Theory and Applications of the Cluster Variation and Path Probability Methods

Theory and Applications of the Cluster Variation and Path Probability Methods

Edited by

J. L. Morán-López

Universidad Autónoma de San Luis Potosí
San Luis Potosí, Mexico

and

J. M. Sanchez

The University of Texas at Austin
Austin, Texas

Plenum Press • New York and London

Library of Congress Cataloging-in-Publication Data

On file

Proceedings of the International Workshop on Theory and Applications of the Cluster Variation
and Path Probability Methods, held June 19 – 23, 1995, in San Juan Teotihuacan, Mexico

ISBN-13: 978-1-4613-8043-6   e-ISBN-13: 978-1-4613-0419-7
DOI: 10.1007/978-1-4613-0419-7

© 1996 Plenum Press, New York
Softcover reprint of the hardcover 1st edition 1996
A Division of Plenum Publishing Corporation
233 Spring Street, New York, N. Y. 10013

Preface

This volume is a compilation of papers presented at the International Workshop on the Theory and Applications of the Cluster Variation and Path Probability Methods, held in the city of San Juan, Teotihuacan, México, during June 18-22, 1995. The presentations at the workshop provided a state of the art review of the fundamental aspects of the CVM and PPM and their application to a wide range of problems in statistical mechanics and alloy theory. The volume begins with several articles dealing with the study of the kinetics of ordering in Ising sytems and alloys using the PPM and other classical techniques. These articles are followed by the contribution of Professor Masuo Suzuki on the Coherent Anomaly Method which has added a new dimension to mean field theory, and the CVM in particular, in the study of critical phenomena. The remaining of the volume is dedicated to fundamental aspects and specific applications of the CVM in a wide range of subjects ranging from bulk and surface studies to new areas of inquiry such as the problem of image reconstruction.

Since the inception by Prof. Ryoichi Kikuchi of the CVM in 1950 and of the PPM in 1966, the latter after a gestation period of approximately six years, the techniques have found wide acceptance in the physics and materials science communities. Both methods are properly regarded as seminal contributions to equilibrium and non-equilibrium statistical mechanics. In particular, the Cluster Variation Method has become one of the methods of choice in the study of the statistical thermodynamics, physics and materials science of alloys, thus transcending its originally intended application as an approximate mean field method to study Ising lattices. The appeal of the CVM is, of course, based primarily on its remarkable success in describing the complexity of statistical correlations in model and real alloy systems. The work of many of the participants to this workshop, and foremost that of Ryo Kikuchi, has been instrumental in demonstrating the power and wide range of applicability of the CVM and PPM.

Quite appropriately, this workshop is dedicated to Professor Ryoichi Kikuchi as a small measure of recognition of his intellectual leadership in the field and, equally important, as a sincere tribute to his qualities as an exceptional friend, colleague and mentor. Those who have had the privilege of working closely with Ryo have been deeply touched by Ryo's inquisitive and uncompromising scientific mind as well as by his uncommon kindness, modesty, and warmth. Ryo has not only brought us the intellectual challenges of the CVM and the PPM but has also set the most rigorous standards of scientific and personal honesty. Thus, we are most grateful to Ryo for his teachings and unselfish dedication to science.

We will be amiss if we were not to extend our gratitude to Toshi Kikuchi. Both Toshi and Ryo Kikuchi have immeasurably enriched the lives of many of us. We hope that the simplicity of a THANK YOU will suffice to convey the extent of our appreciation.

Professor Ryoichi Kikuchi

We also wish to thank the participants for their enthusiasm and support in the organization of the Workshop. Special thanks got to the other members of the organizing committee, Professor Tetsuo Mohri and Professor Faustino Aguilera Granja. We are particularly grateful to Mr. Alejandro Díaz Ortiz and Dr. Juan Martín Montejano Carrizales for their invaluable assistance in the preparation of this volume. Finally, we acknowledge the financial support of the Centro Latinoamericano de Física (México), Consejo Nacional de Ciencia y Tecnología (México), Secretaría de Educación Pública (México), Sociedad Mexicana de Física and the National Science Foundation (U.S.A.)

January, 1996   J. L. Móran-López
                                                 San Luis Potosí, S.L.P., México

                                                 J. M. Sanchez
                                                 Austin, Texas

Contents

Problems in the Foundation of the Path Probability Method: Path Probability Function and Entropy Production

Ryoichi Kikuchi

Department of Materials Science and Engineering
University of California
Los Angeles, CA 90095-1595
U.S.A.

Abstract

The basic procedure of the Path Probability Method (PPM) is to maximize the Path Probability Function (PPF) to obtain the most probable path. The PPF was originally introduced from the probability reasoning. One goal of this paper is to find the physical meaning of the PPF. The paper works with the simple model of the two level atoms. The two-gate probability is formulated, and the general solution of the two-gate paths is derived, in addition to the previously known anticausal path. In a special case, the PPF is shown to be the entropy production. Based on this information, Recipe II of taking into account the energy contribution to the PPF is upheld. Working with the anticausal path, the concept of extended entropy is introduced as a generalization of the entropy and the free energy. The case when the system is in contact with multiple heat baths is treated, and the extended entropy is shown to be a Lyapunov function, whose maximum leads to the non-equilibrium steady state.

I. Introduction

An extension of the Cluster Variation Method (CVM)[1] to time-dependent processes was proposed around 1959,[2,3] and the method has been called the Path Probability Method (PPM).

The main contribution of the CVM is the recipe of writing the entropy of a system in terms of the variables which describe the probability of occurrence of each configuration of a chosen cluster. These variables are called the *state variables* in the present

Theory and Applications of the Cluster Variation and Path Probability Methods
Edited by J.L. Morán-López and J.M. Sanchez, Plenum Press, New York, 1996

paper. After the entropy and the energy are written in terms of the state variables, the main mathematical task in obtaining the equilibrium state is to minimize the free energy with respect to the state variables, since we know that the free energy is a minimum at equilibrium.

In physics theories, variational principles are regarded as important means of expressing basic laws. Therefore, in generalizing the CVM to time-dependent processes, it is desirable to express the basic postulates in a variational principle as an extension of the free energy minimization. However, in 1959 there was no known function which is to be optimized in describing time-dependent processes. We needed a new variational procedure. The search ended up in introducing a function which describes the changes in time Δt in terms of variables of a new category, to be called the *path variables*. Based on what it describes, the introduced function is called the *path probability function* (PPF).

Originally, the PPF was introduced[2,3] based on probability reasoning of time-dependent processes, but not on its physical meaning, and the latter has been left uninvestigated until this time. The present paper reports the study of the physical meaning of the PPF, including the close tie between the latter and the increase of entropy, or decrease of the free energy, in Δt.

We can count two examples as circumstantial evidence to suggest the close relation between the two. One of them is that the free energy always decreases following the natural path, which is for the maximum of the PPF. The other is that in some model processes[3,4] the fluctuated path leads to the two-gate probability which can be written in terms of the free energy difference of the two states. Nevertheless, the physical nature of the PPF has not been self-evident.

In understanding the nature of the PPF, which is the key ingredient of the PPM, we go back to the meaning of the free energy in equilibrium thermodynamics. In section II, we start with Boltzmann's epitaph equation which connects entropy with the number of ways. Then we go to the interpretation of the energy term in the free energy. This interpretation is used in interpreting the PPF in a later section. In treating the nature of the PPF, we use a simple model of two level atoms. Section III treats systems in an isolated condition. Detailed studies are described including the general solution of the two-gate paths. It is shown that the physical meaning of log of the PPF is the increase of entropy. The concept of extended entropy Σ is introduced. The results of section III are extended in section IV which works with cases with one and more heat baths. The multiple heat bath case leads to a non-equilibrium steady state, and it is shown that the extended entropy is a Lyapunov function.

II. Basics of Statistical Mechanics

II.1 Boltzmann's Epitaph Equation

On Ludwig Boltzmann's tombstone in Vienna, the equation

$$S = k \log W \tag{1}$$

is written as the epitaph. We understand that W is the number of ways a state of specified conditions is constructed. When each state appears with the same probability, W is equal to the probability that the state appears in the system. This equation is monumental in the sense that it connects the physics quantity S to the mathematical quantity W of probability theory, and can be regarded as the most basic equation in statistical physics.

Several comments are in order. This tombstone was erected around 1930, about fifteen years after his death, and it is known[5] that Boltzmann himself did not write the entropy relation in the form of Eq. (1). With a valid reason, he always worked with the difference between two entropies. Since in physics reality, we always work with the entropy difference, it is legitimate to write

$$\Delta S = k \Delta(\log W). \qquad (2)$$

The factor between the probability and the number of ways drops out as long as the states to be compared have the same normalization factor, as between two states at the same temperature.

Boltzmann's idea of forming the entropy difference of two states goes well with the second law of thermodynamics. It says that in an isolated system the increment ΔS is positive in irreversible processes, and vanishes at equilibrium:

$$\Delta S \geq 0. \qquad (3)$$

ΔS as in Eq. (2) has a definite value, although S itself has an undetermined additional constant. Note that this entropy is for an isolated system.

II.2 Helmholtz Free Energy

When the system is not isolated, Eq. (3) needs to be modified. We examine how the modification is done in equilibrium thermodynamics. When a system is in thermal equilibrium with a heat bath, the two compose the combined entire system, which is isolated from the rest of the universe. Then Eq. (3) is modified to

$$\Delta S_{\substack{\text{combined} \\ \text{system}}} \equiv \Delta S_{\text{system}} + \Delta S_{\substack{\text{heat} \\ \text{bath}}} \geq 0. \qquad (4)$$

When heat δQ flows from the heat bath to the system, the second law says that the entropy of the heat bath changes by

$$\Delta S_{\substack{\text{heat} \\ \text{bath}}} = -\frac{\delta Q}{T}, \qquad (5)$$

where T is the absolute temperature for the heat bath and hence for the system. When no mechanical work is done on the system, the conservation of the system energy makes

$$\Delta E_{\text{system}} = \delta Q. \qquad (6)$$

Eliminating δQ from the last three relations, we derive

$$-\frac{\Delta F_{\text{system}}}{T} \equiv \Delta S_{\text{system}} - \frac{\Delta E_{\text{system}}}{T} \geq 0, \qquad (7)$$

Since both ΔS_{system} and ΔE_{system} are the quantities for the system, and T is a constant, we may integrate Eq. (7) and define the function F_{system} as

$$F \equiv E - TS, \qquad (8)$$

when the system is at temperature T. The integration constant is not relevant. Equation (7) is the physical basis of the principle of minimum free energy.

What is to be noted is that the energy term in the Helmholtz free energy originates from the entropy change in the heat bath. When the controlling variable is the chemical potential or the pressure, the same idea holds as will be shown below.

II.3 Grand Potential and Gibbs Free Energy

When a number of species in a system is controlled by the chemical potential μ, the system is in thermal equilibrium with a particle reservoir. When the number ΔN of the species flow from the reservoir to the system, the definition of μ says that the energy of the system increases by $\mu \Delta N$. Together with ΔN, we consider that the heat δQ flows into the system from the heat bath. Then the increase of the energy in the system is

$$\Delta E_{\text{system}} = \delta Q + \mu \Delta N_{\text{system}} . \tag{9}$$

The entropy for the combined system in Eq. (4), and the definition of the increase of entropy in the heat bath (5) hold in the present case also. When we eliminate δQ from Eqs. (4), (5) and (9), the inequality (7) is replaced by

$$-\frac{\Delta \Omega_{\text{system}}}{T} \equiv \Delta S_{\text{system}} - \frac{\Delta E_{\text{system}}}{T} + \frac{\mu \Delta N_{\text{system}}}{T} \geq 0 . \tag{10}$$

Integrating this for the constant temperature and neglecting the integration constant, we derive the function Ω as

$$\Omega \equiv E - TS - \mu N , \tag{11}$$

which we call the grand potential. The relation (10) shows that Ω decreases in irreversible processes and is a minimum at equilibrium. This is the expression of the minimum grand potential as an extension of the minimum free energy.

When we control the pressure p of the system, we treat the entropy increase in a way similar as the chemical potential case. We place the system in contact with a pressure source. When the volume of the system increases by ΔV, the system receives the work $-p\Delta V$, and its energy increases by this amount. Together with the mechanical work, we consider that the heat δQ flows into the system from the heat bath. Then the increase of the energy in the system is

$$\Delta E_{\text{system}} = \delta Q - p\Delta V_{\text{system}} . \tag{12}$$

When we eliminate δQ from relations (4), (5) and (12), the inequality (7) is replaced by

$$-\frac{\Delta G_{\text{system}}}{T} \equiv \Delta S_{\text{system}} - \frac{\Delta E_{\text{system}}}{T} - \frac{p\Delta V_{\text{system}}}{T} \geq 0 . \tag{13}$$

Integrating this for the constant temperature and neglecting the integration constant, we derive the function G as

$$G \equiv E - TS + pV , \tag{14}$$

which we call Gibbs free energy. The inequality (13) shows that G decreases in irreversible processes and is a minimum at equilibrium.

III. PPM of Two Level Atoms-Isolated

In formulating the Path Probability Method (PPM), we consider a state of the system at time t and ask the probability with which the system changes toward a certain direction in a short time interval Δt. It is understood that in macroscopic observation the change corresponding to the direction of maximum probability occurs. The function which writes the probability of change is called the Path Probability Function (PPF). Since it is a probability function, it can be likened to W in Boltzman's epitaph equation (1). However, in considering the analogy with Eq. (1), we do not know the left-hand side quantity which is for physics. To search the latter is the first aim of the present paper.

For this purpose, we work with the simplest example presented in the previous review of PPM,[3] namely an ensemble of two-level atoms. Each atom has the ground level (g) and the excited level (e), the energy difference being ε. The formulation in subsequent sections is similar as the one presented before,[3] but is expanded. This model is used in exploring physical interpretations which have not been examined before.

We consider an ensemble of isolated M atoms. A g atom can receive energy ε from the radiation field and is excited to e. An e atom can fall back to g giving the energy back to the radiation field. However, the atom is not in contact with a heat bath.

III.1 Variables

The fractions of g and e atoms at time t are written in the lower case as $p_g(t)$ and $p_e(t)$, respectively. These are the state variables. The probability of finding an atom which is in the state i at t and in j at $t + \Delta t$ is written as $P_{ij}(t; t + \Delta t)$, which is the path variable. The reduction relations are

$$p_g(t) = P_{gg} + P_{ge},$$
$$p_e(t) = P_{ee} + P_{eg}, \qquad (15a)$$

and

$$p_g(t + \tau) = P_{gg} + P_{eg},$$
$$p_e(t + \tau) = P_{ee} + P_{ge}, \qquad (15b)$$

in which the arguments $(t; t + \Delta t)$ are omitted. When $p_g(t)$ and $p_e(t)$ are given, we choose P_{ge} and P_{eg} as independent and write the *static variables* as

$$P_{gg} = p_g(t) - P_{ge},$$
$$P_{ee} = p_e(t) - P_{eg}. \qquad (16)$$

The change of $p_e(t)$ in Δt is written as $\Delta p_e(t)$, which is derived by forming the difference of Eq. (15b) and Eq. (15a) as

$$\Delta p_e(t) \equiv p_e(t + \Delta t) - p_e(t) = P_{ge} - P_{eg}. \qquad (17)$$

III.2 Path Probability Function (PPF)

The PPF is written as a product of two factors: $P \equiv P_1 P_2$. In the equilibrium theory, including the CVM, each state appears with equal probability. One essential difference of the time dependent theory is that each path in Δt has its own *a priori probability*. The *a priori* probability that a g atom is excited to e is written as θ_g, while the reverse probability for an e atom to go down to g is written as θ_e. The *a priori* probability is taken into account in the first factor P_1 which is written as

$$M^{-1} \ln P_1 = P_{gg} \ln(1 - \Delta t \theta_g) + P_{ge} \ln(\Delta t \theta_g) + P_{eg} \ln(\Delta t \theta_e) + P_{ee} \ln(1 - \Delta t \theta_e), \quad (18)$$

P_2 is the number of ways the P_{ij}'s occur and is conceptually similar to the entropy in equilibrium statistical mechanics.

$$M^{-1} \ln P_2 = M^{-1} \ln \frac{(Mp_g(t))! \, (Mp_e(t))!}{(MP_{gg})! \, (MP_{ge})! \, (MP_{eg})! \, (MP_{ee})!}$$

$$= L(p_g(t)) + L(p_e(t)) - L(P_{gg}) - L(P_{ge}) - L(P_{eg}) - L(P_{ee}), \quad (19)$$

where we define

$$L(x) \equiv x \ln x - x. \quad (20)$$

III.3 The Natural Path

When the ensemble starts from a given state at t and is left by itself, it makes a change in a time interval Δt in the most probable direction. It is the direction which maximizes the PPF for given initial state $p_g(t)$ and $p_e(t)$. We differentiate $\ln P(t; t + \Delta t)$ with respect to the independent variables.

$$\frac{\partial \ln P(t; t + \Delta t)}{M \partial P_{ge}} \equiv \ln \left(\frac{P_{gg}}{P_{ge}} \right) + \ln(\Delta t \, \theta_g) - \ln(1 - \Delta t \, \theta_g) = 0, \quad (21a)$$

$$\frac{\partial \ln P(t; t + \Delta t)}{M \partial P_{eg}} \equiv \ln \left(\frac{P_{ee}}{P_{eg}} \right) + \ln(\Delta t \, \theta_e) - \ln(1 - \Delta t \, \theta_e) = 0, \quad (21b)$$

which are rewritten as

$$P_{ge} = \Delta t \, \theta_g \frac{P_{gg}}{1 - \Delta t \, \theta_g}, \quad (22a)$$

$$P_{eg} = \Delta t \, \theta_e \frac{P_{ee}}{1 - \Delta t \, \theta_e}. \quad (22b)$$

When Δt is small, Eq. (22) can be expanded as

$$P_{ge} = \Delta t \, \theta_g \, p_g(t) + \mathcal{O}(\Delta t)^2, \quad (23a)$$

$$P_{eg} = \Delta t \, \theta_e \, p_e(t) + \mathcal{O}(\Delta t)^2. \quad (23b)$$

When the maximization in Eq. (21) holds, the PPF $P(t; t + \Delta t)$ is simplified to

$$\ln P(t; t + \Delta t) = \ln P(t; t + \Delta t) - P_{ge} \frac{\partial \ln P(t; t\Delta t)}{\partial P_{ge}} - P_{eg} \frac{\partial \ln P(t; t\Delta t)}{\partial P_{eg}}$$

$$= p_e(t) \ln p_e(t) + p_g(t) \ln p_g(t) - p_e(t) \ln P_{ee}(t, t + \Delta t)$$

$$- p_g(t) \ln P_{gg}(t, t + \Delta t) + p_g(t) \ln(1 - \Delta t \, \theta_g)$$

$$+ p_e(t) \ln(1 - \Delta t \, \theta_e) = 0. \quad (24)$$

The differential equation for the natural path is derived by using Eq. (23) in (17) as

$$\frac{dp_e(t)}{dt} = \theta_g p_g(t) - \theta_e p_e(t). \quad (25)$$

Using $p_g = 1 - p_e$, this differential equation is solved as

$$p_e^*(t) = \frac{\theta_g}{\theta_g + \theta_e} \pm \frac{\exp[-(\theta_g + \theta_e)(t - t_0)]}{\theta_g + \theta_e}. \quad (26)$$

or

$$p_e^*(t) = \frac{\theta_g}{\theta_g + \theta_e} + \left(p_e^*(t_1) - \frac{\theta_g}{\theta_g + \theta_e} \right) \exp[-(\theta_g + \theta_e)(t - t_1)], \quad (27)$$

where $p_e(t_1)$ is the initial value at t_1. Note that in Eq. (27), the difference $(p_e^*(t_1) - \frac{\theta_g}{\theta_g + \theta_e})$ can be either plus or minus. Both equations go to the correct limit:

$$\lim_{t \to \infty} p_e^*(t) = \frac{\theta_g}{\theta_g + \theta_e} . \tag{28}$$

III.4 Constrained Path

When we need the constraint that both $p_e(t)$ and $p_e(t + \Delta t)$ are fixed, we use a Lagrange multiplier $\lambda(t)$ and write

$$\ln P(t; t + \Delta t) = \ln(P_1 P_2) + M\lambda(t)\big[p_e(t + \Delta t) - p_e(t) + P_{ge} + P_{eg}\big] . \tag{29}$$

When this is maximized, Eq. (21) is changed to

$$\frac{\partial \ln P(t; t + \Delta t)}{M \partial P_{ge}} \equiv \ln\left(\frac{P_{gg}}{P_{ge}}\right) + \ln(\Delta t \, \theta_g) - \ln(1 - \Delta t \, \theta_g) - \lambda(t) = 0 , \tag{30a}$$

$$\frac{\partial \ln P(t; t + \Delta t)}{M \partial P_{eg}} \equiv \ln\left(\frac{P_{ee}}{P_{eg}}\right) + \ln(\Delta t \, \theta_e) - \ln(1 - \Delta t \, \theta_e) + \lambda(t) = 0 , \tag{30b}$$

which changes Eq. (23) to

$$P_{ge} = \Delta t \, \theta_g e^{-\lambda(t)} p_g(t) + \mathcal{O}(\Delta t)^2 , \tag{31a}$$

$$P_{eg} = \Delta t \, \theta_e e^{+\lambda(t)} p_e(t) + \mathcal{O}(\Delta t)^2 . \tag{31b}$$

Using these in Eq. (16), we can write P_{gg} and P_{ee} as

$$P_{gg} = p_g(t)\big[1 - \Delta t \, \theta_g e^{-\lambda(t)}\big] ,$$

$$P_{ee} = p_e(t)\big[1 - \Delta t \, \theta_e e^{+\lambda(t)}\big] . \tag{32}$$

When $\ln P(t; t + \Delta t)$ in Eq. (29) is a maximum, it can be simplified by the procedure similar to (24), but leading to a completely different result. We place a superscript II on P as was done previously for the constrained PPF. We also use Eq. (32) in the transformation.

$$(\Delta t \, M)^{-1} \ln P^{II}\big(p_e(t); p_e(t + \Delta t)\big) = p_g(t)\theta_g \left(e^{-\lambda(t)} - 1\right)$$

$$+ p_e(t)\theta_e \left(e^{+\lambda(t)} - 1\right) + \lambda(t)\frac{\Delta p_e(t)}{\Delta t} , \tag{33a}$$

or

$$(\Delta t \, M)^{-1} \ln P^{II}\big(p_e(t); p_e(t + \Delta t)\big) = \left(e^{-\lambda(t)} - 1\right)\left(p_g(t)\theta_g - p_e(t)\theta_e \, e^{+\lambda(t)}\right)$$

$$+ \lambda(t)\frac{\Delta p_e(t)}{\Delta t} . \tag{33b}$$

We can determine $\lambda(t)$ by substituting (31) in the constraint equation (17), or by differentiating (33) with respect to λ as

$$\frac{\Delta p_e(t)}{\Delta t} = \exp(-\lambda)\theta_g p_g(t) - \exp(+\lambda)\theta_e p_e(t) . \tag{34}$$

III.5 Two-Gate Probability

Using Eq. (33), we can derive the two-gate probability for a finite time interval to go from $p_e(t_{\text{initial}} \equiv t_1)$ to $p_e(t_{\text{final}} \equiv t_f)$. Noting that (33) writes $(\Delta t)^{-1} \ln P^{\text{II}}$, the probability of a path following $p_e(t_n)$ from t_1 to t_f is written as

$$P(p(t); t_1 \leq t \leq t_f) = \exp \left\{ \sum_{n=1}^{(t_f - t_1)/\Delta t} \left[\frac{\ln P^{\text{II}}(p(t_n); p(t_n + \Delta t))}{\Delta t} \right] \Delta t \right\}$$

$$= \exp \left\{ \int_{t_1}^{t_f} dt \, (\Delta t)^{-1} \ln P^{\text{II}}(p(t)) \right\}, \tag{35}$$

where the integrand is written in (33) as

$$(M\Delta t)^{-1} \ln P^{\text{II}}(p_e(t)) = p_g(t)\theta_g \left(e^{-\lambda(t)} - 1 \right)$$

$$+ p_e(t)\theta_e \left(e^{+\lambda(t)} - 1 \right) + \lambda(t) \frac{dp_e(t)}{dt}, \tag{36a}$$

or

$$(M\Delta t)^{-1} \ln P^{\text{II}}(p_e(t)) = \left(e^{-\lambda(t)} - 1 \right)$$

$$\times \left(p_g(t)\theta_g - p_e(t)\theta_e \, e^{+\lambda(t)} \right) + \lambda(t) \frac{dp_e(t)}{dt}. \tag{36b}$$

Unless p_g and p_e appear in an equation we simply write

$$p(t) \equiv p_e(t). \tag{37}$$

When only the initial $p(t_1)$ and the final $p(t_f)$ are specified, the two gate probability can be written by a path integral as

$$P[p(t_1); p(t_f)] = \int_{p(t_1)}^{p(t_f)} P(p(t); t_1 \leq t \leq t_f) D[p(t)]$$

$$= \int_{p(t_1)}^{p(t_f)} \exp \left(\int_{t_1}^{t_f} dt \, (\Delta t)^{-1} \ln P^{\text{II}}[p(t)] \right) D[p(t)]. \tag{38}$$

When M is large we can replace the integral in Eq. (38) by the path which makes the integrand a maximum:

$$P[p(t_1); p(t_f)] = \exp \left(\max \int_{t_1}^{t_f} dt \, (\Delta t)^{-1} \ln P^{\text{II}}[p(t)] \right)_{p(t_1); p(t_f)} \tag{39}$$

This maximum path is derived by varying $p(t)$ in (36). Although $\lambda(t)$ is a function of $p(t)$, the variation of $\ln P^{\text{II}}$ with respect to $\lambda(t)$ vanishes when $\lambda(t)$ satisfies the constraint equation. Thus the variation is done as

$$\delta \int_{t_1}^{t_f} (\Delta t)^{-1} \ln P^{\text{II}}[p(t)] \, dt = \int_{t_1}^{t_f} \left(\frac{\partial \left((\Delta t)^{-1} \ln P^{\text{II}}[p(t)] \right)}{\partial p(t)} \right) \delta p(t) \, dt. \tag{40}$$

In evaluating the derivative with respect to $p(t)$, we first use (36) and partially integrate as

$$\int_{t_1}^{t_f} dt\,(\Delta t)^{-1} \ln P^{\mathrm{II}}\left[p_e(t)\right]$$

$$\equiv M \int_{t_1}^{t_2} \left[\left(e^{-\lambda(t)} - 1\right)\left(p_g(t)\theta_g - p_e(t)\theta_e\, e^{+\lambda(t)}\right) - \frac{d\lambda(t)}{dt}p_e(t)\right] dt$$

$$+ M\left(\lambda(t_2)p_e(t_2) - \lambda(t_1)p_e(t_1)\right). \tag{41}$$

Then the variation of this with respect to $p_e(t)$ leads to

$$\frac{d\lambda(t)}{dt} = -\left(e^{-\lambda(t)} - 1\right)\left(\theta_g + \theta_e\, e^{+\lambda(t)}\right). \tag{42}$$

III.6 General Solution

The most probable path to connect the two gates is determined by the two functions $p(t)$ and $\lambda(t)$. They satisfy the constraint equation (34) and the most probable path equation (42). To simplify the formulation, we introduce

$$a \equiv \frac{\theta_g}{\theta_e}; \quad b \equiv \theta_g + \theta_e, \tag{43a}$$

$$d\tau \equiv b\,dt, \tag{43b}$$

$$\Lambda(\tau) \equiv e^{+\lambda(t)}. \tag{43c}$$

Then the constraint equation (34) is written as

$$(a+1)\frac{dp}{d\tau} = a\Lambda^{-1}(1-p) - \Lambda p, \tag{44}$$

and the most probable path equation (42) becomes

$$(a+1)\frac{d\Lambda}{d\tau} = (\Lambda - 1)(\Lambda + a). \tag{45}$$

We can immediately see that one possible solution of (45) is

$$\Lambda = 1; \quad \lambda(t) = 0, \tag{46}$$

which says that no constraint is imposed on $p(t+\Delta t)$, and is the natural path (25). In addition to this obvious solution, a special solution was found before[3] by inspection, and led to the anticausal path. It is

$$\Lambda = a\frac{1-p^*}{p^*}, \tag{47a}$$

or

$$p(\Lambda)^* = \frac{a}{\Lambda + a}, \tag{47b}$$

the asterisk indicating a special solution. We can verify that (47) satisfies (45) when the constraint relation (44) holds. Using (47), Eq. (44) becomes

$$(a+1)\frac{dp}{d\tau} = (a+1)p - a, \tag{48}$$

which was called the anticausal path since it is the natural path (25) with the time axis reversed.

In addition to the special solution, we can obtain the general solution. For this purpose, first we solve $\Lambda(x)$ from Eq. (45). Because of the integral $\int \frac{d\Lambda}{\Lambda-1}$, the solution depends of the sign of $(\Lambda - 1)$ as

$$\Lambda(t) = \frac{1 + \text{sign}(\Lambda - 1)\, ae^\tau}{1 - \text{sign}(\Lambda - 1)\, e^\tau}, \tag{49a}$$

where from Eq. (43b)

$$\tau \equiv b(t - t_0). \tag{49b}$$

When we form the ratio (44)/(45), we derive an equation

$$\frac{dp}{d\Lambda} + P(\Lambda)p = Q(\Lambda), \tag{50a}$$

with

$$P(\Lambda) \equiv \frac{\Lambda^2 + a}{\Lambda(\Lambda - 1)(\Lambda + a)}, \tag{50b}$$

$$Q(\Lambda) \equiv \frac{a}{\Lambda(\Lambda - 1)(\Lambda + a)}. \tag{50c}$$

Equations (50) are a standard form of a linear ordinary differential equation with variable coefficients, and can be solved analytically.

However, when a special solution of (50) is known, the equation to be solved becomes even simpler. Knowing a special solution $p(\Lambda)^*$, consider a combination

$$p(\Lambda) = p(\Lambda)^* + C\varphi(\Lambda), \tag{51a}$$

where the special solution satisfies

$$\frac{dp^*}{d\Lambda} + P(\Lambda)p^* = Q(\Lambda). \tag{51b}$$

When $\varphi(\Lambda)$ satisfies

$$\frac{d\varphi}{d\Lambda} + P(\Lambda)\varphi = 0, \tag{51c}$$

then the sum $p(\Lambda)$ (51a) with an arbitrary C can satisfy the general equation (50).

Using $P(\Lambda)$ of (50b), we can solve (51c) with an integration constant φ_0 as

$$\frac{\varphi}{\varphi_0} = \frac{\Lambda}{|\Lambda - 1|\,(\Lambda + a)}. \tag{52}$$

Since φ_0 can be absorbed in the constant C, using the special solution (47b), we can write the general solution as

$$p(\Lambda) = \frac{a}{\Lambda + a} + C\frac{\Lambda}{|\Lambda - 1|\,(\Lambda + a)}. \tag{53}$$

At this time in seeing the behavior of the two-gate paths, we examine the two signs of $\Lambda - 1$ separately.

III.6.1 $\Lambda - 1 < 0$

In this case the anticausal path $p^*(\Lambda)$ in Eq. (47b) is larger than its asymptotic value $a/(a+1)$, and hits unity at some value of t. We note $\Lambda(t)$ in Eq. (49a) becomes

$$\Lambda(\tau) = \frac{1 - ae^\tau}{1 + e^\tau}, \tag{54}$$

where τ in Eq. (49b) is written with an undecided reference time t_0. It is helpful to choose the reference time $t = 0$ at the point where $p^* = 1$. This condition leads to the value

$$\exp(bt_0) = a, \tag{55}$$

so that Eq. (54) changes to

$$\Lambda(t) = \frac{a(1 - e^{bt})}{a + e^{bt}}, \tag{56}$$

and the anticausal path is

$$p^*(t) = \frac{a + e^{bt}}{a + 1}. \tag{57}$$

The full general solution (53) becomes

$$p(t) = \frac{a + e^{bt}}{a + 1} + C\frac{(1 - e^{bt})(ae^{-bt} + 1)}{(a + 1)^2}. \tag{58}$$

III.6.2 $\Lambda - 1 > 0$

In this case the anticausal path $p^*(\Lambda)$ in Eq. (47b) is less than its asymptotic value $a/(a + 1)$, and hits zero at some value of t. We note $\Lambda(t)$ in (49a) becomes

$$\Lambda(t) = \frac{1 + ae^{\tau}}{1 - e^{\tau}}. \tag{59}$$

Instead of the undecided t_0 in τ of Eq. (49b), we choose the reference time $t = 0$ at the point where $p^* = 0$. This condition leads to the value

$$t_0 = 0, \tag{60}$$

so that (54) changes to

$$\Lambda(t) = \frac{1 + ae^{bt}}{1 - e^{bt}}, \tag{61}$$

and the anticausal path is

$$p * (t) = \frac{a}{a + 1}(1 - e^{bt}). \tag{62}$$

The full general solution (53) becomes

$$p(t) = \frac{a}{a + 1}(1 - e^{bt}) + C\frac{(e^{-bt} - 1)(1 + ae^{bt})}{(a + 1)^2}. \tag{63}$$

Examples of the natural paths, anticausal paths and two-gate paths are shown in Figs. 1, 2 and 3. Although the anticausal path was derived before,[3] the general solution in this section is a new contribution.

III.7 Anticausal Path and Extended Entropy

The anticausal path for the two-gate probability has special noteworthy properties. The more familiar way of writing the anticausal path in Eq. (47a) is to write

$$e^{\lambda(t)} = \frac{\theta_y\, p_y(t)}{\theta_e\, p_e(t)}, \quad \text{with} \quad \lambda(t) = \ln\left[\theta_g\, p_g(t)\right] - \ln\left[\theta_e\, p_e(t)\right]. \tag{64}$$

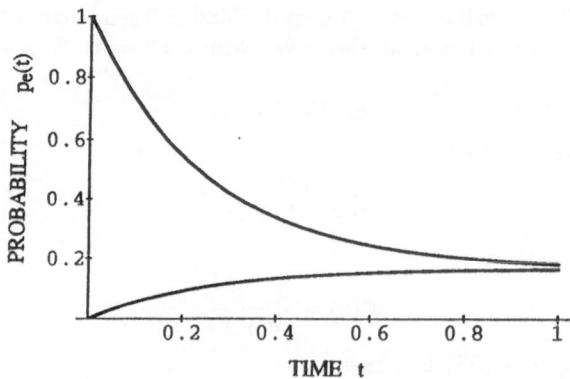

Figure 1. Two branches of the natural path; $a = 0.2$ and $b = 4$.

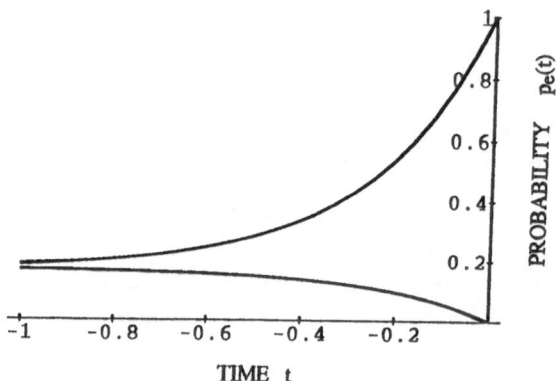

Figure 2. Two branches of the anticausal path; $a = 0.2$ and $b = 4$ and $C = 0$.

We go back to the constrained PPF $\ln P^{II}[p_e(t)]$ in Eq. (36b). It is noted that the anticausal path in (64) makes the first term of (36b) vanish so that

$$(M\Delta t)^{-1} \ln P^{II}[p_e(t)] = \lambda(t)\frac{dp_e(t)}{dt}. \tag{65}$$

Then Eq. (41) can be written as

$$\int_{t_1}^{t_f} dt\, \frac{\ln P^{II}[p_e(t)]}{\Delta t} \equiv M \int_{t_1}^{t_f} \left(\lambda(t)\frac{dp_e(t)}{dt}\right) dt = M \int_{p_e(t_1)}^{p_e(t_f)} \lambda(p_e(t))\, d(p_e(t)), \tag{66}$$

which identifies

$$\lambda(p_e(t)) = \frac{\ln P^{II}[p_e(t)]}{M\Delta p_e(t)}. \tag{67}$$

On the other hand, (64) makes us write

$$\lambda(t) = \ln[\theta_g\, p_g(t)] - \ln[\theta_e\, p_e(t)]$$
$$= \frac{d}{dp_e}\left(-p_g \ln \theta_g - p_e \ln \theta_e - p_g \ln p_g - p_e \ln p_e\right). \tag{68}$$

In the special case of

$$\theta_g = \theta_e, \tag{69}$$

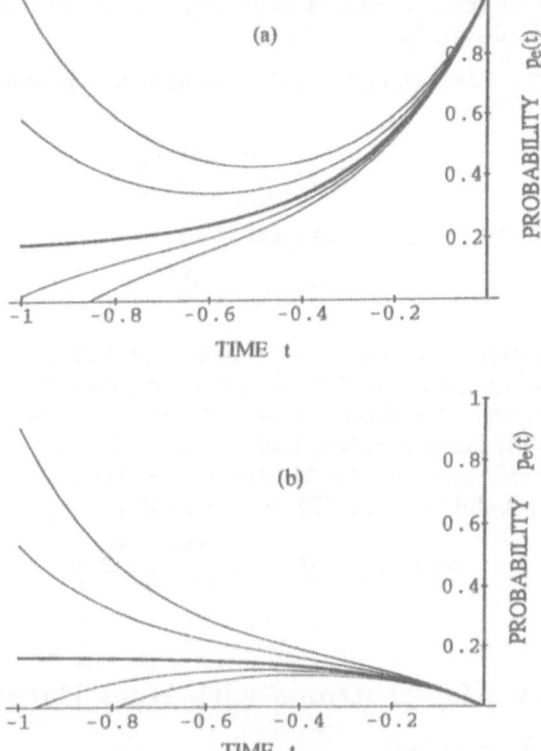

Figure 3. Examples of the anticausal paths (thick curves) and the general solutions (thin curves). (a) $\Lambda - 1 < 0$: $a = 0.2$, $b = 4$; $C = 0.1, 0.05, -0.02$, and -0.04. (b) $\Lambda - 1 > 0$: $a = 0.2$, $b = 4$; $C = 0.02, 0.01, -0.005$, and -0.01.

equation (68) reduces to

$$\lambda\left(p_e(t)\right) = \frac{d}{dp_e}\left(-p_g \ln p_g - p_e \ln p_e\right) = \frac{d}{dp_e}\left(\frac{S}{Mk}\right), \qquad (70)$$

where S is the entropy of the ensemble of M two-level atoms. Combining Eqs. (70) and (67), we can identify

$$\ln P^{II}\left[p_e(t)\right] = \frac{\Delta S}{k}. \qquad (71)$$

Since this is for the maximum path in (39), the two-gate probability in (39) is integrated as

$$P\left[p(t_1); p(t_f)\right] = \exp\left(\int_{t_1}^{t_f} \ln P^{II}[p_e(t)]\, dt\right) = \exp\left(\int_{t_1}^{t_f} \frac{\Delta S}{k}\, dt\right) = \exp\left(\frac{S_f - S_1}{k}\right). \qquad (72)$$

Although this conclusion holds only for a restricted case, it is significant that we can interpret the physical meaning of the PPF, which was originally derived based on the probability reasoning. When we leave aside the kinetic parameters, as far as the thermodynamic properties are concerned, we can now identify the PPF with the entropy production.

In the general case (68), we cannot write $\lambda(t)$ in terms of the entropy, but we can define the extended entropy Σ:

$$\Sigma \equiv kM\left(-p_g \ln \theta_g - p_e \ln \theta_e - p_g \ln p_g - p_e \ln p_e\right), \tag{73}$$

to write Eq. (70) as

$$\lambda\left(p_e(t)\right) = \frac{d}{dp_e}\left(\frac{\Sigma}{Mk}\right). \tag{74}$$

Combining Eqs. (72) and (67), we can identify

$$\ln P^{II}\left[p_e(t)\right] = \frac{\Delta\Sigma}{k}. \tag{75}$$

The PPF $P^{II}\left[p_e(t)\right]$ is the quantity defined in probability theory, while the extended entropy Σ is a physics quantity. This equation connects the quantities of two categories and is regarded significant. It is understandable that Σ in this equality is not purely a thermodynamic quantity, because the PPF is a kinetic concept and can depend on parameters controlling the kinetics of the system.

The two-gate probability in Eq. (72) is now written as the difference of Σ's

$$P\left[p(t_1); p(t_f)\right] = \exp\left(\frac{\Sigma_f - \Sigma_1}{k}\right). \tag{76}$$

IV. PPM of Two Level Atoms-with Heat Baths

IV.1 With a Single Heat Bath

In section II we interpreted the energy term of the free energy expression as the contribution of the entropy change in the heat bath. Since the previous section identifies the PPF as the entropy production when the system is isolated, when it is in contact with a heat bath, the entropy production in the latter is to be taken into account in the same way as the energy increase is treated in the free energy expression.

The model in this section is modified from that of section III, in that the energy ε, to be supplied from the heat bath, is absorbed when the g atom is excited to the e state. However, when an e atom goes down to g, the energy ε is not immediately returned to the heat bath, and thus the path variable P_{eg} is not related to the entropy change of the heat bath.

When the system is in contact with the heat bath, we use the same state variables and path variables as defined in III.1. The difference of this section comes in the PPF, which is now written as a product of three factors: $P \equiv P_1 P_2 P_3$. The first two factors are the same as those written in III.2. The third factor P_3 represents the entropy production in the heat bath, which can be written by equating (5) and (6) as

$$\Delta S_{\substack{\text{heat} \\ \text{bath}}} = \frac{\Delta E_{\text{system}}}{T}. \tag{77}$$

Thus we can write P_3 in the path variable P_{ge} as

$$\ln P_3 = -\frac{\varepsilon}{kT}MP_{ge}. \tag{78}$$

This procedure was called Recipe II in a 1966 report.[3] Recipe I divides ε by $2kT$ in the energy factor, and also takes into account the final state after the jump. Although both Recipes lead to the same equilibrium state, the present paper justifies Recipe II rather than I. This is the second contribution of the paper.

When we combine Eqs. (18), (19) and (78), the entire PPF P is written as

$$M^{-1} \ln P = P_{gg} \ln(1 - \Delta t\, \theta_{gg}) + P_{ge} \ln(\Delta t\, \theta_g) + P_{ee} \ln(1 - \Delta t\, \theta_{ee})$$
$$+ P_{eg} \ln(\Delta t\, \theta_e) + L(p_g(t)) + L(p_e(t)) - L(P_{gg})$$
$$- L(P_{ge}) - L(P_{eg}) - L(P_{ee}) - \frac{\varepsilon}{kT} P_{ge} . \tag{79}$$

In writing this equation, we added the last P_{ge} term and also made changes in θ_{gg} and θ_{ee} in the P_{gg} and P_{ee} terms of P_1. θ_{gg} is the probability that a g atom makes a transition. Different from section III, θ_{gg} is different from θ_g since we now have the energy term contribution. θ_{gg} is defined using the natural path as

$$\Delta t\, \theta_{gg} = \frac{P_{ge}^{(\text{nat})}}{p_g(t)} . \tag{80a}$$

Similarly

$$\Delta t\, \theta_{ee} = \frac{P_{eg}^{(\text{nat})}}{p_e(t)} . \tag{80b}$$

When we maximize $\ln P$ with respect to P_{ge} and P_{eg}, and expand the results for small Δt, Eq. (23) is changed to

$$P_{ge}^{(\text{nat})} = \Delta t\, \theta_g \exp\left(\frac{-\varepsilon}{kT}\right) p_g(t) + \mathcal{O}(\Delta t)^2 , \tag{81a}$$

$$P_{eg}^{(\text{nat})} = \Delta t\, \theta_e p_e(t) + \mathcal{O}(\Delta t)^2 . \tag{81b}$$

When we use these in (80), we derive

$$\theta_{gg} = \theta_g \exp\left(\frac{-\varepsilon}{kT}\right) , \tag{82a}$$

$$\theta_{ee} = \theta_e . \tag{82b}$$

In the constrained path treatment, we use the same $\lambda(t)$ term as in Eq. (29). Maximization of PPF leads to the modification of (81) as

$$P_{ge} = \Delta t\, \theta_g \exp\left(\frac{-\varepsilon}{kT}\right) \exp(-\lambda(t)) p_g(t) + \mathcal{O}(\Delta t)^2 , \tag{83a}$$

$$P_{eg} = \Delta t\, \theta_e \exp(+\lambda(t)) p_e(t) + \mathcal{O}(\Delta t)^2 . \tag{83b}$$

Also, (32) is changed to

$$P_{gg} = p_g(t)\left[1 - \Delta t\, \theta_g \exp\left(\frac{-\varepsilon}{kT}\right) \exp(-\lambda(t))\right] , \tag{84a}$$

$$P_{ee} = p_e(t)\left[1 - \Delta t\, \theta_e \exp(+\lambda(t))\right] , \tag{84b}$$

where the second equation remains the same.

When we compare Eqs. (83) and (84) with Eqs. (31) and (32), we see that the change we need is

$$\theta_g \rightarrow \theta_g \exp\left(\frac{-\varepsilon}{kt}\right) . \tag{85}$$

With this replacement, all equations in section III remain valid for the case of one heat bath. One comment is in order. The natural path equation (25) is now

$$\frac{dp_e}{dt} = \theta_g \exp\left(\frac{-\varepsilon}{kT}\right) p_g(t) - \theta_e p_e(t).\tag{86}$$

In the steady state this leads to

$$\frac{p_e}{p_g} = \frac{\theta_g}{\theta_e} \exp\left(\frac{-\varepsilon}{kT}\right).\tag{87}$$

In case the system is in equilibrium in the steady state, this requires

$$\theta_g = \theta_e.\tag{88}$$

The particular relation to be pointed out is the extended entropy Σ in (73), which now changes to

$$\Sigma \equiv kM\left[-p_g\left(\ln\theta_g - \frac{\varepsilon}{kT}\right) - p_e\ln\theta_e - p_g\ln p_g - p_e\ln p_e\right],\tag{89}$$

which are used in Eqs. (74) and (75):

$$\lambda\left(p_e(t)\right) = \frac{d}{dp_e}\left(\frac{\Sigma}{Mk}\right),\tag{90}$$

and

$$\ln P^{\mathrm{II}}\left[p_e(t)\right] = \frac{\Delta\Sigma}{k}.\tag{91}$$

When the system can become equilibrium, we require (88), so that (89) changes into

$$\Delta\left(\frac{\Sigma}{k}\right) \equiv \Delta M\left(+p_g\left(\frac{\varepsilon}{kT}\right) - p_g\ln p_g - p_e\ln p_e\right) = -\Delta\left(\frac{F}{kT}\right),\tag{92}$$

with the free energy F of the system. This changes (91) to

$$\ln P^{\mathrm{II}}\left[p_e(t)\right] = -\Delta\left(\frac{F}{kT}\right).\tag{93}$$

As is written in Eqs. (38) and (39), the two-gate probability is the integral of this quantity

$$P\left[p(t_1); p(t_f)\right] = \exp\left(\int_{p(t_1)}^{p(t_f)} \ln P^{\mathrm{II}}\left[p_e(t)\right] dp_e(t)\right) = \exp\left(-\frac{F_f}{kT} + \frac{F_1}{kT}\right).\tag{94}$$

This is the result which was obtained in Eq. (4.20) of Ref. 3. When we use the general expression (91), it is to be modified to

$$P\left[p(t_1); p(t_f)\right] = \exp\left(\frac{\Sigma_f}{k} - \frac{\Sigma_1}{k}\right).\tag{95}$$

IV.2 With Multiple Heat Baths

We consider the system of two-level atoms interacting with more than one heat bath, $i = 1, 2, \ldots$. The system cannot achieve thermal equilibrium, but can stay in a non-equilibrium stationary state.

The probability of finding an atom whose g state changes into e by absorbing energy from the heat bath i is written as $P_{ge(i)}$. Then Eq. (15a) is changed to

$$p_g(t) = P_{gg} + \sum_i P_{ge(i)},$$

$$p_e(t) = P_{ee} + P_{eg}. \tag{96a}$$

For the e to g change, our model is that the heat does not go to any specific heat bath, and thus no (i) subscript is identified. Eqs. (15)–(17) change to

$$p_g(t + \tau) = P_{gg} + P_{eg},$$

$$p_e(t + \tau) = P_{ee} + \sum_i P_{ge(i)}, \tag{96b}$$

$$P_{gg} = p_g(t) - \sum_i P_{ge(i)},$$

$$P_{ee} = p_e(t) - P_{eg}, \tag{97}$$

and

$$\Delta p_e(t) \equiv p_e(t + \Delta t) - p_e(t) = \sum_i P_{ge(i)} - P_{eg}. \tag{98}$$

In the PPF in (79), P_{ge} appears in three terms, corresponding to P_1, P_2 and P_3. P_1 is for the *a priori* probability for jumps. We assume the jump $P_{ge(i)}$ occurs with the *a priori* probability $\theta_{ge(i)}$. P_2 is the number of ways the path in Δt can be formed, and the $L(P_{ge})$ term is changed into $\sum_i L(P_{ge(i)})$. For the entropy change in the heat baths, P_3, we follow the reasoning of IV.1 so that the contribution of the entropy change in the ith heat bath is written as $(-\varepsilon/kT_i)P_{ge(i)}$. These changes make us write the PPF P as

$$M^{-1} \ln P = P_{gg} \ln(1 - \Delta t\, \theta_{gg}) + \sum_i P_{eg(i)} \ln(\Delta t\, \theta_{g(i)})$$

$$+ P_{eg} \ln(\Delta t\, \theta_e) + P_{ee} \ln(1 - \Delta t\, \theta_{ee}) + L(p_g(t)) + L(p_e(t))$$

$$- L(P_{gg}) - \sum_i L(P_{ge(i)}) - L(P_{eg}) - L(P_{ee}) - \sum_i \frac{\varepsilon}{kT_i} P_{ge(i)}, \tag{99}$$

where θ_{gg} and θ_{ee} are determined from the natural path by

$$\Delta t\, \theta_{gg} = \frac{1}{p_g(t)} \sum_i P_{ge(i)}^{(\mathrm{nat})}, \tag{100a}$$

$$\Delta t\, \theta_{ee} = \frac{P_{eg}^{(\mathrm{nat})}}{p_e(t)}. \tag{100b}$$

When we maximize $\ln P$, we derive relations corresponding to (81) as

$$P_{ge(i)}^{(\mathrm{nat})} = \Delta t\, \theta_{g(i)} \exp\left(\frac{-\varepsilon}{kT_i}\right) p_g(t) + \mathcal{O}(\Delta t)^2, \tag{101a}$$

$$P_{eg}^{(\mathrm{nat})} = \Delta t\, \theta_e p_e(t) + \mathcal{O}(\Delta t)^2. \tag{101b}$$

When we use (101) in (100), we derive

$$\theta_{gg} = \sum_i \theta_{g(i)} \exp\left(\frac{-\varepsilon}{kT_i}\right),$$

(102a)

$$\theta_{ee} = \theta_e.$$

(102b)

When we write the PPF for a constrained path, (29) is changed to

$$\ln P(t; t + \Delta t) = \ln(P_1 P_2) + M\lambda(t)\left(p_e(t + \Delta t) - p_e(t) - \sum_i P_{ge(i)} + P_{eg}\right).$$

(103)

We note that all $P_{ge(i)}$'s have the same $\lambda(t)$, so that (83) and (84) are changed to

$$P_{ge(i)} = \Delta t\, \theta_{g(i)} \exp\left(\frac{-\varepsilon}{kT_i}\right) \exp(-\lambda(t)) p_g(t) + \mathcal{O}(\Delta t)^2,$$

(104a)

$$P_{eg} = \Delta t\, \theta_e \exp(+\lambda(t)) p_e(t) + \mathcal{O}(\Delta t)^2,$$

(104b)

$$P_{gg} = p_g(t)\left[1 - \Delta t \exp(-\lambda(t)) \sum_i \theta_{g(i)} \exp\left(\frac{-\varepsilon}{kT_i}\right)\right],$$

(105a)

$$P_{ee} = p_e(t)\left[1 - \Delta t \exp(+\lambda(t)) \theta_e\right].$$

(105b)

When we compare Eqs. (104) and (105) with Eqs. (31) and (32), we observe that the following replacement leaves the equations in section III unchanged:

$$\theta_g \to \sum_i \theta_{g(i)} \exp\left(\frac{-\varepsilon}{kT_i}\right).$$

(106)

As a consequence of this replacement, the distribution of e and g in the non-equilibrium steady state is

$$\frac{p_e}{p_g} = \frac{1}{\theta_g} \sum_i \theta_{g(i)} \exp\left(\frac{-\varepsilon}{kT_i}\right),$$

(107)

which is the generalization of the well-known equilibrium distribution.

When we make the replacement of Eq. (106) in (73), we see that the extended entropy Σ is written as

$$\Sigma \equiv kM\left[-p_g \ln\left(\sum_i \theta_{g(i)} \exp\left(\frac{-\varepsilon}{kT_i}\right)\right) - p_e \ln \theta_e - p_g \ln p_g - p_e \ln p_e\right].$$

(108)

This is to be used in the two-gate probability (76):

$$P[p(t_1); p(t_f)] = \exp\left(\frac{\Sigma_f - \Sigma_1}{k}\right).$$

(109)

IV.3 Lyapunov Function

A function whose extremum gives a non-equilibrium steady state is called a Lyapunov function. In the mathematics community, even the free energy is called a Lyapunov function, although its extremum gives an equilibrium state.

We verify that a maximum of Σ in (108) with respect to p_e leads to (107) which is the non-equilibrium steady state. This can identify Σ as a Lyapunov function.

Since Σ reduces to the free energy in special cases, the procedure to maximize Σ is an extension of minimizing the free energy.

Previously we proposed a function whose maximum gives a non-equilibrium steady state, and called the function persistency.[7] It is defined as a special case of the constrained PPF, such that the state does not change in Δt. It may be written as $P^{III}(t; t + \Delta t)$. Using the present notation, we can show

$$(M\Delta t)^{-1} \ln P^{III}(t; t + \Delta t) = -\left(\sqrt{\theta_g p_g} - \sqrt{\theta_e p_e}\right)^2 . \tag{110}$$

Although this does become a maximum at the steady state, and the concept of the persistency is physically sound, $\ln P^{III}$ is not completely satisfactory as a Lyapunov function, because $\ln P^{III}$ always becomes zero at its maximum, and cannot be used in comparing more than one non-equilibrium steady states.

Since the extended entropy Σ can be constructed as a modification of the free energy, it is probably more useful than the persistency, and it deserves further studies.

V. Summary and Discussions

For kinetics of two level atoms, the two-gate probability was treated before,[3] and the anticausal path was proposed as a special solution to describe fluctuated paths. In section III of the present paper, where the same model is taken up, it is shown that general solutions can be obtained in addition to the anticausal path, and numerical examples are shown in Figs. 3.

By treating special cases of the two level atom kinetics in section III, we come to the interpretation that the Path Probability Function (PPF) represents the entropy production as is shown in (71)

$$\ln P^{II}[p_e(t)] = \frac{\Delta S}{k} .$$

When the system does not go to equilibrium in the steady state, we generalize the entropy and introduce the concept of "extended entropy", which we write Σ in this paper. Examples are shown in (73), (86) and (109).

The identification (71) leads us to decide that Recipe II of Ref. 3 is the consistent way of treating the energy inflow from a heat bath as in (78). The third factor of the PPF is to be written as:

$$\ln P_3 = -\frac{\varepsilon}{kT} M P_{ge} .$$

The extended entropy is written in (89):

$$\Sigma \equiv kM \left[-p_g \left(\ln \theta_g - \frac{\varepsilon}{kT} \right) - p_e \ln \theta_e - p_g \ln p_g - p_e \ln p_e \right] ,$$

which reduces to the free energy when the system can go to equilibrium as in (92):

$$\Delta \left(\frac{\Sigma}{k} \right) \equiv \Delta M \left(+p_g \left(\frac{\varepsilon}{kT} \right) - p_g \ln p_g - p_e \ln p_e \right) = -\Delta \left(\frac{F}{kT} \right) .$$

This case was studied before[3] and led to the two-gate probability of (94):

$$P[p(t_1); p(t_f)] = \exp \left(-\frac{F_f}{kT} + \frac{F_i}{kT} \right) .$$

The case of the single heat bath in (78) is generalized to the treatment of multiple heat baths. (78) is then generalized to the last term in (99):

$$\ln P_3 = - \sum_i \frac{\varepsilon}{kT_i} P_{ge(i)} .$$

This leads to the extended entropy in (108):

$$\Sigma \equiv kM \left[-p_g \ln \left(\sum_i \theta_{g(i)} \exp \left(\frac{-\varepsilon}{kT_i} \right) \right) - p_e \ln \theta_e - p_g \ln p_g - p_e \ln p_e \right] .$$

It is noted that the non-equilibrium steady state corresponds to a maximum of Σ. This indicates an important function of Σ that it is a Lyapunov function for non-equilibrium steady state.

The multiple heat bath case is a typical example of the system which does not come to equilibrium in the steady state. The knowledge obtained in this study can be applied to any problems of this category.

References

1. R. Kikuchi, *Phys. Rev.* **81**, 988 (1951).
2. R. Kikuchi, *Annals of Physics (N.Y.)* **10**, 127 (1960).
3. R. Kikuchi, *Prog. Theor. Phys.* Suppl. **35**, 1 (1966).
4. K. Wada and T. Uchida, *Phys. Letters A* **168**, 353 (1992).
5. K. Wada and M. Kaburagi, *Prog. Theor. Phys.* Suppl. **115**, 273 (1994).
6. S. Brush, private communication, 1995.
7. R. Kikuchi, *Phys. Rev.* **124**, 1682 (1961).

Ising Model and Kinetic Mean Field Theories

F. Ducastelle

Direction des Matériaux
Office National d'Etudes et de Recherches Aérospatiales
BP 72, 92322 Châtillon Cedex
FRANCE

Abstract

We discuss the mean field treatments of the kinetic Ising model. We review the general properties of the equations describing the evolution of the local concentrations (or magnetizations) as well as the short-range order parameters. We also discuss the properties of the corresponding mesoscopic master equations.

I. Introduction

The path probability method devised by Prof. Kikuchi is now recognized as an invaluable tool to study ordering or disordering kinetics in alloys. For some applications, for example to treat strongly inhomogeneous systems, its practical implementation however is not quite easy and the simplest point approximation, *i.e.* the usual mean field approximation, is still quite useful and fruitful. Furthermore the mean field approximation provide us with simple explicit analytical formulae which can easily be related to those provided by the phenomenological Ginzburg-Landau or Cahn-Hilliard theories.

Depending on the assumed microscopic mechanism (spin-flip, atomic exchange, etc.) different types of kinetics are expected. Let then η be the (long-range) order parameter characterizing an order-disorder transformation. For simplicity we consider here a scalar order parameter. In general we will have in mind order-disorder transitions of alloys on a fixed underlying lattice.[1]

The order parameter can be defined for the whole system or locally at point r assuming some coarse-grained average or some ensemble average. The order parameter is said to be conserved if its value for the whole system $\eta = \int dr \, \eta(r)$ is fixed. This is the case when the order parameter is the concentration $c(r)$ of an alloy $A_{1-c}B_c$ and when we study phase separation. Otherwise the ordered parameter is said to be

Theory and Applications of the Cluster Variation and Path Probability Methods
Edited by J.L. Morán-López and J.M. Sanchez, Plenum Press, New York, 1996

21

a non-conserved order parameter. This is the case of the order parameter involved in genuine order-disorder transformations. For example if an alloy on a *bcc* lattice orders according to the B2 (CsCl-type) ordered structure, $(1 + \eta)/2$ is simply the concentration in A atoms of one of the two simple cubic sublattices of the B2 ordered structure.

Let now $F(\eta)$ be the Landau free energy functional. The time dependent Ginzburg-Landau evolution equation for the non-conserved order parameter η is

$$\frac{d\eta}{dt} = -\Gamma \frac{\partial F}{\partial \eta}. \tag{1}$$

According to the general classification of kinetic models given by Halperin and Hohenberg[2] this is also called model A. Model B deals with a conserved order parameter with an evolution equation of the Cahn-Hilliard type

$$\frac{d\eta}{dt} = M \nabla^2 \frac{\partial F}{\partial \eta}. \tag{2}$$

More complicated and realistic models can be considered. For example model C deals with two order parameters of both types.

Let us now summarize the content of this brief review. Much more can be found in several review articles.[3-11] We first apply the usual mean field treatments to the microscopic kinetic Ising model. This provides us with equations similar (but not exactly identical) to the above phenomenological equations for the evolution of the long-range order parameter. Two-point correlation functions can also be studied and we shall derive the kinetic equivalent of the Krivoglaz-Clapp-Moss equation for the time evolution of the short-range order parameters. Finally it is also possible to write a master equation for the evolution of the probability distribution $P(\eta, t)$ of the order parameter at time t and we will discuss the general properties of this equation as well as its equivalence with Fokker-Planck or Langevin equations.

II. Kinetic Ising Model and Mean Field Approximations

We start from the Ising model

$$H = -\frac{1}{2} \sum_{n,m} J_{nm} \sigma_n \sigma_m - \sum_n h_n \sigma_n , \tag{3}$$

where H is the hamiltonian or energy of the configuration of the spins or atoms characterized by the variables $\sigma_n = \pm 1$. h_n is in general independent of n and plays the part of a magnetic field for a magnetic system or of a chemical potential difference in the case of alloys.

Let now $\rho(\sigma)$ be the probability of having the configuration denoted here by the global variable $\sigma = \{\sigma_n\}$. The time evolution is assumed to be given by the following microscopic master equation

$$\frac{d\rho(\sigma)}{dt} = - \text{Tr}_{\sigma'} \, \rho(\sigma) W(\sigma, \sigma') + \text{Tr}_{\sigma'} \, \rho(\sigma') W(\sigma', \sigma), \tag{4}$$

where $W(\sigma, \sigma')$ is the transition probability per unit time from configuration σ to configuration σ' and $\text{Tr}_{\sigma'}$ indicates a sum over all configurations σ'. The time derivative of any observable $\langle O \rangle$ is therefore given by

$$\frac{d\langle O\rangle}{dt} = -\,\mathrm{Tr}_{\sigma,\sigma'}\,\rho(\sigma)W(\sigma,\sigma')\,[O(\sigma) - O(\sigma')]\,. \tag{5}$$

In order to insure that the equilibrium distribution

$$\rho_{\mathrm{eq}}(\sigma) = \frac{\exp(-\beta H)}{\mathrm{Tr}\exp(-\beta H)}\,; \qquad \beta = 1/k_B T\,, \tag{6}$$

is the stationary solution of this equation, $W(\sigma',\sigma)$ is generally assumed to satisfy the detailed balance condition

$$\rho_{\mathrm{eq}}(\sigma)W(\sigma,\sigma') = \rho_{\mathrm{eq}}(\sigma')W(\sigma',\sigma)\,. \tag{7}$$

Several choices for $W(\sigma,\sigma')$ are still possible. Let ΔE be the energy difference between the final and initial states, $\Delta E = H(\sigma') - H(\sigma)$, the following choices are frequently used

$$W = \theta\,(1 - \tanh(\beta\Delta E)/2)/2 \qquad \text{Glauber}$$
$$W = \theta\,\min\,(1, \exp(-\beta\Delta E)) \qquad \text{Metropolis}$$
$$W = \theta\,\exp(-\beta\Delta E/2) \qquad \text{Kikuchi}\,, \tag{8}$$

where θ^{-1} is a time scale. This is not sufficient. We have still to define the allowed transitions. The spin-flip mechanism consists in assuming that the elementary events are single spin flips: $\sigma'_n = -\sigma_n$; $\sigma'_m = \sigma_m$ for $m \neq n$. This is what is frequently called the Glauber dynamics. It can be used when the order parameter is not conserved. Another more realistic model for alloys is the Kawasaki dynamics in which the elementary events are the exchange of atoms. First neighbour exchange is generally assumed: $\sigma'_n = \sigma_m$; $\sigma'_m = \sigma_n$ if n and m are first neighbours. Since only the configurations where $\sigma_n \neq \sigma_m$ are modified it is generally assumed that $W = 0$ when $\sigma_n = \sigma_m$ but this is not a necessity. Other more realistic mechanisms involving vacancies or trying to simulate them are also used. It is clear for example that θ in the case of alloys is a thermally activated jump frequency which therefore depends on temperature and which might also introduce a further dependence on the atomic configuration.[12]

II.1 Spin-Flip Dynamics

In the case of spin flips, $W(\sigma,\sigma') = \sum_n W^n(\sigma)$ with

$$W^n(\sigma) = \theta\exp\left(-\sigma_n \beta h_n^{\mathrm{eff}}\right)\,; \qquad h_n^{\mathrm{eff}} = h_n + \sum_m J_{nm}\,\sigma_m\,, \tag{9}$$

so that

$$\frac{d\langle\sigma_n\rangle}{dt} = -2\,\mathrm{Tr}\,\rho(\sigma)W^n(\sigma)\,\sigma_n\,. \tag{10}$$

In a mean field approximation it is quite natural to replace the local effective field h_n^{eff} by its average over the occupancy of sites $m \neq n$. We therefore write

$$h_n^{\mathrm{eff}} \simeq h_n + \sum_m J_{nm}\,m_m\,; \qquad m_m = \langle\sigma_m\rangle\,. \tag{11}$$

The trace over $\rho(\sigma)$ yields $\rho_n = (1 + m_n\sigma_n)/2$ and finally

$$\frac{dm_n}{dt} = -2\theta\left(m_n\cosh(\beta h_n^{\text{eff}}) - \sinh(\beta h_n^{\text{eff}})\right). \tag{12}$$

This equation is satisfied at equilibrium by the well-known mean field equation $m_n = \tanh\beta h_n^{\text{eff}}$. Introducing the concentration $c_n = (1 + m_n)/2$, we can also write

$$\frac{dc_n}{dt} = -\theta\left(c_n\exp(-\beta h_n^{\text{eff}}) - (1 - c_n)\exp(\beta h_n^{\text{eff}})\right). \tag{13}$$

Another form of this equation can still be derived if we notice that

$$\rho_n\exp(-\sigma_n\beta h_n^{\text{eff}}) = \left(\sqrt{1 - m_n^2}/2\right)\exp(\sigma_n\beta\nu_n), \tag{14}$$

with

$$\beta\nu_n = -\beta h_n^{\text{eff}} + \frac{1}{2}\ln\left(\frac{1 + m_n}{1 - m_n}\right) = \frac{\partial\beta F}{\partial m_n}, \tag{15}$$

where F is here the standard mean field free energy functional

$$\beta F = -\frac{1}{2}\sum_{nm}\beta J_{nm}\,m_n\,m_m - \sum_n\beta h_n\,m_n$$
$$+ \frac{1 + m_n}{2}\ln\left(\frac{1 + m_n}{2}\right) + \frac{1 - m_n}{2}\ln\left(\frac{1 - m_n}{2}\right), \tag{16}$$

so that

$$\frac{dm_n}{dt} = -2\theta\sqrt{1 - m_n^2}\,\sinh(\beta\nu_n). \tag{17}$$

If we now use notations appropriate to alloys, the chemical potential $\mu_n = \partial F/\partial c_n$ is equal to $2\nu_n$ and

$$\frac{dc_n}{dt} = -2\theta\sqrt{c_n(1 - c_n)}\,\sinh(\beta\mu_n/2). \tag{18}$$

In the limit of weak ν_n we see that we recover the Ginzburg-Landau form $dm_n/dt = -\Gamma_n\,\partial F/\partial m_n$ with $\Gamma_n = 2\theta\beta\sqrt{1 - m_n^2}$, but in the case of inhomogeneous sytems the dependence on c_n of Γ_n can be important.

Let us now calculate the time derivative of the free energy functional calculated for m_n satisfying the evolution equation Eq. (17)

$$\frac{dF}{dt} = \sum_n\frac{\partial F}{\partial m_n}\frac{dm_n}{dt} = -2\theta\sum_n\sqrt{1 - m_n^2}\,\nu_n\sinh(\beta\nu_n) \le 0. \tag{19}$$

This means that F is a so-called Lyapunov function and that the m_n tend to a local minimum of the free energy if the initial value does not correspond to an unstable stationary position where $\nu_n = 0$.[13]

Another interesting quantity is the function measuring the number of spin flips per unit time and more generally the number of events per unit time. Let Ψ be this function (it is denoted G by Wada et al.[14])

$$\Psi = \mathrm{Tr}_{\sigma,\sigma'} \, \rho(\sigma) W(\sigma, \sigma') = \sum_n \mathrm{Tr} \, \rho_n W^n$$

$$= \theta \sum_n \sqrt{1 - m_n^2} \, \cosh(\beta \nu_n), \qquad (20)$$

Ψ plays the part of a generating function for kinetic processes.[14-16] We have in particular

$$\frac{dm_n}{dt} = 2 \frac{\partial \Psi}{\partial \beta h_n} \equiv R_n . \qquad (21)$$

II.2 Spin-Exchange Dynamics

We assume that the allowed processes are exchanges of different neighboring atoms (spins). It is not so obvious in this case to obtain a proper mean field approximation whose equilibrium solution is the usual equilibrium mean field or Bragg-Williams solution, because we want to write a single site approximation for a "pair" process. This means we should not take into account too precisely the correlation between the occupancies of the two concerned sites. We shall instead assume in some way that the exchange is the succession of two spin flips, so that each spin is flipped within its local mean field. The energy variation in this process is then given by

$$\Delta E_{nm} = \left(h_n^{\mathrm{eff}} - h_m^{\mathrm{eff}} \right) (\sigma_n - \sigma_m) . \qquad (22)$$

Using now equation (5) we obtain

$$\frac{d\langle \sigma_n \rangle}{dt} = - \sum_p{}' \mathrm{Tr} \, \rho(\sigma) W^{np}(\sigma)(\sigma_n - \sigma_p);$$

$$W^{np}(\sigma) = \theta \exp(-\beta \Delta E_{np}/2), \qquad (23)$$

where the sum is over the first neighbors p of site n. If we now approximate the pair density matrix ρ_{np} by $\rho_n \rho_p$, we obtain

$$\frac{dm_n}{dt} = 2 \frac{dc_n}{dt}$$

$$= 2\theta \sum_p{}' \left\{ c_n(1 - c_p) \exp\left[\beta \left(h_p^{\mathrm{eff}} - h_n^{\mathrm{eff}} \right) \right] - c_p(1 - c_n) \exp\left[\beta \left(h_n^{\mathrm{eff}} - h_p^{\mathrm{eff}} \right) \right] \right\}$$

$$= \theta \sum_p{}' \sqrt{(1 - m_n^2)(1 - m_p^2)} \, \sinh\left(\beta(\nu_p - \nu_n) \right) . \qquad (24)$$

Since the concentration is conserved we have obviously $\nu_n = \nu_p$ and $dc_n/dt \equiv 0$ for homogeneous systems. For inhomogeneous systems we also see that the equilibrium solution reduces to the usual mean field equation $\nu_p = \nu_n = -h = \mathrm{constant}$. Consider again the limit of weak chemical potential variations; then $\sinh \beta(\nu_p - \nu_n) \simeq \beta(\nu_p - \nu_n)$. If furthermore the concentration itself is slowly varying we see that dm_n/dt becomes proportional to $\sum_p{}'(\nu_p - \nu_n)$ which is nothing but the discrete Laplacian of ν_n. In this limit we therefore recover model B but even then, the mobility M depends on the local concentrations;[12,17] see also Ref. 18.

Here again the free energy functional is a Lyapunov function since

$$\frac{dF}{dt} = -\frac{1}{2}\theta \sum_{n,m}{}' \sqrt{(1-m_n^2)(1-m_m^2)}\,(\nu_n - \nu_m)\sinh\left(\beta(\nu_n - \nu_m)\right) \leq 0. \qquad (25)$$

Thus our mean field treatment is completely consistent. Other treatments can lead to some difficulties.[13] Finally the function Ψ is given here by

$$\Psi = \frac{1}{2}\theta \sum_{\langle nm \rangle} \sqrt{(1-m_n^2)(1-m_m^2)}\,\cosh\left(\beta(\nu_n - \nu_m)\right), \qquad (26)$$

where the sum is over all first neighbour pairs $\langle nm \rangle$. We can also verify that Eq. (21) is also satisfied here.

III. Fluctuations and Short-Range Order

Since the mean field approximation is a single site approximation, there is no fully consistent way of calculating the pair correlation function $\langle \sigma_n \sigma_m \rangle$. In a genuine mean field approximation, it reduces to $\langle \sigma_n \rangle \langle \sigma_m \rangle$, but using the fluctuation-dissipation theorem an improved formula can be obtained at equilibrium, called the Krivoglaz-Clapp-Moss formula (see e.g., Ref 1). Let us precise our notations. We define the pair correlation function S_{nm} (or cumulant) through

$$\begin{aligned} S_{nm} &= \langle \sigma_n \sigma_m \rangle_c = \langle \sigma_n \sigma_m \rangle - \langle \sigma_n \rangle \langle \sigma_m \rangle \\ &= \langle \sigma_n \sigma_m \rangle - m_n\,m_m\,. \end{aligned} \qquad (27)$$

This function is also related to the susceptibility $\chi_{nm} = \partial m_n / \partial h_m$ or to the so-called Warren-Cowley short-range-order parameter α_{nm} through

$$S_{nm} = k_{\mathrm{B}}T\,\chi_{nm} = 4c_n(1-c_m)\alpha_{nm}\,. \qquad (28)$$

Now, starting from the general equation (5) we obtain, for the spin-flip dynamics,

$$\frac{dS_{nm}}{dt} = -2\,\mathrm{Tr}\,\rho W^n \sigma_n(\sigma_m - m_m) - 2\,\mathrm{Tr}\,\rho W^m \sigma_m(\sigma_n - m_n) + 4\delta_{nm}\,\mathrm{Tr}\,\rho W^n. \qquad (29)$$

A mean field treatment consistent with the general theory of Gaussian fluctuations,[19,20] (see also Section IV) consists in performing in the above formula the substitution[15]

$$W_n\,\sigma_n \to \sum_p \sigma_p \frac{\partial \langle W^n \sigma_n \rangle}{\partial m_p}\,. \qquad (30)$$

We then obtain a closed formula for the time evolution of the pair correlation function

$$\frac{dS_{nm}}{dt} = \sum_p \frac{\partial R_n}{\partial m_p} S_{pm} + \sum_p \frac{\partial R_m}{\partial m_p} S_{pn} + R_{nm}\,, \qquad (31)$$

where

$$R_n = 2\frac{\partial \Psi}{\partial \beta h_n}\,; \qquad R_{nm} = 4\frac{\partial^2 \Psi}{\partial \beta h_n\, \partial \beta h_m}\,. \qquad (32)$$

This is a set of linear differential equations whose coefficients depend on time through the $m_n(t)$.

III.1 Spin-Flip Dynamics

For the spin-flip dynamics,

$$R_n = -2\theta\sqrt{1 - m_n^2}\,\sinh(\beta\nu_n); \qquad R_{nm} = 4\theta\delta_{nm}\sqrt{1 - m_n^2}\,\cosh(\beta\nu_n). \qquad (33)$$

At equilibrium $dS_{nm}/dt = 0$, $\nu_n = 0$ so that

$$\frac{\partial R_n}{\partial m_p} = -2\theta\sqrt{1 - m_n^2}\,\frac{\partial^2\beta F}{\partial m_n\,\partial m_p}; \qquad R_{nm} = 4\theta\delta_{nm}\sqrt{1 - m_n^2}, \qquad (34)$$

and S_{nm} is, as expected, equal to the inverse of the matrix of the second derivatives $\partial^2\beta F/\partial m_n\partial m_m$.

If we are interested in the behavior of the correlation function in the high temperature disordered phase, we can profit from translational invariance and solve the above equation using Fourier transforms. Let $S(q)$ be this Fourier transform, $S(q) = \sum_{n,m} S_{nm}\exp(iq\cdot(m-n))/N$ where N is the number of lattice sites, we obtain

$$\frac{dS(q)}{dt} = -4\theta\left(-\frac{m}{\sqrt{1 - m^2}}\,\sinh(\beta\nu) + \sqrt{1 - m^2}\,D(q)\cosh(\beta\nu)\right)S(q)$$

$$+ 4\theta\sqrt{1 - m^2}\,\cosh(\beta\nu), \qquad (35)$$

where m is the (time-dependent) uniform value of the magnetization and $D(q)$, the Fourier transform of $\partial^2\beta F/\partial m_n\,\partial m_m$, is given by

$$D(q) = \frac{1}{1 - m^2} - \beta J(q), \qquad (36)$$

$J(q)$ being the Fourier transform of J_{nm}. At equilibrium we therefore recover the Krivoglaz-Clapp-Moss formula $S_{eq}(q) = 1/D_{eq}(q)$. Close to equilibrium we can replace m by its equilibrium value and the correlation function is found to relax exponentially towards equilibrium with a relaxation time τ_q, $1/\tau_q = 4\theta\sqrt{1 - m^2}\,D(q)$ and it is easy to check that this is twice the value corresponding to the relaxation of the Fourier component m_q of the magnetization, which is a classical result.[19,20]

III.2 Spin-Exchange Dynamics

In this case,

$$R_n = \theta{\sum_p}'\sqrt{(1 - m_n^2)(1 - m_p^2)}\,\sinh\left(\beta(\nu_p - \nu_n)\right),$$

$$R_{nm} = 2\theta{\sum_p}'(\delta_{nm} - \delta_{pm})\sqrt{(1 - m_n^2)(1 - m_p^2)}\,\cosh\left(\beta(\nu_p - \nu_n)\right). \qquad (37)$$

For uniform systems, $\nu_n = $ constant and we obtain

$$\frac{dS(q)}{dt} = -2\theta(1 - m^2)\gamma(q)\left[D(q)S(q) - 1\right], \tag{38}$$

with

$$\gamma(q) = \sum_{R}' \left(1 - e^{iq \cdot R}\right). \tag{39}$$

where the sum over R is over the first neighbors of a given site (we assume that all lattice sites are equivalent).

We then find a relaxation time such that $1/\tau_q = 2\theta(1 - m^2)\gamma(q)D(q)$. Since $\gamma(q) \sim q^2$ when $q \to 0$ we obtain the expected diffusive mode in the long-wave length limit. It should also be noticed that since the concentration or magnetization is fixed here, the mean field free energy functional is not necessarily convex and $D(q = 0)$ is negative below the spinodal temperature $k_B T = (1 - m^2)J(q = 0)$. Then $S(q)$ grows exponentially for small values of q; the fluctuations are no longer Gaussian and more elaborate treatments are required. There is an important literature devoted to this problem of spinodal decomposition[3-8] and we shall not continue here in this direction, but similar problems will be encountered in the next section.

Finally, let us recall that, as mentioned in section II, the Ginzburg-Landau (or model A) and Cahn-Hilliard (or model B) are recovered if the concentration is uniform and if we keep the lowest terms in the expansion of the hyperbolic sine functions. Then the factor Γ of model A is equal to $4\beta\theta\sqrt{c(1 - c)}$ and the mobility M of model B is equal to $4\beta\theta c(1 - c)$ if model B is defined with a discrete "Laplacian" Δ_d such that its action on any function of n, is given by $\Delta_d f_n = \sum_R'(f_{n+R} - f_n)$. For a simple cubic lattice, $\Delta_d \to 3a^2\nabla^2$ in the continuum limit, where a is the lattice parameter.

The corresponding equations for the correlation function become

$$\frac{dS(q)}{dt} = -2\Gamma k_B T\left[D(q)S(q) - 1\right] \qquad \text{Model A}$$

$$\frac{dS(q)}{dt} = -2M k_B T\gamma(q)\left[D(q)S(q) - 1\right] \qquad \text{Model B}. \tag{40}$$

IV. Master Equation, Gaussian and Non-Gaussian Fluctuations

For simplicity we shall confine the following discussion to uniform systems with the spin-flip dynamics. We know that the magnetization, which we denote here $\overline{m}$ for reasons that will be clear below, varies according to

$$\frac{d\overline{m}}{dt} = -2\theta\sqrt{1 - \overline{m}^2}\sinh(\beta\nu),$$

$$\nu(\overline{m}, h) = \frac{\partial f}{\partial \overline{m}}$$

$$= -\frac{pJ\overline{m}^2}{2} - \overline{m}h + \frac{1 + \overline{m}}{2}\ln\left(\frac{1 + \overline{m}}{2}\right) + \frac{1 - \overline{m}}{2}\ln\left(\frac{1 - \overline{m}}{2}\right). \tag{41}$$

where $f = F/N$. In fact we will be mostly interested in the zero field situation $h = 0$, but it is useful to keep the field dependence of the generating function $\psi(h, \overline{m}) = \Psi(h, \overline{m})/N$ and to set $h = 0$ after derivatives with respect to h are taken

$$\psi(\overline{m}, h) = \theta\sqrt{1 - m^2} \cosh\left(\beta\nu(\overline{m}, h)\right),$$

$$\frac{d\overline{m}}{dt} = 2\frac{\partial\psi}{\partial\beta h}\bigg|_{h=0}. \tag{42}$$

Now in the standard mean field theory we can also calculate the probability $P(m)$ that a system of N spins has the mean magnetization $\overline{m}$. When $N \to \infty$, we have $P(m) \sim C \exp(-N\beta f(m))$ where C is a normalization constant. The mean magnetization $\overline{m} = \int dm\, P(m)\, m$ is asymptotically equal to the value of m minimizing $f(m)$. Expanding then $f(m)$ around this value provides us with a Gaussian approximation for $P(m)$

$$P(m) \sim C \exp\left(-N\beta f''(\overline{m})(m - \overline{m})^2/2\right). \tag{43}$$

The variance $\langle(m - \overline{m})^2\rangle$ is equal to $(N\beta f'')^{-1}$ as it should. Actually this variance should be equal to $S_{eq}(q = 0)/N$, hence the result after the discussion given in section III.1.

IV.1 Master Equation and Kubo Ansatz

The problem now is to calculate $P(m, t)$ for non-equilibrium states. Using the techniques developed for deriving the PPM one can show[14,16,21] that $P(m, t)$ satisfies the equation

$$\frac{\partial P(m, t)}{\partial t} = N\left\{\exp\left(-\frac{2}{N}\frac{\partial}{\partial m}\frac{\partial}{\partial\beta h}\right) - 1\right\}\psi(m, h)\, P(m, t). \tag{44}$$

Using the above form of $\psi(m, h)$, this equation is identical to the following master equation

$$\frac{\partial P(m, t)}{\partial t} = -\sum_{m'} P(m, t)W(m, m') + \sum_{m'} P(m', t)W(m', m), \tag{45}$$

where $m' = m \pm \frac{2}{N}$ and

$$W(m, m' = m \pm \tfrac{2}{N}) = \theta N\frac{\sqrt{1 - m^2}}{2}\exp(\pm\beta\nu). \tag{16}$$

This is just the so-called Weiss kinetic model.[10,20] Since this is a master equation for the total magnetization per site, $m = (\sum_n m_n)/N$, we call it a mesoscopic master equation. We see therefore that this mean field mesoscopic equation can be derived from a mean field treatment of the initial microscopic master Eq. (4). The form Eq. (44) of this equation is very useful. It can be extended to PPM cluster expansions of any order by replacing ψ by its appropriate form as a function of cluster variables m_α and h_α.[14-16] Furthermore the standard Kramers-Moyal expansion is directly obtained by expanding the exponential in the right hand side of this equation.

From the general properties of master equations and Markov processes, we know that $P(m, t) \to P_{eq}(m) = C \exp(-N\beta f(m))$ when $t \to \infty$, but we would like to know more precisely what happens, and how this depends on the initial conditions.

We will generally have in mind an initial Gaussian distribution with a mean value $\overline{m}_0$ and a variance of the order of $1/N$ which we denote σ_0/N. Now, consider for example a ferromagnetic Ising model with positive first neighbor coupling J. Mean field theory tells us that the critical temperature is equal to $k_B T_c = ZJ$ where Z is the number of first neighbors. Above this temperature the equilibrium free energy $f(m)$ is convex and we expect that any initial distribution will relax rapidly towards the equilibrium distribution and will remain approximately Gaussian. There are several methods to prove that. One of them is the large N expansion of Van Kampen.[19] An equivalent method is to use Kubo ansatz, i.e. to postulate that $P(m,t)$ is of the form $C \exp(-N\phi(m,t))$ and to keep only in ϕ the terms of order zero in $1/N$.[20,22] If we apply this to Eq. (44), we obtain

$$\frac{\partial \phi}{\partial t} = \psi(m,h) - \psi(m, h + 2\frac{\partial \phi}{\partial m}) = -H(m, \frac{\partial \phi}{\partial m}). \tag{47}$$

This is a partial differential equation of the Hamilton-Jacobi (H-J) type which can be solved using techniques borrowed from mechanics, ϕ playing the part of the action whereas $m, \partial\phi/\partial m$ and H play the part of the position, of the momentum and of the Hamiltonian, respectively.[20,23] The corresponding mechanical trajectories in phase space $(m, p = \partial\phi/\partial m)$ are given by

$$\frac{dm}{dt} = \frac{\partial H}{\partial p}; \qquad \frac{dp}{dt} = -\frac{\partial H}{\partial m}. \tag{48}$$

Similar equations are also obtained by starting from the path integral form of the master equation.[16,23] Since H is here is a constant of motion, it is fairly easy to represent the trajectories and therefore to solve qualitatively the H-J equation.[20] If we confine ourselves to a Gaussian approximation, which is completely equivalent to the large-N expansion of Van Kampen, we approximate ϕ by

$$\phi(m,t) \simeq \frac{1}{2\sigma(t)} [m - \overline{m}(t)]^2, \tag{49}$$

and expanding the H-J equation up to second order in the fluctuation $m - \overline{m}(t)$, we find

$$\frac{d\overline{m}}{dt} = 2\frac{\partial \psi}{\partial \beta h}$$

$$\frac{d\sigma}{dt} = 4\frac{\partial^2 \psi}{\partial m \, \partial \beta h}\sigma + 4\frac{\partial^2 \psi}{\partial h^2}, \tag{50}$$

which is in complete agreement with the equations derived in section III since σ is nothing but the correlation function $S(q = 0)$. If $f(m)$ is convex, it can be checked that σ never diverges and that both $\overline{m}$ and σ relax towards their equilibrium values with time scales $\tau_{q=0}$ and $\tau_{q=0}/2$ respectively. Of course fluctuations of higher-order beyond the Gaussian approximation can be calculated as well.

So, in the presence of a single minimum of the mean field free energy, the situation is fairly simple: The system evolves towards equilibrium with finite relaxation times, i.e. independent of the size N of the system.

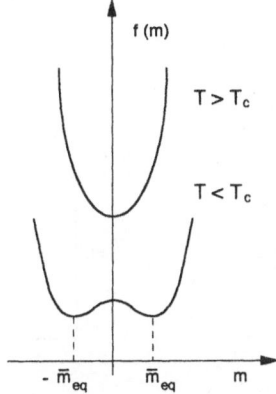

Figure 1. Free energy $f(m)$ in zero field above and below T_c.

IV.2 Relaxation in Unstable and Metastable Systems

The situation is quite different below the transition temperature. Consider first the situation in zero field. The free energy now has the familiar double well shape (Fig. 1). What happens if we quench the system from above to below the transition temperature? Our homogeneous model cannot be realistic since we know that the long time behaviour is governed by domain coarsening, *i.e.* by the evolution of spatial fluctuations. The mean field approximation itself with this non-convex free energy is certainly inaccurate but as we shall see the model remains very instructive.

It is sufficient for this discussion to consider the phenomenological model A. All previous equations apply provided we use for ψ the appropriate expression obtained by expanding the hyperbolic cosine function. This gives

$$\psi = \frac{k_B T\Gamma}{2} \left(1 + (\beta \nu)^2/2\right). \tag{51}$$

For simplicity, we also set $k_B T = \Gamma = 1$. Since ψ is now a quadratic function of h, the Kramers-Moyal expansion of the master equation stops at the second order term and reduces to a Fokker-Planck equation

$$\frac{\partial P}{\partial t} = \frac{\partial}{\partial m}\left(\nu(m)\,P\right) + \frac{1}{N}\frac{\partial^2 P}{\partial m^2}. \tag{52}$$

This is a Fokker-Planck of a famous type whose properties have been investigated in detail and we will just summarize the results.[24,25] Let us first apply the Kubo ansatz. The Fokker-Planck becomes

$$\frac{\partial \phi}{\partial t} = \phi'\left(\nu - \phi'\right) + \left(\phi'' - \nu'\right)/N, \tag{53}$$

where the primes denote derivatives with respect to m. Neglecting the terms of order $1/N$ we obtain again a simple H-J equation. We assume that the initial state is paramagnetic and Gaussian. Then $\overline{m}_0 = 0$, and $\overline{m}(t)$ does not vary although this corresponds to a metastable situation. Now, we also approximate $f(m)$ by the usual simple Landau form

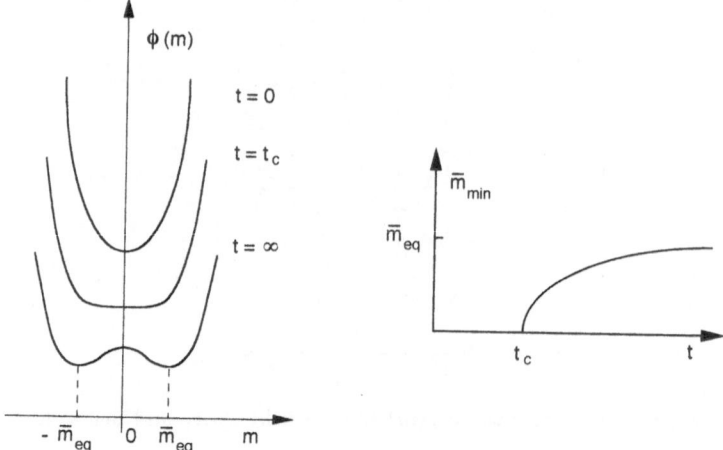

Figure 2. Evolution of $\phi(m)$ and of the position $\bar{m}_{\min}$ of the (positive) minimum as a function of time; t_c is of the order of $\ln N$.

$$f(m) = \tfrac{1}{2}r_{eq}m^2 + \tfrac{1}{4}u_{eq}m^4; \qquad \nu(m) = r_{eq}m + u_{eq}m^3, \qquad (54)$$

where $r_{eq} < 0$. The stationary solutions $\nu(m) = 0$ are the unstable solution $m = 0$ or the two symmetric stable solutions $m = \pm\bar{m}_{eq} = \pm(-r_{eq}/u_{eq})^{1/2}$. Let us write ϕ in a similar form

$$\phi = \tfrac{1}{2}rm^2 + \tfrac{1}{4}um^4 + \cdots \qquad (55)$$

Inserting this expression into the H-J equation yields, in particular,

$$\frac{1}{2}\frac{dr}{dt} = r(r_{eq} - r), \qquad (56)$$

where r is simply the inverse of the variance σ. Since $r_{eq} < 0$, it is clear that $r(t) \to 0$ when $t \to \infty$ for any positive initial value of $r, r_0 = \sigma_0^{-1}$, $r(t) \sim r_0\, e^{-2|r_{eq}|t}$. An investigation of the H-J equation shows that in fact ϕ' tends to 0 when $|m| \le \bar{m}_{eq}$ and to ν otherwise. This is allowed since $\phi' = 0$ and $\phi' = \nu$ are both solutions of the H-J equation. On the other hand $\phi' = \nu$, i.e., $\phi = f$ is the only stationary solution of the Fokker-Planck equation or of its equivalent, Eq. (53). The point is that when $\phi' \to 0$ the terms of order $1/N$ in this equation are no longer negligible. In particular the term $-\nu'/N$ is positive for small m and induces an increase of ϕ and finally the appearance of new symmetric minima of $\phi(m)$. This happens for times t_c of the order of $|r_{eq}|^{-1}\ln N$. The position $m_{\min}$ of the minima then increases as $(t - t_c)^{1/2}$. One can say that there is a dynamic (second order) phase transition at time t_c. If $N \to \infty$ before $t \to \infty$, the analysis based on the H-J equation is correct and we never reach t_c. In practice however $\ln N$ is never very large for realistic macroscopic systems and we therefore expect a transient regime at time t_c describing the escape from the metastable position. It is not easy to develop a quantitative theory of this regime, but a powerful scaling theory has been developed by Suzuki.[24,25]

The global scenario is however clear (Fig. 2): we have first a fast diffusive regime where the fluctuations, as measured by $\sigma = r^{-1}$, increase very rapidly with a time scale

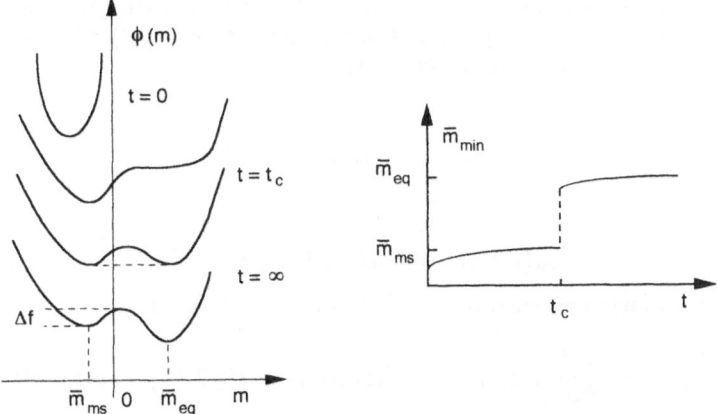

Figure 3. Evolution of $\phi(m)$ and of the position $\overline{m}_{\min}$ of the minimum when the initial minimum lies in the metastable well; t_c is here of the order of $\exp N\Delta f$.

related to the inverse of the curvature r_{eq} of $f(m)$ at the origin. These fluctuations eventually diverge at time t_c. During the corresponding transient regime, new minima of $\phi(m,t)$, ($i.e.$, new maxima of $P(m,t)$) appear at finite and symmetric values $\pm\overline{m}_{\min}$ of m; this is followed by a final relaxation regime controlled by the curvature of $f(m)$ at the equilibrium values $\pm\overline{m}_{eq}$. This final regime can also be studied using other techniques, by studying the lowest eigenvalues of a Schrödinger equation equivalent to the master or Fokker-Planck equation.[19,26,27]

These methods as well as "first-passage" theories[19] are very powerful to study what happens if the initial system has a finite mean magnetization $i.e.$, when the initial distribution is already preferentially in one well. Here again the initial regime can be studied using the H-J equation. Since the symmetry is broken we find of course that the mean magnetization and the fluctuations relax first towards the equilibrium values corresponding to the prefered well (but of course there is still a transient regime with enhanced fluctuations if the initial magnetization is small). This is true even in the presence of a magnetic field. Then, the wells are no longer symmetrical and the initial well may be metastable. On intuitive grounds it is clear that the system has a probability to escape from the metastable well ($\overline{m} = \overline{m}_{ms}$) proportional to $\exp(-N\Delta f)$ where Δf is the free energy difference between the metastable and the unstable positions. The techniques mentioned above allow us to justify and precise this guess. So, in that case the transient regime (this is frequently called the Kramers regime), separating the regime of fast relaxation towards the metastable and towards the final stable states (first order dynamical transition) occurs at times t_c of the order of $\exp N\Delta f$, which is now really very large (Fig. 3). Recent numerical studies of the master Eq. (45) confirm these analyses.[14,28]

IV.3 Langevin Approach and Summary

The approaches described previously based on the study of master or Fokker-Planck equations are well suited to analytic discussions, but equivalent Langevin approaches are also frequently used, and are particularly convenient for numerical simulations. Actually the phenomenological equations of models A, B, etc., are usually written in the form of Langevin equations, with random forces satisfying the fluctuation-dissipation theorem. In this approach, the magnetization $m_n(t)$ becomes a stochastic variable. For example for model A we write

$$\frac{dm_n}{dt} = -\Gamma \frac{\partial F}{\partial m_n} + \xi_n(t), \tag{57}$$

where $\xi_n(t)$ is a Gaussian white noise

$$\langle \xi_n(t)\, \xi_m(t') \rangle = 2k_B\, T\Gamma \delta_{nm} \delta(t - t'). \tag{58}$$

In the case of uniform systems, $m = (\sum_n m_n)/N$, and this gives

$$\frac{dm}{dt} = -\Gamma \frac{\partial f}{\partial m} + \xi(t); \qquad \langle \xi(t)\, \xi(t') \rangle = 2k_B T\, (\Gamma/N)\, \delta(t - t'), \tag{59}$$

and this Langevin equation is perfectly equivalent to the "quasilinear" Fokker-Planck equation (52).[19]

The advantage of this Langevin approach is that it allows us to perform numerical simulations for inhomogeneous situations, and many interesting results have been obtained recently,[9] but then one should be careful with the definition of m_n which in a mean field theory is already a coarse-grained or ensemble average. This poses more generally the problem of relating the microscopic and mesoscopic scales in a consistent scheme.

To summarize, the behaviour of the distribution $P(m, t)$ in the mean field approximation and for homogeneous systems is fairly well understood, but some work is needed still in the case of inhomogeneous systems to make contact with the usual theories of domain coarsening (model A) and of nucleation and growth (model B).

Acknowledgements

The author is indebted to Prof. M. Suzuki for his fruitful comments concerning the relaxation of unstable systems which have, hopefully, improved the final version of this short review. The interested reader should consult his very complete review on the subject.[25]

References

1. For a review, see e.g. F. Ducastelle, *Order and Phase Stability in Alloys*, Eds. F. R. de Boer and D. G. Pettifor, Cohesion and Structure (North-Holland, Amsterdam, 1991), Vol. 3.
2. P. C. Hohenberg and B. I. Halperin, *Rev. Mod. Phys.* **49**, 435 (1977).
3. J. S. Langer, in *Solids Far from Equilibrium*, edited by C. Godrèche (Cambridge University Press, 1992), p. 297.
4. A. J. Bray, *Adv.Phys.* **43**, 357 (1994).
5. J. D. Gunton, M. San Miguel, and P. S. Sahni, in *Phase Transitions and Critical Phenomena*, edited by C. Domb and J. Lebowitz (Academic Press, New York, 1983), p. 267.
6. J. D. Gunton and M. Droz, *Introduction to the Theory of Metastable and Unstable States*, Lecture Notes in Physics 183 (Springer, Berlin, 1983).

7. K. Binder, *Rep. Prog. Phys.* **50**, 783 (1987), and in *Materials Science and Technology*, edited by P. Haasen (VCH, Weinheim, 1991), Vol. 5, p. 405.

8. Furakawa, H., *Adv. Phys.* **34**, 703 (1985).

9. A. G. Khachaturyan, *Theory of Structural Transformations in Solids* (Wiley, New York, 1983); A. G. Khachaturyan, Y. Wang, and H. Y. Wang, *Materials Science Forum*, Vols. 155-156 (Trans.Tech.Pub., Aedermannsdorf, 1994), p. 345; Y. Wang, H. Y. Wang, L.-Q. Chen, and A. G. Khachaturyan, in *Solid → Solid Phase Transformations*, edited by W. C. Johnson, J. M. Howe, D. E. Laughlin, and W. A. Soffa (The Minerals, Metals and Materials Society, Warrendale, 1994), p. 245.

10. K. Kawasaki, in *Phase Transitions and Critical Phenomena*, edited by C. Domb and M.S. Green (Academic Press, New York, 1972), p. 443.

11. H. Yamauchi and D. de Fontaine, in *Order-Disorder Transformations in Alloys*, edited by H. Warlimont (Springer, Berlin, 1974).

12. G. Martin, *Phys. Rev. B* **41**, 2279 (1990).

13. O. Penrose, *J. Stat. Phys.* **63**, 975 (1991).

14. K. Wada, A. Kawada, and Y. Kabasawa, in *Theory and Application of the Cluster Variation and Path Probability Methods*, edited by J. L. Morán-López and J. M. Sanchez (Plenum, New York, 1996), p. 53.

15. F. Ducastelle, *Prog. Theor. Phys.* Suppl. **115**, 255 (1994).

16. K. Wada and M. Kaburagi, *Prog. Theor. Phys.* Suppl. **115**, 273 (1994).

17. G. Martin, *Phys. Rev. B* **50**, 12362 (1994).

18. J. -F. Gouyet, *Europhys. Lett.* **21**, 335 (1993), and *Phys. Rev. E* **51**, 1695 (1995).

19. N. G. Van Kampen, *Stochastic Processes in Physics and Chemistry*, revised and enlarged edition (North-Holland, Amsterdam, 1992).

20. R. Kubo, K. Matsuo, and K. Kitahara, *J. Stat. Phys.* **9**, 51 (1973).

21. K. Wada, T. Ishikawa, and A. Yamashita, *Phys. Letters A* **110**, 355 (1985).

22. M. Suzuki, *Prog. Theor. Phys.* **53**, 1657 (1975), **55**, 383 and 1064 (1976).

23. K. Wada, T. Ishikawa, and H. Tsuchiniga, *Physica A* **142**, 38 (1987).

24. M. Suzuki, *Prog. Theor. Phys.* **56**, 77, 477 (1976) 77, 477 and **57**, 380 (1977); *J. Stat. Phys.* **16**, 11 and 477 (1977).

25. M. Suzuki, *Adv. Chem. Phys.* **46**, 195 (1981).

26. B. Caroli, C. Caroli, and B. Roulet, *J. Stat. Phys.* **21** (1979) and **22**, 515 (1980).

27. J. -F. Gouyet, *J. Stat. Phys.* **45**, 267 (1986); A. Bunde and J. -F. Gouyet, *Physica A* **132**, 357 (1985).

28. W. Paul, D. W. Heermann, and K. Binder, *J. Phys. A: Math. Gen.* **22**, 3325 (1989).

Kinetic Path and Fluctuations Calculated by the Path Probability Method

Tetsuo Mohri, Yoichi Ichikawa, Takayuki Nakahara* and Tomoo Suzuki

Division of Materials Science and Engineering
Graduate School of Engineering
Hokkaido University
Kita-13 Nishi-8, Kita-ku, Sapporo 060
JAPAN

Abstract

The Path Probability Method of non-equilibrium statistical mechanics is employed to investigate the ordering kinetics for the disorder-$L1_0$ and the disorder-$L1_2$ transitions. For both the cases, nucleation-growth character and spinodal character of ordering reaction are clarified from the kinetic point of view. The derived kinetic path deviates significantly from the steepest descent direction of the free energy contour surface. A preliminary study of fluctuation analysis is attempted for the ordering relaxation process in the disordered phase. The calculated fluctuation spectrum suggests that the most probable path of the kinetic evolution is characterized by the trace of the easiest direction of fluctuation at each time.

I. Introduction

The Path Probability Method[1] (hereafter PPM) of statistical mechanics has been recently recognized as one of the powerful tools to study the non-equilibrium time evolution process in an alloy system. An advantage of the PPM is due to the fact that relevant quantities derived for the long time limit correctly converge to the equilibrium ones predicted by the Cluster Variation Method (hereafter CVM).[2] Hence the combination of the CVM provides an invaluable tool to study the thermodynamic properties of a given system starting from non-equilibrium to equilibrium states. Since the PPM is recognized as the natural extension of the CVM to the time domain, we begin with a brief review of the CVM in the following.

Theory and Applications of the Cluster Variation and Path Probability Methods
Edited by J.L. Morán-López and J.M. Sanchez, Plenum Press, New York, 1996

In the CVM, the atomic correlations in a given alloy is approximated by a selected set of finite clusters, which is most efficiently described by correlation functions,[3-5] and the free energy functional is symbolically written as

$$F = F(T, \{\xi_i\}), \tag{1}$$

where ξ_i represents the atomic correlation function for an i-cluster, defined as

$$\xi_i = \langle \sigma_1 \sigma_2 \cdots \sigma_n \cdots \sigma_i \rangle, \tag{2}$$

where σ_n is the spin operator which takes values $+1$ or -1 depending on wheter an A or a B atom, respectively, is located at a lattice point n, and $\langle \cdots \rangle$ denotes the ensemble average. It is noted that the correlation functions form an independent set of configurational variables. The maximum cluster $i_{\max}$ explicitly considered in the free energy functional F specifies the level of the approximation. The common practice for an fcc-based system is the tetrahedron approximation[6] in which the nearest-neighbor tetrahedron cluster is taken as the maximum cluster. Hence, within the tetrahedron approximation, the free energy expression, Eq. (1), is rewritten as

$$F = F(T, \xi_1, \xi_2, \xi_3, \xi_4), \tag{3}$$

where ξ_1, ξ_2, ξ_3 and ξ_4 represents, respectively, point, nearest-neighbor pair, nearest-neighbor triangle and nearest-neighbor tetrahedron clusters. It should be noted that the breakdown of symmetry of a disordered phase associated with an ordering transition induces sublattices, and the number of correlation functions required to describe an ordered phase increases. For the L1$_0$ and L1$_2$ phases, which are the main concerns of the present study, the free energies are expressed as

$$F^{\mathrm{L1_0}} = F(T, \xi_1^{\alpha}, \xi_1^{\beta}, \xi_2^{\alpha\alpha}, \xi_2^{\alpha\beta}, \xi_2^{\beta\beta}, \xi_3^{\alpha\alpha\beta}, \xi_3^{\alpha\beta\beta}, \xi_4^{\alpha\alpha\beta\beta}), \tag{4}$$

and

$$F^{\mathrm{L1_2}} = F(T, \xi_1^{\alpha}, \xi_1^{\beta}, \xi_2^{\alpha\beta}, \xi_2^{\beta\beta}, \xi_3^{\alpha\beta\beta}, \xi_3^{\beta\beta\beta}, \xi_4^{\alpha\beta\beta\beta}), \tag{5}$$

respectively, where α and β designate sublattices. Minimization of the free energies followed by the common tangent construction (or an equivalent method) determines the phase equilibria among the three phases. The calculated disorder-L1$_0$ -L1$_2$ phase diagram is shown in Fig. 1.

In the present expression of the free energy of a disordered phase, Eq. (3), the second order derivative with respect to the deviation of the correlation functions from their equilibrium values provides information on the stability of a solid solution,

$$F'' = \frac{\partial^2 F}{\partial \Delta \xi_i \, \partial \Delta \xi_j}, \tag{6}$$

where $\Delta \xi_i$ is defined as

$$\Delta \xi_i = \xi_i - \xi_i^{\mathrm{eq}}, \tag{7}$$

where superscript 'eq' indicates the equilibrium state. The vanishing condition of the above equation is the stability criterion of a homogeneous solid solution against an excitation and the amplification of a concentration wave, which is most efficiently

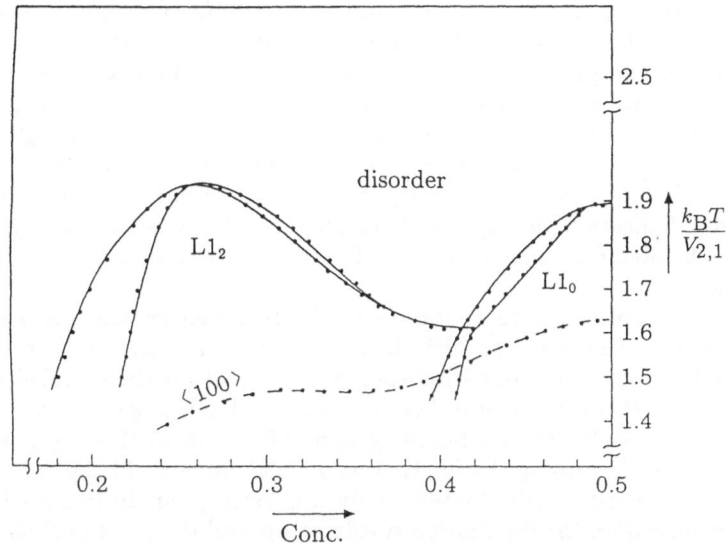

Figure 1. Disorder-L1$_0$ -L1$_2$ phase diagram. The broken line indicates the $\langle 100 \rangle$ spinodal ordering locus. The temperature axis is normalized with respect to the nearest-neighbor pair interaction energy $v_{2,1}$.

described by the Fourier transformation of $\Delta \xi_i$. The broken line in Fig. 1 is the stability locus, which is termed spinodal ordering,[7-10] for the $\langle 100 \rangle$ concentration wave. It is expected that the ordering reaction takes place by nucleation-growth mechanism above the locus, whereas the $\langle 100 \rangle$ concentration wave destabilizes the homogenous solid solution and leads to spontaneous decomposition of an ordered phase below the locus.

It should be emphasized that various fruitful features of the CVM stem from the fact that the short-range order atomic correlations, which play an important role in the ordering reaction, are explicitly incorporated in the free energy expression. Inherited from the CVM, the PPM is able to derive the time evolution of not only the long-range order but also the short-range order correlations, which is the noticeable advantage over other kinetic theories. However, the number of variables involved in the fundamental equation of the PPM is fairly large as will be described shortly. This is more pronounced for an alloy system for which the atomic *exchange* process drives the transition. In fact, contrary to the equilibrium thermodynamics, a clear distinction should be made for the kinetic study between an alloy system and a spin system. The exchange mechanism[11] which dominates the former system strictly conserves the composition, while the flipping mechanism[12] for the latter one does not necessarily observe the conservation law except for a special case. The additional constraints induced by the conservation law further complicates numerical computations, and PPM studies for the alloy systems have been limited to *bcc*-based systems[13] for which primitive pair approximation provides fairly accurate results. For the *fcc*-based systems, on the other hand, it has been recognized that the pair approximation is not sufficient and, at least, the tetrahedron approximation is required. Although the formulation for an *fcc* alloy system within the tetrahedron approximation has been done, the actual calculation is not fully tractable yet. Following the previous studies[14-22] for the disorder-L1$_0$ transition, the present study is also limited to a spin system which is believed to provide a precursor for the more complex alloy system.

The aim of the present study is two-fold. One is the extension of the previous studies to the disorder-L1$_2$ transition. In fact, various intriguing features of the kinetic behavior observed experimentally have been centered around the disorder-L1$_2$ transition, for instance, of the Al-Li system.[23] The key to understanding these phenomena is believed to be the interplay among instabilities. The present study based on the phase diagram with a spinodal ordering instability calculated by the CVM and the kinetic behavior analyzed by the PPM is expected to facilitate the theoretical investigation.

The second aim is the fluctuation analysis. It is well known that a system undergoes incessant fluctuations by which the actual phase transition is driven and assisted. The fluctuation is, therefore, an indispensable subject for the detailed analysis of kinetic behavior. However, to the best of authors' knowledge, no systematic studies have been reported for the *fcc*-based systems. Following Kikuchi's prescription,[1] an additional factor describing the fluctuation which is missing in the conventional PPM formula is introduced and fluctuation analysis is attempted. In this preliminary study, the focus is placed on the fluctuation spectrum around the most probable kinetic path for an ordering relaxation process in the single disordered phase region.

The organization of the present report is as follows. In the next section, a brief description of the theoretical framework of the PPM is provided. In the third section, for the sake of completeness, we reproduce the previous results for the disorder-L1$_0$ transition. In the fourth section, the extension to the disorder-L1$_2$ transition is described. And, in the last section, the fluctuation analysis is reported.

II. Theoretical Formulation of the PPM

Unlike the case of the CVM, the body of the reports of the PPM have been centered around fundamental studies of statistical mechanics aspects. Among them is the celebrated paper by Kikuchi[1] who devised, developed and applied the PPM. The one by Ducastelle[24] is also noteworthy. The latter clarifies the essential features of the PPM by comparing with the CVM from a different angle. Also, more recent basic studies can be found in this proceedings. Hence, the major focus of this section is not on the same kind of basic arguments but to provide the essential formulas which are necessary for the discussions in the subsequent sections

A noticeable feature of the PPM is the fact that state variables (cluster probabilities or correlation functions of the CVM) and their time derivatives are not explicitly dealt with, which is in marked difference with other kinetic theories. Instead, the central quantities of the PPM are the path variables which describe transition among state variables. Since, as was mentioned in the previous section, the PPM is the natural extension of the CVM to the time domain, the essential feature can be grasped well by comparing with the CVM as follows.

The counterpart of the grand potential (or equivalently the free energy) of the CVM is the *Path Probability Function* (hereafter *PPF*), $P(t; t + \Delta t)$, which is an explicit function of time and is defined as the product of three factors P_1, P_2 and P_3. Each factor is provided in the following in the logarithmic expression,

$$\ln P_1 = \tfrac{1}{4} \sum_\delta N_1^\delta \left\{ \left(X_{1,2}^\delta + X_{2,1}^\delta \right) \ln \left(\theta \Delta t \right) + \left(X_{1,1}^\delta + X_{2,2}^\delta \right) \ln \left(1 - \theta \Delta t \right) \right\}, \qquad (8)$$

$$\ln P_2 = -\frac{\Delta E}{2k_B T},\tag{9}$$

and

$$\ln P_3 = -\tfrac{5}{4}\sum_\delta N_1^\delta \left(\sum_{i,j} L(X_{i,j}^\delta) - \sum_i L(x_i^\delta(t)) \right)$$

$$+ \sum_{\delta\delta'} N_2^{\delta\delta'} \left(\sum_{ij,kl} L(Y_{ij,kl}^{\delta\delta'}) - \sum_{ij} L(y_{ij}^{\delta\delta'}(t)) \right)$$

$$- 2 \left(\sum_{ijkl,mnop} L(W_{ijkl,mnop}) - \sum_{ijkl} L(w_{ijkl}(t)) \right),\tag{10}$$

where δ and δ' denote sublattices, $L(z)$ represents $z \ln z - z$, θ is the spin flip probability per unit time, $X_{i,j}^\delta$, $Y_{ij,kl}^{\delta\delta'}$ and $W_{ijkl,mnop}$ are the path variables for the flipping from one spin configuration to another designated by the subscript(s) (up spin or down spin) before (at time t) and after (at time $t + \Delta t$) the comma sign, on a point, pair and tetrahedron clusters over the specified sublattice(s), respectively; $x_i^\delta(t)$, $y_{ij}^{\delta\delta'}(t)$ and $w_{ijkl}(t)$ are the state variables (cluster probabilities) which describe the probability of finding a spin configuration designated by the subscript(s) at time t on a point, pair and tetrahedron clusters, respectively, over the specified sublattices, and ΔE is the change of the internal energy before and after the flipping events. The subscripts $i, j, k, \ldots, o, p$ take values $+1$ or -1 for up or down spins, respectively. N_1^δ and $N_2^{\delta\delta'}$ are determined by the symmetry of the phase. It is readily shown that, for the $L1_0$ ordered phase, $N_1^\alpha = N_1^\beta = 2$, $N_2^{\alpha\alpha} = N_2^{\beta\beta} = 1$ and $N_2^{\alpha\beta} = 4$ and, for the $L1_2$ ordered phase, $N_1^\alpha = 1$, $N_1^\beta = 3$, $N_2^{\alpha\beta} = N_2^{\beta\beta} = 3$ and $N_2^{\alpha\alpha} = 0$. For the disordered phase, since there is no distinction between the sublattices the summation operations with respect to δ and δ' are eliminated and $N_1(= N_1^\delta) = 4$ and $N_2(= N_2^{\delta\delta'}) = 6$.

Within the nearest-neighbor pair interaction model, the description of $\Delta E = (E(t + \Delta t) - E(t))$ also depends on the symmetry of the phase and, with N_1^δ and $N_2^{\delta\delta'}$ specified above, one can readily derive the following expression,

$$E(t) = \frac{1}{2}\omega \sum_{\delta\delta'} \frac{N_2^{\delta\delta'}}{6} \sum_{ij} e_{ij} y_{ij}^{\delta\delta'}(t) + \sum_\delta \frac{N_1^\delta}{4} \sum_i \mu_i^\delta x_i^\delta(t),\tag{11}$$

where ω is the coordination number, e_{ij} is the nearest-neighbor interaction energy between i and j, and μ_i^δ is the chemical potential of i-species on sublattice δ. It is noted that the effective nearest-neighbor pair interaction energy $v_{2,1}$ is related to the pair interaction energies e_{ij} through

$$v_{2,1} = \tfrac{1}{2}\{(e_{ii} + e_{jj}) - 2e_{ij}\},\tag{12}$$

by which the temperature axis in Fig. 1 is normalized.

In the PPF, the first factor P_1 describes the statistical average of non-correlated spin flip events over the entire lattice points, and the second factor P_2 is the conventional thermal activation factor. Hence, the product of P_1 and P_2 corresponds to the Boltzmann factor in the free energy and gives the probability that one of the paths specified by a set of path variables occurs. The third factor P_3 characterizes the PPM. One may see the similarity with the configurational entropy term of the CVM, which gives the multiplicity, i.e. the number of equivalent states. In a similar sense, P_3

can be viewed as the number of equivalent paths, *i.e.* the degrees of freedom of the microscopic evolution from one state to another.

The path variables of the PPM correspond to the cluster probabilities of the CVM by which the grand potential is minimized to obtain the most probable state. Likewise, the *PPF* is maximized with respect to the path variables for each time step, which yields the optimized set of path variables. Since a set of path variables, $\Xi_{\psi,\phi}(t; t+\Delta t)$, relates cluster probabilities $\chi_\psi(t)$ at time t and $\chi_\phi(t+\Delta t)$ at time $t+\Delta t$ through

$$\chi_\phi(t+\Delta t) = \chi_\psi(t) + \sum \Xi_{\psi,\phi}(t : t+\Delta t), \qquad (13)$$

where ψ and ϕ represent clusters, one can pursue the time evolution of cluster probabilities χ (equivalently correlation functions) with a given initial set of cluster probabilities.

The time evolution derived by the above formula traces the most probable kinetic path in the configuration space spanned by the state variables. In order to analyze the fluctuations around the most probable path, it is necessary to introduce an additional factor, P_4 in the *PPF*.[1] When the configuration space is spanned by tetrahedron cluster probabilities as a set of state variables, the fluctuations are most conveniently defined as deviations from their most probable values, w^*_{ijkl}, in the following manner,

$$S_{ijkl}(t; t+\Delta t) = w_{ijkl}(t+\Delta t) - w^*_{ijkl}(t). \qquad (14)$$

Then, with pre-assigned values of S_{ijkl}, the additional factor given as

$$\ln P_4 = \sum_{ijkl} \lambda_{ijkl} \left\{ w_{ijkl}(t+\Delta t) - w^*_{ijkl}(t) - S_{ijkl}(t; t+\Delta t) \right\}, \qquad (15)$$

where λ_{ijkl} is a Lagrange multiplier, describes a constrained *Path Probability Function*. Following the variational procedure, the probability of the path towards an assigned state is derived. And the operation of the above procedure for various values of S_{ijkl} constitute a fluctuation spectrum.

III. Kinetics of Disorder-L1$_0$ Transition

In Figure 1, one can find that the calculated transition temperature, T_t, for disorder-L1$_0$ at 1:1 stoichiometric composition is 1.89 and the spinodal ordering temperature, T_s, is 1.6333. The present PPM calculations are performed for the two kinds of temperature 1.70 between T_t and T_s and 1.60 below T_s at which the system quenched from $T = 2.5$ in the disordered region is subject to the aging operation.

Prior to the kinetic study, it is interesting to clarify the thermodynamic implication of the spinodal ordering temperature by examining the free energy contour surface in the thermodynamic configuration space. Shown in Figs. 2a and 2b (Ref. 22) are the contour surface at 1:1 stoichiometry for $T = 1.70$ and 1.60, respectively, spanned by the point correlation function, ξ_1^α, and the tetrahedron correlation function, $\xi_4^{\alpha\alpha\beta\beta}$. The contour surface is calculated by minimizing the free energy function given by Eq. (4) with respect to the pair correlations $\xi_2^{\alpha\alpha}(= \xi_2^{\beta\beta})$ and $\xi_2^{\alpha\beta}$, and the triangle correlations $\xi_3^{\alpha\alpha\beta}(= -\xi_3^{\alpha\beta\beta})$, under a given set of temperature, point correlation $\xi_1^\alpha(= -\xi_1^\beta)$, long-range order parameter, and tetrahedron correlation function $\xi_4^{\alpha\alpha\beta\beta}$. The null value of the point correlation function corresponds to the disordered state, hence both surfaces are in the vicinity of disordered state. In both cases, the lowest energy corresponding to the L1$_0$ ordered state is found near unity of ξ_1^α which is off-scale in these figures. One sees that there exists a saddle point configuration for $T = 1.70$, while the free energy surface varies monotonously for $T = 1.60$. It is noted that the contour surface is symmetric around $\xi_1^\alpha = 0$ for 1:1 stoichiometric composition. Hence, the saddle point near $\xi_1^\alpha = 0.175$ can be also found near $\xi_1^\alpha = -0.175$. The two saddle points approach with decreasing temperature and merge together at the spinodal ordering temperature, T_s. The local minimum of the free energy found near $\xi_4^{\alpha\alpha\beta\beta} = 0.267$ in the disordered state ($\xi_1^\alpha = 0$) at $T = 1.70$ indicates that the disordered phase is metastable above T_s. The local minimum and absolute minimum corresponding to the L1$_0$ ordered state reach the same level at the transition temperature, $T_t = 1.89$. These features confirm that the disorder-L1$_0$ transition is of first order nature.

In summarizing the above discussions, the second order derivative of the free energy, Eq. (6), describes the curvature of the hypersurface of configuration space, and the critical value corresponding to the appearance and disappearance of the saddle point configuration mathematically determines the spinodal ordering temperature and physically distinguishes the nucleation-growth and spinodal ordering reactions. In such an analysis based on thermodynamic arguments, however, actual kinetic behavior of the system are not well investigated. Without kinetic factors, for instance, the kinetic evolution path should precisely follow the steepest descent direction, which may be rationalized for near-equilibrium transitions but not for far-from-equilibrium transitions. This is the subject of kinetics. In the following, the PPM studies for the ordering transition are demonstrated.

As was mentioned in the previous section, the fluctuations are not explicitly incorporated in the conventional PPM. In order to trigger the transition, a perturbation is imposed on the chemical potential in the ΔE term. Shown in Figs. 3a (Ref. 22), by solid and broken lines, are the time evolution of x_A^α and x_A^β, the concentration of A-atoms (up spin) on α and β sublattices, respectively, which serve as the long-range order parameters, when the system of 50% is quenched from the temperature of 2.5 down to 1.7 and is subject to the aging treatment. Note that the time axis is normalized by the spin flip probability θ throughout this study. The initial separations of x_A^α and x_A^β are due to the imposed fluctuation. One can see that with sufficient amount of initial fluctuation (solid line), the system decomposes into the α and β sublattices, which is confirmed by the agreement of the steady state values of the x_A^α and x_A^β with the equilibrium ones of the L1$_0$ ordered phase at $T = 1.7$ independently obtained by the CVM. However, the initial fluctuation decays and returns to the disordered solid solution for a small amount of fluctuation (broken line). These two results clearly indicates that there exists a critical amount of fluctuation at the temperature 1.7, which is an indication of a nucleation-growth type ordering. The kinetic paths of the two cases are demonstrated by solid and broken lines in Fig. 2a. One clearly sees that the perturbed chemical potential induces the fluctuation in concentration on a sublattice.

On the other hand, as is shown in Fig. 3b, (Ref. 22) at $T = 1.60$ below T_s an infinitesimal fluctuation is amplified and a spontaneous ordering reaction takes place. The corresponding kinetic path is drawn by the solid line in Fig. 2b. One notices that the kinetic path does not necessarily follow the steepest descent direction, which is also the case for the nucleation-growth mode. It is noteworthy that the system spends most time at the saddle point state ($T = 1.70$) and at the unstable disordered state ($T = 1.60$), as are indicated by Figs. 2 and 3. This may be understood as the loss of the driving force manifested by the flatness of the free energy surface at these states.

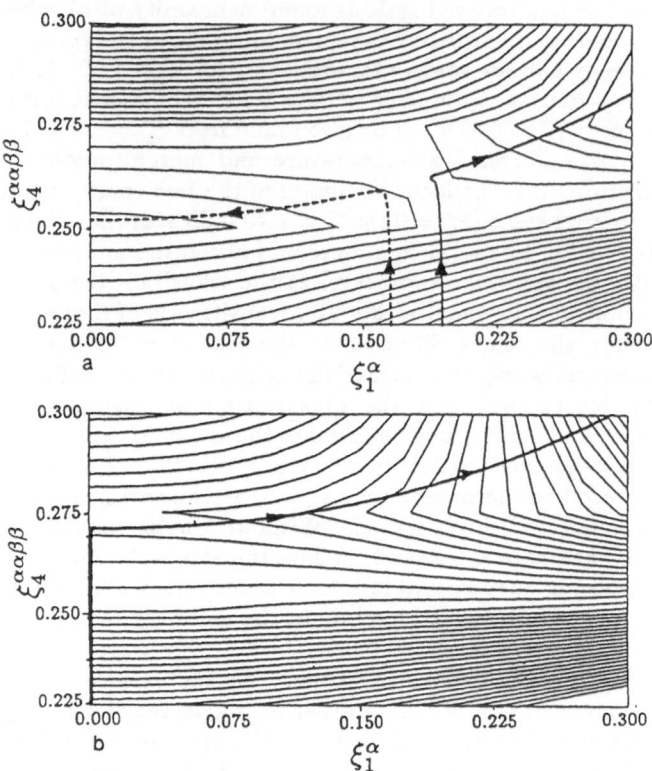

Figure 2. Free energy surface at 1:1 stoichiometry in the vicinity of disordered state ($\xi_1^\alpha = 0.0$) and kinetic path for $T = 1.70$ (Fig. 2a) and 1.6 (Fig. 2b).[22] The solid line in each figure indicates the kinetic path evolving towards the $L1_0$ ordered phase and the broken line in Fig. 2a is the one devolving towards the disordered phase. The arrow designates the direction of time evolution or devolution.

It is recalled that the advantage of the PPM is that the short-range order kinetics is clarified, which enables one to trace the kinetic path in configuration space spanned by the long-range and short-range order ($\xi_4^{\alpha\alpha\beta\beta}$) parameters as is demonstrated above. However, a shortcoming common to the PPM and CVM is the fact that the correlations beyond the specified maximum cluster size are not easily grasped. In order to subsidize such a shortcoming, a computer simulation is carried out to synthesize the configuration in a crystal lattice based on the derived point and pair correlation functions, ξ_1^α, $\xi_2^{\alpha\alpha}$ and $\xi_2^{\alpha\beta}$. Shown in Fig. 4 (Refs. 22, 25) are the series of snap shots of spin configurations on a (100) plane of a model crystal which consists of $25 \times 24 \times 21$ *fcc* lattice points. The left hand snapshot corresponds to equilibrium at $T = 2.5$ before the quenching operation, and the others are at time $t = 20,000$, $43,000$ and $50,000$, as are indicated in the figure, during the aging process at $T = 1.7$. The calculated long range and short range order parameters by the PPM are indicated by solid and broken lines, respectively, and the open circle and triangle correspond to the resultant ones in the simulated lattice. One confirms that the PPM results are satisfactorily reproduced.

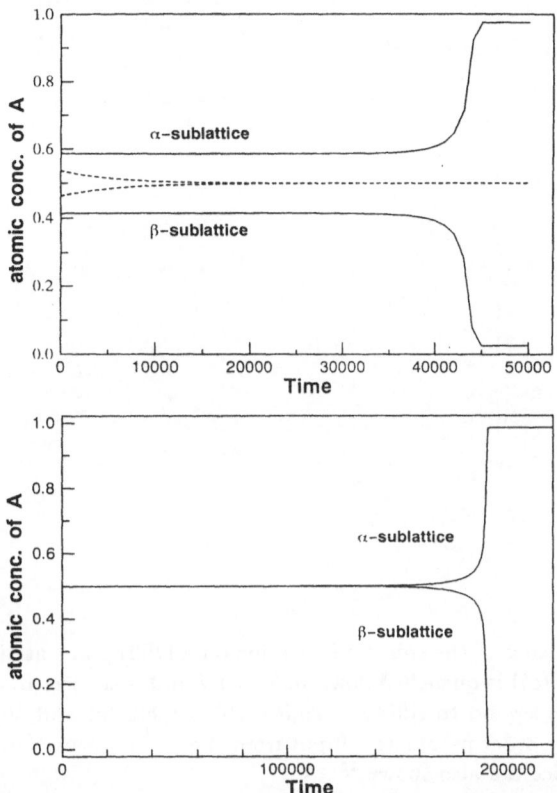

Figure 3. Time evolution of the concentration of A atom (*up* spin) on $\alpha(x_A^\alpha)$ and $\beta(x_A^\beta)$ sublattices at 50% when the system is quenched from $T = 2.5$ to $T = 1.7$ (a) and to $T = 1.6$ (b) with the spin flip probability $\theta = 0.001$. The solid and broken lines indicate the ones with and without sufficient initial fluctuation, respectively.[22]

The long-range order does not exist in the initial configuration at $T = 2.5$ while short-range order remains. At $T = 1.70$, one observes that with aging process the lattice symmetry of the disordered phase is broken and the alternative lattice points along the [010] and [001] directions form α and β sublattices of the L1$_0$ ordered phase. The evolution of the three dominant pair cluster probabilities obtained by the simulation, $y_{AA}^{\alpha\alpha}$; 0.297 at $t = 1$ to 0.489 at $t = 43,000$, $y_{AB}^{\alpha\beta}$; 0.382 to 0.556 and $y_{BB}^{\beta\beta}$; 0.300 to 0.489, indicates the ordering reaction proceeding among the nearest neighbor sites in the intermediate time up to $t = 43,000$. Then, one observes that the ordered configuration considerably evolves at $t = 43,000$ which corresponds to the steeply rising (declining) portion of $x_A^\alpha(x_A^\beta)$ in Fig. 3a and long-range order parameter (LRO) in Fig. 4 prior to the completion at $t = 50,000$. It is pointed out that the many body correlation functions, $\xi_3^{\alpha\alpha\beta}$ and $\xi_4^{\alpha\alpha\beta\beta}$, obtained by the PPM are ignored in the present simulation since an efficient algorithm to incorporate all the short range order correlations is not yet at our disposal. Together with the analyses for $T = 1.60$, the details of the present simulation will be reported in a separate issue.

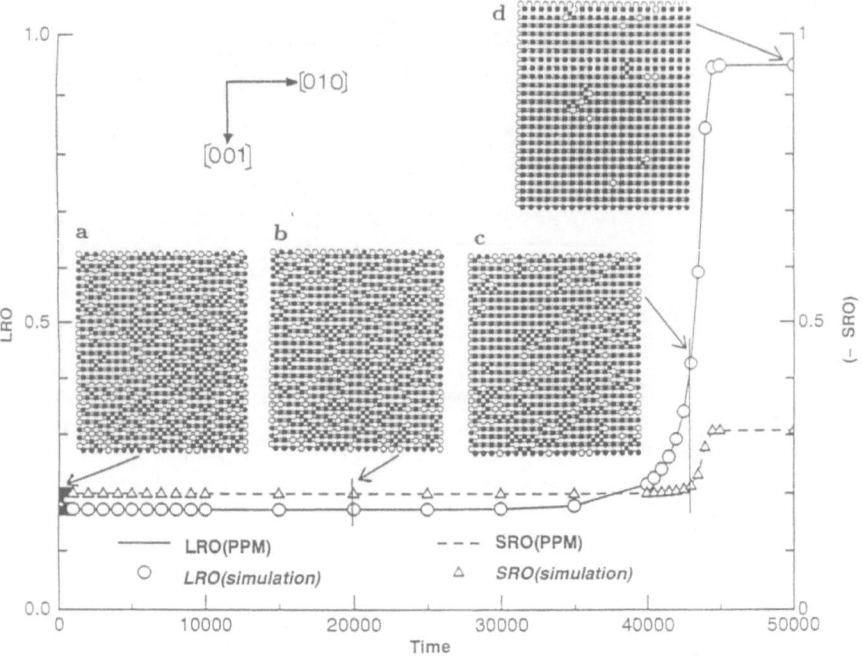

Figure 4. Time sequence of the spin configuration on a (100) plane at 50% when the system at $T = 2.5$ (snapshot (a)) is quenched down to $T = 1.7$ and is subject to an isothermal aging. Snapshots shown correspond to time $t = 20,000$ (b), $43,000$ (c) and $50,000$ (d). The long-range and short-range order parameters input from the PPM calculations and resultant ones in the simulated lattice are also shown.[25]

IV. Kinetics of Disorder-L1$_2$ Transition

Recent high resolution experimental studies have detected various intriguing features in the transition behavior involving the L1$_2$ ordered phase. Among them is the phase separation with ordering phenomenon found in a Al-Li system.[23] It has been pointed out that the key to understanding such phenomenon is the interplay among various instabilities in the disorder+L1$_2$ two-phase field. Hence, the extension of the PPM studies to the two-phase field involving the L1$_2$ phase is desirable. Although a preliminary attempt of PPM calculations to the two-phase field was made for the disorder+L1$_0$ system,[22] the analysis was not trivial since the synthesis of two results separately derived for disordered and L1$_0$ phases into a single framework leaves uncertainties. This is also anticipated for other cases of two-phase decomposition process. In the present study, therefore, the focus is placed only on the single phase decomposition of L1$_2$ ordered phase.

Shown in Figs. 5a and 5b are the free energy surfaces for $T = 1.60$ and 1.20, respectively, at 1:3 stoichiometric composition. Note that the long range order parameter of the L1$_2$ phase, in the present study, is defined as the difference of the point correlation functions of the two sublattices, $\xi_1^\alpha - \xi_1^\beta$. In order to facilitate the detailed analysis, the contour surface in the vicinity of disordered state is also shown in each figure. These are derived by minimizing the free energy, Eq. (5), in a similar manner carried out for the L1$_0$ ordered phase. As are the cases at 1:1 stoichiometry, a saddle

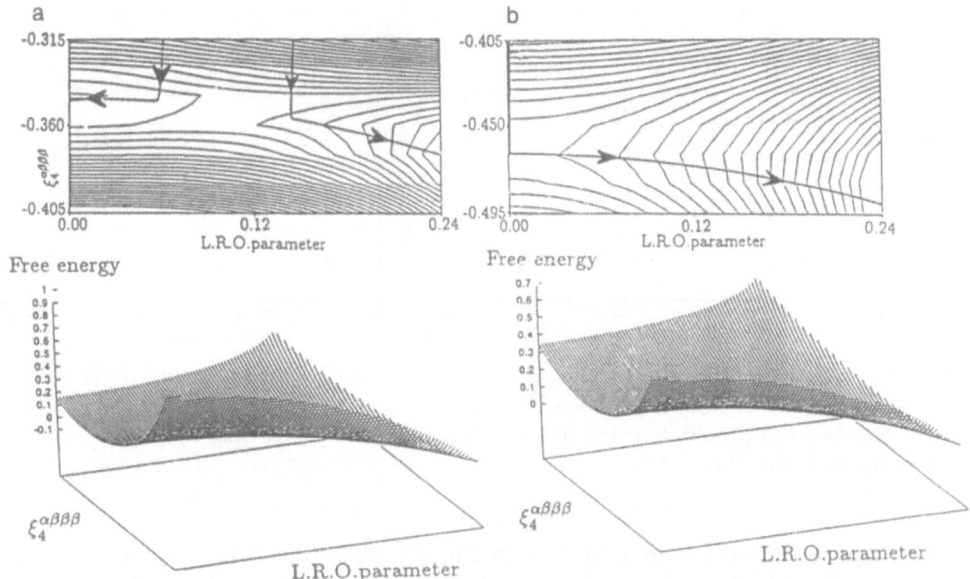

Figure 5. Free energy as a function of long-range order parameter ($\xi_1^\alpha - \xi_1^\beta$) and short-range order parameter ($\xi_4^{\alpha\beta\beta\beta}$) for $T = 1.60$ (a) and for $T = 1.20$ (b) at 1:3 stoichiometric composition. Shown above are the free energy contour surfaces projected on the two dimensional $\xi_1^\alpha - \xi_4^{\alpha\beta\beta\beta}$ plane.

point is observed at $T = 1.60$ above the spinodal ordering temperature, $T_s = 1.4$ at this stoichiometry (see Fig. 1), whereas the free energy behaves monotonously at $T = 1.20$ below T_s.

By applying the same procedure for the $L1_0$ phase, the time evolution process of the long-range order parameter and the short-range order parameter ($\xi_4^{\alpha\beta\beta\beta}$) is investigated. It is recalled that except the case at 1:1 stoichiometry the spin system generally does not conserve the concentration, and it is more convenient to work with the grand canonical ensemble with a fixed chemical potential. In fact, the former study[22] of the two-phase decomposition process of disorder+$L1_0$ was performed in such a scheme. In the present study, however, a slightly different attempt is made. The initial chemical potential is fixed at $\mu = -8.4725$ which realizes exactly 1:3 stoichiometry in the equilibrium state at the pre-quenching temperature of $T = 3.0$. Then, right after quenching to an aging temperature, the chemical potential is switched to a one which yields the 1:3 stoichiometry in the equilibrium state at that temperature. Those chemical potentials are $\mu = -8.4090$ for $T = 1.60$ and $\mu = -8.3263$ for $T = 1.20$, respectively. The spin flip probability employed in the present calculation is $\theta = 0.001$.

The results are shown in Figs. 6a and 6b for $T = 1.60$ and 1.20, respectively. The final steady state values of the concentration for each case are again confirmed to be the equilibrium ones independently obtained by the CVM. Like the case for the $L1_0$ transition shown in Fig. 3a, the existence of a critical amount of fluctuation is indicated for the $T = 1.60$ above the spinodal ordering temperature, T_s. While a barrierless ordering reaction is suggested at $T = 1.20$ below T_s. Interestingly, the incubation period is less pronounced at $T = 1.60$ as compared with the spinodal ordering at $T = 1.20$ and nucleation-growth at $T = 1.60$ at 1:1 stoichiometry. As hinted by the

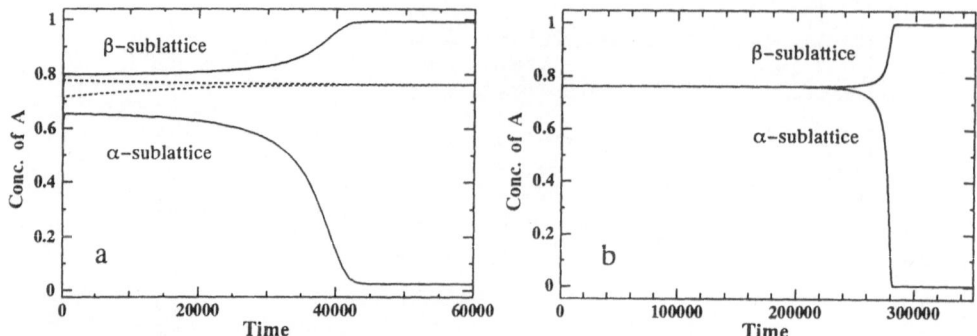

Figure 6. Time evolution of concentration of A atoms (*up* spin) on $\alpha(x_A^\alpha)$ and $\beta(x_A^\beta)$ sublattices when the system is quenched from $T = 3.0$ to $T = 1.6$ (a) and to $T = 1.20$ (b) with the spin flip probability $\theta = 0.001$. The solid and broken lines in (a) have the similar meaning as in Fig. 3a.

kinetic path drawn by the solid line in Fig. 5a, this difference may be attributed to the considerable deviation of the initial portion of the path from the saddle point configuration. One should, however, be careful that the detailed discussion of the kinetic paths shown in Figs. 5a and 5b is not fully rationalized. This is because the concentration is fixed for the calculated contour surface while incessant variations are present along the path.

V. Fluctuation Spectrum in the Ordering Relaxation Process

In the previous sections, we observed that a kinetic path does not necessarily follow the steepest descent direction. Although the deviation may be attributed to the effects of kinetic factors (more precisely the spin flip probability) which modify the thermodynamic driving force given by the gradient of the free energy, such an argument is, in a strict sense, not rationalized for the far-from-equilibrium transition.

It is recalled that a system undergoes incessant microscopic fluctuations to search for the most stable state and the shown kinetic paths are nothing but a resultant trace of such states. Hence, in order to clarify the direction of the kinetic path, one may seek the underlying necessity in the fluctuation spectrum obtained through *PPF* with an additional fluctuation factor, P_4. In the present section, we study the fluctuation around the kinetic path for the ordering relaxation process in a single disordered phase field at 50%.

At 50%, the free energy of the disordered phase within the tetrahedron approximation is further symmetrized to include only two independent variables. This is because of the fact that the point and triangle correlation functions, ξ_1 and ξ_3, vanish at the fixed composition of 50%. The construction of the free energy contour surface is, therefore, quite straightforward, and pair and tetrahedron correlation functions, ξ_2 and ξ_4 uniquely span the configuration space. Since it can be shown that the tetrahedron cluster probability, w_{ijkl}, is represented in terms of pair and triangle correlation functions as

$$w_{ijkl} = \frac{1}{2^4}\left[1 + (ij + ik + il + jk + jl + kl)\xi_2 + ijkl\xi_4\right], \qquad (16)$$

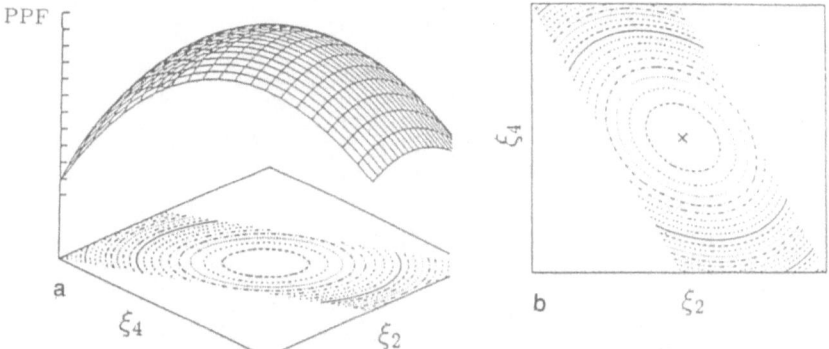

Figure 7. Schematic representation of fluctuation spectrum, PPF, as a function of pair correlation (ξ_2) and tetrahedron correlation (ξ_4) (a). The projection onto the $\xi_2 - \xi_4$ plane is also shown (b). The cross mark indicates the most probable state.[25]

an assigned fluctuation $S_{ijkl}(t; t+\Delta t)(= w_{ijkl}(t+\Delta t) - w^*_{ijkl})$ can also be transformed to the fluctuations of pair and the tetrahedron correlation functions, .

$$\Delta\xi_2 = \xi_2 - \xi_2^*, \tag{17}$$

and

$$\Delta\xi_4 = \xi_4 - \xi_4^*, \tag{18}$$

where * denotes the most probable state. The maximized PPF under a given constraint of S_{ijkl} (or equivalently given set of $\Delta\xi_2$ and $\Delta\xi_4$) constitutes the fluctuation spectrum which takes the maximum value along a kinetic path which is amply demonstrated in the previous section for the ordering transitions. A schematic representation is shown in Fig. 7a and 7b.[25] The latter is the projection of the PPF onto the $\xi_2 - \xi_4$ plane and the cross mark in the center indicates the most probable state corresponding to the maximum of PPF in Fig. 7a.

The calculation of the fluctuation spectrum is carried out for four kinds of times during the aging process at $T = 2.5$ quenched from $T = 5.0$ with an infinite cooling rate. The results are shown in Fig. 8 (Ref. 25) together with the kinetic path traced on the free energy contour surface at $T = 2.5$. For the sake of convenience, the scale of the ξ_2 and ξ_4 axes for contour surface and that of $\Delta\xi_2$ and $\Delta\xi_4$ axes for the spectrum are taken to be equivalent. One realizes that the kinetic path deviates significantly from the steepest descent direction for the contour surface.

Each fluctuation spectrum is well characterized by an ellipsoidal shape. One observes that the intensity distribution of the spectrum does not coincides with that of the free energy contour except at time $t = 1,000$ for which the equilibrium state is nearly attained. The careful observation of the spectra indicates that the major axis of the ellipsoid coincides with the tangential direction of the kinetic path at each time. This suggests that the kinetic path of the system is not the steepest descent direction but is characterized by the trace of the easiest direction of fluctuation. Also the sharper spectrum in the latter stage implies that the directionality of the fluctuation is pronounced with aging process. Together with the temperature variation of the spectrum, more details will be reported in a separate publication.

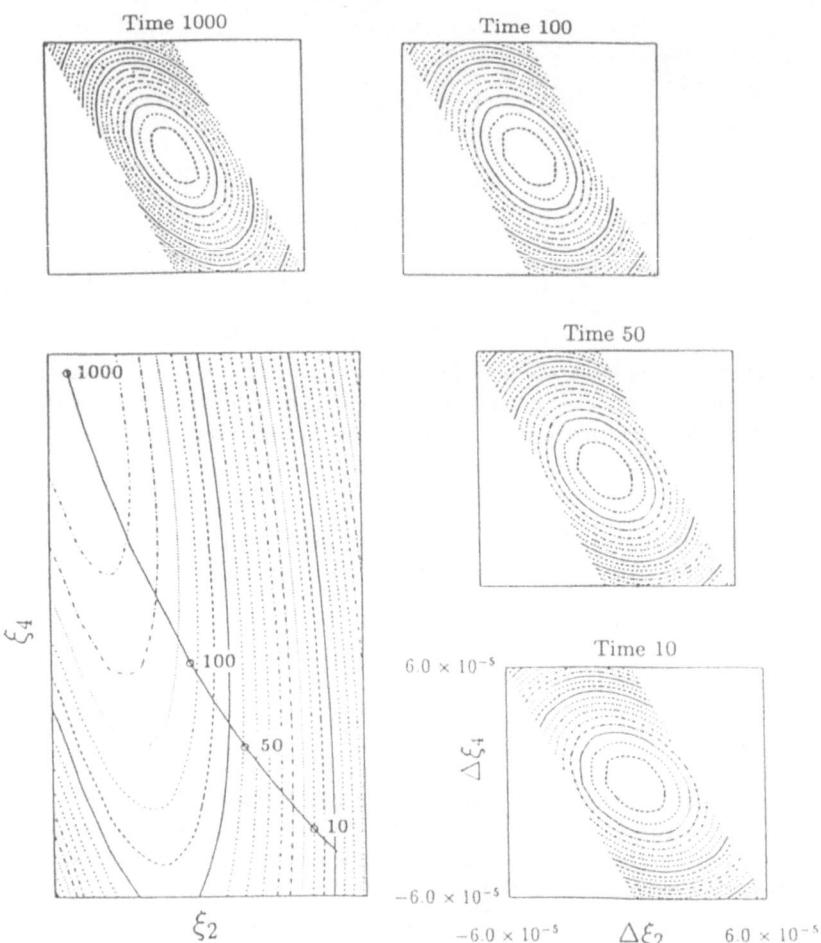

Figure 8. Fluctuation spectrum for four kinds of time during the aging period at $T = 2.50$. The system is maintained at $T = 5.0$ prior to the quenching to the aging temperature. In each figure, the horizontal and vertical axes are, respectively, the deviation of the pair and the tetrahedron correlation functions. The free energy contour surface at this temperature and the kinetic path during the aging process are also shown. The number along the kinetic path indicates the time. Note that the scale of all the axes are equivalent.[25]

Before closing the present report, it is emphasized that a conventional phase diagram merely provides a clue and a guideline to clarify features of the phase transition process. Additional information of various instabilities facilitates the detailed analysis, and a reliable kinetic theory which is able to manage far-from-equilibrium transition is indispensable. The combination of the CVM and PPM is a unique tool satisfying such requirements. We suggest that further efforts should be directed towards the extension of the present study to a two-phase alloy system.

References

* Present address: Nisshin Steel Co. Ltd,. Kure Works, Kure, Hiroshima 737, Japan.
1. R. Kikuchi, *Prog. Theor. Phys.* Suppl. **35**, 1 (1966).
2. R. Kikuchi, *Phys. Rev.* **81**, 998 (1951).
3. J. M. Sanchez and D. de Fontaine, *Phys. Rev. B* **17**, 2926 (1978).
4. J. M. Sanchez, F. Ducastelle, and D. Gratias, *Physica A* **128**, 334 (1984).
5. T. Mohri, J. M. Sanchez, and D. de Fontaine, *Acta Metall.* **33**, 1171 (1985).
6. R. Kikuchi, *J. Chem. Phys.* **60**, 1071 (1974).
7. D. de Fontaine, *Acta Metall.* **23**, 553 (1975).
8. J. M. Sanchez, *Physica A* **111**, 200 (1982).
9. T. Mohri, J. M. Sanchez, and D. de Fontaine, *Acta Metall.* **33**, 1463 (1985).
10. T. Mohri, Y. Sugawara, K. Watanabe, and J. M. Sanchez, *Mat. Trans., JIM.* **33**, 558 (1992).
11. K. Kawasaki, *Phys. Rev.* **145**, 224 (1966).
12. R. J. Glauber, *J. Math. Phys.* **4**, 294 (1953).
13. H. Sato and R. Kikuchi, *Acta Metall.* **24**, 797 (1976).
14. T. Mohri and T. Ikegami, *Deffects and Diffusion Forum* **95-98**, 119 (1993).
15. T. Mohri, in *Structural and Phase Stability of Alloys*, edited by J. L. Morán-López *et al.* (Plenum Press, New York 1992), p. 87.
16. T. Mohri and Y. Sugawara, in *Proc. Intn'l. Conf. and Exhit. on Computer Appl. to Mat. Sci. and Engr.-CAMSE '90*, edited by M. Doyama, (Elsevier 1991), p. 853.
17. T. Mohri, *Proc. Intn'l. Symp. on Intermetallic Compounds*, in *Structure and Mechanical Properties* JIMIS-6, edited by O. Izumi, (The Japan Institute of Metals 1991), p. 209.
18. T. Mohri, in *Statistics and Dynamics of Alloy Phase Transformations*, edited by P. E. A. Turchi and A. Gonis, (Plenum, New York 1994), p. 665.
19. T. Mohri, in *Interatomic Potential and Phase Stability*, Springer Series in Solid-State Sciences **114**, edited by K. Terakura and H. Akai, (Springer-Verlag, 1993), p. 168.
20. T. Mohri and T. Ikegami, *TMS EMPMD Monograph Series* **3**, in Diffusion in Ordered Alloys and Intermetallic Compounds edited by B. Fultz *et al.* (TMS Publication, Warrendale, PA 1993), p. 79.
21. T. Mohri, *Acta Metall.* **38**, 2445 (1990).
22. T. Mohri, in *Solid-Solid Phase Transformations*, edited by W. C. Johnson, J. M. Howe, D. E. Laughlin, and W. A. Soffa, (The Minerals, Metals and Materials Society, 1994), p. 53.
23. M.-S. Yu and H. Chen, in *Kinetics of Ordering Transformations in Metals*, edited by H. Chen and V. Vasudevan, (TMS Publication, Warrendale, PA 1992), p. 307.
24. F. Ducastelle, *Prog. Theor. Phys.* in press.
25. T. Nakahara, M.E. Thesis, Graduate School of Engineering, Hokkaido University, (1995).

Time Development of Fluctuations in the Path Probability Method

Koh Wada, Akiyoshi Kawada, and Yuki Kabasawa

Department of Physics
Faculty of Science
Hokkaido University
Sapporo 060
JAPAN

Abstract

The fluctuation properties inherent in the variation principle of the Path Probability Method (PPM) devised by Kikuchi are extensively studied. First, as a prototype of a kinetic process the Becker-Döring nucleation theory is reformulated from the viewpoint of the PPM. Second, a general formulation of the PPM is reviewed with a spin flip model being taken as a model with interaction among particles. Third, the general formulation is applied to an inhomogeneous spin system in the point and the pair approximations. Not only the master equation for inhomogeneous magnetization but also the path integral representation is also obtained. Along a path called anticausal path the path integral representation is shown to give the creation of a free energy to be expected from the Cluster Variation Method (CVM) in equilibrium statistical mechanics. Concerning the magnetic susceptibility, the validity of the well known fluctuation dissipation theorem is confirmed. These facts show the consistency between the PPM and the CVM. Finally the master equation in the homogeneous system is numerically integrated in comparison with the Becker-Döring nucleation theory to study the relaxation from the metastable to stable states in the point and pair approximations. The nucleation rates are in good agreement with theoretical expectations.

I. Introduction

About thirty years ago the Path Probability Method (PPM)[1] was devised by Kikuchi as a dynamical version of his Cluster Variation Method (CVM)[2] in equilibrium statistical mechanics. Since then the PPM has been applied to various phase transitions and transport phenomena successfully.[3] The merits of the PPM lie in the facts that

Theory and Applications of the Cluster Variation and Path Probability Methods
Edited by J.L. Morán-López and J.M. Sanchez, Plenum Press, New York, 1996

the PPM is a systematic approximation scheme parallel to an approximation series of the CVM and that the equation derived from the PPM has a static solution connected to the free energy expected from the CVM. Actually it has been recognized recently that the approximation scheme in the PPM goes completely parallel to that of the CVM when a system itself in the ensemble is chosen as a basic cluster.[4,5] However, the merit of the PPM is more than that. The PPM can treat not only the most probable motion of the system but also fluctuation from its most probable path owing to the variation principle of the PPM,[6] though the actual applications are limited to the former one until recently.

In the general formulation of the PPM, an Ising type spin flip model is the most convenient. We first construct the transition probability of an ensemble of equivalent systems in a short time interval from t to $t + \Delta t$ called the path probability function which is the core ingredient of the PPM. The path probability function in the PPM corresponds to the probability of finding a free energy of the ensemble in equilibrium. The maximum condition of the path probability function gives the most probable path of the ensemble when a state of the ensemble at time t is specified. However, there are many other paths in addition to the most probable path which represent fluctuations in the path probability function. Thus, by making use of information of fluctuations contained in the path probability function, we can open a way for the new applications of the PPM. One application is the derivation of the master equation in each approximation.[6] The master equation describes a time development of a distribution function of some physical quantity such as order parameter. Another application of fluctuations is the path integral representation. In the Euler-Lagrange equation of the path integral representation there are at least two solutions easily found by inspection.[1,7,8] One is a causal path corresponding to the most probable motion discussed up to now. Another solution is called an anticausal path along which the free energy to be expected from the CVM is created. Though the anticausal path may be the most improbable path of the system, the path creating the free energy exists owing to fluctuations. We can also see the effect of fluctuations in the well known fluctuation-dissipation relation.[13] In each approximation level we can actually show the validity of the fluctuation-dissipation theorem between the magnetic susceptibility and the time correlation function of magnetization in the PPM frame work.

Other characteristics to be noted is that the PPM is a kinetic method for which the kinetics is to be supplemented from outside. In an Ising spin flip model, once the spin kinetics along with Ising Hamiltonian is specified, the PPM can do everything for the time evolution of the system. From the viewpoint of the kinetics we note that the birth and death process is best suited for the PPM formulation. Transport phenomena and crystal growth are also considered as one of the birth and death processes.[9]

In the present article, as an introduction to the PPM we first discuss the Becker-Döring theory[10] as a typical example of a birth and death process from the viewpoint of the PPM. This part is also meant as an introduction to nucleation theory which is related to a relaxation phenomena from metastable to stable states in section VI. In section II a general formulation of the PPM is presented using a spin flip Ising model for later convenience. In section III the general formulation is applied to an inhomogeneous spin flip model in the point approximation. The master equation and the path integral representation for the inhomogeneous system are discussed. Along the anticausal path the creation of the free energy expected from the CVM is actually derived. Moreover, the validity of the fluctuation-dissipation theorem is shown for an inhomogeneous system in the point approximation. In section IV the results of section III are extended to the pair approximation. In section V the master equation for a homogeneous system is numerically solved to discuss the relaxation from metastable to stable states in the pair approximation as well as in the point approximation. The nucleation rate is qualitatively in good agreement with a theoretical evaluation. A summary and discussions are given in section VII.

II. Reformulation of the Becker-Döring Nucleation Theory

First let us reconsider the Becker-Döring nucleation theory[10] from the viewpoint of the PPM. Initially a spin system was in thermal equilibrium at a temperature T below the critical temperature with an external magnetic field in the negative z-direction. When the field direction is suddenly reversed to the positive z-direction at time $t = 0$, the spin system would be placed in a metastable state. Let the number of up spin clusters of size k at time t be $N_k(t)$ $(k = 1, 2, \dots)$. In time $t \to t + \Delta t$ some of the up spin clusters of size k change to $k + 1$ or $k - 1$ due to spin flip process by thermal fluctuation such that

$$N_k(t) = Q_{k,k+1}(t, t + \Delta t) + Q_{k,k-1}(t, t + \Delta t) + Q_{k,k}(t, t + \Delta t), \tag{1}$$

where $Q_{k,k+1}(t, t + \Delta t)$ is the number of up spin clusters which have a size k at time t and has changed to $k + 1$ at time $t + \Delta t$, $Q_{k,k-1}(t, t + \Delta t)$ is the number of up spin clusters which have a size k at time t and have changed to $k - 1$ at $t + \Delta t$ and $Q_{k,k}(t, t + \Delta t)$ is the number of up spin clusters which are of size k at t and remain constant k until $t + \Delta t$. In the PPM terminology $N_k(t)$ is called a state variable and $Q_{k,k'}(t, t + \Delta t)$ is called a path variable, respectively. We also see, by definition, that the number of clusters of size k at $t + \Delta t$ can be written in terms of the same path variables as

$$N_k(t + \Delta t) = Q_{k-1,k}(t, t + \Delta t) + Q_{k+1,k}(t, t + \Delta t) + Q_{k,k}(t, t + \Delta t). \tag{2}$$

In the following the arguments of path variables are often omitted when no confusion occurs. The rate of increase and decrease of up spin clusters of size k is taken to be g_k and r_k, respectively. In the present model g_k and r_k are phenomenological kinetic coefficients to be determined later. When a configuration of the system at time t is specified by the $N_k(t)$'s, the transition probability of the system in time $t \to t + \Delta t$ changing according to the relation (1) is obtained as

$$T(t, t + \Delta t) = \prod_k (g_k \Delta t)^{Q_{k,k+1}} (r_k \Delta t)^{Q_{k,k-1}} \left(1 - (g_k + r_k)\Delta t\right)^{Q_{k,k}}$$

$$\times \frac{N_k(t)!}{Q_{k,k+1}! \, Q_{k,k-1}! \, Q_{k,k}!}, \tag{3}$$

where the second combinatorial factor represents a multiplicity factor of the first term which is the probability of one of the possible transitions of k size clusters. The most probable transition of the system in $t \to t + \Delta t$ is realized by maximizing $T(t, t + \Delta t)$ with respect to independent path variables under the condition of given $N_k(t)$'s at time t. To this end the logarithmic form of Eq. (3) is convenient:

$$\ln T(t, t + \Delta t) = \sum_k (Q_{k,k+1} \ln(g_k \Delta t) + Q_{k,k-1} \ln(r_k \Delta t) + Q_{k,k} \ln(1 - (g_k + r_k) \Delta t))$$

$$+ \sum_k (\mathcal{L}(N_k(t)) - \mathcal{L}(Q_{k,k+1}) - \mathcal{L}(Q_{k,k-1}) - \mathcal{L}(Q_{k,k}))$$

$$+ \sum_k s_k (N_k(t) - Q_{k,k+1} - Q_{k,k-1} - Q_{k,k}), \tag{4}$$

where Stirling's formula $\ln N! \simeq N(\ln N - 1)$ is used with a definition $\mathcal{L}(x) = x(\ln x - 1)$ and s_k is a Lagrange multiplier to be determined by Eq. (1). Differentiation of $\ln T(t, t + \Delta t)$ with $Q_{k,k'}$'s yields

$$\hat{Q}_{k,k+1} = e^{-s_k} g_k \Delta t,$$
$$\hat{Q}_{k,k-1} = e^{-s_k} r_k \Delta t,$$
$$\hat{Q}_{k,k} = e^{-s_k} (1 - (g_k + r_k) \Delta t), \tag{5}$$

where the Lagrange multiplier is $e^{-s_k} = N_k(t)$ and a caret denotes the most probable path value. Equation (5) represents the expected result that, for example, $\hat{Q}_{k,k+1}$ is obtained as the number $N_k(t)$ at time t multiplied by the transition probability of increase in Δt. When inserting Eq. (5) into Eq. (4), we immediately obtain

$$\ln \hat{T}(t, t + \Delta t) = 0. \tag{6}$$

This result means that the transition of $T(t, t + \Delta t)$ in $t \to t + \Delta t$ is realized with probability 1 along the most probable path $\hat{Q}_{k,k'}$'s in the macroscopic system.

Now the most probable evolution of the number of up spin clusters of size k is obtained by making a difference between $N_k(t + \Delta t)$ and $N_k(t)$ from Eqs. (1) and (2) in the limit of $\Delta t \to 0$ as

$$\frac{dN_k(t)}{dt} = J_{k-1}(t) - J_k(t), \tag{7}$$

with

$$J_k(t) = g_k N_k(t) - r_{k+1} N_{k+1}(t), \tag{8}$$

where $J_k(t)$ is the net flow by which clusters of size k grow to size $k + 1$. Since g_k and r_k are phenomenological kinetic coefficients containing all the details of the kinetic process, we have to determine their contents based on physical considerations. First we note that the stationary solution of Eq. (7) is obtained by the relation

$$J_{k-1}(t) - J_k(t) = 0, \tag{9}$$

of which the equilibrium solution is obtained by requiring the detailed balance condition, *i.e.*, $J_k = 0$,

$$\frac{g_k}{r_{k+1}} = \frac{N_{k+1}^e}{N_k^e} = \exp[-\beta(\varepsilon_{k+1} - \varepsilon_k)], \tag{10}$$

where ε_k is a free energy of an up spin cluster of size k and Boltzmann's distribution $N_k^e \propto \exp(-\beta \varepsilon_k)$ is assumed in equilibrium. Hereafter an inverse temperature β is sometimes used instead of $1/k_\mathrm{B}T$. Here the free energy is composed of bulk and surface energy

$$\varepsilon_k = -2hk + \sigma k^{(d-1)/d}, \tag{11}$$

where h is a magnetic field, σ is a surface tension due to the boundary between up and down spins and d is the dimension of the space. The critical size beyond which the cluster grows even without fluctuation is given by the maximum condition of Eq. (11), $k_c = \left(\frac{d-1}{d}\frac{\sigma}{2h}\right)^d$. Thus the flow term is now given by

$$J_k(t) = g_k \exp(-\beta\varepsilon_k)\left[\exp(\beta\varepsilon_k)N_k(t) - \exp(\beta\varepsilon_{k+1})N_{k+1}(t)\right]. \tag{12}$$

To complete the theory, we have to specify the kinetic coefficient g_k. The Becker-Döring assumption is that the growth rate of an up spin cluster of size k is proportional to its surface area such that $g_k = Ck^{(d-1)/d}(C > 0)$.

In nucleation theory we seek a non-equilibrium, steady state solution of Eq. (7) with $J_k = I = $ const., while the condition $J_k = 0$ corresponds to the equilibrium solution. Here the quantity I is called the nucleation rate which measures the net growth rate of up spin clusters larger than the critical size in a non-equilibrium steady state. This non-equilibrium steady state is supposed to be a time independent solution N_k^s of Eq. (7). In order to find the solution we take the following source and sink boundary conditions $N_1^s = N\exp(-\beta\varepsilon_1)$ and $N_k^s \to 0$ $(k \to \infty)$. Summing up Eq. (8) under the sink-source boundary condition we easily have the nucleation rate I as

$$I = 1 \bigg/ \sum_{k=1}^{\infty} \frac{1}{g_k \bar{N}_k}, \qquad \bar{N}_k = N\exp(-\beta\varepsilon_k). \tag{13}$$

The numerical results are shown in Fig. 1 and Fig. 2.

Moreover, in a continuous limit an expansion of the integral term about the critical size leads to the well-known Becker-Döring nucleation rate:

$$I = I_0 \exp\left(\frac{-\varepsilon_{k_c}}{k_\mathrm{B}T}\right), \tag{14}$$

where ε_{k_c} is the free energy of the cluster at critical size and I_0 is given by

$$I_0 = N g_{k_c} \left(\left\|\left(\frac{\partial^2\varepsilon}{\partial k^2}\right)_{k_c}\right\| \bigg/ 2\pi k_\mathrm{B}T\right)^{1/2}.$$

In order to check the validity of the above formula, we show $\ln(I/I_0)$ calculated from Eq. (13) in Fig. 3. We see that the Becker-Döring formula (14) is in good agreement with the direct numerical calculation.

By the way, Eq. (7) with Eq. (8) is a simple and plausible evolution equation and we might derive this equation intuitively without recourse to the PPM. However, that is just what we want for the PPM, because the PPM yields a very natural evolution equation when we apply it to the simple system while we can apply the PPM to more complicated systems systematically once a Hamiltonian and the kinetics are specified. Moreover, since the PPM formulation contains information of fluctuations from the most probable motion owing to the variational principle, we can also derive from the PPM the evolution equation of the distribution function of the numbers of up spin clusters. For that purpose, first we derive a transition probability of the system taking N_k''s at time $t + \Delta t$ when N_k's at t are specified:

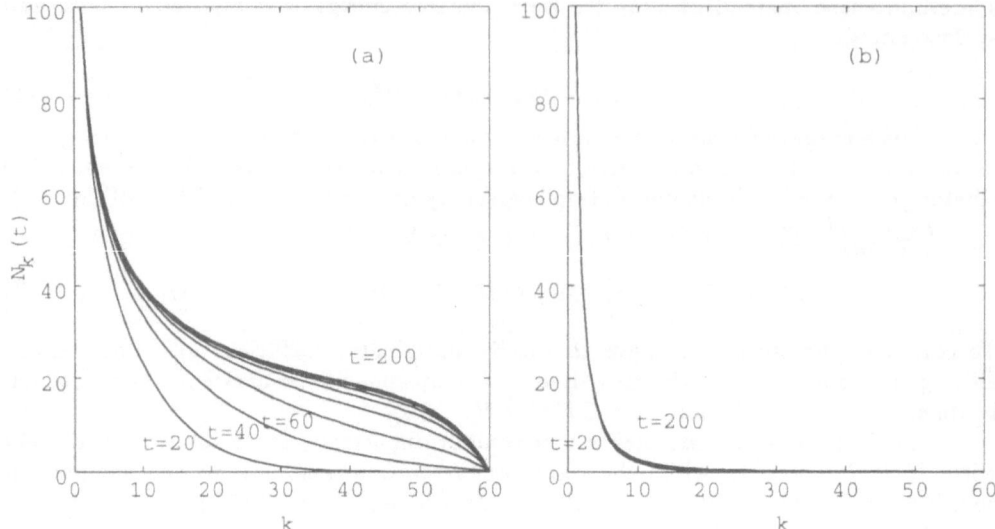

Figure 1. Time development of the numbers $N_k(t)$ of clusters at time $t = 20, 40, \ldots 200$. k stands for the size of a cluster. Parameters are chosen so that $d = 2$, $\beta h = 0.1$ and the number of source clusters $N_1 = 100$ and the size of sink cluster is $k_s = 60$ in common, and that (a) $\beta\sigma = 0.5$ and (b) $\beta\sigma = 2.5$, respectively. Critical size k_c is given by (a) $k_c = 1.5625$ and (b) $k_c = 39.0625$, respectively. When the critical size is large, clusters stop growing.

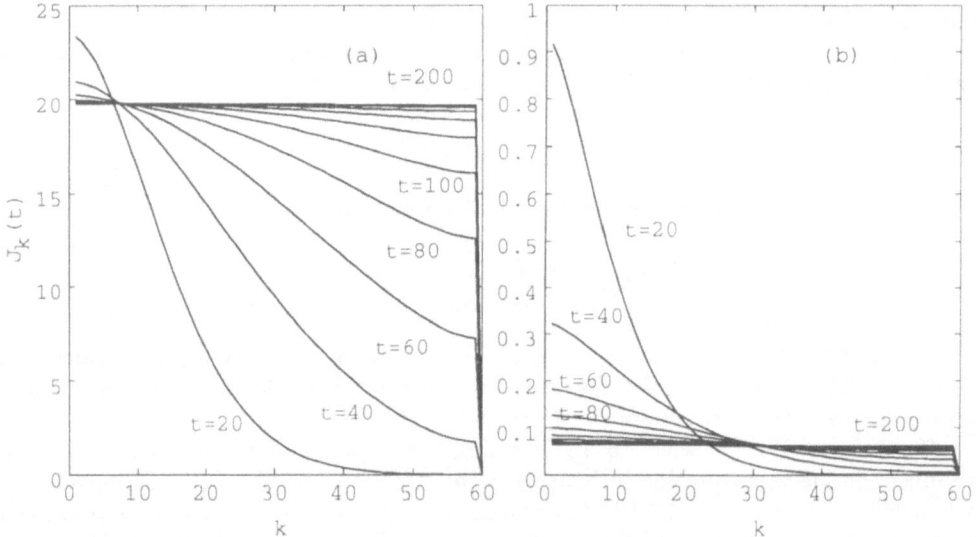

Figure 2. Time development of net flow $J_k(t)$ of cluster size k at time $t = 20, 40, \ldots, 200$. The same parameter values as those in Fig. 1 are chosen ($d = 2$, $\beta h = 0.1$, the sink cluster size $= 60$, and $\beta\sigma = 0.5$ (a), 2.5 (b)). A finite constant nucleation rate $I = J_k$ is obtained in case (a) but $I \to 0$ in case (b). The scale in ordinate is different between (a) and (b).

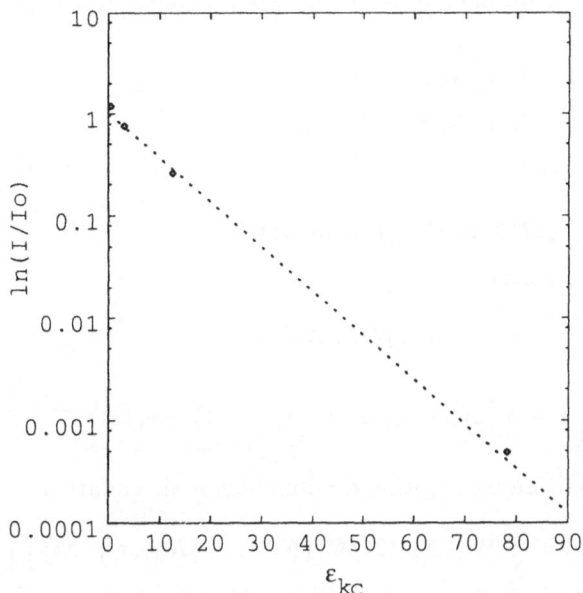

Figure 3. Dependence of nucleation rate upon critical sizes. Square marks are obtained by numerically integrating a kinetic equation and making use of a relation $I = 1/\sum_{k=1}^{\infty} \frac{1}{g_k N_k}$. The dotted line is obtained from $I = I_o \exp(-\beta \varepsilon_{k_c})$. Parameters are chosen so that $d = 2$, $\beta h = 0.1$ and the sink cluster size $= 60$ in common, and $\beta \sigma = 2$ (a), 5 (b), 10 (c), 25 (d), respectively, while the corresponding critical size is given by $k_c = 0.25$ (a), 1.5625 (b), 6.25 (c), 39.0625 (d), respectively.

$$\Psi(\{N_k\}, t \mid \{N_k'\}, t + \Delta t)$$
$$= \prod_k \int T(t, t + \Delta t) \delta[\Delta N_k - (N_k(t + \Delta t) - N_k(t))] \prod_{k'} dQ_{k',k'+1} \, dQ_{k',k'-1}. \quad (15)$$

where $\Delta N_k = N_k' - N_k$. Eq. (15) is further transformed by using (1), (2) and (4) with a delta function expression as

$$\Psi(\{N_k\}, t \mid \{N_k'\}, t + \Delta t)$$
$$= \prod_k \frac{1}{2\pi} \int_{-\infty}^{\infty} dL_k \int \prod_{k'} dQ_{k',k'+1} \, dQ_{k',k'-1} \exp(iL_k \Delta N_k)$$
$$\times \exp\Big[Q_{k,k+1} \ln\big(g_k \Delta t e^{iL_k - iL_{k+1}}\big) + Q_{k,k-1} \ln\big(r_k \Delta t e^{iL_k - iL_{k-1}}\big)$$
$$+ Q_{k,k} \ln\big(1 - (g_k + r_k)\Delta t\big) + N_k(t) \ln N_k(t)$$
$$- \big(Q_{k,k+1} \ln Q_{k,k+1} + Q_{k,k-1} \ln Q_{k,k-1} + Q_{k,k} \ln Q_{k,k}\big)\Big], \quad (16)$$

The maximum term approximation of the argument of the exponent yields up to $\mathcal{O}(\Delta t)$

$$\widehat{Q}_{k,k+1} = N_k g_k \Delta t e^{iL_k - iL_{k+1}},$$
$$\widehat{Q}_{k,k-1} = N_k r_k \Delta t e^{iL_k - iL_{k-1}},$$
$$\widehat{Q}_{k,k} = N_k \left[1 - (g_k \Delta t e^{iL_k - iL_{k+1}} + r_k \Delta t e^{iL_k - iL_{k-1}})\right]. \tag{17}$$

Thus substituting Eq. (17) into Eq. (16), we obtain

$$\Psi(\{N_k\}, t \mid \{N_k'\}, t + \Delta t)$$
$$= \prod_k \frac{1}{2\pi} \int_{-\infty}^{\infty} dL_k \exp(iL_k \Delta N_k)$$
$$\times \left\{1 + \sum_s N_s \Delta t \left[g_s(e^{iL_s - iL_{s+1}} - 1) + r_s(e^{iL_s - iL_{s-1}} - 1)\right]\right\}. \tag{18}$$

Finally under a Markovian assumption the Smoluchowski equation

$$W(\{N_k\}, t + \Delta t) = \int W(\{N_k'\}, t) \Psi(\{N_k'\}, t \mid \{N_k\}, t + \Delta t) \prod_k dN_k', \tag{19}$$

leads us to the master equation of the system,

$$\frac{\partial}{\partial t} W(\{N_k\}, t)$$
$$= \sum_k \left[(N_k + 1)g_k W(\ldots, N_{k-1}, N_k + 1, N_{k+1} - 1, N_{k+2}, \ldots) - N_k g_k W(\{N_k\}, t)\right]$$
$$+ \sum_k \left[(N_k + 1)r_k W(\ldots, N_{k-1} - 1, N_k + 1, N_{k+1}, \ldots) - N_k r_k W(\{N_k\}, t)\right]. \tag{20}$$

Since the master equation of the system is an evolution equation of distribution function of the numbers of up spin clusters and information of fluctuations around the most probable motion of the system is contained in Eq. (20), we can derive the average motion of $\langle N_k(t) \rangle$ from Eq. (20) by defining as $\langle N_k(t) \rangle = \int \prod_l dN_l N_k W(\{N_l\})$:

$$\frac{d\langle N_k(t) \rangle}{dt} = [g_{k-1}\langle N_{k-1}(t) \rangle - r_k \langle N_k(t) \rangle] - [g_k \langle N_k(t) \rangle - r_{k+1}\langle N_{k+1}(t) \rangle]. \tag{21}$$

As is expected, the average motion is identical to the most probable motion of the system Eq. (7).

As an introduction to the PPM we demonstrated that we can derive not only the most probable evolution of the system but also the evolution equation of the distribution function in the birth and death type process.

III. Ising Spin Flip Model

Let us take up as a concrete example an Ising spin flip model with interaction among them for general formulation. The Hamiltonian of a ferromagnetic Ising model with N spins is given by

$$H(\{\sigma\}) = -J \sum_{\langle i,j \rangle} \sigma_i \sigma_j - \mu_0 h \sum_i \sigma_i, \tag{22}$$

where σ_i is a spin variable taking two values ± 1, $J(> 0)$ is an exchange integral, $\langle i, j \rangle$ runs over all nearest-neighbor pairs and the last term is the Zeeman energy. In order to introduce the spin kinetics it is assumed that each spin can flip with a rate θ when there is no interaction among them. In order to formulate the PPM, first, we consider an ensemble of L equivalent systems with the Hamiltonian $H(\{\sigma\})$ in contact with a heat reservoir of temperature T. To describe the change of a state of the ensemble in a short time interval from t to $t + \Delta t$, in addition to the number of systems $Lp_N(\{\sigma\}, t)$ taking a configuration $\{\sigma\}$ at time t, we introduce $LQ_N(\{\sigma\}, t; \{\sigma'\}, t + \Delta t)$ which is defined by the number of systems taking a configuration $\{\sigma\}$ at t and $\{\sigma'\}$ at $t + \Delta t$. From these definitions it is clear that $p_N(\{\sigma\}, t)$ is the probability of finding a configuration $\{\sigma\}$ at t and $Q_N(\{\sigma\}, t; \{\sigma'\}, t + \Delta t)$ is the joint probability of finding a configuration of $\{\sigma\}$ at t and $\{\sigma'\}$ at $t + \Delta t$, respectively. And thus two kind of probabilities are connected by

$$p_N(\{\sigma\}, t) = \text{Tr}_{\{\sigma'\}} Q_N(\{\sigma\}, t; \{\sigma'\}, t + \Delta t),$$

$$p_N(\{\sigma\}, t + \Delta t) = \text{Tr}_{\{\sigma'\}} Q_N(\{\sigma'\}, t; \{\sigma\}, t + \Delta t), \qquad (23)$$

where Tr is a trace operator defined by $\text{Tr}_{\{\sigma\}} = \sum_{\{\sigma = \pm 1\}}$ and in the PPM terminology $p_N(\{\sigma\}, t)$ and $Q_N(\{\sigma\}, t; \{\sigma'\}, t + \Delta t)$ are called a state and a path variable, respectively. Now when a state $\{Lp_N(\{\sigma\}, t)\}$ of the ensemble at time t is specified, the transition probability of the ensemble in the time interval Δt is given by

$$T(t, t + \Delta t) = \frac{\prod_\sigma (Lp_N(\{\sigma\}, t))!}{\prod_\sigma \prod_{\sigma'} (LQ_N(\{\sigma\}, t; \{\sigma'\}, t + \Delta t))!}$$

$$\times \prod_\sigma \prod_{\sigma'} (W_N(\{\sigma\}, t \mid \{\sigma'\}, t + \Delta t))^{LQ_N(\{\sigma\}, t; \{\sigma'\}, t + \Delta t)}, \qquad (24)$$

where $W_N(\{\sigma\}, t \mid \{\sigma'\}, t + \Delta t)$ is the transition probability per system to a configuration $\{\sigma'\}$ at $t + \Delta t$ when a configuration $\{\sigma\}$ at t is given and each factor in Eq. (24) has completely the same interpretation as in the section II. The transition probability $T(t, t + \Delta t)$ of the ensemble in $t \to t + \Delta t$ is called the path probability function, which corresponds to a realizable probability of a free energy in equilibrium statistical mechanics. In contrast to a free energy it is often convenient to work with $\ln T(t, t + \Delta t)$ because a free energy per system is given by

$$F_N = \sum_{\{\sigma\}} [H(\{\sigma\}) + k_B T \ln p_N(\{\sigma\})] p_N(\{\sigma\}). \qquad (25)$$

While the maximum condition of $\exp(-L\beta F_N)$ with respect to $p_N(\{\sigma\})$'s gives the equilibrium distribution function with $Z = \text{Tr}_\sigma \exp[-\beta H(\{\sigma\})]$

$$p_N^e(\{\sigma\}) = \frac{1}{Z} \exp[-\beta H(\{\sigma\})], \qquad (26)$$

the maximum condition of the path probability function with respect to path variables under given state variables at t leads us to a relation

$$\hat{Q}_N(\{\sigma\}, t; \{\sigma'\}, t + \Delta t) = p_N(\{\sigma\}, t) W_N(\{\sigma\}, t \mid \{\sigma'\}, t + \Delta t). \qquad (27)$$

Then the most probable motion of the system is determined with the help of Eq. (23) when the transition probability per system is specified. Since Δt is short enough, we may rewrite the transition probability using the transition rate of a system as

$$W_N(\{\sigma\}, t \mid \{\sigma'\}, t + \Delta t) = w_N(\{\sigma\} \mid \{\sigma'\}) \Delta t (1 - \delta_{\{\sigma\},\{\sigma'\}})$$
$$+ [1 - \text{Tr}'_{\{\sigma''\}} w_N(\{\sigma\} \mid \{\sigma''\}) \Delta t] \, \delta_{\{\sigma\},\{\sigma'\}} , \quad (28)$$

where Tr' denotes a configuration $\{\sigma\}$ being omitted in the sum over $\{\sigma''\}$. Substituting Eq. (28) into Eq. (23) and making a difference of $p_N(\{\sigma\}, t)$ at t and $t + \Delta t$, we get

$$\frac{d}{dt} p_N(\{\sigma\}, t) = - \text{Tr}_{\{\sigma'\}} w_N(\{\sigma\} \mid \{\sigma'\}) \, p_N(\{\sigma\})$$
$$+ \text{Tr}_{\{\sigma'\}} w_N(\{\sigma'\} \mid \{\sigma\}) \, p_N(\{\sigma'\}) . \quad (29)$$

The most probable motion of the system with N spins is just what we call a microscopic master equation of the system in contrast to the coarse-grained master equation at each approximation discussed later.

Now let us determine the transition probability $W_N(\{\sigma\}, t \mid \{\sigma'\}, t + \Delta t)$ in the present spin flip model. For the system to tend to thermal equilibrium, the transition probability follows the detailed balance condition:

$$p_N^e(\{\sigma\}) \, w_N(\{\sigma\} \mid \{\sigma'\}) = p_N^e(\{\sigma'\}) \, w_N(\{\sigma'\} \mid \{\sigma\}) . \quad (30)$$

Though there remains some arbitrariness in the determination of $w_N(\{\sigma\} \mid \{\sigma'\})$, in the relaxation kinetics the PPM usually chooses

$$w_N(\{\sigma\} \mid \{\sigma'\}) \propto \exp\left\{ -\frac{\beta}{2} \Big[H(\{\sigma'\}) - H(\{\sigma\}) \Big] \right\}, \quad \text{for } \{\sigma\} \neq \{\sigma'\}. \quad (31)$$

Thus, with the above transition rate in mind we choose the transition probability per system as[7]

$$W_N(\{\sigma\}, t \mid \{\sigma'\}, t + \Delta t) = \prod_i (\theta \Delta t)^{1 - \delta(\sigma_i, \sigma_i')} \left(1 - \Theta(\sigma_i) \Delta t \right)^{\delta(\sigma_i, \sigma_i')}$$
$$\times \exp\left\{ -\frac{\beta}{2} \Big[H(\{\sigma'\}) - H(\{\sigma\}) \Big] \right\}, \quad (32)$$

where $1 - \Theta(\sigma_i)\Delta t$ is the residual probability of the ith spin in $t \to t + \Delta t$, which is introduced for conservation of the transition probability $\text{Tr}_{\{\sigma'\}} W_N(\{\sigma\}, t \mid \{\sigma'\}, t + \Delta t) = 1$. Here it is noteworthy to stress that only the terms up to $\mathcal{O}(\Delta t)$ contribute to the PPM calculation.

For practical use of the PPM we usually have to proceed to an approximation scheme. One of the most valuable advantages of the PPM is that a systematic approximation treatment was devised by Kikuchi as a natural extension of the CVM. A choice of the basic cluster to represent a system determines an approximation level in which the PPM versus the CVM has a one to one correspondence. It has been discussed that an equilibrium solution in the PPM yields a free energy expected from the corresponding approximation in the CVM.

Now, in a similar fashion leading to Eq. (23), let a state variable of n spin cluster at time t be denoted by $p_n(\{\sigma\}, t)$ and a path variable of n spin cluster at t and

$t + \Delta t$ by $Q_n(\{\sigma\}, t; \{\sigma'\}, t + \Delta t)$, respectively. In the CVM, it is assumed that the probability $p_n(\{\sigma\}_n)$ of finding any n-spin cluster $(n = 1, 2, \dots)$ in the state $\{\sigma\}_n$ be systematically decomposed as

$$p_n(\{\sigma\}_n) = \prod_{i=1}^{n} p_1(\sigma_i) \prod_{\langle i,j \rangle} g_2(\sigma_i, \sigma_j) \prod_{\langle i,j,k \rangle} g_3(\sigma_i, \sigma_j, \sigma_k)$$

$$\times \cdots \times g_n(\sigma_1, \sigma_2, \cdots, \sigma_n), \tag{33}$$

where $g_k(\{\sigma\}_k)$ is called the kth spin correlation function.[11] The essence of the CVM lies in that starting with $n = 2$ the correlation function can be successively determined in terms of cluster probability functions up to kth cluster. Then the nth order cluster approximation assumes $g_m(\{\sigma\}_m) = 1$ for all $m(> n)$ in which approximation all quantities in the system can be expressed using $p_k(\{\sigma\}_k)$'s up to $k = n$. For example, the point approximation uses only $p_1(\sigma_i)$'s and the pair approximation uses $p_1(\sigma_i)$'s and $p_2(\sigma_i, \sigma_j)$'s. On the other side, in the PPM we apply the same reasoning not only to the states variable $p_k(\{\sigma\}_k)$ but also to the path variable $Q_k(\{\sigma\}_k, t; \{\sigma'\}_k, t + \Delta t)$. Then in the nth order approximation of the PPM the path probability function is expressed in terms of $p_k(\{\sigma\}_k)$'s and $Q_n(\{\sigma\}, t; \{\sigma'\}, t + \Delta t)$'s up to $n(k = 1, 2, \dots, n)$.[4]

IV. Site Dependent Point Approximation

For illustration purpose, let us apply the general formulation of the PPM to an inhomogeneous system in the point approximation. The path probability function in the point approximation is given by

$$T(t, t + \Delta t) = \prod_{k} \prod_{\sigma_k = \pm 1} \frac{(Lp_1(\sigma_k, t))!}{\prod_{\sigma_k' = \pm 1}(LQ_1(\sigma_k; \sigma_k'))!}$$

$$\times \left\{ \theta \Delta t \exp\left[-\left(K \sum_{\rho} m_{k+\rho}(t) + B_k \right) \sigma_k \right] \right\}^{LQ_1(\sigma_k; -\sigma_k)}$$

$$\times (1 - \Theta(\sigma_k)\Delta t)^{LQ_1(\sigma_k; \sigma_k)}$$

$$\times \exp\left[Ls_k(p_1(\sigma_k, t) - \mathrm{Tr}_{\sigma_k = \pm 1} Q_1(\sigma_k; \sigma_k')) \right], \tag{34}$$

where $m_k(t)$ denotes a magnetization at k site, $\sum_{\rho}$ stands for a sum over nearest neighboring sites and s_k is the Lagrange multiplier. For notational simplicity we use k for a site instead of vector notation k, and we also write $p_1(\sigma_k, t)$, though the probability function $p_1(\sigma_k, t)$ has a site dependence such as

$$p_1(\sigma_k, t) = \tfrac{1}{2}(1 + m_k(t)\sigma_k), \tag{35}$$

and the same is true for a path variable $Q_1(\sigma_k, t; \sigma_k', t + \Delta t)$ when no confusion occurs. There are some geometrical relations between state and path variables;

$$p_1(\sigma_k, t) = Q_1(\sigma_k, t; -\sigma_k, t + \Delta t) + Q_1(\sigma_k, t; \sigma_k, t + \Delta t),$$

$$p_1(\sigma_k, t + \Delta t) = Q_1(-\sigma_k, t; \sigma_k, t + \Delta t) + Q_1(\sigma_k, t; \sigma_k, t + \Delta t). \tag{36}$$

Differentiation of the above expression with respect to path variables $Q_1(\sigma_k, t; \sigma_k', t + \Delta t)$ with the state variables at t being given leads us to

$$Q_1(\sigma_k, t; \sigma_k, t + \Delta t) = \exp(-s_k)(1 - \Theta(\sigma_k)\Delta t),$$

$$Q_1(\sigma_k, t; -\sigma_k, t + \Delta t) = \exp(-s_k)\theta\Delta t \exp\left[-\sum_\rho (Km_{k+\rho} + B_k)\sigma_k\right]. \qquad (37)$$

Moreover, the substitution of Eq. (37) into Eq. (36) yields

$$\exp(-s_k) = p_1(\sigma_k, t), \qquad \Theta(\sigma_k)\Delta t = \theta\exp\left[-\sum_\rho (Km_{k+\rho} + B_k)\sigma_k\right]\Delta t. \qquad (38)$$

Thus $\hat{Q}_1(\sigma_k; -\sigma_k) = \theta\Delta t p_1(\sigma_k)\exp\left[-\sum_\rho (Km_{k+\rho} + B_k)\sigma_k\right]$ is obtained. The evolution of magnetization is obtained by making a difference of two relations in Eq. (36) as

$$\frac{d}{dt}m_k(t) = 2\frac{\partial}{\partial B_k}G_k(m_k, B_k),$$

$$G_k(m_k, B_k) = \mathrm{Tr}_k\,\theta p_1(\sigma_k)\exp\left[-\sum_\rho (Km_{k+\rho} + B_k)\sigma_k\right]. \qquad (39)$$

Next, we proceed to derive the master equation for magnetization and the path integral representation of the system in the present site dependent point approximation. We first note that the transition probability from $\{M_k \equiv Lm_k\}$ at time t to $\{M_k' \equiv Lm_k'\}$ at $t + \Delta t$ is obtained by making use of the path probability function $T(t, t + \Delta t)$:

$$\Psi(\{M_k\}, t \mid \{M_k'\}, t+\Delta t) = \prod_k \int d(LQ(\sigma_k; -\sigma_k))\,\delta[M_k' - M_k - \Delta M_k(t)]\,T(t, t+\Delta t),$$
$$(40)$$

with $\Delta M_k(t) = -2L\,\mathrm{Tr}_k\,\sigma_k Q(\sigma_k; -\sigma_k)$. Use of a delta function expression $\delta(x) = \frac{1}{2\pi}\int_{-\infty}^\infty dq\,\exp(iq\cdot x)$ and the saddle point method owing to the size of ensemble L give the relation with $\mathbf{M} = \{M_k\}$

$$\Psi(\mathbf{M}, t|\mathbf{M}', t + \Delta t) = \frac{1}{(2\pi)^N}\int d^N\mathbf{q}\,\exp\{L[i\mathbf{q}\cdot(\mathbf{m}' - \mathbf{m}) - H(\mathbf{m}, i\mathbf{q})\Delta t]\}, \qquad (41)$$

where

$$H(\mathbf{m}, i\mathbf{q}) = \left[1 - \exp\left(2i\mathbf{q}\cdot\frac{\partial}{\partial\mathbf{B}}\right)\right]G(\mathbf{m}, \mathbf{B}),$$

$$G(\mathbf{m}, \mathbf{B}) = \sum_k G_k(m_k, B_k). \qquad (42)$$

First, with the help of this transition probability we arrive at the master equation by making use of Smoluchowski equation

$$\varepsilon\frac{\partial W(\mathbf{m}, t)}{\partial t} = \sum_k \left[\exp\left(-2\varepsilon\frac{\partial}{\partial B_k}\cdot\frac{\partial}{\partial m_k}\right) - 1\right]G_k(m_k, B_k)W(\mathbf{m}, t), \qquad (43)$$

where $\varepsilon = 1/L$ is a smallness parameter of the ensemble. This master equation is a Landau-Ginzburg type evolution equation of the ensemble. When we treat a homogeneous system, that is, $m_k = m$ for any site k, we have a master equation for a homogeneous system in the point approximation with $\varepsilon = 1/NL$:

$$\varepsilon \frac{\partial W(m,t)}{\partial t} = \left[\exp\left(-2\varepsilon \frac{\partial}{\partial B} \cdot \frac{\partial}{\partial m} \right) - 1 \right] G(m,B) W(m,t) ,$$

$$G(m,B) = \text{Tr}\, \theta p_1(\sigma) e^{-(zKm+B)\sigma} . \tag{44}$$

Second, when we connect together the transition probability Eq. (41) over all possible paths in a finite interval from t to t', we finally have a path integral expression of the ensemble as

$$W(\mathbf{M}_0, t_0 \mid \mathbf{M}, t) = \int \cdots \int D[\mathbf{M}(t)]\, D[\boldsymbol{\lambda}(t)]$$

$$\times \exp\left\{ L \int_{t_0}^{t} dt\, [\boldsymbol{\lambda}(t) \cdot \dot{\mathbf{m}}(t) - H(\mathbf{m}(t), \boldsymbol{\lambda}(t))] \right\}, \tag{45}$$

with

$$D[\mathbf{M}(t)]\, D[\boldsymbol{\lambda}(t)] = \lim_{n \to \infty} \prod_{j=1}^{n-1} d^N \mathbf{M}_j \frac{d^N \boldsymbol{\lambda}_j}{(2\pi)^N} ,$$

where $\mathbf{M}(t) = \mathbf{M}, \mathbf{M}(t_0) = \mathbf{M}_0$ and $\boldsymbol{\lambda}_j = i\mathbf{q}_j$.

As an application of the above path integral representation we show that an integral along the anticausal path follows the creation of a free energy of the system. Since the number of systems in the ensemble is very large, the most important contribution to the path integral comes from a maximum term of the integrand which gives an Euler-Lagrange equation

$$\frac{d}{dt} \mathbf{m}(t) = \frac{\partial}{\partial \boldsymbol{\lambda}} H(\mathbf{m}, \boldsymbol{\lambda}), \tag{46}$$

$$\frac{d}{dt} \boldsymbol{\lambda}(t) = -\frac{\partial}{\partial \mathbf{m}} H(\mathbf{m}, \boldsymbol{\lambda}). \tag{47}$$

More explicitly, one component of Eq. (46) is rewritten as

$$\frac{d}{dt} m_k(t) = 2 \frac{\partial}{\partial B_k} \text{Tr}_k\, \theta p_1(\sigma_k, t) \exp\left[-\left(K \sum_{\rho} m_{k+\rho} + B_k - 2\lambda_k \right) \sigma_k \right]. \tag{48}$$

While $\lambda_k = 0$ for any k gives a causal solution of Eq. (39), corresponding to the most probable motion, we can find another solution by inspection which corresponds to an anticausal path. That is, when we choose the following relation

$$p_1(\sigma_k, t) \exp\left[-\left(K \sum_{\rho} m_{k+\rho} + B_k - 2\lambda_k \right) \sigma_k \right] = p_1(-\sigma_k, t) \exp\left[\left(K \sum_{\rho} m_{k+\rho} + B_k \right) \sigma_k \right],$$
$$\tag{49}$$

for all k, this condition also satisfies Eq. (47) since Eq. (49) is equivalent to a relation

$$H(\mathbf{m}(t), \boldsymbol{\lambda}(t)) = G(\mathbf{m}(t), B) - G(\mathbf{m}(t), B - 2\lambda) = 0. \tag{50}$$

As the present solution leads to an evolution equation

$$\frac{d}{dt}m_k(t) = -2\frac{\partial}{\partial B_k}G_k(m_k, B_k), \tag{51}$$

we can say that the solution represents an anticausal path against that of Eq. (39). Next let us evaluate the contribution in the path integral equation (45) along the anticausal path by making use of Eq. (49) and Eq. (50). We can easily obtain an integral value as

$$\int_{t_0}^{t} dt \left(\boldsymbol{\lambda}(t) \cdot \frac{d\mathbf{m}(t)}{dt} - H(\mathbf{m}, \boldsymbol{\lambda})\right) = \int_{t_0}^{t} -\frac{d}{dt}\frac{F_N(\mathbf{m}(t))}{k_B T}dt$$

$$= -\frac{1}{k_B T}(F_N(\mathbf{m}) - F_N(\mathbf{m}_0)), \tag{52}$$

where $F_N(\mathbf{m})$ is given by

$$\frac{F_N(\mathbf{m})}{k_B T} = -K\sum_{\langle i,j\rangle}m_i m_j - \sum_i B_i m_i + \sum_k \mathrm{Tr}_k \frac{1 + m_k \sigma_k}{2}\ln\frac{1 + m_k \sigma_k}{2}. \tag{53}$$

Thus along the anticausal path we obtain a contribution to the path integral as

$$W(\mathbf{m}_0, t_0 \mid \mathbf{m}, t) = \exp\left[-\frac{L}{k_B T}(F_N(\mathbf{m}) - F_N(\mathbf{m}_0))\right], \tag{54}$$

with $\mathbf{m}_0 = \mathbf{m}(t_0)$ and $\mathbf{m} = \mathbf{m}(t)$. The above function $F_N(\mathbf{m})$ is just the same free energy as that obtained from the site dependent point approximation of the CVM. It should be noted that along the anticausal path we always have an expression of the free energy creation which is expected from the CVM in the corresponding approximation.[7] This fact shows the consistency between the PPM and the CVM.

The consistency between the PPM and the CVM can be also seen through the fluctuation-dissipation theorem. First let us find the magnetic susceptibility $\chi(q, \omega)$ by applying a small magnetic field $h_k(t) = h(\omega, q)\exp\{i(\omega t - q \cdot k)\}$ to the system. Since Eq. (39) is rewritten to linear order of magnetic field for a temperature above the critical temperature T_C as

$$\frac{1}{2\theta}\frac{dm_k(t)}{dt} = -\left(m_k(t) - K\sum_\rho m_{k+\rho}(t)\right) + B_k(t), \tag{55}$$

the susceptibility defined by $\chi(q, \omega) = N\mu_0 m(q, \omega)/h(q, \omega)$ is obtained as

$$\chi(q, \omega) = \frac{N\mu_0^2}{k_B T}\frac{2\theta}{i\omega + 2\theta(1 - K\gamma(q))}, \tag{56}$$

where $\gamma(q) = \sum_\rho e^{iq\cdot\rho}$ and $m_k(t) = m(q, \omega)\exp[i(\omega t - q \cdot k)]$ is used. Next, consider a retarded spin correlation function in equilibrium without external magnetic field defined by

$$\langle m_j(0)m_k(t)\rangle_e = \int d\mathbf{m}\, p_N^e(\{m_i\})\, m_j\, W_N(\{m_i\}, 0 \mid \{m_i'\}, t)\, m_k', \tag{57}$$

where $W_N(\{m_i\}, 0 \mid \{m_i'\}, t)$ is the transition probability from a spin configuration $\{m_i\}$ to another one $\{m_i'\}$ in time t and $\langle \cdots \rangle_e$ denotes an thermal average in equilibrium. The expression is further rewritten as

$$\langle m_j(0)m_k(t)\rangle_e = \int dm \; p_N^e(\{m_i\}) \, m_j \, \langle m_k(t)\rangle_{\{m_i\}}, \tag{58}$$

where $\langle m_k(t)\rangle_{\{m_i\}}$ is the expectation of $m_k(t)$ for the given configuration $\{m_i\}$ at $t = 0$. Since in no magnetic field the Fourier transform of Eq. (55) is given, with a definition $M(q,t) = \mu_0 \sum_k m_k(t) \exp(iq \cdot k)$, by

$$\frac{1}{2\theta}\frac{dM(q,t)}{dt} = -[1 - K\gamma(q)]M(q,t), \tag{59}$$

from the definition, Eq. (58), the retarded spin correlation function must follows the same equation Eq. (59) as a linear relaxation of the magnetization deviated from equilibrium due to fluctuations:

$$\frac{1}{2\theta}\frac{d}{dt}\langle M(-q,0)M(q,t)\rangle_e = -[1 - K\gamma(q)]\langle M(-q,0)M(q,t)\rangle_e. \tag{60}$$

This equation is easily solved to yield

$$\langle M(-q,0)M(q,t)\rangle_e = \langle M(-q,0)M(q,0)\rangle_e \exp[-2\theta(1 - K\gamma(q))t]. \tag{61}$$

As $p_N^e(\{m_i\})$ is proportional to $\exp[-\beta F_N(\{m_i\})]$, in the Fourier transformed representation we have from Eq. (53) to second order in the magnetization

$$p_N^e(\{m_i\}) \propto \exp\left[-\frac{1}{2N\mu_0^2}\sum_q (1 - K\gamma(q))M(-q)M(q)\right]. \tag{62}$$

From this expression the static correlation function becomes

$$\langle M(-q,0)M(q,0)\rangle_e = N\mu_0^2\frac{1}{1 - K\gamma(q)}. \tag{63}$$

Comparing Eq. (56) with Eq. (63) we have a general relation for the static response function and fluctuations of the magnetization

$$\chi(q,0) = \frac{1}{k_B T}\langle M(-q,0)M(q,0)\rangle_e. \tag{64}$$

It is shown that by combining Eq. (56), Eq. (61) and Eq. (64) we can immediately extend the above relation to a general relation between the dynamical response function and fluctuations of the magnetization.

$$\chi(q,\omega) = \chi(q,0) - \frac{i\omega}{k_B T}\int_0^\infty dt \langle M(q,0)M(-q,t)\rangle_e \exp(-i\omega t). \tag{65}$$

This relation is called the fluctuation-dissipation theorem whose general validity Suzuki and Kubo[13] proved in the Glauber kinetic model.[12] The fact that the fluctuation-dissipation theorem holds in the site dependent point approximation is also considered as evidence of the consistency of the PPM and the CVM. In the next section we will show that the fluctuation-dissipation theorem holds even in the site dependent pair approximation.

V. Fluctuation-Dissipation Theorem in the Pair

Approximation

Let us apply the general formulation of the PPM to the spin flip Ising model in the pair approximation. The logarithmic form of the path probability function for an inhomogeneous system is given by

$$\ln T(t, t + \Delta t) = \sum_i \mathrm{Tr}_i \, Q_1(\sigma_i; -\sigma_i) \ln(\theta \Delta t) + \mathrm{Tr}_i \, Q_1(\sigma_i; \sigma_i) \ln(1 - \Theta(\sigma_i) \Delta t)$$

$$+ \frac{1}{2} \sum_{\langle i,j \rangle} K \, \mathrm{Tr}_{i,j} \, \sigma_i \sigma_j \Big\{ [Q_2(-\sigma_i, \sigma_j; \sigma_i, \sigma_j) - Q_2(\sigma_i, \sigma_j; -\sigma_i, \sigma_j)]$$

$$+ [Q_2(\sigma_i, -\sigma_j; \sigma_i, \sigma_j) - Q_2(\sigma_i, \sigma_j; \sigma_i, -\sigma_j)] \Big\}$$

$$+ \frac{1}{2} \sum_i B_i \, \mathrm{Tr}_i \, \sigma_i [Q_1(-\sigma_i; \sigma_i) - Q_1(\sigma_i; -\sigma_i)]$$

$$+ \Big[-(z-1) \sum_i \mathrm{Tr}_i \, \mathcal{L}(p_1(\sigma_i)) + \sum_{\langle i,j \rangle} \mathrm{Tr}_{i,j} \, \mathcal{L}(p_2(\sigma_i, \sigma_j)) \Big]$$

$$- \Big\{ -(z-1) \sum_i \mathrm{Tr}_i \, \mathcal{L}(Q_1(\sigma_i; -\sigma_i)) + \mathrm{Tr}_i \, \mathcal{L}(Q_1(\sigma_i; \sigma_i))$$

$$+ \sum_{\langle i,j \rangle} [\mathrm{Tr}_{i,j} \, \mathcal{L}(Q_2(\sigma_i, \sigma_j; -\sigma_i, \sigma_j)) + \mathrm{Tr}_{i,j} \, \mathcal{L}(Q_2(\sigma_i, \sigma_j; \sigma_i, -\sigma_j))$$

$$+ \mathrm{Tr}_{i,j} \, \mathcal{L}(Q_2(\sigma_i, \sigma_j; \sigma_i, \sigma_j))] \Big\}, \tag{66}$$

where the state variables are related to the path variables by the following relations

$$p_2(\sigma_i, \sigma_j, t) = Q_2(\sigma_i, \sigma_j; -\sigma_i, \sigma_j) + Q_2(\sigma_i, \sigma_j; \sigma_i, -\sigma_j) + Q_2(\sigma_i, \sigma_j; \sigma_i, \sigma_j),$$

$$p_2(\sigma_i, \sigma_j, t + \Delta t) = Q_2(-\sigma_i, \sigma_j; \sigma_i, \sigma_j) + Q_2(\sigma_i, -\sigma_j; \sigma_i, \sigma_j) + Q_2(\sigma_i, \sigma_j; \sigma_i, \sigma_j), \tag{67}$$

$$p_1(\sigma_i, t) = \mathrm{Tr}_j \, p_2(\sigma_i, \sigma_j, t) = Q_1(\sigma_i; -\sigma_i) + Q_1(\sigma_i; \sigma_i),$$

$$p_1(\sigma_i, t + \Delta t) = \mathrm{Tr}_j \, p_2(\sigma_i, \sigma_j, t + \Delta t) = Q_1(-\sigma_i; \sigma_i) + Q_1(\sigma_i; \sigma_i). \tag{68}$$

The time t in the path variable $Q_2(\sigma_i, \sigma_j, t; \sigma_i', \sigma_j', t + \Delta t)$ is omitted without confusion. In order to find most probable path variables it is convenient to rewrite the above path probability function as

$$\ln T(t, t + \Delta t) = \sum_i \mathrm{Tr}_i \, Q_1(\sigma_i; -\sigma_i) \ln(\theta \Delta t) + \mathrm{Tr}_i \, Q_1(\sigma_i; \sigma_i) \ln(1 - \Theta(\sigma_i)\Delta t)$$

$$+ \sum_i \sum_\rho K \, \mathrm{Tr}_{i,i+\rho} \, \sigma_i \sigma_{i+\rho} Q_2(\sigma_i, \sigma_{i+\rho}; -\sigma_i, \sigma_{i+\rho})$$

$$- \sum_i B_i \, \mathrm{Tr}_i \, \sigma_i Q_1(\sigma_i; -\sigma_i)$$

$$+ \left[-(z-1) \sum_i \mathrm{Tr}_i \, \mathcal{L}(p_1(\sigma_i)) + \frac{1}{2} \sum_i \sum_\rho \mathrm{Tr}_{i,i+\rho} \, \mathcal{L}(p_2(\sigma_i, \sigma_{i+\rho})) \right]$$

$$+ \left\{ (z-1) \sum_i [\mathrm{Tr}_i \, \mathcal{L}(Q_1(\sigma_i; -\sigma_i)) + \mathrm{Tr}_i \, \mathcal{L}(Q_1(\sigma_i; \sigma_i))] \right\}$$

$$- \sum_i \sum_\rho \left[\mathrm{Tr}_{i,i+\rho} \mathcal{L}(Q_2(\sigma_i, \sigma_{i+\rho}; -\sigma_i, \sigma_{i+\rho})) \right.$$

$$\left. + \tfrac{1}{2} \, \mathrm{Tr}_{i,i+\rho} \, \mathcal{L}(Q_2(\sigma_i, \sigma_{i+\rho}; \sigma_i, \sigma_{i+\rho})) \right]$$

$$+ \sum_i \sum_{\rho=1}^z \mathrm{Tr}_i \, s_\rho(\sigma_i) [Q_1(\sigma_i; -\sigma_i)$$

$$- \mathrm{Tr}_{i+\rho} \, Q_2(\sigma_i, \sigma_{i+\rho}; -\sigma_i, \sigma_{i+\rho})] \,, \tag{69}$$

where $s_\rho(\sigma_i)$ is a Lagrange multiplier which guarantees the consistency among path variables. By differentiating the path probability function with respect to independent path variables $Q_1(\sigma_i; -\sigma_i)$ and $Q_2(\sigma_i, \sigma_{i+\rho}; -\sigma_i, \sigma_{i+\rho})$ with state variables at t specified, we obtain two kinds of relations

$$1 = \frac{\theta \Delta t \, e^{-B_i \sigma_i}}{1 - \Theta(\sigma_i)\Delta t} \left(\frac{Q_1(\sigma_i; -\sigma_i)}{Q_1(\sigma_i; \sigma_i)} \right)^{z-1} \exp \left(\sum_\rho s_\rho(\sigma_i) \right), \tag{70}$$

$$e^{s_\rho(\sigma_i)} = \frac{Q_2(\sigma_i, \sigma_{i+\rho}; \sigma_i, \sigma_{i+\rho}) \exp(-K \sigma_i \sigma_{i+\rho})}{Q_2(\sigma_i, \sigma_{i+\rho}; -\sigma_i, \sigma_{i+\rho})}, \qquad \rho = 1, \dots, z. \tag{71}$$

These equations are easily solved up to $\mathcal{O}(\Delta t)$ as

$$\hat{Q}_1(\sigma_i; -\sigma_i) = \theta \Delta t p_1(\sigma_i, t) e^{-B_i \sigma_i} \prod_\rho \frac{\mathrm{Tr}_{i+\rho} \, p_2(\sigma_i, \sigma_{i+\rho}) e^{-K \sigma_i \sigma_{i+\rho}}}{p_1(\sigma_i)}, \tag{72}$$

$$\hat{Q}_2(\sigma_i, \sigma_{i+\rho}; -\sigma_i, \sigma_{i+\rho}) = \hat{Q}_1(\sigma_i; -\sigma_i) \frac{p_2(\sigma_i, \sigma_{i+\rho}) e^{-K \sigma_i \sigma_{i+\rho}}}{\mathrm{Tr}_{i+\rho} \, p_2(\sigma_i, \sigma_{i+\rho}) e^{-K \sigma_i \sigma_{i+\rho}}}. \tag{73}$$

Here, let us define two kinds of order parameters, the magnetization and the short range order parameter, as

$$m_i(t) = \mathrm{Tr}_i \, \sigma_i p_1(\sigma_i, t), \qquad s_{i,j}(t) = \mathrm{Tr}_{i,j} \, \sigma_i \sigma_j p_2(\sigma_i, \sigma_j). \tag{74}$$

Since use of Eq. (66) $\sim$ Eq. (68) with definitions of order parameters yields

$$m_i(t + \Delta t) - m_i(t) = -2 \, \mathrm{Tr}_i \, \sigma_i Q_1(\sigma_i; -\sigma_i), \tag{75}$$

$$s_{ij}(t + \Delta t) - s_{ij}(t) = -2 \operatorname{Tr}_{i,j} \sigma_i \sigma_j (Q_2(\sigma_i, \sigma_j; -\sigma_i, \sigma_j) + Q_2(\sigma_i, \sigma_j; \sigma_i, -\sigma_i)), \quad (76)$$

we obtain a set of evolution equations for the average motion of the system

$$\frac{dm_i(t)}{dt} = 2 \frac{\partial}{\partial B_i} \operatorname{Tr}_i \theta p_1(\sigma_i, t) e^{-B_i \sigma_i} \prod_\rho \lambda_\rho(\sigma_i), \quad (77)$$

$$\frac{ds_{i,j}(t)}{dt} = 2 \left[\operatorname{Tr} r_i \theta p_1(\sigma_i) e^{-B_i \sigma_i} \frac{\partial \lambda_j(\sigma_i)}{\partial K} \prod_{\rho \neq j} \lambda_\rho(\sigma_i) \right.$$

$$\left. + \operatorname{Tr}_j \theta p_1(\sigma_j) e^{-B_j \sigma_j} \frac{\partial \lambda_i(\sigma_j)}{\partial K} \prod_{\rho \neq i} \lambda_\rho(\sigma_j) \right], \quad (78)$$

where i, j in the above equation is a nearest-neighboring pair and $\lambda_\rho(\sigma_i)$ is given by

$$\lambda_\rho(\sigma_i) = \frac{\operatorname{Tr}_{i+\rho} p_2(\sigma_i, \sigma_{i+\rho}) e^{-K \sigma_i \sigma_{i+\rho}}}{p_1(\sigma_i)}. \quad (79)$$

In a homogeneous system this set of equations reduces to[6]

$$\frac{dm(t)}{dt} = 2 \frac{\partial}{\partial B} G(m, s; B, K), \quad (80)$$

$$\frac{ds(t)}{dt} = 2 \frac{\partial}{\partial K} G(m, s; B, K), \quad (81)$$

where the generating function $G(m, s; B, K)$ of the pair approximation becomes

$$G(m, s; B, K) = \operatorname{Tr}_i \theta e^{-B_i \sigma_i} p_1(\sigma_i) \prod_\rho \frac{\operatorname{Tr}_{i+\rho} p_2(\sigma_i, \sigma_{i+\rho}) e^{-K \sigma_i \sigma_{i+\rho}}}{p_1(\sigma_i)}. \quad (82)$$

Now let us discuss whether the fluctuation-dissipation theorem holds in the present system. A weak inhomogeneous magnetic field deviates the system from equilibrium. Then for a temperature above the critical temperature T_C the evolution equation of the system becomes, up to the linear order of a magnetic field,

$$\frac{dm_k(t)}{dt} = \frac{1}{\tau(T)} \left[\frac{1}{2} \left(z - 2 - z \cosh 2K \right) m_k + \frac{1}{2} \sinh 2K \sum_\rho m_{k+\rho} + \frac{\mu_0 h_k(t)}{k_B T} \right], \quad (83)$$

where $\tau(T) = (\cosh K)^z / 2\theta$ and $s_{i,i+\rho} = s = \tanh K$ are used in the derivation. Consequently, with a definition $M(q, t) = \mu_0 \sum_k m_k(t) \exp(iq \cdot k)$ the magnetization follows the equation

$$\tau(T) \frac{dM(q, t)}{dt} = -\frac{1}{2} \left[z \cosh 2K - (z - 2) - \gamma(q) \sinh 2K \right] M(q, t)$$

$$+ \frac{N \mu_0^2 h(q, \omega) \exp(i\omega t)}{k_B T} \quad (84)$$

from which the dynamical susceptibility is obtained as

$$\chi(q, \omega) = \frac{N \mu_B^2}{k_B T} \frac{1}{i\omega \tau(T) + \frac{1}{2} \left[z \cosh 2K - (z - 2) - \gamma(q) \sinh 2K \right]}, \quad (85)$$

where $\gamma(q) = \sum_\rho e^{iq\cdot\rho}$. On the other hand, we note that in no magnetic field the retarded spin correlation function in equilibrium follows the same equation as a linear relaxation of the magnetization deviated from equilibrium owing to fluctuation

$$\tau(T)\frac{d}{dt}\langle M(q,0)M(-q,t)\rangle_e = -\left(\tfrac{1}{2}\left[z\cosh 2K - (z-2)\right] - \tfrac{1}{2}\gamma(q)\sinh 2K\right)$$
$$\times \langle M(q,0)M(-q,t)\rangle_e. \tag{86}$$

This equation is easily solved to yield,

$$\langle M(-q,0)M(q,t)\rangle_e = \langle M(-q,0)M(q,0)\rangle_e$$
$$\times \exp\left[-\frac{t}{2\tau(T)}(z\cosh 2K - (z-2) - \gamma(q)\sinh 2K)\right]. \tag{87}$$

Now we can evaluate the static correlation function $\langle M(-q,0)M(q,0)\rangle_e$ in equilibrium by using the static free energy in the pair approximation which is obtained by calculating the path integral along the anticausal path or by directly applying the CVM. The site dependent free energy in the pair approximation is given by

$$\beta F_N(\{m_k\}) = -K\sum_{\langle i,j\rangle} s_{i,j} - \sum_i m_i B_i - (z-1)\sum_i \mathrm{Tr}_i \frac{1+\sigma_i m_i}{2}\ln\frac{1+\sigma_i m_i}{2}$$
$$+ \sum_{\langle i,j\rangle} \mathrm{Tr}_{i,j}\frac{1+m_i\sigma_i+m_j\sigma_j+s_{ij}\sigma_i\sigma_j}{4}\ln\frac{1+m_i\sigma_i+m_j\sigma_j+s_{ij}\sigma_i\sigma_j}{4}. \tag{88}$$

Since, in order to have the correlation function of magnetization, it is enough to get the free energy up to second order in the magnetization, we have

$$\beta F_N(\{m_k\}) = \frac{zN}{2}\left[Ks + 2\left(\frac{1+s}{4}\ln\frac{1+s}{4} + \frac{1-s}{4}\ln\frac{1-s}{4}\right)\right]$$
$$- \sum_i m_i B_i - \frac{1}{2}\sum_i m_i^2\left(z-1-\frac{z}{1-s^2}\right)$$
$$- \frac{1}{2}\sum_i\sum_\rho \frac{s}{1-s^2}m_i m_{i+\rho}, \tag{89}$$

where $s = \tanh K$ is the zeroth order of s_{ij}. Since $p_N^e(\{m_i\})$ is proportional to $\exp[-\beta F_N(\{m_k\})]$, we have in the representation of the Fourier transform

$$P_N^e(\{m_k\}) \propto \exp\left\{-\frac{1}{2N\mu_0^2}\sum_q\left[\left(\frac{z}{1-s^2}-(z-1)\right) - \frac{s\gamma(q)}{1-s^2}\right]M(-q)M(q)\right\}. \tag{90}$$

From this expression we have a static correlation function

$$\langle M(-q,0)M(q,0)\rangle_e = N\mu_0^2\frac{1}{\frac{1}{2}\{[z\cosh 2K - (z-2)] - \gamma(q)\sinh 2K\}}. \tag{91}$$

Comparing Eq. (85) with Eq. (91) we have a general relation in statistical mechanics between the response function and fluctuations of magnetization

$$\chi(q,0) = \frac{\langle M(-q,0)M(q,0)\rangle_e}{k_B T}. \tag{92}$$

Moreover, we can show by combining Eqs. (85), (87) and (91) that the next relation holds

$$\chi(q,\omega) = \chi(q,0) - \frac{i\omega}{k_\text{B}T} \int_0^\infty dt \, \langle M(q,0)M(-q,t)\rangle_e \exp(i\omega t) . \tag{93}$$

The similar relation is also shown to hold

$$\Im(\chi(q,\omega)) = -\frac{\omega}{2k_\text{B}T} \int_{-\infty}^\infty dt \, \langle M(q,0)M(-q,t)\rangle_e \exp(-i\omega t) . \tag{94}$$

The present derivation for $z = 2$ reduces to an exact result of a one dimensional model treated by Suzuki and Kubo.[13] This fact shows the consistency of the degree of approximations in the CVM and the PPM. From the results of the point and the pair approximation it is expected that the fluctuation-dissipation theorem generally holds in any approximation in the present Ising spin flip model.

VI. Relaxation Phenomena by the Master Equation

We have been stressing the importance of fluctuations from the most probable motion of the system. There are some problems which cannot be treated by studying the most probable motion of the system such as the relaxation from metastable to stable states. Even the relaxation from an unstable state cannot be treated on the basis of the most probable motion when the system is quenched from a paramagnetic to a ferromagnetic phase. In section IV, based on the variational principle of the PPM we have shown the derivation of the master equation in the point approximation. In a similar fashion we can derive the master equation not only in the pair approximation but also in the triangular and the square approximation for a homogeneous system.[14] However, since we hope to treat the relaxation process from metastable to stable states and the time evolution of the quenched system mainly from the numerical point of view, we first list a summary for a homogeneous system in the point and the pair approximation for later convenience. Here it should be noted that the formula of a homogeneous case can be derived as a special case of an inhomogeneous case.

1. The point approximation:
   (a) Equation of motion

$$\frac{dm(t)}{dt} = 2\frac{\partial}{\partial B}G(m, B) . \tag{95}$$

   (b) Master equation

$$\varepsilon\frac{\partial W(m,t)}{\partial t} = \left[\exp\left(-2\varepsilon\frac{\partial}{\partial B}\cdot\frac{\partial}{\partial m}\right) - 1\right]G(m, B)W(m, t), \tag{96}$$

   with

$$G(m, B) = \text{Tr}\,\theta p_1(\sigma)e^{-(zKm+B)\sigma} \tag{97}$$

$$p_1(\sigma) = \tfrac{1}{2}(1 + m\sigma) . \tag{98}$$

   (c) Free energy

$$\frac{F(m)}{k_\text{B}T} = N\left(-\frac{z}{2}Km^2 - mB + \sum_{\sigma=\pm 1}\frac{1+m\sigma}{2}\ln\frac{1+m\sigma}{2}\right) . \tag{99}$$

2. The pair approximation:

  (a) Equation of motion

$$\frac{dm(t)}{dt} = 2\frac{\partial}{\partial \mathbf{B}}G(\mathbf{m}, \mathbf{B}).\tag{100}$$

  (b) Master equation

$$\varepsilon\frac{\partial W(\mathbf{m}, t)}{\partial t} = \left[\exp\left(-2\varepsilon\frac{\partial}{\partial \mathbf{B}}\cdot\frac{\partial}{\partial \mathbf{m}}\right) - 1\right]G(\mathbf{m}, \mathbf{B})W(\mathbf{m}, t),\tag{101}$$

with $\mathbf{m} = (m, s)$, $\mathbf{B} = (B, K)$ and

$$G(\mathbf{m}, \mathbf{B}) = \mathrm{Tr}\,\theta p_1(\sigma)e^{-B\sigma}\lambda(\sigma)^z,\tag{102}$$

$$\lambda(\sigma) = \left[e^{-K}p_2(\sigma, \sigma) + e^{K}p_2(\sigma, -\sigma)\right]/p_1(\sigma),\tag{103}$$

$$p_1(\sigma) = \tfrac{1}{2}(1 + m\sigma),\quad p_2(\sigma, \sigma') = \tfrac{1}{4}\left[1 + m(\sigma + \sigma') + \tfrac{2}{z}s\sigma\sigma'\right].\tag{104}$$

  (c) Free energy

$$\frac{F(m)}{k_B T} = N\left[-Ks - mB - (z - 1)\sum_{\sigma=\pm 1} p_1(\sigma)\ln p_1(\sigma)\right.$$

$$\left. + \frac{z}{2}\sum_{\sigma=\pm 1}\sum_{\sigma'=\pm 1} p_2(\sigma, \sigma')\ln p_2(\sigma, \sigma')\right].\tag{105}$$

In the above list, z is the number of nearest neighboring spins and m and s are the magnetization and the short range order parameter defined by $s = \frac{z}{2}\langle\sigma\sigma'\rangle$, respectively. At this point one comment is in order. Even in the triangular and the square approximation, the same structure of the above expressions, such as Eq. (100) and Eq. (101), are conserved, though the content of a generating function $G(\mathbf{m}, \mathbf{B})$ is changed in each approximation, especially, the content of $\lambda(\sigma)$ representing a molecular field effect about a specified spin being changed.

Now we proceed to the relaxation process from metastable to stable states which is equivalent to the nucleation problem treated in section II. First, let us treat the relaxation process in the point approximation. Griffiths et al.[15] already studied essentially the same model for the similar purpose. Thus, the analytical discussion in the point approximation is mainly a guide to that in the pair approximation since in the present article the analytical discussion of the pair approximation is omitted. The physical situation is completely the same as in section II. Initially a spin system was in thermal equilibrium at a temperature T below the critical temperature T_C with an external magnetic field in the negative z-direction. When the field direction is suddenly reversed to the positive z-direction at time $t = 0$, the spin system would be placed in a metastable state.

To treat Eq. (96) analytically or numerically it is more convenient to rewrite it as

$$\varepsilon\frac{\partial}{\partial t}W(m, t) = \sum_{\sigma=\pm 1}\left[-\theta\frac{1 + m\sigma}{2}e^{-(zKm+B)\sigma}W(m, t)\right.$$

$$\left. + \theta\frac{1 - (m - 2\varepsilon\sigma)\sigma}{2}e^{(zK(m-2\varepsilon\sigma)+B)\sigma}W(m - 2\varepsilon\sigma, t)\right],\tag{106}$$

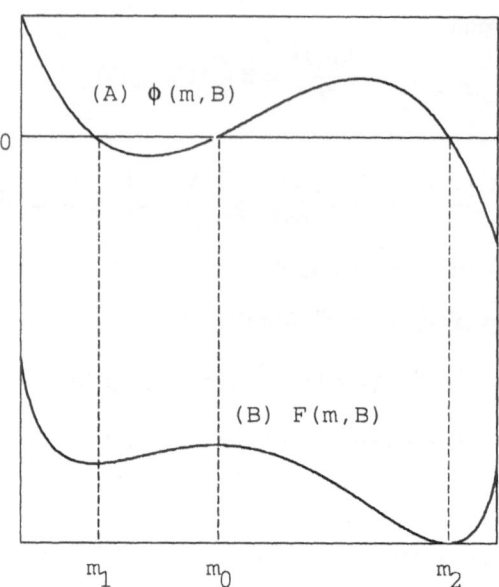

Figure 4. (A) Force term of a kinetic equation and (B) free energy for magnetization are schematically depicted. The zero points m_1, m_0 and m_2 of $\phi(m, B) = 2\frac{\partial G(m,B)}{\partial B}$ correspond to a metastable, a saddle and a stable point in the free energy $F(m, B)$, respectively.

where a smallness parameter ε is considered as the inverse of the number of spins in the system. The above equation can be also written in terms of probability current as

$$\frac{\partial}{\partial t} W(m, t) = J(m - 2\varepsilon, t) - J(m, t),\tag{107}$$

where the probability current $J(m, t)$ is defined as

$$J(m, t) = N \left[\theta \frac{1 - m}{2} e^{(zKm + B)} W(m, t) \right.$$

$$\left. - \theta \frac{1 + (m + 2\varepsilon)}{2} e^{-(zK(m+2\varepsilon)+B)} W(m + 2\varepsilon, t) \right].\tag{108}$$

The points m_0, m_1 and m_2 where $2\frac{\partial}{\partial B} G(m, B)$ in the average motion of the magnetization, Eq. (95), is zero, are the extreme points in the free energy Eq. (99), the last two corresponding to a metastable and a stable state, respectively. The point m_0 is the point to be called a saddle point for the more general case (See Fig. 4). Since the probability of finding a metastable state a time t is defined by $S(t)(= \sum_{m < m_0} W(m, t))$, we define the nucleation rate as the rate of decrease of probability of the metastable state

$$I(t) = -\frac{dS(t)}{dt} = J(m_0 - 2\varepsilon, t),\tag{109}$$

where the sum of Eq. (107) is made use of. To extract the characteristics of metastable state from the master equation, we change it into an equation for $Q(m, t)$ by a transformation

$$W(m, t) = p_N^e(m) Q(m, t), \qquad (110)$$

where $p_N^e(m)$ is an equilibrium distribution probability at $t \to \infty$. Then, by making use of the detailed balance condition $J(m - 2\varepsilon, t) = J(m, t) = 0$, we have an equation for $Q(m, t)$

$$\frac{\partial}{\partial t} Q(m, t) = -N \sum_{\sigma = \pm 1} \theta \frac{1 + m\sigma}{2} e^{-(zKm + B)\sigma} \left[Q(m, t) - Q(m - 2\varepsilon\sigma, t) \right], \qquad (111)$$

where $Q(m, t)$ is expected to behave like a step function about a saddle point m_0 of the free energy. Up to $\mathcal{O}(\varepsilon)$ this equation becomes

$$\frac{\partial}{\partial t} Q(m, t) = 2 \frac{\partial}{\partial B} G(m, B) \frac{\partial}{\partial m} Q(m, t) + 2\varepsilon G(m, B) \frac{\partial^2}{\partial m^2} Q(m, t). \qquad (112)$$

Noting a step function like behavior of $Q(m, t)$ and the sink-source boundary condition, we have approximately the following static solution for Eq. (112)

$$Q(m) = \frac{1}{Z_0} \sqrt{\frac{a}{2b\varepsilon\pi}} \int_m^\infty dm' \exp\left[-\frac{a}{2b\varepsilon} (m' - m_0)^2 \right], \qquad (113)$$

$$Z_0 = \int_{-\infty}^{m_0} dm \exp\left[-\beta F_N(m) \right], \qquad (114)$$

where $a = \left(\frac{\partial}{\partial m} \frac{\partial}{\partial B} G(m, B) \right)_{m_0}$ and $b = (G(m, B))_{m_0}$ and Z_0 is chosen so that $\int_{-\infty}^\infty dm W(m, t) = 1$ is conserved under the source-sink boundary condition. Expanding the free energy about the point m_1 in Eq. (114) and evaluating the nucleation rate, Eq. (109), up to the lowest order in ε, we finally have the time independent nucleation rate under the source-sink boundary condition

$$I = I_0 \exp\left(-\frac{F_N(m_0) - F_N(m_1)}{k_B T} \right), \qquad (115)$$

$$I_0 = \theta N \frac{1 + m_0}{2} \exp\left[-(zKm_0 + B) \right] \sqrt{\frac{\left(\frac{\partial^2}{\partial m^2} F_N(m) \right)_{m_1}}{8\pi k_B T}}. \qquad (116)$$

Now we follow the time development of Eq. (106) by numerical integration in various system sizes at appropriate temperatures. The time evolution of relaxation from a metastable to a stable state is shown in Fig. 5. When we take τ the relaxation time of a metastable state, the dependence of $\ln(\theta\tau)$ corresponding to $-\ln(I/I_0)$ upon the number of spins N is calculated. We see that the tangent is approximately equal to that expected from Eq. (115) in Fig. 7.

It is possible to make a similar analysis in the pair approximation. In the present article we just present numerical results using the master equation, Eq. (101), which is rewritten in a more convenient form for numerical calculations as

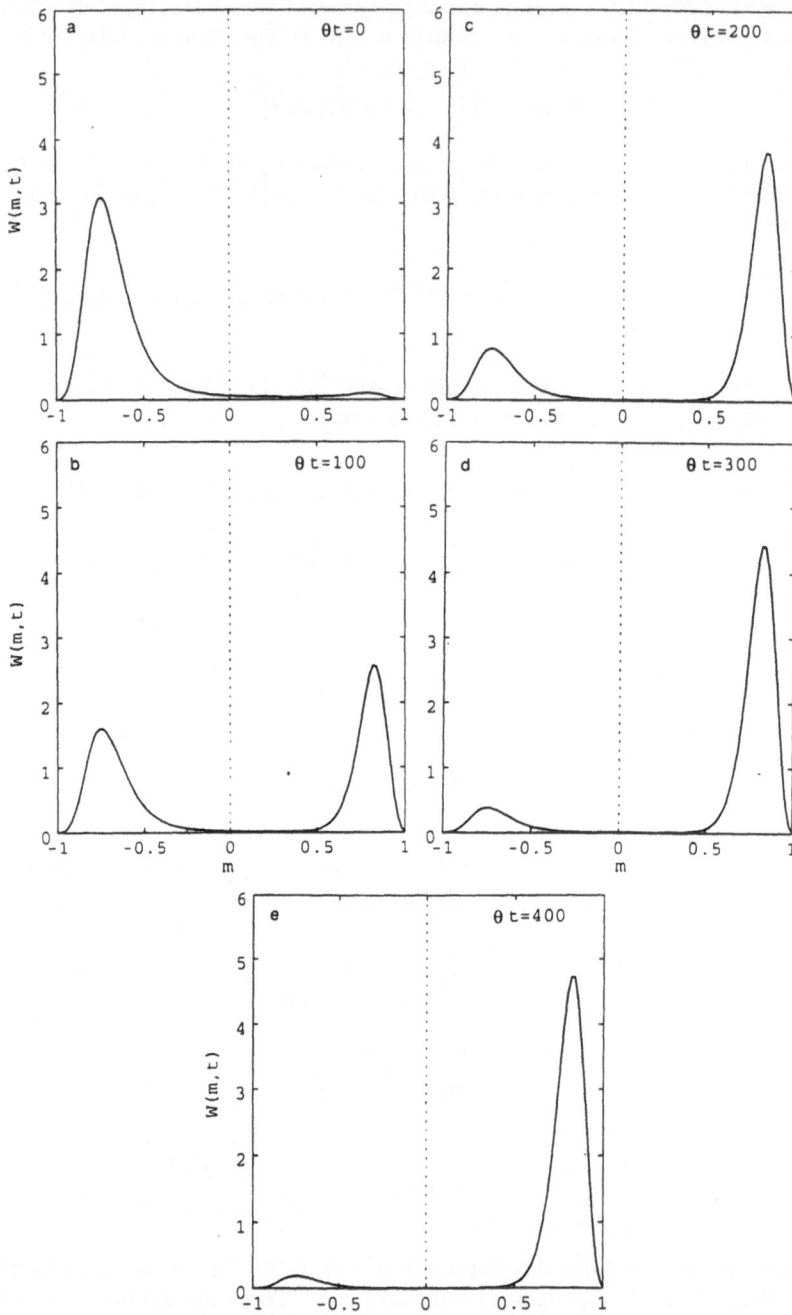

Figure 5. Relaxation from metastable to stable states in field inverted system in the point approximation. The time development of a distribution function $W(m,t)$ of magnetization m is followed at time $\theta t = 0, 100, 200, 300, 400$. Parameters are chosen as $N = 100$, $z = 6$, $B = -0.05$ (initial equilibrium) $\rightarrow 0.05$ (inverted field), and $k_B T/J = 0.747\, k_B T_C/J$ where $k_B T_C/J = z$ is a critical temperature in the point approximation.

$$\varepsilon \frac{\partial}{\partial t} W(m, s, t)$$

$$= -\theta \left[\mathrm{Tr}_{\sigma=\pm 1} \frac{1 + m\sigma}{2} e^{-B\sigma} \right.$$

$$\times \prod_{\rho=1}^{z} \sum_{\sigma_\rho} \frac{\frac{1}{4}[(1 + m(\sigma + \sigma_\rho) + (2/z)s\sigma\sigma_\rho)e^{-K\sigma\sigma_\rho}]}{\frac{1}{2}(1 + m\sigma)} W(m, s, t) \right]$$

$$+ \theta \left[\mathrm{Tr}_{\sigma=\pm 1} \frac{1 + (m + 2\varepsilon\sigma)\sigma}{2} e^{-B\sigma} \right.$$

$$\times \prod_{\rho=1}^{z} \mathrm{Tr}_{\sigma_\rho} \frac{\frac{1}{4}\left\{[1 + (m + 2\varepsilon\sigma)(\sigma + \sigma_\rho) + (2/z)(s + 2\varepsilon\sigma \sum_\rho \sigma_\rho)\sigma\sigma_\rho]e^{-K\sigma\sigma_\rho}\right\}}{\frac{1}{2}[1 + (m + 2\varepsilon\sigma)\sigma]}$$

$$\left. \times W\left(m + 2\varepsilon\sigma, s + 2\varepsilon\sigma \sum_\rho \sigma_\rho, t\right) \right]. \tag{117}$$

We follow the time development of Eq. (117) by numerical integration in various system sizes at appropriate temperatures. Time evolution of relaxation from metastable to stable states are shown in Fig. 6. Dependence of $\ln(\theta\tau)$ upon the system size N in the pair approximation are shown in Fig. 7.

VII. Summary and Discussions

As an introduction to the path probability method, first, we studied the Becker-Döring nucleation theory from the viewpoint of the PPM. The content of the theory is a typical birth and death type kinetic process. Though the Becker-Döring nucleation theory is simple because of no interaction among nucleated clusters, all ingredients required for the PPM formulation is contained in it. When we see the PPM in a new light of birth and death type process, we expect to find various fields of applications for the PPM. The crystal growth by MBE (molecular beam epitaxy) recently studied by us is one of them.[9] In the study of the Becker-Döring nucleation theory we showed that we can derive not only the evolution equation of the average number of clusters of size k, but also the evolution equation of the distribution function of the numbers of various cluster sizes. The evolution of the distribution function is considered just as a representation of the time development of fluctuations. Second, after the general formulation of the PPM was constructed on the basis of the Ising spin flip model, we applied it to an inhomogeneous system in the point and the pair approximation. By making use of fluctuation properties of the transition probability of the ensemble called the path probability function in the PPM terminology, we derived a master equation for an inhomogeneous system. We also derived the path integral representation as one of the applications of fluctuations inherited in the PPM. In the Euler-Lagrange equation of the path integral representation we can find at least two solutions by inspection. One is a causal solution which corresponds to the most probable path tending toward equilibrium. Another is an anticausal path along which the path integral leads to a free energy creation. The free energy is the one to be expected from the CVM in the corresponding approximation level. Onsager and Machlup[16]

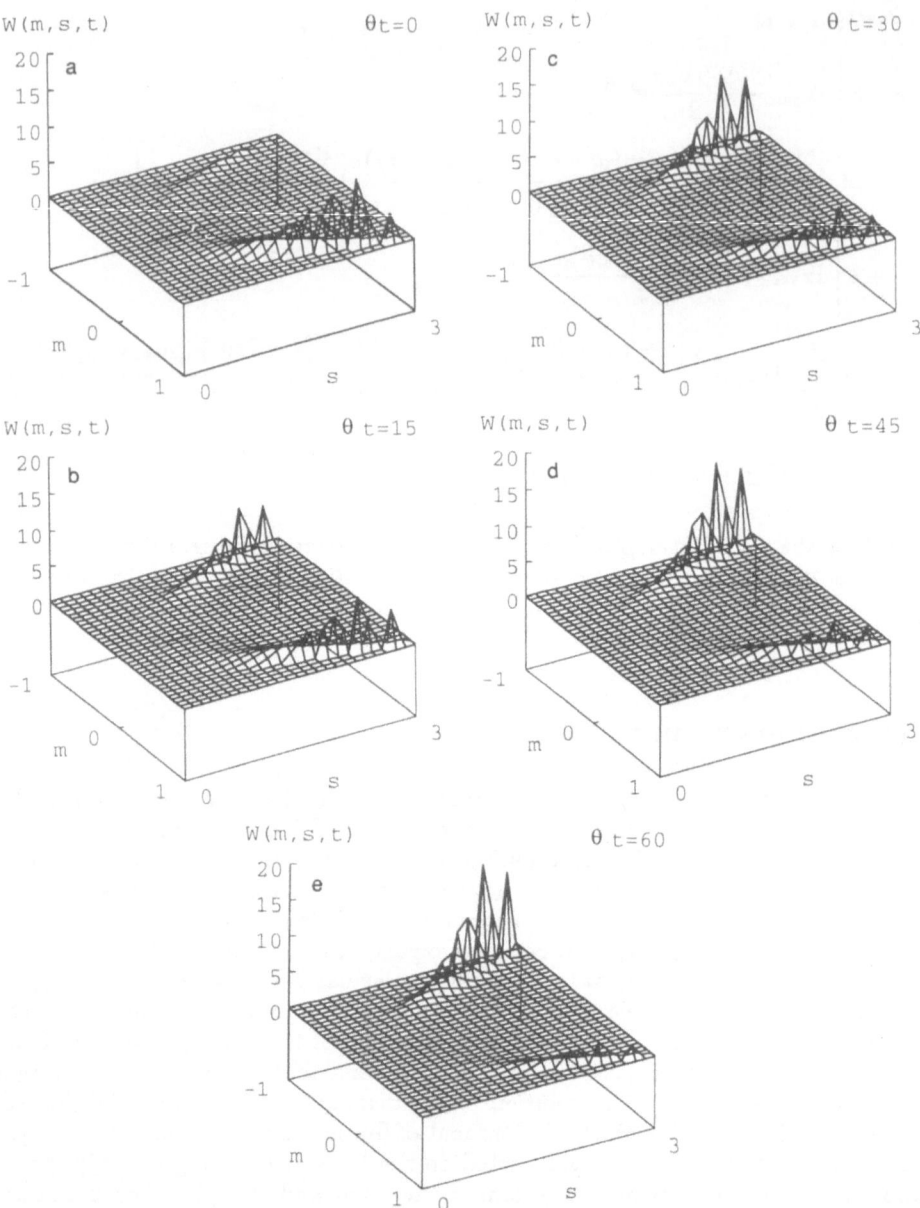

Figure 6. Relaxation from metastable to stable states in field inverted system in the pair approximation. The time development of a distribution function $W(m, s, t)$ of magnetization m and short range order parameter s is follwed at time $\theta t = 0, 15, 30, 45, 60$. Parameters are chosen as $N = 32$, $z = 6$, $B = -0.05$ (initial equilibrium) $\rightarrow 0.05$ (inverted field), and $k_B T/J = 0.71\ k_B T_C/J$ where $k_B T_C/J = 2/\ln(z/(z-2))$ is a critical temperature in the pair approximation.

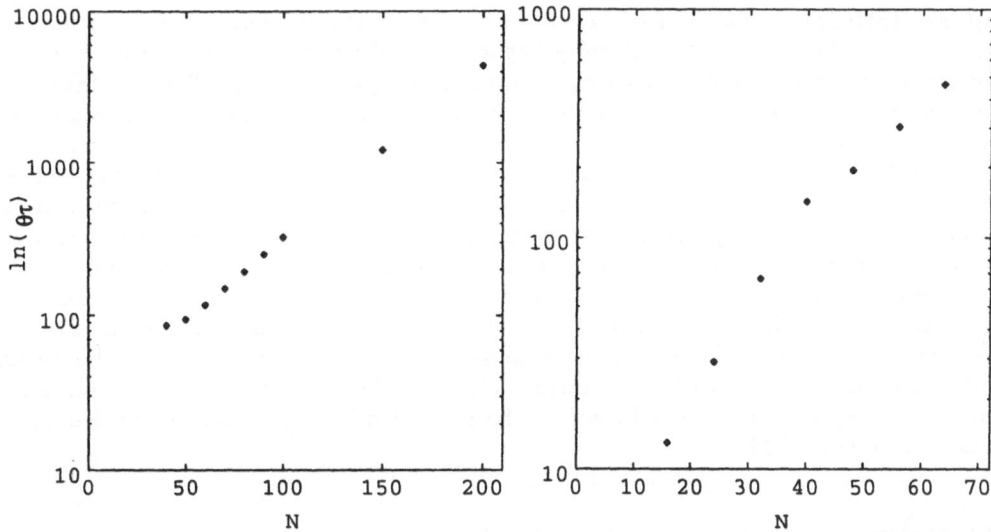

Figure 7. Dependence of relaxation time τ upon spin numbers. Relaxation time τ is evaluated by numerically calculating a decrease of the metastable state in a free energy space with the help of the master equation. $\tau \sim \exp[-N(\Delta f)/k_B T]$ is expected theoretically, where N is the spin numbers in the system and Δf is an activation free energy from a metastable to a saddle point in the free energy space. Parameters are chosen so that $z = 6$, $B = -0.05$ (initial equilibrium) $\rightarrow 0.05$ (inverted field) in common, and $k_B T/J = 0.747\ k_B T_C/J$ in the point approximation and $k_B T/J = 0.71\ k_B T_C/J$ in the pair approximation, respectively.

introduced the concept of the entropy decrease along an anticausal path in their stochastic theory. The present result is considered as a generalization of the Onsager-Machlup theory to an interacting system in contact with a heat bath at a temperature T. Previously, though in the homogeneous system, we confirmed the free energy creation even in the triangular and square approximation. In the present Ising spin flip model the property that the creation of the free energy along the anticausal path seems to be valid in any approximation level, though Kikuchi[1] noted its property in a simple model for the first time. Moreover, when we choose a system itself in an ensemble as a basic cluster, we can prove that the exact free energy creation of the system is realized along the anticausal path.[7] However, the general proof that the present statement holds in any approximation level is an open question. Next, as a confirmation of the consistency of the PPM and the CVM, we discussed the validity of the fluctuation-dissipation theorem in the point and the pair approximation for an inhomogeneous system. There are two ways for deriving the dynamical susceptibility $\chi(q, \omega)$ of a given system. One is to calculate directly the linear response of the magnetization under a given inhomogeneous magnetic field. Another is to calculate the time correlation functions of the magnetization in equilibrium without magnetic field and to use the general fluctuation-dissipation theorem. As long as the approximations in the PPM and the CVM are consistent, the two results should coincide. Though this consistency seems to be valid in any approximation which we could treat analytically, the general proof is also an open question.

In section IV and V we were interested mainly in an inhomogeneous treatment of the PPM. Recently we have been studying the crystal growth by molecular beam epitaxy (MBE) on a vicinal surface.[9] In this problem, the site dependence is essential. However, we find sometimes difficulty beyond our efforts even in the pair approximation, since the number of the participating variables grows rapidly. Thus, for the same reason as above, in order to study the relaxation from metastable to stable states we integrated numerically the master equation obtained for a homogeneous treatment of the PPM in the point and the pair approximation. Since the most probable motion without fluctuation does not work in a first order transition such as the nucleation problem, the master equation plays an important role in such a fluctuation related situation. The activation due to fluctuations from a metastable to a saddle point state is essential in nucleation of the stable state. However, as we have shown numerically as well as theoretically, the activation energy becomes proportional to the number of spins. Thus the nucleation in the homogeneous system becomes impossible in an infinite system by the PPM treatment. Whether the master equation for an inhomogeneous system in section IV solves the present difficulty or not is another open question in the PPM.

References

1. R. Kikuchi, *Prog. Theor. Phys.* Suppl. **35**, 1 (1966).
2. R. Kikuchi, *Phys. Rev.* **81**, 988 (1951).
3. D. E. Temkin, *Sov. Phys. -Crystallogr.* **14**, 344 (1969); R. Kikuchi and H. Sato, *J. Chem. Phys.* **55**, 702 (1971); T. Ishii, H. Sato, and R. Kikuchi, *Phys. Rev.* **34**, 8335 (1986); T. Uchida and K. Wada, *J. Stat. Phys.* **64**, 605 (1991).
4. K. Wada, M. Kaburagi, T. Uchida, and R. Kikuchi, *J. Stat. Phys.* **53**, 1081 (1988).
5. F. Ducastelle, *Prog. Theor. Phys.* Suppl. **115**, 255 (1994).
6. T. Ishikawa, K. Wada, H. Sato, and R. Kikuchi, *Phys. Rev. A* **33**, 4164 (1986); K. Wada, T. Ishikawa, H. Sato, and R. Kikuchi, *Phys. Rev. A* **33**, 4171 (1986).
7. K. Wada and M. Kaburagi, *Prog. Theor. Phys.* Suppl. **115**, 273 (1994).
8. K. Wada and TE. Uchida, *Phys. Lett. A* **168**, 353 (1992).
9. T. Uchida and K. Wada, *Appl. Surf. Sci.* **60/61**, 346 (1992); T. Uchida, K. Wada, and T. Ishikawa, *Phys. Lett.* **154**, 264 (1991).
10. R. Becker and W. Döring, *Ann. der. Phys.* **24**, 719 (1935).
11. T. Morita, *J. Math. Phys.* **13**, 115 (1972).
12. R. J.Glauber, *J. Math. Phys.* **4**, 294 (1963).
13. M. Suzuki and R. Kubo, *J. Phys. Soc. Jpn.* **24**, 51 (1968).
14. K. Wada, T. Ishikawa, and H. Tsuchinaga, *Physica A* **142**, 38 (1987).
15. R. B. Griffiths, C. Y. Weng, and J. S. Langer, *Phys. Rev.* **149**, 301 (1966).
16. L. Onsager, and S. Machlup, *Phys. Rev.* **91**, 1505 (1953).

Universal Dynamic Response in Solid Electrolytes: Formalism of the Path Probability Method as Applied to Transport Problems

Hiroshi Sato,[1] Anuradha Datta,[1] and Takuma Ishikawa[2]

[1] School of Materials Engineering
Purdue University
W. Lafayette, IN 47907-1289
U.S.A.

[2] Faculty of Engineering
Tokyo Institute of Polytechnics
Kanagawa 243-02
JAPAN

Abstract

The relaxation process of hopping ionic conduction in interacting lattice gas systems has been treated analytically from first principles in order to gain understanding of the origin of the "universal dynamic response" in solid electrolytes. The derivation is based on the combination of the pair approximation of the Path Probability Method and the Bethe method. The result indicates that the relaxation motion of a single particle in a lattice gas is intrinsically non-Debye and follows the Kohlrausch-Williams-Watts relaxation behavior even without any interactions among mobile particles. This conclusion is in apparent contrast to the current view that the anomaly is created by interaction among mobile particles and their disorder. It can be shown that the appearance of the KWW effect is due to a memory effect arising from the fact that it takes time for information to propagate over a distance by hopping which has been overlooked so far. A Monte Carlo simulation of the relaxation process was made and the results support the above concept.

Theory and Applications of the Cluster Variation and Path Probability Methods
Edited by J.L. Morán-López and J.M. Sanchez, Plenum Press, New York, 1996

I. Introduction

The major interest of this treatment is to understand the "universal dynamic response" observed in the dispersion process or the relaxation characteristics in the ionic conductivity in solid electrolytes, and to examine the formalism of the Path Probability Method (PPM) as applied to transport problems at the same time.

The phenomenon "universal dynamic response" is generally represented by the following two empirical relations.[1,2] The first indicates that the frequency dependent ionic conductivity $\sigma(\omega)$ is represented by the Jonscher's power law

$$\sigma(\omega) \cong \omega^p, \qquad 0 < p < 1, \tag{1}$$

with p limited between 0 and 1. The second is that the relaxation process in the induced current, when a constant external field is switched on at $t = 0$, is non-Debye (non-exponential), with the behavior to reach the stationary state represented by the Kohlrausch-Williams-Watts (KWW) stretched exponential function $\Xi(t)$[2]

$$\Xi(t) \approx \exp\left[-\left(\frac{t}{\lambda}\right)^\beta\right], \qquad 1 - p = \beta, \quad 0 < \beta < 1. \tag{2}$$

Equation (2) indicates that, in an expression of the form $\exp[-(t/\lambda)]$, λ increases with time.

In the hopping process, a particle which jumps out to a nearest neighboring vacancy leaves behind a vacancy in its wake, and there is always a high probability for that particle to jump back into the vacancy which had been left behind. This jump back process thus constitutes the relaxation process in the hopping motion and, hence, the problem is reduced to the question of whether this jump back process is exponential with time or not.

The task involved here is to derive the relaxation curve for the induced current upon the application of a constant external field at $t = 0$ in the lattice gas model. Here, we limit ourselves to the derivation of flow of a single particle in an assembly, rather than that of the assembly, because the treatment of the former is far simpler. The relation between the two is now well known,[3] but, as far as the mechanism of the relaxation process is concerned, the former is satisfactory as well.

Typically, the relaxation curve can be divided into three time regions: adiabatic, intermediate and isothermal[4] as shown in Fig. 1. If the relaxation curve is normalized, the curve represents the time dependent correlation factor $f(t)$ as introduced by Funke[2] and the relaxed value in the isothermal region for flow of a single particle is commonly called the tracer correlation factor. The difficulty in deriving the relaxation behavior theoretically is to deal with all the three time regions with equal rigor. The adiabatic region and the isothermal region are manageable because, at $t \leq 0$, the system is in equilibrium up to the adiabatic state and, in the isothermal region, a highly sophisticated linear response theory can be used. However, in the intermediate region, it is necessary to rely on kinetics because the reference state is not the same equilibrium state. If the same equilibrium state persists from the adiabatic range through the isothermal range, the jump back process should be exponential, because this reduces the relaxation process to the Markovian process.[5] Therefore, prior theoretical efforts emphasize that the KWW effect should arise from interactions among mobile particles and their disordered arrangements.[6] This means that the origin of the non-linear relaxation process is ascribed to a coupling with the relaxation process of

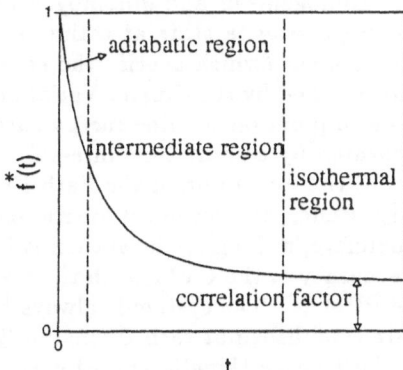

Figure 1. Normalized relaxation curve. The quantity $f^*(t)$ represents the time dependent correlation factor.

the surroundings through interactions with neighbors. Several phenomenological theories have been developed for the appearance of the KWW behavior by constructing models to account for the difference in the equilibrium state in the adiabatic range and in the isothermal range.[2,7,8] For this, the models require interactions among mobile ions and their disordered arrangements. Also, highly sophisticated simulation models in disordered systems, such as glass, are being developed to account for the universal dynamic response. However, in order to understand the real origin of the universal response, it is necessary to develop an analytical treatment based on first principles rather than resorting to phenomenological models.

We have developed such an analytical theory for interacting lattice gas models by combining the pair approximation of the Path Probability Method (PPM) and the Bethe method. Here, the PPM is utilized to describe the adiabatic state correctly, while the Bethe method is used to lead this adiabatic state to the isothermal state correctly through the intermediate state in the form of the propagation of order. The background of why the relaxation process can be analytically described with the pair approximation from the adiabatic state to the isothermal state will be given briefly. By applying this treatment to the jump back process of the tagged particle in the assembly of particles of the same species (without any interaction) with a single vacant site, we realize that the existence probability of the vacancy at the left behind site increases with time from the value expected at $t = 0$ due to its return after the escape. This leads the jump back process to the KWW behavior. In these treatments, the result for a one dimensional lattice is confirmed to be exact. A Monte Carlo simulation of the time dependent correlation factor is made for the assemblies of particles of the same species. The agreement with analytical calculations is found to be excellent.

The behavior which represents the universal dynamic response is thus extracted from the analytical results.

II. Formalism of the PPM

In order to extract important conclusions of the PPM as applied to transport problems, it is convenient to review general characteristics of the PPM briefly with respect to how it handles the kinetic problems. For this purpose, we divide cases to be treated into two general categories: homogeneous and inhomogeneous. If the system is to be kept always statistically homogeneous, a state at a time instant t to be used in the PPM is defined with state variables (which specify the energy of the system) with a corresponding entropy term specified by the Cluster Variation Method (CVM). Hence, the state at t is determined as a point on the free energy surface defined by the CVM. Each consecutive state separated by a short time interval Δt defined by the PPM is connected by the condition of the maximum of the Path Probability function (PPF) as the transition probability. Hence, the resultant kinetic path is defined as a line on the free energy surface. Therefore, as long as the system is homogeneous, the kinetics can be described with the accuracy of the CVM, with the transition probability which is strictly prescribed in the PPM.[9,10] The system is always kept under the isothermal condition and new states are established at each Δt instantly. This indicates that the treatment of the PPM in the t space (kinetic space) corresponds to the mean field approximation, but as long as the system is kept homogeneous, the PPM gives an excellent representation of the kinetics.

In dealing with transport problems, systems include gradients in the state variables and are not homogeneous. However, in the linear range, the flow is represented by the motion of particles in the reference state under equilibrium, multiplied by the driving force. Therefore, in dealing with transport problems by the PPM, the motion of particles in the same equilibrium state, rather than in the state on the free energy surface, at each time instant is discussed irrespective of whether it is in the adiabatic, intermediate or isothermal time range. This also means that the equilibrium state which the PPM represents is that of the adiabatic state, because this is the state connected to the equilibrium state of the system at $t \leq 0$.

If the nearest neighbor distance jump of particles is assumed for the kinetic process, at the adiabatic stage, only particles at the nearest neighbor of the vacancy can contribute to the change of state. Therefore, the pair approximation rather than higher approximations of the PPM represents the kinetics at the adiabatic state well. Further, if one considers the assembly of particles of the same species and no interactions among particles is taken into account, the pair approximation of the PPM represents the adiabatic condition exactly as long as the concentration of vacancies is negligibly small.

Because the PPM corresponds to the mean field approximation in t-space, it is required to improve the degree of approximation in t-space to that of pair in order to be consistent with the treatment of the adiabatic state. For this purpose, the application of the Bethe concept is appropriate for the reason stated below.

The original Bethe method[11] is devised to deal with the order-disorder transformation in alloys and is known to correspond to the pair approximation of the CVM on which the PPM is based. Here, the consistency relation[11] in the Bethe method which connects the thermodynamical properties of a small cluster to that of its outside gives a change from the local order to the long range order with distance in the form of the propagation of order.[12] This concept can be extended in the present treatment in t-space as follows. In Fig. 2, we show a relation of the tagged particle which has just jumped out, leaving a vacancy at the center at $t = 0$ (adiabatic state) which is derived by the pair approximation of the PPM. The distribution of particles around the vacancy, Q_j, is given by the pair approximation of the PPM. The jump back probability of the tagged particle in competition with the surrounding particles at the adiabatic state, τ at $t = 0$, is thus defined. The change of the jump back probability τ (defined in Eq. (7)) with t can then be calculated by the Bethe method in the form

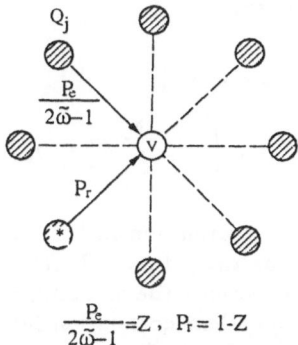

Figure 2. The jump back process of the tagged particle (represented by an asterisk) in a disordered binary alloys by means of the pair approximation of the PPM.

of the propagation of order as t changes from 0 to ∞. The value of τ at $t \to \infty$ corresponds to the value of τ in the isothermal state in the pair approximation.

The combination of the pair approximation of the PPM and the Bethe method thus treats the relaxation of current analytically from the adiabatic state to the isothermal state with the same rigor.

III. Formal Treatment of the Relaxation Process by the PPM

Here, we deal with the calculation of the relaxation of flow when a constant driving force for the tagged particle, $\dot{\alpha}_i$, is switched on at $t = 0$ in a binary system consisting of A and B (and of tracer particle of B or B^*) indicated by i and j which specify A, B and B^* as 1, 2 and 3, respectively in a lattice gas with pairwise, nearest-neighbor interactions, ϵ_{AA}, ϵ_{BB} and $\epsilon_{AB} = \epsilon_{BA}$ among mobile particles A and B by means of the pair approximation of the PPM. A detailed derivation of the tracer correlation factor in the range $t \to \infty$ has been carried out and is shown in the Appendix of Ref. 13.

The flow equation for a small number of tracer ions in the disordered distribution in the linear approximation at a time instant t is written down as

$$\Psi_i(t) = -\dot{\alpha}_i + \sum_{j=1}^{3} Q_j \psi_{ji}(t), \qquad i, j = 1, 2, 3. \tag{3}$$

Here, i is meant to be the tagged species. Therefore, $\Psi_i(t)$ indicates the normalized flow at t of the tagged species, Q_j specifies the equilibrium distribution of particles represented by the concentration of the jth species around the vacancy, $\psi_{ji}(t)$ indicates the gradient in the distribution of particles created by the driving force for the ith species $\dot{\alpha}_i$. $\Psi_i(t)$ is evaluated in a short time period between t and $t + \Delta t$ under the same reference state under equilibrium Q_j at any t.

Based on Eq. (3), we can rewrite

$$\Psi_i(t) = -f_i(t)\,\dot{\alpha}_i$$

$$= -\left(1 - \sum_{j=1}^{3} Q_j \psi_{ji}/\dot{\alpha}_i\right)\dot{\alpha}_i. \tag{4}$$

This defines the time dependent correlation factor. In this expression, the first term represents the jumping out of the tagged particle from the central site (at $t = 0$) (see Fig. 2) and the second term represents the jumping back of the tagged particle into the vacancy. Here, the time dependent correlation factor takes the form of $1 - A_i(t)$ as in Eq. (4) and $A_i(t)$ takes the form of an exponential function because $\psi_{ji}(t)$ in Eq. (3) is a function of t only. In other words, in the disordered state, the relaxation process is expressed by the PPM in terms of a single exponential mode even in binary systems. Our major interest is to examine $1 - A_i(t)$ with the aid of the Bethe method with an improved degree of approximation through the use of τ_i.

If mutual interactions among mobile ions are taken into account, there should be four independent state variables (even if the distribution of vacancies is considered to be random) and there should then be four relaxation modes (in case the long-range order exists) in binary alloys.[14,15] In such a case, the time dependent correlation factor should be composed of four exponential modes and, hence, the relaxation mode becomes non-Debye even in the mean field approximation. However, our concern here is to show that each relaxation mode thus defined is non-Debye. The result that, in the disordered state, the number of modes is reduced to one is due to the fact that the PPM is a mean field approximation in t-space.[16]

Equation (3) can be most easily handled in the time region $t \to \infty$ where $\psi_{ij}(t)$ saturates and does not change with time.[13] In this time region, the stationary state value of Ψ_i can be solved in terms of a time independent function Z_i as

$$\Psi = -\frac{(2\tilde{\omega} - 1)Z_i}{2 + (2\tilde{\omega} - 3)Z_i}\,\dot{\alpha}_i, \tag{5}$$

where Z_i represents

$$Z_i = \sum_j \frac{Q_j \hat{w}_j}{\hat{w}_i + \hat{w}_j}, \tag{6}$$

and $2\tilde{\omega}$ represents the coordination number. In Eq. (6), $\hat{w}_i$, etc. are defined as the jump frequency of the ith species including the bond breaking factor. The meaning of Z_i is clear from Fig. 2 which represents the second term of Eq. (3) and the jump back probability of the tagged particle into the vacancy left behind in competition with other particles around the vacancy under the equilibrium condition is being calculated. Here, the quantity $1 - Z_i$ represents the jump back (return) probability P_r of the tagged particle whereas Z_i represents the escape probability P_e (the jumping-in probability of a particle other than the tagged particle). The normalized jump back probability is then defined as

$$\tau_i = \frac{1 - Z_i}{(1 - Z_i) + (2\tilde{\omega} - 1)Z_i}, \tag{7}$$

It should be realized that, although τ_i in Eq. (7) can be interpreted as a normalized jump back probability of a particle of the ith species under equilibrium, it actually is the time dependent jump back probability $\tau_i^*(t)$ of the tagged particle evaluated at $t = t_0$ in the time range $t \rightarrow \infty$. In terms of $\tau_i = \tau_i^*(t_0)$, the tracer correlation factor f_i (at t_0) is defined as

$$f_i = \frac{1 - \tau_i}{1 + \tau_i}. \tag{8}$$

Since the time dependence of the reference state is dealt with, and since τ_i depends on the reference state, τ_i is indirectly time dependent as mentionated in Section II. Equation (3) has the same form at any t except for the value of $\psi_{ji}(t)$. Because of this, at any t, the same type of solution for $\tau_i^*(t)$ should exist. Then, the time dependent correlation factor for the ith species $f_i^*(t)$ is determined from the relation

$$f_i^*(t) = \frac{1 - \tau^*(t)}{1 + \tau_i^*(t)}, \tag{9}$$

as long as $\tau_i^*(t)$ can be obtained (we put an asterisk to $f_i(t)$ in order to conform to the previous treatment), rather than as $1 - A_i(t)$. With the aid of Eq. (7), $\tau_i^*(t)$ can be obtained as

$$\tau_i^*(t) = \tau_i \left[1 - \exp\left(-\frac{\hat{w}_i t}{\tau_i} \right) \right]. \tag{10}$$

Here, $\tau_i^*(t)$ tends to τ_i as $t \rightarrow \infty$ as expected.

In the PPM, the background equilibrium state is the same from the adiabatic state to the isothermal state and, hence, τ_i remains the same. Therefore, in the PPM, the relaxation of $\tau_i^*(t)$ is exponential. However, if the reference state changes from the adiabatic state to the isothermal state and τ_i increases with time, the relaxation of $\tau_i^*(t)$ becomes KWW, and so does $f^*(t)$. The essential part of the present work is to investigate the dependence of τ_i with t from the adiabatic state to the isothermal state by means of the Bethe method.

The treatment of the PPM combined with the Bethe method is based on the assumption of a "stochastic model" for the jump of constituent particles in the lattice gas model. We observe the system and when a particle exchanges with a vacancy, we set $t = 0$ and tag the particle, and the correlation of the tagged particle with the vacancy is closely followed from $t = 0$ to $t \rightarrow \infty$. The whole system is kept under equilibrium condition.

IV. Conversion of the Averaging Process

In the PPM, the flow of the tagged particle is calculated at each time instant (separated by Δt) in the reference state under equilibrium in competition with particles around the vacancy with the averaged distribution Q_j and with its (averaged) jump frequency $\hat{w}_j$ specified by the Cluster Variation Method (CVM). In this respect, in Eq. (3), the values of Q_j and $\hat{w}_j$ do not fluctuate with time and take the same averaged values (ensemble average) at any time. In other words, in the original PPM, a flow in a binary system is treated as a flow in a homogeneous system consisting of averaged particles with averaged jump frequencies. In the adiabatic or in the intermediate time range, however the averaging under equilibrium such as in the PPM is not allowed. In other words, the distribution of surrounding particles, Q_j, cannot remain the same, but should fluctuate from time to time between two values which represent the short range order distribution Q_j^α around an A particle and Q_j^β around a B particle, respectively, which is specified by the pair approximation of the CVM (instantaneous distribution conversion process).[17,18] At the same time, the jump frequency of a particle at the surrounding of the vacancy left behind is not that of the averaged particle $Q_j \hat{w}_j$, but should fluctuate between that of an A particle and of a B particle with time (time conversion process).[13,17,18] Since the system as a whole is kept under equilibrium, the ensemble average is the same as the time average and fluctuations are allowed within this limit. In other words, in the pair approximation,

1. Instantaneous Distribution Conversion Process

$$Q_j \rightarrow x_A Q_j^\alpha + x_B Q_j^\beta . \tag{11}$$

2. Time Conversion Process

$$Q_j \hat{w}_j (1 - \tau_j) \rightarrow Q_A \hat{w}_A (1 - \tau_A) + Q_B \hat{w}_B (1 - \tau_B) , \text{etc.} \tag{12}$$

for each subcomponent of Q_i.

The inclusion of the factor $(1 - \tau_j)$, etc. in Eq. (12) indicates that particles on the average are moving with a jump back probability τ_j as specified in the Bethe method.[11]

With the instantaneous distribution conversion process, the number of modes in disordered binary alloys increases from one to four according to the number of independent state variables.[16] Based on Eq. (11), we eventually deal with two types of flow each for A and B particles with specific surroundings Q_j^α and Q_j^β in converting to the Bethe method. This conversion can be handled by the concept of microscopic balance.[16] In this paper, however, we do not proceed further on this concept.

V. Assembly of Particles of the Same Species with a Negligible Amount of Vacancies

Here, we treat the problem of the motion of a single particle in the assembly of particles of the same species without any mutual interaction with a negligible amount of vacant sites in a lattice gas. Without any interactions among mobile particles, and with no species of particles to be distinguished, the pair approximation of the PPM represents the exact adiabatic condition.

From Eq. (6) of the pair approximation of the PPM, without the need of specifying the different jump frequency of the tagged species

$$Z = \frac{w}{(w + w)} , \tag{13}$$

where w represents the jump frequency of the particle. From this, the adiabatic value of τ is obtained using Eq. (7) as $\tau = w/\{w + (2\tilde{w} - 1)w\}$. For $t \geq 0$, utilizing the time conversion relation Eq. (12)

$$Z = \frac{w(1 - \tau)}{w + w(1 - \tau)} = \frac{1 - \tau}{2 - \tau} . \tag{14}$$

The consistency relation in τ is then obtained with the use of Eq. (7)

$$\tau = \frac{1}{2\tilde{w} - (2\tilde{w} - 1)\tau} , \tag{15}$$

in the form of an iteration formula. Eq. (15) can be written in a differential form as

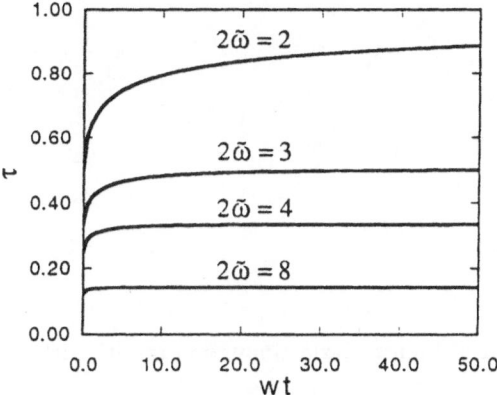

Figure 3. τ–t relations for structures with different coordination numbers, $2\tilde{\omega}$.

$$\frac{d\tau}{dN} = \frac{2\left(2\tilde{\omega} - (2\tilde{\omega} - 1)\tau\right)\left(1 - (2\tilde{\omega} - 1)\tau\right)(1 - \tau)}{(2\tilde{\omega} - 1) + (2\tilde{\omega} - (2\tilde{\omega} - 1)\tau)^2}, \qquad (16)$$

where N represents the number of iterations. As the number of iterations, N, increases, τ approaches the isothermal value $\tau \to 1/(2\tilde{\omega} - 1)$ $(N \to \infty)$ from the adiabatic value $\tau = 1/(2\tilde{\omega})$ $(N = 1)$ in Eq. (15). Therefore, if N is related to time t, Eq. (16) represents the propagation of order, $d\tau/dt$, in t space through $d\tau/dN$.

The number of iterations N represents the distance of propagation of information, and, hence, represents the number of jumps n which effectively covers N unit distance by the vacancy which makes a random walk. If so, N and the number of jumps n are related as $N = \sqrt{n}$ and $t = n/w$. Because $N = 1$ corresponds to $t = 0$, $N - 1 = \sqrt{wt}$.[19] This is the relation between N and t. The propagation of order in Eq. (15) means that information at the center of the group in Fig. 2 is accumulated over a distance with time by hopping of particles[20] represented by the motion of the vacancy. Here, specifically, the increase of τ indicated in Fig. 3 represents the return of the escaped vacancy with time. The presence of interactions among conduction ions, especially those which tend to create an ordering of conduction ions enhances the KWW behavior.

In Fig. 3, τ–t relations are plotted for different values of $2\tilde{\omega}$ with $2\tilde{\omega} = 2$ for the one dimensional lattice, 3 for the two dimensional honeycomb lattice, 4 for the two dimensional square net and 8 for the body-centered cubic lattice. The increase of τ with t means that $\tau^*(t)$ and, hence, $f^*(t)$, also have the KWW behavior (Eq. (10)). In other words, the non-Debye behavior is inherent even in the single mode case although the effect is almost negligible except for the case, $2\tilde{\omega} = 2$.

Based on Eq. (16), the increase of τ as $t \to \infty$ for cases with $2\tilde{\omega} > 2$ is exponential, but for the one dimensional case ($2\tilde{\omega} = 2$), where the percolation limit is reached as $t \to \infty$, the decay becomes proportional to $t^{-1/2}$. The change of the rate of approach from exponential to $t^{-1/2}$ is observed in Fig. 4 which shows $d(\log f^*(t))/d(\log t)$ at the value of $f^*(t)$ around 0.1. The extensive linearity in the relation $\log f(t) - \log t$, such as that observed for the case $2\tilde{\omega} = 2$, is also observed in systems where $f(t)$ decreases with time to a small enough value such as in mixed alkali systems.[21] This is due to the existence of a scaling law for the decrease of $f(t)$ with time near the percolation limit. The

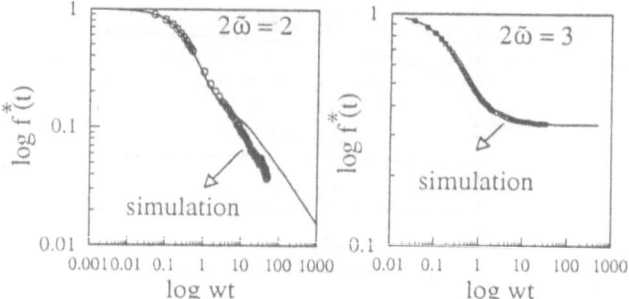

Figure 4. Comparison of Monte Carlo simulation results with the analytical theory for $2\widetilde{\omega} = 2$ and for $2\widetilde{\omega} = 3$.

value of p is obtained from the slope of this curve as $d(\log f(t))/d(\log t) = -p$ and are 0.12, 0.22, 0.32 and 0.5, respectively for $2\widetilde{\omega} = 6, 4, 3$ and 2. The value 0.5 for the one dimensional lattice is proven to be the exact value for this case. The relation $\log f^*(t) - \log t$ can be converted to $\log f^*(\omega) - \log \omega$ by simply converting t into $2\pi/\omega$. The Jonscher's power law exponent, p, can also be obtained from $d(\log f^*(\omega))/d(\log \omega) = p$.

As mentioned earlier, Monte-Carlo simulation of the relaxation process can be made in a straightforward fashion. The simulation calculates $f^*(t) = -\int_0^t \langle v(0)v(t') \rangle / \langle v^2(0) \rangle dt'$ directly, assuming a single vacancy in the system which makes the random walk starting from an instant when the vacancy has exchanged places with the tagged particle. The agreement of the analytical calculation of $f^*(t)$ obtained from $\tau^*(t)$ (Eq. (11)) with the Monte-Carlo simulation of $f^*(t)$ is excellent. In the case $2\widetilde{\omega} = 3$, where the tracer correlation factor obtained by the pair approximation coincides with the exact value of $1/3$, an almost exact agreement of the results is observed (Fig. 4). This shows that the pair approximation is quite acceptable for the calculation of $f^*(t)$. In the one dimensional case, a change in slope in the $\log f^*(t) - \log t$ curve is observed at the value of $f^*(t) \approx 0.1$ (Fig. 4), in agreement with the theory. With respect to the comparison with Monte-Carlo simulations, a small, but a systematic deviation of the simulation curve from that of the analytical value is observed in the range where t is large. Monte-Carlo simulation indicates a slightly smaller value of $f^*(t)$ for the same value of t, although all the qualitative behavior is the same. At this moment, the origin of such a systematic deviation is unknown to us.

It is to be mentioned here that universal response is a property of the total assembly of charged particles and not of the motion of a single particle in the assembly. In the case of the assembly of particles of the same species with a negligible amount of vacancies as treated here, the correlation effect for the assembly vanishes in the completely homogeneous case due to the indistinguishability among particles. In order to regain the correlation effect for the assembly of particles, it is necessary to deal with more general cases with mutual interactions among particles and fluctuation in the distribution is necessary.

VI. Universal Dynamic Response

So far, we have derived analytically that the relaxation motion of a single particle in a lattice gas is intrinsically non-Debye and follows a typical KWW relaxation behavior even without any mutual interactions among mobile particles. Based on these analytical expressions, we would like to extract some rules with respect to the universal dynamic response.

In the previous section, for the assembly of particles of the same species with a negligible amount of vacancies, we saw that, although the KWW behavior is inherent, the effect is practically negligible except for the one dimensional case. The major cause for this is that $f^*(t)$ does not decrease enough for those cases. Solid electrolytes which follow the universal dynamic response commonly have very small values of tracer correlation factor. This indicates that more general models with interactions among mobile particles have to be introduced where $f^*(t)$ can decrease sufficiently. In our earlier publication,[20] we discussed such general models and concluded that two kinds of generalized lattice gas models are sufficient to cover general cases; one with a large number of vacant sites with repulsive interactions among mobile particles, and the other, a mixture of two kinds of mobile ions with interactions which tends to create ordered distribution between the two kinds of particles. Using the latter model, we are able to discuss the origin of the mixed alkali effect.[21] In this treatment, due to the necessity of introducing fluctuations (Eq. (11)), the number of relaxation modes increases. In the pair approximation, the former requires two independent relaxation modes while the latter requires four independent relaxation modes.

The Jonscher's power exponent, p, is obtained as

$$-p = \frac{d(\log f^*(t))}{d(\log wt)}$$

$$= \frac{f^{*'}(t)wt}{f^*(t)}. \tag{17}$$

Here, the prime indicates the derivative with respect to t, and w is the jump frequency and serves as a scaling factor. As discussed in the previous section, there are two ways for τ in approaching its limiting value. When the correlation factor is not too small (the limiting value of τ is not too close to 1), τ approaches its limiting value exponentially, whereas when τ is close to 1, τ approaches $\tau(\infty)$ extremely slowly. In the range where τ approaches $\tau(\infty)$ exponentially, $f^*(t)$ has an exponential tail. Therefore, in the time range where t is relatively large, $f^*(t)$ in Eq. 17) can be treated as if it is an exponential function of the form given in Eq. (10) with a constant τ. Then it is easy to obtain some general conclusions from Eq. (17). In such a case Eq. (17) reduces to $\approx \frac{-2\tau w}{(1+\tau)(1-\tau)}$. This relation indicates that, as τ becomes larger (or the correlation factor becomes smaller), p becomes larger, and w serves as the scaling factor (Fig. 5).

Even in the case of multiple relaxation modes, the situation is not too different. Here, the $f^*(t)$–t relation is a superposition of these modes and the mode with the smallest partial f plays the decisive role in determining p. This is because in the mode which has the smallest partial f, the jump frequency and, hence, the time scale is the smallest, and, hence, this mode determines the value of p. This serves as another scaling factor. In general, when the difference in the time scale between different modes is small, p is usually determined by the behavior in the mode with the smallest partial correlation factor. If the value of f is not close to zero in this mode, p is found

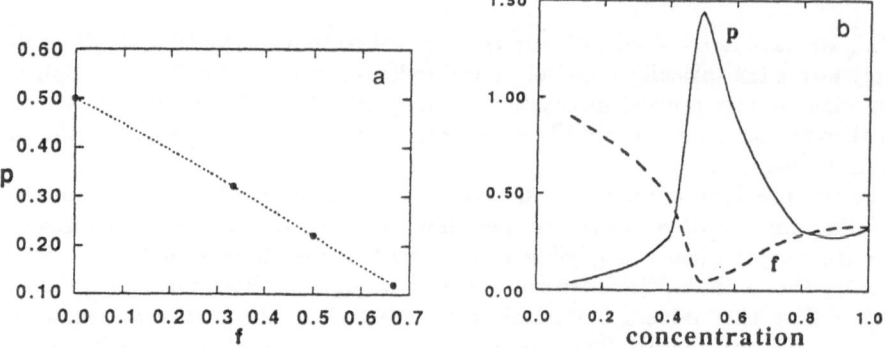

Figure 5. The relation between p and f^*. (a) Single mode; p and f^* for $2\tilde{\omega} = 2, 3, 4$, and 6. (b) Two mode case; p and f for a large concentration of vacant sites.

to be inversely proportional to f^* (Fig. 5), while in the range near the percolation limit, the scaling law appears and the value of p becomes relatively small. However, no specific relation seems to be found to exist in the scaling factor to make the value of p within one. Indeed, cases with p values larger than 1 are found.[24]

VII. Conclusion

Relaxation of a current when a constant field is switched on at $t = 0$, in particular that of a single particle in the assembly, has been calculated analytically in a lattice gas model from the adiabatic state to the isothermal state with the same degree of rigor. The pair approximation of the PPM is utilized for the description of the adiabatic state, and its change with time is followed by the Bethe method in the form of the propagation of order. Our results show clearly that the relaxation motion of a single particle in a lattice gas is intrinsically non-Debye and follows the KWW relaxation behavior even without any interaction among mobile particles. From this treatment, we can conclude the following:

1) The calculation can be extended easily to more general cases with interactions among mobile particles.

2) From such analytical expressions, the origin of the KWW behavior, and, hence, the universal dynamic response of assembly of particles can be extracted.

3) The origin of the KWW behavior is explained as being due to the return of the escaped vacancy with time while the jumping back process of the tagged particle into the vacancy it has left behind is being considered. More generally, the KWW behavior is due to a memory effect arising from the fact that it takes time for information to propagate over a distance by hopping.

4) The same relaxation process can be followed by a Monte Carlo simulation and the results support the above statements.

Acknowledgements

This work is a result of cooperation with K. Funke of the University of Münster, Germany. The authors also owe valuable discussions to R. Kikuchi of UCLA and to G. L. Liedl of Purdue University. The work was supported by the U.S. Department of Energy under grant number DE-EG02-84ER45133.

References

1. A. K. Jonscher, *Nature* **267** 673 (1977).
2. K. Funke, in *Prog. Solid State Chem.* **22**, 1 (1993).
3. G. E. Murch and J. C. Dyre, *CRC Crit. Rev.* **15**, 348 (1989).
4. C. Zener, *Elasticity and Anelasticity of Metals*, University of Chicago Press (1948).
5. J. C. Dyre, *J. Appl. Phys.* **64**, 2456 (1988).
6. J. P. Bouchard and A. Georges, *Phys. Rep.* **195**, 127 (1990).
7. K. L. Ngai, A. K. Rajagopal, and S. Teitler, *J. Chem. Phys.* **88**, 5086 (1988).
8. S. R. Elliott and A. P. Owens, *Phys. Rev. B* **44**, 47 (1991).
9. T. Ishikawa, K. Wada, H. Sato, and R. Kikuchi, *Phys. Rev. A* **33**, 4164 (1986).
10. K. Wada, T. Ishikawa, H. Sato, and R. Kikuchi, *Phys. Rev. A* **33**, 4170 (1986).
11. H. A. Bethe, *Proc. Roy. Soc.* **150**, 552 (1935).
12. F. Zernike, *Physica* **1**, 565 (1940).
13. H. Sato and R. Kikuchi, *Phys. Rev. B* **28**, 648 (1983).
14. K. Gschwend, H. Sato, and R. Kikuchi, *J. Chem. Phys.* **69**, 5006 (1978).
15. Y. Ozeki and T. Ishikawa, *J. Phys. Soc. Jpn.* **55**, 3931 (1986).
16. H. Sato and H. Zhang, *Atomistic Approach to Diffusion in Ordered Alloys*, edited by B. Fultz, R. W. Cahn, and D. Gupta, (The Minerals, Metals and Materials Society, 1993), p. 21.
17. H. Sato, *Nontraditional Methods in Diffusion*, edited by G. E. Murch, H. K. Birnbaum, and J. R. Cost, (The Metallurgical Society of AIME, 1984), p. 203.
18. H. Sato, S. A. Akbar, and G. E. Murch, *Diffusion in Solids: Recent Developments*, edited by M. A. Dayananda and G. E. Murch, (Conference Proceedings, The Metallurgical Society of AIME, 1985), p. 67.
19. H. Sato and A. Datta, *Solid State Ionics* **72**, 19 (1994).
20. H. Sato, T. Ishikawa, and K. Funke, *Solid State Ionics* **53-56**, 907 (1992).
21. H. Sato, *Ceramic Transactions Vol. 20, Glasses for Electronic Applications*, edited by K. M. Nair, (The American Ceramic Society, 1991), p. 19.
22. R. Kikuchi and H. Sato, *J. Chem. Phys.* **55**, 677 (1971).
23. R. Kikuchi and H. Sato, *J. Chem. Phys.* **57**, 4962 (1972).
24. K. Funke, T. Mane, D. Wilmer, C. Cramer, and T. Saatkamp, in *Ionic and Mixed Conducting Ceramics*, edited by T. A. Ramanerayanon, W. L. Worrell, and H. L. Tuller, (The Electrochemical Society, Inc., 1994), p. 564.

Cluster Variational Approach to Non-Equilibrium Lattice Models

Makoto Katori

Department of Physics
Faculty of Science and Engineering
Chuo University
Kasuga, Bunkyo-ku, Tokyo 112
JAPAN

Abstract

We report our recent attempt to study *non-equilibrium stationary states* following the spirit of the cluster variation method (CVM) and the path probability method of Kikuchi. We consider two kinds of non-equilibrium lattice models, the sandpile model of Bak, Tang and Wiesenfeld (BTW) and the basic contact process (CP).

Computer simulations show that starting from an arbitrary initial state the system evolves into a unique stationary state called the *self-organized critical* (SOC) state, in which characteristic time or length scales vanish. We introduce a single-site system and a two-site system as subsystems in the two-dimensional BTW model and derive approximate expressions for the height distribution functions in the SOC state. By imposing a *reducibility condition* of the distribution functions, we obtain an approximation, which shall be called a pair approximation for the SOC state. The results are in good agreement with the values estimated by the Monte Carlo simulation.

The basic CP is a simple mathematical model for the spread of infection of a contagious disease. Though it has only nearest-neighbor interactions of individuals, a phase transition occurs and critical phenomena are observed at a critical value of infection rate $\lambda = \lambda_c$ even in the one-dimensional system. We study the non-equilibrium stationary state which appears in the supercritical phase $\lambda > \lambda_c$ using cluster approximations. This procedure can be regarded as a trial to approximate non-equilibrium stationary states by the Gibbs states with potentials which are appropriately chosen following the CVM.

Theory and Applications of the Cluster Variation and Path Probability Methods
Edited by J.L. Morán-López and J.M. Sanchez, Plenum Press, New York, 1996

I. Introduction

The Cluster Variation Method (CVM)[1] is a general scheme for making approximate expressions for free energies and has been widely used to study thermal equilibrium states in a variety of systems.[2] This method was extended to treat time-dependent phenomena and the Path Probability Method (PPM)[3] is now a useful theory to study relaxation phenomena to thermal equilibrium states and other stochastic processes.[4]

In the present paper, we will report our recent attempt to study *non-equilibrium stationary states* following the spirit of the CVM and the PPM. We will treat two kinds of non-equilibrium lattice models, the "sandpile" model by Bak, Tang and Wiesenfeld (BTW)[5,6] and the basic contact process.[7,8]

BTW introduced a simple automaton model for dissipative dynamical systems with spatially extended degrees of freedom. It displays the remarkable property that starting from an arbitrary initial state, it evolves into a unique stationary state characterized by power-law correlations both in space and time. Power-law correlations imply lack of characteristic time or length scales and thus this stationary state is called a *self-organized critical* (SOC) state.[5,6] Extensive simulations have been performed by Kadanoff *et al.*[9] and Manna.[10] Though distribution functions of many quantities were given exactly for the BTW model on the Bethe lattice by Dhar and Majumdar,[11] exact results for the SOC states on the usual d-dimensional hypercubic lattices are still limited.[12,13]

The basic contact process (CP) is a simple mathematical model for the spread of infection of a contagious disease.[7] Though it has only nearest-neighbor interactions, phase transitions occur and critical phenomena are observed even in the one-dimensional system and non-trivial stationary states appear in the supercritical phase.[8,14,15] This implies that detailed balance is violated in the supercritical stationary states and they can not be described by the ordinary Gibbs states with finite-ranged potentials.

Can we apply the CVM and the PPM to investigate such non-equilibrium stationary states which are far from thermal equilibrium states? This question is related with the fundamental problem: whether non-equilibrium stationary states can be well-approximated by the Gibbs states with appropriately chosen potentials or not.

II. BTW Model

The BTW model[5,6] is a stochastic cellular automata and its definition on a square lattice is given as follows. We consider an $L \times L$ finite region Λ_L in a square lattice $\mathbf{Z}^2$. To each site $(x, y) \in \Lambda_L$ an integer variable $z(x, y) \in \{0, 1, 2, \ldots\}$ is assigned. All of the variables on the site outside of Λ_L are fixed to be 0. The variables in Λ_L evolve by the following rules. (i) We select a site at random with equal probability for each site and increase the variable at the site by 1, $z(x, y) \rightarrow z(z, y) + 1$. Other values $z(u, v)$, $(u, v) \neq (x, y)$, are unchanged. (ii) Let the maximum stable value $z_m = 3$. If $z(x, y) > z_m$ at a site (x, y), then we change the variables at (x, y) and its four nearest-neighbors as

$$z(x, y) \rightarrow z(x, y) - 4,$$
$$z(x \pm 1, y \pm 1) \rightarrow z(x \pm 1, y \pm 1) + 1. \tag{1}$$

This process continues until every site in Λ_L is stable; $z(x, y) \leq z_m$ for all $(x, y) \in \Lambda_L$.

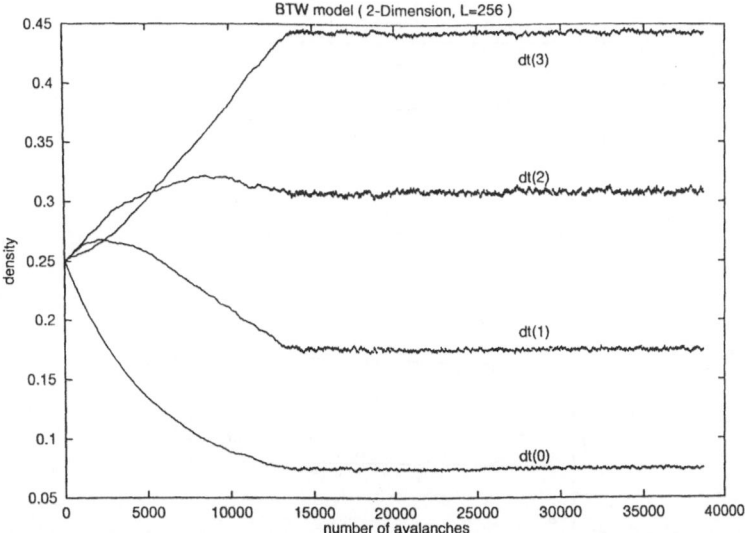

Figure 1. A typical result of the Monte Carlo simulation of the two-dimensional BTW model with the lattice size $L = 256$. Starting from a random configuration with equal probability $1/4$ for each $n = 0, 1, 2, 3$, we measure the time-dependence of the density of sites having a value n, $d_t(n)$. In this case we have the SOC state after about 15,000 avalanches. We define $p_L(n)$ as a time average of the density $d_t(n)$ in the SOC state.

The BTW model can be regarded as a simple model for sandpiles, where the variable $z(x, y)$ represents a local "height" in sandpiles. The process (i) is a random addition of a grain of sand and the process (ii) is for toppling. In many cases toppling at some site causes toppling at the nearest-neighbor sites, some of which will cause other toppling at their surrounding sites. Such a series of toppling processes is called an *avalanche*. After an avalanche process ceases, we repeat the process (i). It should be remarked that the condition for toppling, $z(x, y) > z_m$, depends only on a local height and it follows that the configuration of variables $\{z(x, y)\}$ obtained after an avalanche does not depend on the order of toppling processes which occur in the avalanche. Dhar[13] proposed to refer to this model as the Abelian model to distinguish it from other sandpile models in which the toppling condition depends on gradients of height and thus the toppling procedures are not commutable.

Starting from an arbitrary initial state, the system evolves into a SOC state. There various seizes and lifetimes of avalanches are observed and the probability distributions of them follow power laws:[5,6] $P(\text{avalanche size} = s) \sim s^{-\tau}$ and $P(\text{lifetime of avalanche} = t) \sim t^{-b'}$ with $\tau \simeq 1.22$ and $b' \simeq 0.85$.[10] At the SOC state, the averaged value of $z(x, y)$ per lattice site attains a steady value $\langle z \rangle = z_c$, which can be recognized as a critical value when we draw an analogy between the self-organized criticality and critical phenomena of second-order phase transitions.[16]

Let $p_L(n), n = 0, 1, 2, 3$ be the probability for $z(x, y) = n$ in the SOC state. Fig. 1 shows a typical result of computer simulation for the system with $L = 256$, which gives $p_{L=256}(0) = 0.074$, $p_{L=256}(1) = 0.175$, $p_{L=256}(2) = 0.308$, $p_{L=256}(3) = 0.443$ and $\langle z \rangle_{L=256} \equiv \sum_{n=0}^{3} n p_{L=256}(n) = 2.12$. We performed the simulation[17] for $L = 32, 64, 128, 256$ and 512 and found that these values slightly depend on the size L

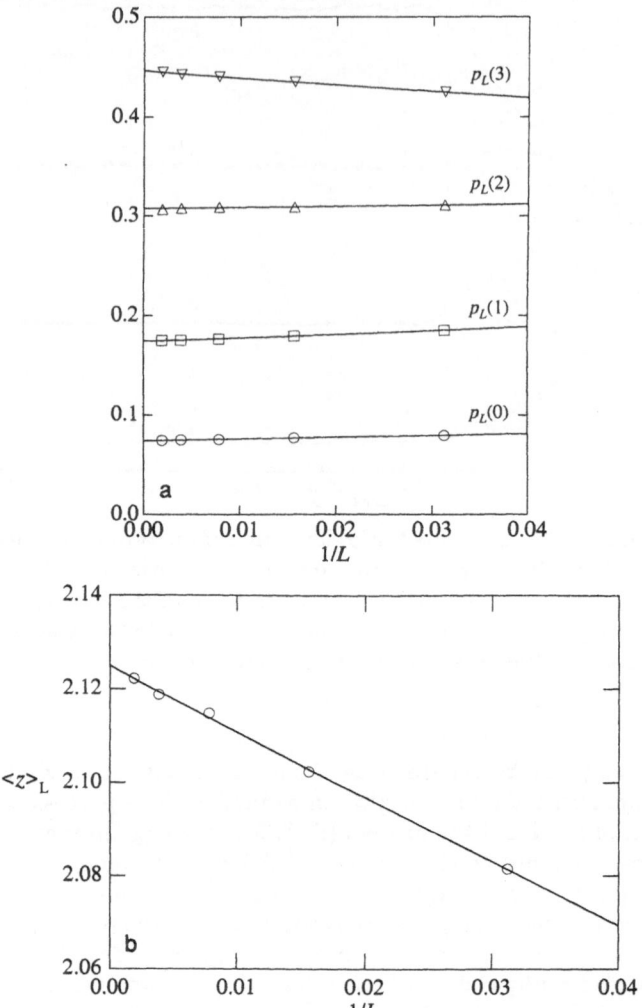

Figure 2. Lattice size L dependence of (a): $p_L(n)$, $n = 0, 1, 2, 3$ and (b): $\langle z \rangle_L$. The lines are obtained by the least-squares fitting of data to the equations $p_L(n) = p(n) + c_n L^{-a_n}$, $n = 0, 1, 2, 3$ and $\langle z \rangle_L = z_c + cL^{-a}$.

as shown in Fig. 2a and 2b. Extrapolation of the $L \to \infty$ limit gives $p(0) = 0.074$, $p(1) = 0.174$, $p(2) = 0.306$, $p(3) = 0.446$ and $z_c = 2.124$, as already reported by Manna.[10]

Majumdar and Dhar[13] obtained the exact value $p(0) = 2(1 - 2/\pi)/\pi^2 = 0.07363\ldots$, but there are no exact results for other values. Simple mean-filed-type theory gives $p^{mf}(n) = 1/4$ for all $0 \le n \le 3$ and $z_c^{mf} = 1.5$,[18,19] which are far from the values estimated by simulations. In the next section, I report the results obtained by a pair approximation.

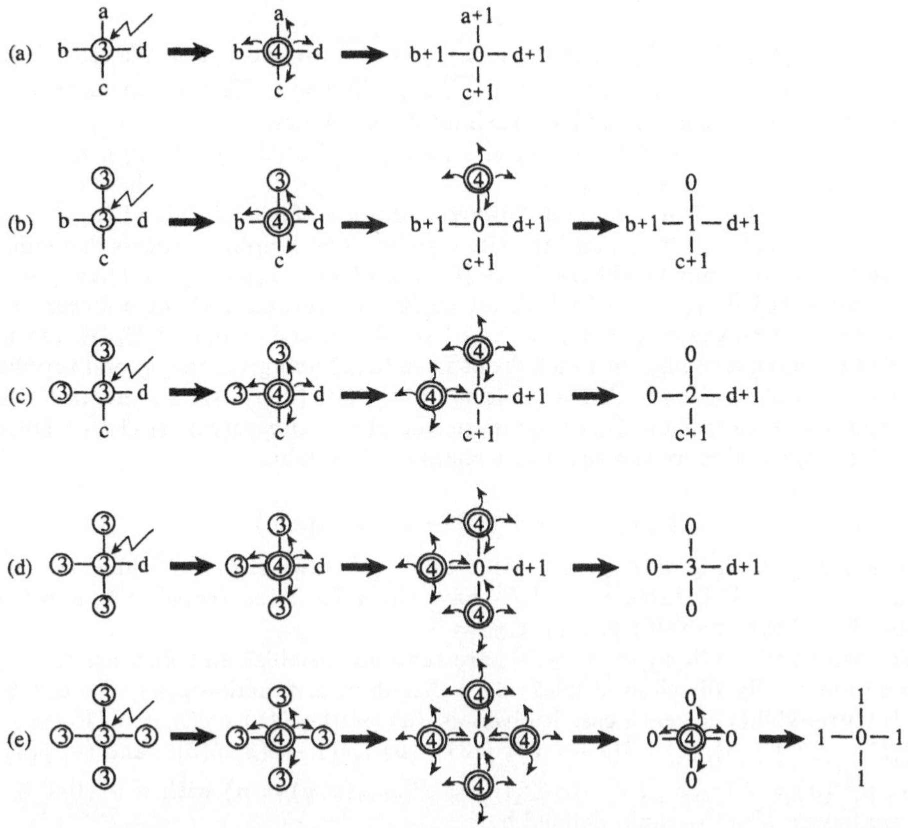

Figure 3. Avalanche processes in the single-site system $\Omega_1 \cup \mathcal{N}_1$. The value 3 is the maximum stable value z_m and the sites with the value 4 are unstable and topple. The values a, b, c and d are in $\{0, 1, 2\}$.

III. Pair Approximation for the SOC State

The CVM is based on the variational principle of free energy. We prepare a set of sites called basic clusters and make a trial free energy function using density matrices defined on the basic clusters, which is a function of a set of parameters. It is assumed that the best approximation is obtained when we choose the parameters so that the trial free energy is reduced to its minimum.[1,2]

We do not know how to define a function of states which corresponds to the free energy for non-equilibrium lattice models such as the BTW model and the CP. We know, however, that in the CVM the stationary conditions of a trial free energy with respect to variational parameters can be regarded as the self-consistency equations among different basic clusters, or the *reducibility* conditions of the system.[20-23]

In this section, we consider a single-site system Ω_1 and a two-site system Ω_2 and assume the reducibility condition between them. This procedure will give an approximation which shall be called the *pair approximation* for the BTW model.

III.1 Single-Site System

Let $\Omega_1 = \{(x, y)\} \subset \mathbf{Z}^2$ and $p^{(1)}(n) = P(z(x, y) = n)$ for $n = 0, 1, 2, 3$. We assume that the system is in the SOC state and that $\sum_{n=0}^{3} p^{(1)}(n) = 1$. The height distribution function $p^{(1)}(n)$ can be calculated approximately as follows.

Suppose that $z_t(x, y) = n$ and the site (x, y) is selected for adding a new grain of sand at time t. If $n \in \{0, 1, 2\}$, then $z_{t+1}(x, y) = z_t(x, y) + 1$ and there occurs no changes at other sites. If $n = z_m = 3$, however, addition of a sandgrain at (x, y) makes the site unstable, $z(x, y) > z_m$, and the site topples. The toppling process depends on the variables on the four neighbors $\mathcal{N}_1 \equiv \{(x + 1, y), (x - 1, y), (x, y + 1), (x, y - 1)\}$ at that time t. (a) If $z_t(u, v) \in \{0, 1, 2\}$ for all $(u, v) \in \mathcal{N}_1$, each of them increases by 1 and the avalanche ceases (see Fig. 3a). (b) If $z_t(u, v) = 3$ at one of $\mathcal{N}_1$, it becomes unstable by adding a sandgrain which drops from (x, y) and gives the second toppling. This second toppling gives back one sandgrain to the site (x, y). We assume that other sandgrains which drop from (u, v) dissipate out of the subsystem $\Omega_1 \cup \mathcal{N}_1$. During this avalanche, we observe the successive changes of variable

$$z_t(x, y) = 3 \rightarrow 4 \rightarrow 0 \rightarrow 1 = z_{t+1}(x, y), \tag{2}$$

as shown in Fig. 3b. Fig. 3 shows also other cases, (c) if two sites of $\mathcal{N}_1$ have the value 3, $z_{t+1}(x, y) = 2$, (d) if three sites of $\mathcal{N}_1$ have the value 3, $z_{t+1}(x, y) = 3$, and (e) if all sites of $\mathcal{N}_1$ have the value 3, $z_{t+1}(x, y) = 0$.

We assume that $z(u, v), (u, v) \in \mathcal{N}_1$, are random variables and they are independent and identically distributed following a distribution function $\tilde{p}(n), n = 0, 1, 2, 3$. Then the probability for each case is given as, (a) $(\tilde{p}(0) + \tilde{p}(1) + \tilde{p}(2))^4 = (1 - \tilde{p}(3))^4$, (b) $\binom{4}{1}(1 - \tilde{p}(3))^3 \tilde{p}(3)$, (c) $\binom{4}{2}(1 - \tilde{p}(3))^2 \tilde{p}(3)^2$, (d) $\binom{4}{3}(1 - \tilde{p}(3))\tilde{p}(3)^3$, and (e) $\tilde{p}(3)^4$.

Let $p_t^{(1)}(n) = P(z_t(x, y) = n), p_{t+1}^{(1)}(n) = P(z_{t+1}(x, y) = n)$ with $n \in \{0, 1, 2, 3\}$. Then we have a Markov chain defined by

$$\left(p_{t+1}^{(1)}(0) \quad p_{t+1}^{(1)}(1) \quad p_{t+1}^{(1)}(2) \quad p_{t+1}^{(1)}(3) \right) = \left(p_t^{(1)}(0) \quad p_t^{(1)}(1) \quad p_t^{(1)}(2) \quad p_t^{(1)}(3) \right)$$

$$\times \begin{pmatrix} 0 & 1 & 0 & 0 \\ 0 & 0 & 1 & 0 \\ 0 & 0 & 0 & 1 \\ (1 - \tilde{p})^4 + \tilde{p}^4 & 4(1 - \tilde{p})^3 \tilde{p} & 6(1 - \tilde{p})^2 \tilde{p}^2 & 4(1 - \tilde{p})\tilde{p}^3 \end{pmatrix}, \tag{3}$$

where $\tilde{p} = \tilde{p}(3)$.

The height distribution function $p^{(1)}(n)$ should be the stationary solution of this Markov chain (3). It is given by

$$p^{(1)}(0) = \frac{(1 - \tilde{p})^4 + \tilde{p}^4}{4(1 - \tilde{p} + \tilde{p}^4)},$$

$$p^{(1)}(1) = \frac{(1 - \tilde{p})^4 + 4(1 - \tilde{p})^3 \tilde{p} + \tilde{p}^4}{4(1 - \tilde{p} + \tilde{p}^4)},$$

$$p^{(1)}(2) = \frac{(1 - \tilde{p})^4 + 4(1 - \tilde{p})^3 \tilde{p} + 6(1 - \tilde{p})^2 \tilde{p}^2 + \tilde{p}^4}{4(1 - \tilde{p} + \tilde{p}^4)},$$

$$p^{(1)}(3) = \frac{1}{4(1 - \tilde{p} + \tilde{p}^4)}. \tag{4}$$

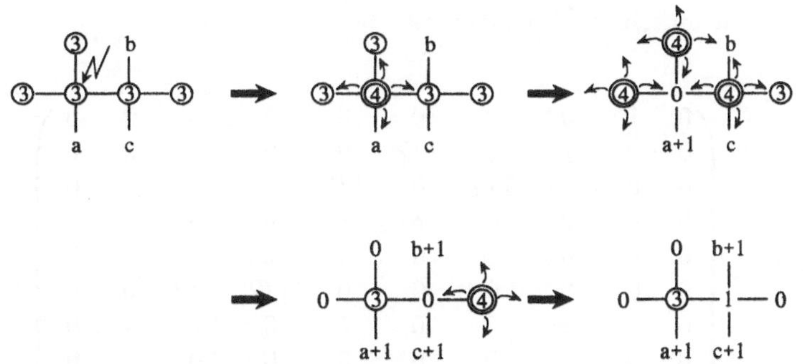

Figure 4. An example of an avalanche in the two-site system. Assume that $a, b, c \in \{0, 1, 2\}$.

This gives

$$\langle z \rangle^{(1)} \equiv \sum_{n=0}^{3} n p^{(1)}(n)$$

$$= \frac{3(1 - \tilde{p}^2 + \tilde{p}^4)}{2(1 - \tilde{p} + \tilde{p}^4)}. \tag{5}$$

It should be remarked here that we will obtain an approximation, which we call the *single-site approximation*, by imposing the simplest self-consistency condition

$$\tilde{p}(3) = p^{(1)}(3). \tag{6}$$

More details are given in the appendix.

III.2 Two-Site System

Next we consider a two-site system $\Omega_2 = \{(x, y), (x + 1, y)\}$ and its neighbor $\mathcal{N}_2 = \{(x - 1, y), (x, y + 1), (x + 1, y + 1), (x + 2, y), (x + 1, y - 1), (x, y - 1)\}$. We assume that one of the two sites in Ω_2 is selected for adding a new sandgrain at time t. Again we suppose that the variables in $\mathcal{N}_2$ are independently and identically distributed random variables with distribution $\tilde{p}(n)$ and that sandgrains which fall outside of $\Omega_2 \cup \mathcal{N}_2$ dissipate and will not return to the region $\Omega_2 \cup \mathcal{N}_2$. Fig. 4 shows the toppling process in the case in which $z_t(x, y) = z_t(x + 1, y) = 3$, the site (x, y) is selected out of two sites Ω_2 for adding a sandgrain, and two of the neighbor sites of (x, y) and one of the neighbor sites of $(x + 1, y)$ have the value 3 and other sites in $\mathcal{N}_2$ have the values less than 3. In this case, $(z_t(x, y), z_t(x + 1, y)) = (3, 3) \rightarrow (z_{t+1}(x, y), z_{t+1}(x + 1, y)) = (3, 1)$ and the transition probability is $\frac{1}{2} \binom{3}{2}(1 - \tilde{p})\tilde{p}^2 \binom{3}{1}(1 - \tilde{p})^2 \tilde{p}$.

We define the joint distribution function $p_t(n, m) = P(z_t(x, y) = n, z_t(x + 1, y) = m)$ and obtain a 16×16 transition matrix. By the symmetry $(x, y) \leftrightarrow (x + 1, y)$, this matrix is reduced to a 10×10 matrix. Let

$$\mathbf{p}_t = (p_t(0, 0) \quad 2p_t(0, 1) \quad 2p_t(0, 2) \quad 2p_t(0, 3) \quad p_t(1, 1)$$
$$2p_t(1, 2) \quad 2p_t(1, 3) \quad p_t(2, 2) \quad 2p_t(2, 3) \quad p_t(3, 3)). \tag{7}$$

Then the transition matrix $T^{(2)}$ is given as

$$T^{(2)} = \begin{pmatrix}
0 & 1 & 0 & 0 & 0 & 0 & 0 & 0 & 0 & 0 \\
0 & 0 & 1/2 & 0 & 1/2 & 0 & 0 & 0 & 0 & 0 \\
0 & 0 & 0 & 1/2 & 0 & 1/2 & 0 & 0 & 0 & 0 \\
0 & a & 0 & 0 & b & c & d & 0 & 0 & 0 \\
0 & 0 & 0 & 0 & 0 & 1 & 0 & 0 & 0 & 0 \\
0 & 0 & 0 & 0 & 0 & 0 & 1/2 & 1/2 & 0 & 0 \\
0 & 0 & a & 0 & 0 & b & 0 & c & d & 0 \\
0 & 0 & 0 & 0 & 0 & 0 & 0 & 0 & 1 & 0 \\
0 & 0 & 0 & a & 0 & 0 & b & 0 & c & d \\
0 & e_1 & e_2 & e_3 & e_4 & e_5 & e_6 & e_7 & e_8 & e_9
\end{pmatrix}, \tag{8}$$

with

$$a = \tfrac{1}{2}(1-\bar{p})^3, \quad b = \tfrac{3}{2}(1-\bar{p})^2\bar{p}, \quad c = \tfrac{3}{2}(1-\bar{p})\bar{p}^2, \quad d = \tfrac{1}{2}(1+\bar{p}^3),$$

$$e_1 = \bar{p}^3(1-\bar{p})^3 + (1-\bar{p})^6 + \bar{p}^6, \qquad e_2 = 3(1-\bar{p})^2\bar{p}^4 + 3(1-\bar{p})^5\bar{p},$$

$$e_3 = 3(1-\bar{p})\bar{p}^5 + 3(1-\bar{p})^4\bar{p}^2, \qquad e_4 = 3(1-\bar{p})^5\bar{p},$$

$$e_5 = 12(1-\bar{p})^4\bar{p}^2, \qquad\qquad\qquad e_6 = 10(1-\bar{p})^3\bar{p}^3,$$

$$e_7 = 9(1-\bar{p})^3\bar{p}^3, \qquad\qquad\qquad e_8 = 12(1-\bar{p})^2\bar{p}^4,$$

$$e_9 = 3(1-\bar{p})\bar{p}^5. \tag{9}$$

An approximate joint distribution function $p^{(2)}(n,m)$ is obtained as a stationary solution of the Markov chain defined by $T^{(2)}$.

At a first glance at $T^{(2)}$ we find that

$$p^{(2)}(0,0) = 0. \tag{10}$$

In other words, the subconfiguration $\{0,0\}$ is forbidden in the SOC state. Dhar[12] gave a general criterion for such a *forbidden subconfiguration* (FSC) in the SOC states for any finite subset in a lattice. The FSC on Ω_2 is only $\{0,0\}$. The present procedure for giving approximations naturally excludes the FSC's on subsystems.[17] Using Maple V, a mathematical package, we obtain the stationary joint distribution of the Markov chain $T^{(2)}$ and it gives the reduced distribution function which is defined as

$$p^{(2)}(n) = \sum_{m=0}^{3} p^{(2)}(n,m), \tag{11}$$

$n \in \{0,1,2,3\}$. We obtain the following results,

$$p^{(2)}(0) = \frac{-1}{3D^{(2)}}(6\tilde{p}^{14} - 45\tilde{p}^{13} + 132\tilde{p}^{12} - 219\tilde{p}^{11} + 267\tilde{p}^{10} - 288\tilde{p}^{9} + 281\tilde{p}^{8}$$
$$- 251\tilde{p}^{7} + 119\tilde{p}^{6} + 66\tilde{p}^{5} - 258\tilde{p}^{4} + 399\tilde{p}^{3} - 420\tilde{p}^{2} + 240\tilde{p} - 57),$$

$$p^{(2)}(1) = \frac{1}{3D^{(2)}}(12\tilde{p}^{14} - 72\tilde{p}^{13} + 156\tilde{p}^{12} - 222\tilde{p}^{11} + 261\tilde{p}^{10} - 225\tilde{p}^{9} + 200\tilde{p}^{8}$$
$$- 104\tilde{p}^{7} - 73\tilde{p}^{6} + 234\tilde{p}^{5} - 267\tilde{p}^{4} + 258\tilde{p}^{3} - 50\tilde{p}^{2} - 130\tilde{p} + 76),$$

$$p^{(2)}(2) = -\frac{1}{3D^{(2)}}(6\tilde{p}^{14} - 36\tilde{p}^{13} + 78\tilde{p}^{12} - 117\tilde{p}^{11} + 99\tilde{p}^{10} - 69\tilde{p}^{9} + 61\tilde{p}^{8}$$
$$- 16\tilde{p}^{7} - 35\tilde{p}^{6} - 12\tilde{p}^{5} - 18\tilde{p}^{4} + 9\tilde{p}^{3} + 5\tilde{p}^{2} + 73\tilde{p} - 76),$$

$$p^{(2)}(3) = \frac{1}{3D^{(2)}}(6\tilde{p}^{11} - 24\tilde{p}^{10} + 21\tilde{p}^{9} - 35\tilde{p}^{8} + 41\tilde{p}^{7} - 17\tilde{p}^{6} + 45\tilde{p}^{5}$$
$$- 9\tilde{p}^{4} - 21\tilde{p}^{3} + 52\tilde{p}^{2} - 73\tilde{p} + 76), \tag{12}$$

with

$$D^{(2)} = 3\tilde{p}^{13} - 18\tilde{p}^{12} + 40\tilde{p}^{11} - 43\tilde{p}^{10} + 51\tilde{p}^{9} - 59\tilde{p}^{8}$$
$$+ 68\tilde{p}^{7} - 58\tilde{p}^{6} + 75\tilde{p}^{5} - 57\tilde{p}^{3} + 139\tilde{p}^{2} - 172\tilde{p} + 95, \tag{13}$$

and the averaged height is given as

$$\langle z \rangle^{(2)} \equiv \sum_{n=0}^{3} n p^{(2)}(n)$$
$$= \frac{1}{D^{(2)}}(10\tilde{p}^{11} - 3\tilde{p}^{10} - 8\tilde{p}^{9} - 9\tilde{p}^{8} + 17\tilde{p}^{7} - 18\tilde{p}^{6} + 131\tilde{p}^{5}$$
$$- 86\tilde{p}^{4} + 59\tilde{p}^{3} + 32\tilde{p}^{2} - 165\tilde{p} + 152). \tag{14}$$

III.3 Reducibility Condition and Results

We impose the following self-consistency condition,

$$\langle z \rangle^{(1)} = \langle z \rangle^{(2)}. \tag{15}$$

By the definitions (5), (11) and (14), it is regarded as the reducibility condition of distribution functions,[23]

$$\sum_{n=0}^{3} n p^{(1)}(n) = \sum_{n=0}^{3} n \sum_{m=0}^{3} p^{(2)}(n, m). \tag{16}$$

This condition gives an equation for $\tilde{p}$ and we find a following root in $[0,1]$,

$$\tilde{p}_c^{(2)} = 0.487307. \tag{17}$$

Let $\tilde{p} = \tilde{p}_c^{(2)}$ in (5) or (14). Then we have a result,

$$z_c^{\text{pair}} = \langle z \rangle^{(1)} \big|_{\tilde{p} = \tilde{p}_c^{(2)}} = \langle z \rangle^{(2)} \big|_{\tilde{p} = \tilde{p}_c^{(2)}}$$
$$= 2.1585. \tag{18}$$

Table I

Results of the single-site approximation and the pair approximation
for the SOC state of the two-dimensional BTW model.
Values estimated by the Monte Carlo simulation are also listed.

	Single-site approximation	Pair approximation	Monte Carlo simulation
$p(0)$	0.0621	0.0493	0.074
$p(1)$	0.2004	0.1767	0.174
$p(2)$	0.3382	0.3401	0.306
$p(3)$	0.3993	0.4339	0.446
z_c	2.0746	2.1585	2.124

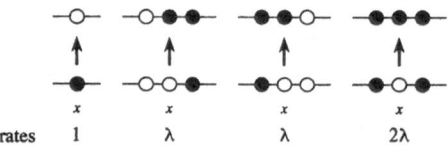

Figure 5. The elementary processes of the one-dimensional basic CP. The full (open) circles denote particles (vacancies).

Let $\tilde{p} = \tilde{p}_c^{(2)}$ in (12) with (13) and we obtain the approximate values for $p^{(2)}(n), n = 0, 1, 2, 3$. The results are listed in Table I. In this table, the values obtained by the single-site approximation, which is explained in the appendix, are also listed with the numerical values estimated by the Monte Carlo simulation.[17] The value z_c^{pair} obtained by the present pair approximation is in good agreement with the simulation result.

IV. Basic Contact Process

The basic contact process (CP) was introduced by Harris[7] as a simple mathematical model of epidemics. It is a continuous-time Markov process on a d-dimensional hypercubic lattice $\mathbf{Z}^d$. Each site x takes one of two states, $\eta_t(x) = 0$ (vacancy) and $\eta_t(x) = 1$ (particle). The system evolves by the following single-spin-flip dynamics,

(i) If $\eta_{t-}(x) = 1$, then $\eta_t(x) = 0$ at a rate 1.

(ii) If $\eta_{t-}(x) = 0$, then $\eta_t(x) = 1$ at the rate $\lambda \sum_{y:|y-x|=1} \eta_{t-}(y)$,

where λ is a non-negative parameter.

Figure 5 illustrates the above elementary processes for the one-dimensional case, where the full circles denote particles and the open circles denote vacancies. This process can be viewed as a simple model of the spread of infection of a contagious disease. An individual is infected if $\eta_t(x) = 1$ and healthy if $\eta_t(x) = 0$ at each time

t. Recovery process (i) occurs independently. The parameter λ is the infection rate in the case that only one of the neighbors is infected. The total infection rate is proportional to the total number of infected individuals at the nearest-neighbor sites.

It should be remarked that if there is no infected individuals in the neighborhood, the infection does not take place. Therefore if all individuals recover from the disease at some time s, then the disease is exterminated and there will be no infection for all $t \geq s$. For each initial state, there are two possibilities for long-term behavior: *extinction* or *survival*. The former means that the disease is exterminated and the infection process ceases. The latter means that the disease is spreading and the infection process *survives* even for $t \to \infty$.

It is proven that there is a unique critical value λ_c so that any process becomes extinct with probability 1 for $\lambda < \lambda_c$, but all processes starting from initial states with at least one infected individual have positive probabilities of survival for $\lambda > \lambda_c$.[8] Since the total infection rate is proportional to the total infected individuals in neighboring sites and the recovery rate is independent of the surrounding configuration, the process starting from an initial state with more infected individuals has higher probability for survival. It follows that the process starting from the state with all individuals infected, which we write as δ_1, has the highest probability for survival. The stationary distribution function of configuration $\{\eta(x) : x \in Z^d\}$ of the process starting from δ_1 is called the upper invariant measure and is denoted by ν_λ. Let δ_0 be the state with $\eta(x) = 0$ for all $x \in Z^d$. Then the precise definition of the critical value λ_c is given as[8]

$$\lambda_c = \inf\{\lambda \geq 0 : \nu_\lambda \neq \delta_0\}$$
$$= \sup\{\lambda \geq 0 : \nu_\lambda = \delta_0\}. \tag{19}$$

Best rigorous lower[24] and upper bounds[25] of the critical value of the one-dimensional basic CP are given as

$$1.539 \leq \lambda_c \leq 1.942. \tag{20}$$

Though the existence is proved, neither the exact value of λ_c nor the explicit form of the distribution ν_λ are yet known. Mathematical physicists have been seeking the precise description of this non-trivial distribution ν_λ, since processes which are essentially the same as the CP have appeared in a wide variety of fields of theoretical physics. It is known as a Reggeon quantum spin model for the strong interaction of hadrons.[26-28] We can consider the CP as a lattice model version of Schlögl's first model[29] of autocatalytic chemical reactions.[30,31] As mentioned above, it is also viewed as a simple epidemic model in which immunization is not taken into account.[32] The appearance of the non-trivial stationary state ν_λ at $\lambda = \lambda_c$ can be regarded as a phase transition. The bounds (20) means that there occurs a phase transition even in the one-dimensional system. Moreover, we observe critical phenomena at $\lambda = \lambda_c$ also in the one-dimensional system.[28,30,33,34] Since the CP has only nearest-neighbor interactions, this implies that ν_λ can not be described by the ordinary Gibbs state with finite-ranged potentials. In other words, detailed balance is violated in the survival states which appear in the supercritical phase $\lambda > \lambda_c$.

V. Cluster Approximations for Contact Process

V.1 Correlation Equalities

We consider the non-equilibrium stationary state ν_λ of the one-dimensional CP. It is a stationary distribution of configuration $\{\eta(x) : x \in \mathbf{Z}\}$ of the CP starting from the state δ_1 in which $\eta(x) = 1$ (infected) for all $x \in \mathbf{Z}$. As the system evolves, infected individuals recover from the disease. Since we do not take immunization into account in this simple model, they become infected again by contact interaction with nearest-neighbor infected individuals. We concentrate on the distribution of the healthy individuals in the state ν_λ. Let A be an arbitrary finite subset of $\mathbf{Z}$ and define

$$\bar{\rho}_\lambda(A) = \nu_\lambda\{\eta : \eta(x) = 0 \text{ for all } x \in A\}. \qquad (21)$$

That is, $\bar{\rho}_\lambda(A)$ is a probability that we observe the configuration such that $\eta(x) = 0$ for all $x \in A$ in the non-trivial stationary state ν_λ. If we can determine (21) for any finite $A \subset \mathbf{Z}$, the state ν_λ is completely determined.

It is proven that the following equality holds for any finite subset $A \subset \mathbf{Z}$,[8]

$$\lambda \sum_{x \in A} \sum_{y:|y-x|=1} [\bar{\rho}_\lambda(A \cup \{y\}) - \bar{\rho}_\lambda(A)] + \sum_{x \in A} [\bar{\rho}_\lambda(A \setminus \{x\}) - \bar{\rho}_\lambda(A)] = 0. \qquad (22)$$

Therefore we can say that the stationary state ν_λ is a non-trivial solution of Eqs. (22). Here the trivial solution, which corresponds to the state δ_0, is $\bar{\rho}_\lambda(A) \equiv 1$ for any A for all $\lambda \geq 0$. Let $A = \{x\}$ and $A = \{x, x+1\}$ in (22). We obtain the following equations.

$$1 - (2\lambda + 1)\bar{\rho}_\lambda(\{x\}) + 2\lambda\bar{\rho}_\lambda(\{x, x+1\}) = 0, \qquad (23)$$

$$\bar{\rho}_\lambda(\{x\}) - (\lambda + 1)\bar{\rho}_\lambda(\{x, x+1\}) + \lambda\bar{\rho}_\lambda(\{x, x+1, x+2\}) = 0, \qquad (24)$$

where we have used the translational invariance of ν_λ. That is, the density of healthy individuals, $\bar{\rho}_\lambda(\{x\})$, is related to the pair correlation function, $\bar{\rho}_\lambda(\{x, x+1\})$ by (23), and the three-point correlation function, $\bar{\rho}_\lambda(\{x, x+1, x+2\})$, appears in Eq. (24), which is derived from (22) by letting $A = \{x, x+1\}$. We will see that higher-order correlation functions appear when we put $A = \{x, x+1, x+2\}$ in (22). We often call such equations *the correlation equalities*. They follow a kind of BBGKY hierarchy and Eqs. (22) make an infinite set of simultaneous equations. Some truncation procedure is required to obtain approximate solutions.

V.2 Cluster Approximations

For each subset $A \subset \mathbf{Z}$, we introduce an indicator function $I_A(x)$ as

$$I_A(x) = \begin{cases} 1 & \text{if } x \in A \\ 0 & \text{if } x \notin A. \end{cases} \qquad (25)$$

Then we assume that the logarithm of $\bar{\rho}_\lambda(A)$ can be expanded as follows.

$$\bar{\rho}_\lambda(A) = \exp\{K_1 \sum_x I_A(x) + K_2 \sum_x I_A(x)I_A(x+1)$$

$$+ K_3 \sum_x I_A(x)I_A(x+1)I_A(x+2) + K_2' \sum_x I_A(x)I_A(x+2) + \cdots\} \quad (26)$$

In other words, we try to approximate the non-equilibrium stationary state ν_λ by a generalized Gibbs state with many kinds of potentials. The effective potentials $\{K_i\}$ should be determined so that (26) satisfy the correlation equalities (22).

If we take into account only finite terms in (26) and neglect other terms, we will obtain an approximate expression for $\bar\rho_\lambda(A)$. Such a procedure is very similar to the CVM which is used for obtaining approximate density matrices in equilibrium spin systems.[21,22] Of course, there are the following differences. (i) In the usual application of the CVM to equilibrium spin systems, we introduce a constant term $F^{(i)}$ in the logarithm of the approximate density matrix, which should be determined by the normalization condition of the density matrix. Here $\bar\rho_\lambda(A)$ is a probability and thus already normalized to have its value between 0 and 1. (ii) If we treat, for example, the Ising model with the Hamiltonian, $\mathcal{H} = -J\sum_{<i,j>} S_i S_j - h\sum_i S_i$, K_1 and K_2 are the following form: $K_1 = \beta(h + \text{effective fields})$, $K_2 = \beta(J + \text{effective interactions})$ with $\beta = 1/k_B T$ and the effective fields and effective interactions are expressed by variational parameters. In the present situation, however, there is no Hamiltonian for the CP and then K_1 and K_2 are themselves the parameters which should be determined by stationary conditions.

The simplest choice of trial function for $\bar\rho_\lambda(A)$ is

$$\bar\rho_\lambda^{(1)}(A) = \exp\left\{ K_1^{(1)} \sum_x I_A(x) \right\}. \tag{27}$$

We impose the equality (23) for (27) and have the result

$$K_1^{(1)} = -\log(2\lambda). \tag{28}$$

In order to characterize the state ν_λ we introduce an *order parameter* as

$$\rho(\lambda) = \nu_\lambda\{\eta : \eta(x) = 1\}, \tag{29}$$

which represents the density of infected individuals in the stationary state. Since it is independent of the site x by the translational invariance, we find that $\rho(\lambda) = 0$ means $\nu_\lambda = \delta_0$, while $\rho(\lambda) > 0$ means $\nu_\lambda \neq \delta_0$. On the other hand, by definition (21),

$$\rho(\lambda) = 1 - \bar\rho_\lambda(\{x\}). \tag{30}$$

The solution (28) gives

$$\rho(\lambda)^{(1)} = \frac{2\lambda - 1}{2\lambda} \qquad \text{for} \qquad \lambda \geq \lambda_c^{(1)} = \tfrac{1}{2}, \tag{31}$$

which is a *mean-field approximation* of the one-dimensional CP.[35]

Next we assume that

$$\bar\rho_\lambda^{(2)}(A) = \exp\left\{ K_1^{(2)} \sum_x I_A(x) + K_2^{(2)} \sum_x I_A(x) I_A(x+1) \right\}, \tag{32}$$

and impose the equalities (23) and (24). Then we have

$$K_1^{(2)} = -\log(2\lambda - 1),$$
$$K_2^{(2)} = \log\left(\frac{2\lambda - 1}{\lambda}\right). \tag{33}$$

This *pair approximation* gives[35]

$$\rho(\lambda)^{(2)} = \frac{2(\lambda - 1)}{2\lambda - 1} \quad \text{for} \quad \lambda \geq \lambda_c^{(2)} = 1. \tag{34}$$

Now we take into account four terms in (26), $K_1 \neq 0$, $K_2 \neq 0$, $K_3 \neq 0$ and $K_2' \neq 0$, and neglect other terms. We impose the correlation equalities (23), (24) and the following two equations which are derived from (22) by letting $A = \{x, x+1, x+2\}$ and $A = \{x, x+2\}$,[15]

$$2\bar{\rho}_\lambda(\{x, x+1\}) - (2\lambda + 3)\bar{\rho}_\lambda(\{x, x+1, x+2\}) + \bar{\rho}_\lambda(\{x, x+2\})$$
$$+ 2\lambda\bar{\rho}_\lambda(\{x, x+1, x+2, x+3\}) = 0, \tag{35}$$

and

$$\bar{\rho}_\lambda(\{x\}) + \lambda\bar{\rho}_\lambda(\{x, x+1, x+2\}) - (2\lambda+1)\bar{\rho}_\lambda(\{x, x+2\}) + \lambda\bar{\rho}_\lambda(\{x, x+1, x+3\}) = 0. \tag{36}$$

The results are as follows,

$$K_1^{(3)} = \log\left(\frac{\lambda(2\lambda + 3) + \sqrt{D}}{(2\lambda + 1)(6\lambda^2 - 3\lambda - 1)}\right),$$

$$K_2^{(3)} = \log\left(\frac{-(24\lambda^4 + 16\lambda^3 + 8\lambda^2 + \lambda + 1) + (6\lambda - 1)(\lambda + 1)\sqrt{D}}{2\lambda(\lambda - 1)^2}\right),$$

$$K_3^{(3)} = \log\left(\frac{(24\lambda^3 + 16\lambda^2 - 2\lambda - 3) + (4\lambda + 3)\sqrt{D}}{8\lambda(2\lambda + 1)^2}\right),$$

$$K_2'^{(3)} = \log\left(\frac{(\lambda + 1)\{(12\lambda^3 - 2\lambda^2 - \lambda + 1) - (3\lambda - 1)\sqrt{D}\}}{2\lambda^2(\lambda - 1)}\right), \tag{37}$$

and

$$\rho(\lambda)^{(3)} = \frac{4\lambda(3\lambda^2 - \lambda - 3)}{(12\lambda^3 - 2\lambda^2 - 8\lambda - 1) + \sqrt{D}} \quad \text{for} \quad \lambda \geq \lambda_c^{(3)} = \tfrac{1}{6}(1 + \sqrt{37}), \tag{38}$$

where

$$D = 16\lambda^4 + 4\lambda^2 + 4\lambda + 1. \tag{39}$$

V.3 Upper Bounds and Convergence

Figure 6 summarizes the approximations for the order parameter $\rho(\lambda)$. It suggest that the approximations monotonically converge to the unknown $\rho(\lambda)$ from the above. Konno and Katori[34] numerically studied higher approximations using Suzuki's coherent-anomaly method.[36-38] Though the result strongly suggests the convergence, we have not yet given proof for convergence. It should be remarked that, however, we can prove[39,40] that the present three approximations, $\rho(\lambda)^{(i)}, i = 1, 2, 3$ are indeed upper bounds of the true order parameter $\rho(\lambda)$ by using a probability theorem called Harris' lemma.[41]

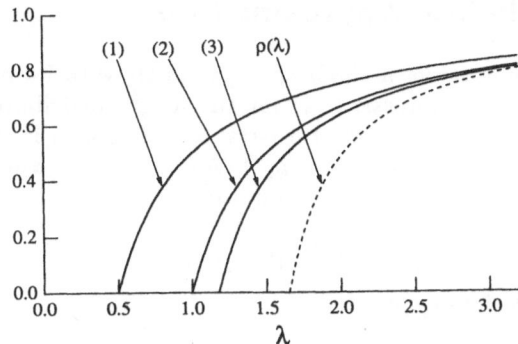

Figure 6. Three approximations for the order parameter $\rho(\lambda)$ of the one-dimensional CP. (1): $\rho(\lambda)^{(1)}$, (2): $\rho(\lambda)^{(2)}$, and (3): $\rho(\lambda)^{(3)}$.

VI. Concluding Remarks

In the present paper, we treated two kinds of non-equilibrium lattice modes, the BTW sandpile model and the basic CP. In section III, we introduced a single-site system and a two-site system with their neighbor sites, and gave an approximate expression to the averaged height $\langle z \rangle$ in the SOC state for each system. Following the idea of the CVM, we imposed a reducibility condition between these two subsystems and obtained a pair approximation for the two-dimensional BTW model. Table I shows that the pair approximation gives better results than the single-site approximation as expected. We find, however, still some differences between the values of distribution function $p(n)$ obtained by the pair approximation and the values estimated by the Monte Carlo simulation. In the pair approximation, we consider only the averaged height $\langle z \rangle$. Of course, we want to know other quantities such as the pair correlation $\langle z(x,y)z(x + 1,y)\rangle$ to characterize the SOC state. For improving the pair approximation and for estimating correlation functions, the square approximation, or the so-called *Kikuchi approximation*, is required for the SOC state. In 55 distinct configurations on a square of four sites, 15 configurations are FSC's. Our procedure excludes these FSC's automatically and the transition matrix, which we have to treat, is reduced to a 40×40 matrix. The calculation is now in progress and the details will be reported in the near future.

In section V, we demonstrated an approximate procedure for the non-equilibrium stationery state ν_λ of the supercritical CP, $\lambda > \lambda_c$. Since we observe critical phenomena at $\lambda = \lambda_c$ even in the one-dimensional CP, the state ν_λ is not the ordinary Gibbs state with short-ranged potentials, even though the elementary interactions of the CP are nearest-neighbor interactions. In other words, we see the disagreement between the range of elementary interactions and that of effective potentials in the stationary state, which is one of the characteristics of non-equilibrium stationary states. We noticed that if we follow the CVM, we can introduce many kinds of effective potentials systematically in making approximations. The obtained approximations seem to converge to the true one monotonically. However, the convergence problem is still open. This procedure can be regarded as approximating the non-equilibrium stationary states by the generalized Gibbs states. Therefore this convergence problem of approximations is very interesting.

Appendix: Single-Site Approximation

In subsection III.1, we consider a single-site system Ω_1 with its nearest-neighbor $\mathcal{N}_1$ and gave an approximate expression (4) for the height distribution function in the SOC state. It is given as a function of a parameter $\tilde{p}$, which is the probability that the site in $\mathcal{N}_1$ has the value $z_{\mathrm{m}} = 3$; $\tilde{p} = \tilde{p}(3)$. A simple approximation is obtained if we impose the following self-consistency condition,

$$p^{(1)}(3) = \tilde{p}, \tag{40}$$

which gives a self-consistency equation of $\tilde{p}$ as

$$1 - 4\left[(1 - \tilde{p}) + \tilde{p}^4\right]\tilde{p} = 0. \tag{41}$$

This equation has a unique root in [0,1] as

$$\tilde{p}_{\mathrm{c}}^{(1)} = 0.3993, \tag{42}$$

and we have the values for $p(n), n = 0, 1, 2, 3$, by letting $\tilde{p} = \tilde{p}_{\mathrm{c}}^{(1)}$ in (4), as shown in Table I. We call this approximation *the single-site approximation*.

It should be remarked that the self-consistency equation (41) is easily generalized for the BTW model on the d-dimensional hypercubic lattice as

$$1 - 2d\left[(1 - \tilde{p}) + \tilde{p}^{2d}\right]\tilde{p} = 0, \tag{43}$$

for $\tilde{p} = \tilde{p}(z_{\mathrm{m}})$ with $z_{\mathrm{m}} = 2d - 1$. Moreover this approximation is also valid for other lattice models which exhibit the SOC states. More details will be reported elsewhere.[17]

References

1. R. Kikuchi, *Phys. Rev.* **81**, 988 (1951).
2. For a review, see D. M. Burley, in *Phase Transitions and Critical Phenomena*, Vol. 2, edited by C. Domb and M. S. Green (Academic Press, London, 1972), p. 329.
3. R. Kikuchi, *Prog. Theor. Phys.* Suppl. **35**, 1 (1966).
4. For recent development of the CVM and the PPM, see *Prog. Theor. Phys.* Suppl. **115** (1994).
5. P. Bak, C. Tang, and K. Wiesenfeld, *Phys. Rev. Lett.* **59**, 381 (1987).
6. P. Bak, C. Tang, and K. Wiesenfeld, *Phys. Rev. A* **38**, 364 (1988).
7. T. E. Harris, *Ann. Probab.* **2**, 969 (1974).
8. T. M. Liggett, *Interacting Particle Systems* (Springer, New York, 1985).
9. L. P. Kadanoff, S. R. Nagel, L. Wu, and S. M. Zhou, *Phys. Rev. A* **39**, 6524 (1989).
10. S. S. Manna, *J. Stat. Phys.* **59**, 509 (1990).
11. D. Dhar and S. N. Majumdar, *J. Phys. A: Math. Gen.* **23**, 4333 (1990).
12. D. Dhar, *Phys. Rev. Lett.* **64**, 1613 (1990).
13. S. N. Majumdar and D. Dhar, *J. Phys. A: Math. Gen.* **24**, L357 (1991).
14. M. Katori and N. Konno, in *Formation, Dynamics and Statistics of Pattern*, Vol.2, edited by K. Kawasaki and M. Suzuki (World Scientific, Singapore, 1993), p. 23.
15. N. Konno, *Phase Transitions of Interacting Particle Systems* (World Scientific, Singapore, 1994).
16. C. Tang and P. Bak, *Phys. Rev. Lett.* **60**, 2347 (1988).

17. M. Katori and H. Kobayashi, to appear in Physica A (1996).
18. C. Tang and P. Bak, *J. Stat. Phys.* **51**, 797 (1988).
19. J. Theiler, in *Modeling Complex Phenomena*, Proceedings of the Third Woodward Conference, San Jose State University, April 12-13, 1991, edited by L. Lam and V. Naroditsky (Springer, New York, 1992), p. 221.
20. T. Morita, *J. Phys. Soc. Jpn.* **12**, 753, 1060 (1957).
21. T. Morita and T. Tanaka, *Phys. Rev.* **145**, 288 (1966).
22. T. Morita, *J. Math. Phys.* **13**, 115 (1972).
23. T. Morita, *J. Stat. Phys.* **34**, 319 (1984).
24. H. Ziezold and A. Grillenberger, *J. Appl. Prob.* **25**, 1 (1987).
25. T. M. Liggett, *Ann. Probab.* **23**, 697 (1995).
26. D. Amati, M. Le Bellac, G. Marchesini, and M. Ciafaloni, *Nucl. Phys.* **B112**, 107 (1976).
27. R. C. Brower, M. A. Furman, and K. Subbarao, *Phys. Rev. D* **15**, 1756 (1977).
28. R. C. Brower, M. A. Furman, and M. Moshe, *Phys. Lett.* **B76**, 213 (1978).
29. F. Schlögl, *Z. Physik* **253**, 147 (1972).
30. P. Grassberger and A. de la Torre, *Ann. Phys.* **122**, 373 (1979).
31. P. Grassberger and M. Scheunert, *Fortschr. Phys.* **28**, 547 (1980).
32. T. Ohtsuki and T. Keyes, *Phys. Rev. A* **33**, 1223 (1986).
33. R. Dickman, *J. Stat. Phys.* **55**, 997 (1989).
34. N. Konno and M. Katori, *J. Phys. Soc. Jpn.* **59**, 1581 (1990).
35. M. Katori and N. Konno, *J. Phys. Soc. Jpn.* **60**, 95 (1991).
36. M. Suzuki, *J. Phys. Soc. Jpn.* **55**, 4205 (1986).
37. M. Suzuki, M. Katori, and X. Hu, *J. Phys. Soc. Jpn.* **56**, 3092 (1987).
38. M. Suzuki, *J. Stat. Phys.* **53**, 483 (1988).
39. M. Katori and N. Konno, *J. Phys. Soc. Jpn.* **60**, 418 (1991).
40. M. Katori and N. Konno, *J. Stat. Phys.* **63**, 115 (1991).
41. T. E. Harris, *Ann. Probab.* **4**, 175 (1976).

Coherent Anomaly Method
and its Applications

Masuo Suzuki and Miroslav Kolesik

Department of Physics
University of Tokyo
Bunkyo-ku, Tokyo 113
JAPAN

Abstract

Various methods for constructing systematic series of mean-field-type approximations are reviewed from the point of view of the coherent-anomaly method.

I. Introduction

Mean-field theory plays an important role in physics. It provides an insight into the properties of the system under investigation without the necessity of large-scale computations, and it is therefore an ideal tool for investigation of new physical systems. In statistical physics, there have been developed many mean-field-type theories which go beyond the ordinary mean-field approximation, namely systematic approaches which give very detailed information about the studied model. For example, one can obtain a series of phase diagrams with the phase boundaries closely approaching the exact ones. In some cases, the mean-field-type approximations can even give the exact results for critical temperatures. However, the mean-field-type approximations have an inherent drawback, as far as the behavior in the critical region is concerned. Namely, they provide always the so-called classical singularity near their critical points, irrespectively of how accurate these approximate critical points could be. Thus, no improvement can be achieved in this respect. Namely the critical exponents calculated from mean-field approximation remain the same when the approximation is improved. Nevertheless, the systematic series of mean-field solutions actually contain much more information concerning the true critical behavior as it was discovered by Suzuki[1] about ten years ago. Such systematic series of approximations exhibit the property called coherent-anomaly which allows to extract the true critical exponents. This is an important finding which turns the mean-field approach into a powerful tool to investigate the phenomena which could originally seem unreachable because of the built-in mean-field

Theory and Applications of the Cluster Variation and Path Probability Methods
Edited by J.L. Morán-López and J.M. Sanchez, Plenum Press, New York, 1996

limitations. One of the strengths of the combination of the mean-field and coherent-anomaly methods lies in the fact that even the simplest approximations can give a fairly good picture of the studied problem. Moreover, nowadays, the accuracy of the estimates of the critical exponents from the coherent-anomaly method approaches the high precision achieved by the conventional methods such as Monte Carlo and series expansions.

The present article is devoted to the interaction between the mean-field approach and the coherent-anomaly method, and to their mutual fertilization leading to the improvements of the mean-field techniques motivated by the coherent-anomaly approach and viceversa. In the following section, we give a brief recapitulation of the coherent-anomaly method, and the subsequent sections are devoted to summarizing several successful mean-field-type approaches as appreciated from the point of view of the Coherent-Anomaly Method.

II. Summary of the Coherent Anomaly Method

The Coherent Anomaly Method is a general approach to study critical phenomena. It is based on the observation that apparently any natural series of mean-field approximations to a given problem exhibits the so-called coherent-anomaly. Let us suppose that we can calculate such a series of mean-field approximate solutions in some systematic way. Naturally, each of these approximations exhibits the classical singularity in the vicinity of its critical point, and has the same set of mean-field critical exponents. However, the prefactors of the mean-field critical diverge as the approximation improves. It is the systematic change of these mean-field critical coefficients, which is called the coherent-anomaly, and which can be used to extract information about the true critical behavior. Series of approximations which exhibits convergence to the exact solution together with the anomalous scaling of critical exponents is called *canonical*. As we have already mentioned, in practice, any set of mean-field approximations is expected to be canonical, provided it is built in a natural and systematic way. The concept of this powerful method was introduced by Suzuki[1] and developed into a selfconsistent theory in a series of following works.[2-7] It is impossible to mention here all of the numerous applications of the CAM, but one can a find review of recent developments concerning the method *e.g.* in Ref. 9, and a comprehensive review of a variety of applications together with original works in Ref. 10.

The source of data for the CAM analysis is thus a set of mean-field approximations for the given problem. For each member of this set labeled by L, the order of approximation, one can calculate its critical temperature T_L^c. Further, for each singular quantity Q of interest one has to calculate the above mentioned mean-field critical coefficient $\overline{Q}_L$. The coefficient $\overline{Q}_L$ characterizes the classical singular behavior of Q in the vicinity of the critical point T_L^c,

$$Q_L \sim \overline{Q}_L (T/T_L^c - 1)^{\omega_{class}}, \tag{1}$$

where ω_{class} denotes the classical value of the corresponding critical index. Suppose that the true critical behavior is described by

$$Q \sim (T/T^* - 1)^\omega, \tag{2}$$

with T^* being the exact critical temperature, and ω the exact critical exponent. Then, the mean-field critical coefficients $\{\overline{Q}_L\}$ can be shown to follow the scaling formula[1-3]

$$\overline{Q}_L \sim (T_L^c - T^*)^{\omega - \omega_{class}}. \tag{3}$$

It characterizes how the mean-field approximations approach the exact solution. Thus, we can obtain an estimate for the exact critical index ω by fitting the set of data $\{T_L^c, \overline{Q}_L\}$ to the formula (3). To be more specific, the critical indices α, β, γ and δ can be estimated from

$$\overline{c}_L \sim (T_L^c - T^*)^{-\alpha}, \tag{4}$$
$$\overline{m}_L \sim (T_L^c - T^*)^{\beta - 1/2}, \tag{5}$$
$$\overline{\chi}_L \sim (T_L^c - T^*)^{1-\gamma}, \tag{6}$$
$$\overline{m}_L^c \sim (T_L^c - T^*)^{-\psi}, \qquad \psi = \gamma(\delta - 3)/3(\delta - 1), \tag{7}$$

where $\overline{c}_L$, $\overline{m}_L$, $\overline{\chi}_L$ and $\overline{m}_L^c$ are the mean-field critical coefficients of the specific heat, magnetization, susceptibility and the critical magnetization, respectively. These mean-field critical coefficients are not independent, but they fulfill the relations[1]

$$\overline{\chi}_L \overline{c}_L / (\overline{m}_L)^2 = \text{const.}, \quad \text{and} \quad (\overline{\chi}_L)^2 \overline{c}_L / (\overline{m}_L^c)^3 = \text{const.}, \tag{8}$$

regardless of the used series of approximations. As a consequence, it is an inherent property of the CAM that the scaling relations $\alpha + 2\beta + \gamma = 2$ and $\gamma = \beta(\delta - 1)$ are always satisfied by the resulting exponents.

It is worth to mention one technical point here. The property of the coherent-anomaly scaling of the critical exponents can be expressed not only using the temperature as the scaling variable, but also in terms of the inverse temperature or yet a different variable. While these schemes coincide in the asymptotic region, from the practical point of view they are different, because usually the mean-field approximations do not provide the critical points with such accuracy. So, it is rather unfortunate that the results of the coherent-anomaly analysis may depend on the choice of the scaling variable. The deviations in the resulting critical exponents are not so large, but if one aims at very precise estimates, then it is necessary to cope with this problem. Several possible ways to avoid this difficulty have been suggested.[3] One of them is to fit the coherent-anomaly data using the scaling variable[11] $\Delta = |x^{(a)} - x^*| / \sqrt{x^{(a)} x^*}$ where $x^{(a)}$ and x^* are the approximate and exact critical point, respectively. The advantage of this variable is the fact that it does not change when one replaces the critical temperature for the inverse critical temperature. Thus, this is a very natural variable to use when extracting critical exponents from the coherent-anomaly data. Moreover, one has also to consider possible corrections to the ideal coherent-anomaly scaling. This may be a difficult problem, because these deviations seem to be dependent on how the particular series of approximation was constructed and it is therefore impossible to know their analytical form. Nevertheless, if the set of approximations gives a nice series of critical coefficients and temperatures which scale smoothly enough, then it may be possible to extract the unwanted corrections and obtain rather accurate results.

The Coherent Anomaly Method was successfully applied to a sheer number of various problems ranging from the critical phenomena in classical spin models, through nonuniversal critical behavior, spin glasses and critical dynamics to quantum transitions and many others. This review article is concerned mainly with the methods for generating systematic series of mean-field approximations suitable for the CAM analysis of the critical phenomena, such as the Cluster Variation Method and other related approaches.

While systematic mean-field approximations themselves can provide often very accurate insight into the properties of studied systems such as its phase structure and phase diagram in combination with the CAM analysis, they represent a powerful tool to understand also the critical phenomena. Usually it is possible to design more types of systematic approximations for a given system, and it seems that the choice of the best scheme for the CAM analysis depends on the specific problem. In what follows we review several methods to obtain such approximations from the point of view of the Coherent Anomaly Method.

III. Systematic Mean-Field Methods

III.1 Cluster Variation Method

The Cluster Variation Method was proposed by Kikuchi[12] about forty-five years ago and since then it has been used as a standard tool in many different fields of statistical physics. The Cluster Variation Method is an approach for constructing approximate solutions for the free energy in the thermodynamic limit from the solutions for finite clusters. In its range of applicability, the method is very general and has been applied to a great variety of systems. In the present article, we concentrate only on the relation between the Cluster Variation Method and the Coherent Anomaly Method, firstly because it is impossible to review all the numerous applications of the Cluster Variation Method, and secondly because the Coherent Anomaly Method gives the Cluster Variation Method a new dimension: the coherent-anomaly analysis of the set of approximations generated by the Cluster Variation Method can go far beyond the scope of mean-field theories. It allows to obtain not only the phase diagram with high accuracy, but also very reliable estimates of critical exponents.

Let us summarize the basic ideas of the Cluster Variation Method. Suppose that we are given a model defined on an infinite lattice, and that we are required to find an approximation for its free energy. As we have already mentioned, one starts with finite clusters. Having specified this set of clusters, one gives the trial function for the free energy constructed from the density matrices of clusters. The free energies of clusters contain the effective fields acting on a single spin and/or on multiplets of spins. With the final trial function for the infinite-lattice free energy, one considers the effective fields as variational parameters: they are fixed by minimizing the resulting free energy. Thus, this scheme provides a mean-field-type solution for the free energy.

To be more specific, let us present concrete formulas for the simplest Kikuchi's approximation. It is based on the set of clusters consisting of just three basic elements of the hyper-cubic lattice, namely singleton, doubleton and square:

$$\Lambda_0 = \{\{x\}, \{x, x + e_i\}, \{x, x + e_i, x + e_j, x + e_i + e_j\}\},$$

where e_i are the unit vectors of the lattice. One constructs the free energies of these clusters such that the influence of the rest of the infinite system is taken into account by introducing the effective fields $\{\lambda_1, \lambda_2, \lambda_3\}$ acting on the spins from outside of the basic cluster

$$F(\{x\}) = -T \ln \left[\operatorname{Tr} \exp \beta(z\lambda_1 + H)s_x \right], \tag{9}$$

$$F(\{x + e_i\}) = -T \ln \left[\operatorname{Tr} \exp \beta \{ (J + \nu_4 \lambda_3) s_x s_{x+e_i} \right.$$
$$+ \left. [(z-1)\lambda_1 + \nu_4 \lambda_2 + H](s_x + s_{x+e_i}) \} \right], \tag{10}$$

$$F(\{x, x + e_i, x + e_j, x + e_i + e_j\}) =$$
$$- T \ln \left[\operatorname{Tr} \exp \beta \{ [J + (\nu_4 - 1)\lambda_3] \right.$$
$$\times (s_x s_{x+e_i} + s_x s_{x+e_j} + s_{x+e_i} s_{x+e_i+e_j} + s_{x+e_j} s_{x+e_i+e_j})$$
$$+ [(z-2)\lambda_1 + 2(\nu_4 - 1)\lambda_2 + H]$$
$$\times \left. (s_x + s_{x+e_i} + s_{x+e_j} + s_{x+e_i+e_j}) \} \right]. \tag{11}$$

Here, $z = 2d$ is the coordination number of the lattice and $\nu_4 = 2(d-1)$ is the number of squares sharing an edge. The parameter λ_1 is the mean-field generated by the nearest neighbors, and λ_2 and λ_3 are the fields acting on a spin and the product of two spins, respectively, both from a square cluster of spins.

Having the free energies of basic clusters, one expresses these as the sum of the proper contributions from the subclusters,

$$F(A) = \sum_{B \subseteq A} f(B). \tag{12}$$

These equations are easily inverted to obtain the approximation for the free energy in the thermodynamic limit

$$f^{\Lambda_0} = \sum_{B_i \in \Lambda_0} n(B_i) f(B_i), \tag{13}$$

where the numbers $n(B_i)$ are the normalization factors which are necessary to obtain the free energy per spin. As a result, one arrives at the final formula of Kikuchi's approximation

$$f^{\Lambda_0} = (1 - z + \tfrac{1}{2} z \nu_4) F(\{x\}) - \tfrac{1}{2} z (\nu_4 - 1) F(\{x + e_i\})$$
$$+ \tfrac{1}{8} z \nu_4 F(\{x, x + e_i, x + e_j, x + e_i + e_j\}). \tag{14}$$

Here, the mean-fields are determined by the selfconsistency equations

$$\partial_{\lambda_1} f^{\Lambda_0} = 0, \quad \partial_{\lambda_2} f^{\Lambda_0} = 0, \quad \partial_{\lambda_3} f^{\Lambda_0} = 0, \tag{15}$$

and from all the solutions of these equations we choose the set of values that give the minimal free energy.

Two important characteristics of the method are worth to mention here: firstly, the method is systematic. Its results can be, in principle, improved by taking into account larger and larger basic clusters. It is supposed that such systematic procedure would eventually lead to the exact solution in the thermodynamic limit. Actually it was proved by Schlijper,[13,14] that the free energies calculated by the Cluster Variation Method indeed converge to the exact one. Similarly, the convergence of entropies has been proven by Cenedese, Sanchez and Kikuchi.[15] The convergence of order parameters remains an open question.

The second important property of the method is the possibility to derive the order parameters and short-range correlations. This is possible due to the presence of effective fields coupled with the relevant order parameters. Consequently, the method allows to study the structure of the phases and the phase diagram in great detail.

The coherent anomaly of the cluster variation approximations was studied by Katori and Suzuki,[16] and by Fujiki, Katori and Suzuki.[17] They addressed the question whether or not the simplified cluster variation approximation can be used to estimate critical exponents by the coherent-anomaly method. Indeed, they found that the coherent-anomaly scaling of the critical mean-field coefficients is exhibited already by the lowest order approximations, whose property allows to estimate reasonable critical exponents from relatively simple computations. It was also observed that it is important that the basic clusters reflect the structure of the lattice properly. For example, for the three-dimensional Ising model, the inclusion of the cube-cluster is essential for a good accuracy of the resulting critical exponents.

Fujiki[18] applied the combination of the CAM and CVM to the two-dimensional 6-clock model exhibiting the Kosterlitz-Thouless (KT) phase transition. However, his comparative study of the Ising model showed that in order to be able to distinguish clearly between the KT and ordinary second-order phase transition, higher-order CVM approximations would be needed.

Katori and Suzuki[19] pointed out that that all the CVM approximations, including the simplest one, exhibit the coherent-anomaly. This is in contrast with the conventional approach to critical phenomena such as Monte Carlo and series expansions, where rather hard work is necessary to obtain even rough estimates of critical exponents. On the contrary, the CAM based on the CVM provides sensible estimates at relatively low cost.

The comprehensive investigation of the canonicality of the CVM approximation was given by Katori and Suzuki.[16,19] They demonstrated that the coherent-anomaly analysis based on the cluster variation approach represent a powerful tool for investigating critical properties. They presented fairly good estimates of the critical exponents for the two- and three-dimensional Ising models.

As it has been already pointed out, the cluster variation approximations can be in principle improved to arbitrary accuracy. However, in practice such a straightforward approach is hampered at a rather early stage by the fact that the number of effective fields needed for large basic clusters grows rapidly, and the problem ceases to be tractable. To weaken this difficulty, simplified CVM approximations were proposed by Morita.[20] These simplified approximations were tested from the point of view of the CAM by Tanaka, Horiguchi and Morita.[21] They obtained very good estimates for the critical exponents of the square lattice Ising model, namely $\gamma = 1.75$, $\nu = 0.977$ and $\eta\nu = 0.2554$. The same authors[22] applied the CAM analysis of the cluster variation approximation to the Ising model with the next-nearest-neighbor antiferromagnetic interactions. They found the breakdown of the universality in accord with some previous findings concerning this interesting system. Their results also gave some indication of the weak universality breaking, but they could not solve this problem conclusively.

Wada and Watanabe[23] studied the dynamical critical exponent of the square Ising model using the pair, square-cactus and square cluster variation approximations. They concluded that $\Delta = 2.125$. In Ref. 24 they applied the same approach to the exponent ν of the D-dimensional Ising model, demonstrating the strength of the CVM method which allowed to study this problem in general dimensions analytically from the beginning to the end. The bond-percolation was studied by these authors[25] by means of the mapping on the $s \to 1$ limit of the s-state Potts model. Using the CAM, they obtained the estimates of critical exponents with satisfactory accuracy, especially for higher dimensions.

Finally, we mention a work which well demonstrates how the Coherent Anomaly Method motivates the improvement of the mean-field-type methods, in the case of the cluster variation method. To avoid the difficulty of generating many CVM approximations needed for the subsequent CAM analysis, Cenedese, Sanchez and Kikuchi[15] developed a new method based on the CVM. Their approach allows to generate the continuous family of CVM approximations starting from just two different CVM approximations. The resulting family gives good critical exponents in spite of the fact that the starting approximations are relatively simple.

III.2 Multi-Effective-Field Theory

Another successful mean-field-type approach to critical phenomena is the Multi-effective-field theory introduced by Suzuki,[26] and further developed by Suzuki, Minami, Katori and Nonomura.[26-28] In this method, one considers a finite cluster with effective fields applied to the spins on its boundary. These fields act on single spins as well as on multiplets of spins. More precisely, the original Hamiltonian $\mathcal{H}$ is modified as follows

$$\mathcal{H} = \mathcal{H}_{cl} - \sum J_i Q_i^{\text{even}} - \sum H_i Q_i^{\text{odd}},$$

$$\mathcal{H}_{cl} = \sum_{\langle ij \rangle \in \Omega} J s_i s_j - \sum_{i \in \Omega} H s_i,$$

where the symbol Ω means the considered cluster, and $\mathcal{H}_{cl}$ is the original Hamiltonian restricted to spins in the cluster. Here, H is the external field acting on all the spins within the cluster. The symbols Q_i^{even} and Q_i^{odd} stand for the sums of spin-products containing even and odd number of spins, respectively. Furtheremore, J_i and H_i are the corresponding effective fields applied to the cluster boundary. The self-consistency equations determining the effective fields are derived from the requirement that the expectation values for products of spins at the boundary are the same as the expectation values of the corresponding products of the spins inside the cluster. To calculate the selfconsistency equations explicitly, one neglects the second- and higher-order terms in the odd effective fields and finally arrives at the following self-consistency equations

$$-\sum_j \langle \Delta Q_i^{\text{odd}} Q_j^{\text{odd}} \rangle^* - H \langle \Delta Q_i^{\text{odd}} \sum_j s_j \rangle^*,$$

and

$$\langle \Delta Q_i^{\text{even}} \rangle^* = 0.$$

In these equations, ΔQ's mean the difference between the spin products at the boundary and their counterparts inside the cluster. The notation $\langle \ldots \rangle^*$ represents the expectation value with respect to the effective Hamiltonian without odd effective fields.

The multi-effective-field theory was tested with the square-lattice Ising model. Minami and Suzuki[28] used 3×3 and 4×4 clusters and applied the coherent-anomaly analysis to the resulting sets of approximations. They formulated the rules for constructing a sequence of approximations suitable for estimating of the critical exponents. The rule states that the canonical series can be constructed when (1) the effective fields which decrease the symmetry, namely the odd fields, are introduced such that they do not break the symmetry if combined with the expected spontaneous symmetry breaking, and (2) the symmetry conserving (even) fields are introduced according to their dominance.

The resulting approximations exhibit extremely good coherent-anomaly scaling. In two dimensions, the multieffective-field-theory seems to provide the most accurate estimates of the critical exponents among other methods based on the coherent-anomaly approach.

III.3 Variational Series Expansion Method

The Variational Series Expansion Method was designed by Kolesik and Šamaj[29−31] for investigation of phase diagrams and critical properties of classical spin systems. It is based on combination of two approaches widely and successfully used in statistical mechanics, namely on the series expansions and on the variational approach.

This method makes use of the fact that a great majority of spin systems can be exactly mapped onto so-called vertex models. Unlike Ising-like models, where the spin variables live in nodes of a lattice with interactions assigned to the bonds, in the vertex models the spin variables are linked to the bonds and the interactions called vertex weights are defined in the vertices of the lattice. Thus, typically the first step in the Variational Series-Expansion Method consists in rewriting the given model in terms of a suitable vertex model (see *e.g.* Ref. 32 for a general method for such a mapping). Then, a series expansion as a function of all vertex weights is generated for the resulting vertex model. This is purely a formal graphical expansion which is calculated for an arbitrary suitable ground-state without considering the convergence properties. This is the most difficult part of the whole calculation, because usually the series consists of thousands of terms. Fortunately, there are effective methods[33] which allow for calculations of a reasonable long series of this type. Thus, having generated the series, one has a formal expression for the free energy as a function of all possible vertex weights:

$$\mathcal{F}_L = -\ln(w_0) - \sum_{n=2}^{L} f_n\left(w_1/w_0, w_2/w_0, \dots\right), \tag{16}$$

where w_0 stands for the vertex weight which corresponds to the formal ground state, and $w_1, w_2, \dots$ denote other weights. The functions f_n are homogeneous polynomials representing the contributions of graphs with n vertices. The parameter L is the maximal size of graphs included and represents the degree of approximation.

The main ingredient of the method is the fact that the vertex models are gauge-invariant: there is always a family of transformations which leaves all physical quantities unchanged while it transforms the vertex weights of the model. These gauge transformations are parameterized by the so-called gauge parameters denoted by y below. One can express the vertex weights w_i as functions of the parameters of the original model and of the gauge parameter(s): $w_i = w_i(\text{temperature, external field,} \dots, y)$. In this way, $\mathcal{F}_L$ turns into a function of the temperature, external field and other quantities characterizing the given model *and* it also depends on the gauge parameter y for any finite approximation degree L. Of course, this dependence on the gauge must disappear in the limit $\lim_{L \to \infty} \mathcal{F}_L$, provided the limit exists. By trying to recover the gauge symmetry for finite L, the following minimal sensitivity condition is imposed:

$$\partial_y \mathcal{F}_L(\text{temperature, external field}, \ldots, y) = 0. \tag{17}$$

This is a self-consistency equation which determines the gauge parameter as a function of the original model parameters. Its justification is the following: for a fixed y, $\mathcal{F}_L$ represents a good approximation usually only in a small region of the model-parameter space, say in the vicinity of the temperature and external field $[t(y), h(y)]$. For a different value y, one obtains an approximation suitable for a different region. Thus, we have a family of *local* approximations with y parameterizing this family. It can be shown easily that Eq. (17) produces an *envelope* to these local solutions, resulting in a *global* approximate solution. Naturally, sometimes there are more than one solutions to the minimal sensitivity condition. In such a case one has to choose the one which provides the minimal free energy. In other words, the free gauge parameters are used as *variational* ones to construct an approximation for each length of the series expansion. In this way, a systematic series of mean-field-type approximate solutions for the free energy and also for order parameter(s) is obtained.

This method was applied to a variety of systems, such as the antiferromagnetic $2D$ Ising model,[29] nonuniversal vertex models,[30,31] Blume-Emery-Griffiths model,[34] and has proven its versatility and a very good accuracy of estimated critical exponents. Recently, it was shown in Ref. 11 that the accuracy of the critical exponents obtained for the $3D$ Ising model, namely

$$\alpha = 0.108(5), \quad \beta = 0.327(4), \quad \gamma = 1.237(4), \quad \delta = 4.77(5), \tag{18}$$

is comparable to the one of the most precise conventional methods such as Monte Carlo simulations and ordinary series expansion analyses. For the $3D$ antiferromagnetic three-state Potts model we recently obtained the critical exponents[35]

$$\alpha = -0.011, \quad \beta = 0.351, \quad \gamma = 1.309, \quad \delta = 4.73, \tag{19}$$

which indicate the XY universality class and strongly disagree with the proposed new universality class for this model.

III.4 Variational Finite-Lattice Method

A new method for constructing a systematic series of mean-field approximations was developed recently.[11] In its spirit, it is similar to the above Variational Series-Expansion Method;[29-37] one prepares a set of approximations for a vertex model equivalent to the original problem, and identifies its gauge parameter(s) as the variational parameter(s). The only, but from the practical point of view crucial difference, is the way how the approximate solution for the the corresponding vertex model is calculated. As the name suggests, this method is based on the finite lattice approach for generating series expansions. However, in the present approach the finite-lattice technique is used to calculate the approximation of the free energy numerically as a combination of free energies of finite lattices. Similarly, all the derivatives of the free energy necessary for derivation of the critical mean-field coefficients are calculated numerically.

In order to be more specific, let us show how to implement the method in the simplest case, namely for the square lattice Ising model. It is not possible to give a detailed justification of the method here, just because of lack of space, but we would like to present at least basic formulas to give the reader the main ideas of the method.

Let us denote the matrix representing the pair interactions between spins as $V_{s_i s_j} = \exp(K s_i s_j + H(s_i + s_j)/4)$ where its indices run over the two spin states. Next, define a new matrix $W(y) = A(y)V^{1/2}$ with A being the orthogonal matrix

$$A(y) = \frac{1}{\sqrt{1+y^2}} \begin{pmatrix} 1 & y \\ -y & 1 \end{pmatrix}. \tag{20}$$

With the matrix W, calculate the partition function Z_{ij} of a rectangle of the size $i \times j$ with the boundary conditions defined through the matrix W: to each external edge of the finite lattice which is connected to a spin in the state s we assign the weight factor $W_{1,s}$. All the finite-lattice calculations are performed at the fixed value $y = 1$ and the partitions functions together with their derivatives with respect the external field are calculated up to the second order in powers of $y - 1$. Having calculated all the necessary partition functions, one can obtain the infinite lattice approximation for the dimensionless free energy by combining the results for finite lattices:[37]

$$-F^{(s)} = \sum_{[i,j] \in B(s)} i\left(\delta_{i,s-j} - 3\delta_{i,s-j-1} + 3\delta_{i,s-j-2} - \delta_{i,s-j-3}\right)\ln(Z_{ij}), \tag{21}$$

with the summation restricted to the rectangles with spans less or equal to the maximal span s (which determines the degree of approximation)

$$B(s) = \{[i,j] : i + j \le s\}. \tag{22}$$

Finally, the mean-field critical coefficient of the susceptibility is evaluated from

$$\overline{\chi}_s = \frac{-(\partial_{yH} F_{(s)})^2}{\partial_{yyK} F_{(s)}}, \tag{23}$$

with the derivatives taken at the critical point

$$\{K = K_c^{(s)}, H = 0, y = 1\}, \quad \partial_{yy} F_{(s)}(K = K_c^{(s)}, H = 0, y = 1) = 0, \tag{24}$$

where the second equation serves for finding the approximate critical point $K_c(s)$ in the approximation characterized by its maximal span s.

Finally, having the mean-field critical coefficients for as many approximations as needed or possible, one can apply the Coherent Anomaly Method to estimate the critical exponent. Naturally, similar formulas can also be given for other critical coefficients as well as for three dimensions. The method can be applied straightforwardly to other models such as, e.g., Potts models or a wide family of vertex models.

This method turns out to be very accurate in two dimensions[36] and is presently tested in three dimensions. For the square Ising model it is possible to obtain the estimates as good as $\gamma = 1.750016$ from the high-order approximations. It is possible to generate up to forty different approximations for two-state models in two dimensions and this series is really systematic, meaning that all the quantities approach their exact values smoothly, without "random" deviations. For example, from just two approximations with spans 29 and 30, one can obtain an extrapolated critical temperature of the square Ising model with a relative error of 2×10^{-5}. The range of applicability of this approach is nearly as wide as with the Variational Series-Expansion Method, while it is considerably simpler to implement, because it avoids the tedious generation of the multivariable expansions. We hope to give the details of this promising method in the near future.

IV. Summary

We have reviewed a few methods for generating mean-field approximations for lattice models studied in statistical mechanics. There has been continuing progress in developing new ways to approach critical phenomena, and the mean-field-type methods combined with the coherent-anomaly approach become competitive with the most powerful conventional methods in this field. We have concentrated on two main streams in attacking the critical region from the mean-field. The first method conceptually similar to Kikuchi's cluster variation method use relatively small clusters, but at the same time give a rather rich palette of mean-fields encompassing also the short-range correlations. The second type of methods is based on rather long graphical expansions including relatively large clusters, but on the other hand it uses only simple "effective fields". Both approaches have already demonstrated their strength and it is a tempting idea to try to find a kind of compromise between these two extreme approaches. It could hopefully lead to approximations searching into the critical region while preserving the better coherent anomalies of the series expansion approach. For a further substantial improvement, it would be also extremely useful to elucidate how the concrete method of building the series of mean-field approximation influences the resulting coherent anomaly, namely what type of corrections to the coherent-anomaly scaling are to be expected. With a proper grasp of these corrections, a significantly improved accuracy of the CAM-based estimates of critical exponents would be on the way.

References

1. M. Suzuki, *J. Phys. Soc. Jpn.* **55**, 4205 (1986).
2. M. Suzuki, M. Katori, and X. Hu, *J. Phys. Soc. Jpn.* **56**, 3092 (1987).
3. M. Katori and M. Suzuki, *J. Phys. Soc. Jpn.* **56**, 3113 (1987).
4. X. Hu, M. Katori,and M. Suzuki, *J. Phys. Soc. Jpn.* **56**, 3865 (1987).
5. X. Hu and M. Suzuki, *J. Phys. Soc. Jpn.* **57**, 791 (1988).
6. M. Suzuki, *Prog. Theor. Phys.* Suppl. **87**, 1 (1986).
7. M. Suzuki, *Phys. Lett. A* **116**, 375 (1986).
8. M. Suzuki, *J. Stat. Phys.* **49**, 977 (1987).
9. M. Suzuki, K. Minami, and Y. Nonomura, *Physica A* **205**, 80 (1994).
10. M. Suzuki, X. Hu, M. Katori, A. Lipowski, N. Hatano, K. Minami, and Y. Nonomura, in *Coherent Anomaly Method-Mean-Field, Fluctuations and Systematics*, (World Scientific, Singapore 1995).
11. M. Kolesik and M. Suzuki, *Physica A* **215**, 138 (1995).
12. R. Kikuchi, *Phys. Rev.* **81**, 988 (1951).
13. A. G. Schlijper, *Phys. Rev. B* **27**, 6841 (1983).
14. A. G. Schlijper, *J. Stat. Phys.* **40**, 1 (1985).
15. P. Cenedese, J. M. Sanchez, and R. Kikuchi, *Physica A* **209**, 257 (1994).
16. M. Katori and M. Suzuki, *J. Phys. Soc. Jpn.* **57**, 3753 (1988).
17. S. Fujiki, M. Katori, and M. Suzuki, *J. Phys. Soc. Jpn.* **59**, 2681 (1990).
18. S. Fujiki, *J. Phys. Soc. Jpn.* **62**, 556 (1993).
19. M. Katori and M. Suzuki, *Prog. Ther. Phys.* Suppl. **115**, 83 (1994).
20. T. Morita, *Physica A* **155**, 73 (1989).
21. K. Tanaka, T. Horiguchi, and T. Morita, *J. Phys. Soc. Jpn.* **60**, 2576 (1991).
22. K. Tanaka, T. Horiguchi, and T. Morita, *Phys. Lett. A* **165**, 266 (1992).

23. K. Wada and N. Watanabe, *J. Phys. Soc. Jpn.* **58**, 4358 (1989).
23. K. Wada and N. Watanabe, *J. Phys. Soc. Jpn.* **59**, 2610 (1990).
24. K. Wada and N. Watanabe, *J. Phys. Soc. Jpn.* **60**, 3289 (1991).
25. M. Suzuki, in *New Trends in Magnetism*, edited by M. D. Coutinho-Filoho and S. M. Rezende (World Scientific, Singapore, (1990). p. 304.
26. K. Minami, Y. Nonomura, M. Katori, and M. Suzuki, Physica A **174**, 479 (1991).
27. K. Minami and M. Suzuki *Physica A* **187**, 282 (1992).
28. M. Kolesik and L. Šamaj *J. Phys. I (France)* **3**, 93 (1993).
29. M. Kolesik and L. Šamaj *J. Stat. Phys.* **72**, 1203 (1993).
30. M. Kolesik and L. Šamaj, *Phys. Lett. A* **177**, 87 (1993).
31. M. Kolesik, *Int. J. Mod. Phys. B* **6**, 1529 (1992).
32. M. Kolesik, *Physica A* **202**, 529 (1994).
33. M. Kolesik, *Mod. Phys. Lett. B* **8**, 113 (1994).
34. M. Kolesik and M. Suzuki, *Physica A* **216**, 469 (1995).
35. M. Kolesik and M. Suzuki, submitted to *J. Phys. A*
36. I. G. Enting, *J. Phys. A* **11**, 563 (1978).
37. A. J. Guttmann and I. G. Enting, *J. Phys.* **A26**, 807 (1993).

Cluster Variation and Cluster Statics

Didier de Fontaine

*Materials Science Department
and Lawrence Berkeley Laboratory
University of California, Berkeley
U.S.A.*

Abstract

Ever since its early formulation by Ryoichi Kikuchi in 1951, the Cluster Variation Method (CVM) has become a versatile tool for Ising Model applications, particularly in complex situations such as those required by phase diagram calculations. Now that first-principles approaches are becoming feasible, the CVM has become virtually indispensable: the deficiencies of the Bragg-Williams can no longer be masked by fudging adjustable interaction parameters. Thanks to delightful tutorials given to us by Ryo Kikuchi at UCLA in the 70's, Juan Miguel Sanchez was soon able to develop a remarkable "Cluster Algebra" which has proved to be the formalism of choice for *ab initio* alloy thermodynamics. I shall review this formalism and its possible applications to alloy phase stability.

I. Introduction

The Cluster Variation Method (CVM) was proposed by Ryoichi Kikuchi in 1951[1,2] primarily as a method for improving the calculation of the Ising model critical temperature T_c. The method rested on the use of a hierarchy of clusters of lattice points (points, pairs, triplets, quadruplets, ...) whose decorations by atoms A, B, C, ... (or spins $+1$, -1, ...), could be described by corresponding "cluster probabilities" having well-defined values at equilibrium. It was found that these cluster probabilities $x_\alpha(\sigma)$, where α is the cluster type, σ is its atomic (spin) configuration, or decoration entered into the configurational entropy functional $[S]$ through terms of the type $x \ln x$ times positive and negative integers which later came to be known as "Kikuchi-Barker coefficients." The ordering, or configurational energy functional (Hamiltonian) $[E]$ was usually taken as a linear combination of cluster probabilities. The free energy F was obtained by minimizing the expression $[E] - T[S]$ with respect to the unknown cluster probabilities $x_\alpha(\sigma)$. Hence the CVM, as its initials imply, is a variational technique which can give an upper bound of the free energy and can come very close indeed to

Theory and Applications of the Cluster Variation and Path Probability Methods
Edited by J.L. Morán-López and J.M. Sanchez, Plenum Press, New York, 1996

the exact one if the cluster approximation is well-chosen. Hence the originality of the CVM manifested itself in the improved expression for the configurational entropy, the essential many-body (cluster) character of which was thus duly emphasized.

As a procedure for the particular study of critical phenomena, the CVM was soon overshadowed by Renormalization Group methods and went into something of a decline. Ryo Kikuchi then turned his attention to the kinetic counterpart of the CVM, the PPM or Path Probability Method,[3] a very successful technique discussed extensively at this symposium, but which not be further mentioned in the present communication.

II. Early CVM phase diagram calculations

What was apparently missing in the early days of the CVM was a recognition of the wide applicability of the CVM to practical problems in phase stability, beyond the strict domain of critical phenomena. The needed breakthrough came with the presentation by van Baal[4] of a tetrahedron-approximation CVM calculation of binary alloy ordering on the *fcc* lattice. This calculation was soon followed by a "fitted" CVM calculation of the solid-state portion of the Cu-Au phase diagram,[5] a calculation emphasizing the necessity of a new free energy functional, as provided by the CVM, since single-site mean-field approximations, such as the Bragg-Williams (or regular solution) model, failed to produce phase equilibria which were even qualitatively correct. The often-quoted result that single-site ("point") mean field become exact in the limit of infinitely long-range pair interactions may be correct for the ferromagnetic Ising model, but fails completely for "ordering" in the presence of frustration, *i.e.*, for antiferromagnetism on the *fcc* lattice.

The CVM treatment of *fcc* ordering was an enormous improvement over the BW calculations of Shockley,[6] but it had one serious limitation: it was not possible in this approximation to introduce second-neighbor (nnn, next-nearest neighbor) pair interactions. The introduction of at least nnn pairs was essential to capture some of the diversity of ordering encountered on *fcc* and *bcc* lattices, for example. Indeed, it had been proved by Kanamori and co-workers[7] and by Cahn and co-workers[8] that many of the important ordered superstructures observed experimentally on those lattices could be stabilized formally by the use of nn (nearest-neighbor) and nnn pair interactions. A more elaborate, though still tractable CVM approximation was therefore needed. It was Juan Miguel Sanchez who came up with an acceptable cluster combination: the tetrahedron-octahedron (TO) approximation.[9] With larger clusters come much larger cluster configurations, soon giving rise to what some have called the "combinatorial explosion." Such is still one of the major practical problems of the CVM: if longer-range interactions are needed in the Hamiltonian for physical reasons, the configurational entropy must include some compact clusters which must contain the required interaction ranges, and hence results in unmanageable numbers of configurations on the largest clusters. The TO combination is still manageable, but minimizing the free energy functional with respect to a large number of cluster probabilities which are not all mutually independent becomes rather awkward. It was then found to be more advantageous to express the free energy functional in terms of linearly independent variables, the so-called correlation functions ξ_α, which are linearly related to the cluster concentrations (or probabilities).[9] A simple rule may be enunciated: for a given (undecorated) cluster of equivalent lattice points, the number of independent correlation functions is equal to the number of distinct sub-clusters which may be extracted from the given cluster, a number which can be much less than the number of distinct configurations on said cluster. Thus was born a veritable "cluster algebra," later given an extremely elegant formulation, valid for binary and multicomponent systems.[10] A brief exposition of this approach will be given in the next section. Most modern expositions of the CVM formalism are based on this approach, which owes much to original work of Morita.[11]

In the late 70's—early 80's, first at UCLA, then at Berkeley, we calculated a series of prototype binary alloy phase diagrams[12-14] which differed from one another only by the ratio $\rho = V_1/V_2$ of nnn-to-nn interaction parameter values. Phase diagrams were calculated, based on the *fcc* ordered ground states discovered by Kanamori and Cahn,[7,8] for selected positive and negative values of ρ. In retrospect, it seems surprising that, in 1980, nobody had yet solved, even approximately, such an apparently simple problem as the phase equilibria of the *fcc* anti-ferromagnetic Ising model. We called these *prototype* phase diagram calculations because the energy parameters were chosen arbitrarily, in this case the nnn/nn interaction ratio ρ, to illustrate the types of phase relations one would expect in simple cases, and to understand how some phase diagram topologies gradually evolve into others as some physical parameter is altered. The *bcc* case was investigated by others, but re-examined in more detail by Sluiter *et al.*[15] The *hcp* case, which is quite distinct from the *fcc* case when nnn pair interactions are included, was studied by McCormack *et al.*,[16] and in references quoted therein.

Another form of phase diagram may be called fitted. In this case, the empirical interaction energies are varied systematically until a good fit is achieved with some well-known features of an experimentally determined phase diagram. The objectives of such an undertaking are, for example, the extraction of thermodynamic properties, such as free energies, entropies, from known phase equilibrium data, the extrapolation of known phase relations into regions of temperature-concentration space where no data is available, or the approximate determination of metastable phase equilibria, often of great practical importance for materials design. Much work along these lines has been carried out by the CALPHAD group, but their exclusive use of the "point" mean field approximation constitutes a severe limitation in "ordering" cases. A very good example of an early CVM "fitted" phase diagram calculation is that of Sigli and Sanchez[17] for the Al-Li system. Some years later and also more recently, the same system was analyzed by means of an *ab initio* phase diagram calculation.[18] The meaning of the term "*ab initio*," strictly speaking, is the following: it refers to a calculation which requires as input only the atomic numbers of the atomic constituents.

While these three types of phase diagram calculations, *prototype*, *fitted*, and *ab initio*, were being carried out, much progress was realized in other aspects of the CVM formalism and applications. Let us cite a few with some of the key references: the implementation of larger-cluster approximations, a k-space formulation of the CVM with applications to short-range order SRO fluctuations,[19] group theory-based computer codes required to determine the relationships between cluster probabilities and correlation functions (as contained in the configuration or so called "V" matrix) and also the Kikuchi-Barker coefficients, valid in principle for arbitrary structural symmetry and cluster choices,[20] the derivation from the V-matrix of inequalities required for ground state analysis,[21-23] the extension of alloy phase equilibria calculations to cases involving magnetism. A good example of the latter is provided by the recent work of Sanchez *et al.*[24] cited at this conference by Dr. Cadeville. In this work, several of the techniques just cited are illustrated: use of large clusters, magnetic interactions, k-space formalism, direct "retrofitting" in k-space to experimental SRO intensity to determine effective pair interactions.

It is apparent that what began rather modestly in the mid-70's as an application of the CVM to alloy phase equilibria has now been developed extensively and is even acquiring a new dimension thanks to the integration of cluster methods with electronic structure calculations, a topic to be succinctly reviewed in the remainder of this article. A very complete treatment of the topics mentioned thus far can be found in the excellent monograph of Ducastelle, entitled *Order and Phase Stability in Alloys*[22] and in the Doctoral Dissertation of Finel.[27] There are also two review articles by the present author in *Solid State Physics*;[25,26] the first one reviews the field before the CVM made its big impact, the second one emphasizes the cluster expansion technique which will now be described.

III. Cluster expansion

The concept of expansion of a function of local configuration in a complete set of cluster functions was suggested in early work of Kawasaki[28] and of Sanchez,[9] but it was only with the publication by Sanchez, Ducastelle, and Gratias (SDG) in 1984 (Ref. 10) that the formalism was given definite shape in the context of the Cluster Variation Method. In this paper, the problem of multicomponent systems was treated quite generally. Here, I shall describe only the simpler case of binary systems. The multicomponent case is reviewed in the articles of Inden and Pitsch[23] and of the present author,[26] for example.

If it were possible to find a complete set of functions $\Phi(\sigma)$ in the space of configurations of some binary alloy, then any function of configuration $f(\sigma)$ could be expanded in such a set as follows

$$f(\sigma) = \sum_{\alpha} F_{\alpha}\Phi(\alpha), \tag{1}$$

where F_{α} are appropriate expansion coefficients (or generalized Fourier coefficients), the index α signifying a given cluster of lattice sites. The summation in Eq. (1) extends over all possible clusters in the system of N lattice points. If the basis functions Φ are also orthogonal, then the coefficients can be calculated by inversion in the usual way

$$F_{\alpha} = \rho_0 \sum_{\sigma} f(\sigma)\Phi_{\alpha}(\sigma) \equiv \langle \Phi_{\alpha}, f \rangle, \tag{2}$$

in which the angle brackets indicate a normalized sum over all configurations, *i.e.* an inner product over the space of all configurations $\{\sigma\}$, the normalization factor ρ_0 being equal to the inverses of the total number 2^N of configurations, in the present case.

As shown by Asta *et al.*[29] it is also possible to define other basis sets of functions such that the summation in Eq. (2) extends only to configurations having some arbitrary average concentration (of B-type atoms, say) $c = (1 - \overline{\sigma})/2$, where $\overline{\sigma}$ is the corresponding average of the "spin variable" $\sigma = \pm 1$. In this case, the inner product (2) consists of a *restricted* summation (*R-Sum*) over configurations of fixed c; in the general case, the summation is *unrestricted* (*U-Sum*). Later, Sanchez[31] (see also the contribution by this author in the present volume) showed that both approaches could be combined into one by defining the inner product of two functions $f(\sigma)$ and $g(\sigma)$ over the space of all configurations in the following way

$$\langle f, g \rangle_\mu = \frac{1}{[2\cosh(\mu)]^N} \sum_\sigma e^{N\mu x} f(\sigma) g(\sigma), \tag{3}$$

thus generalizing Eq. (2). In Eq. (3), μ is an arbitrary constant and x is the average value of σ over all N lattice sites. We shall denote the summation involved in this inner product as the μ-*Sum*, for short. It is of course an unrestricted sum, but with weighting factor involving the parameter μ. In the limit of very large N, the summation in Eq. (3) is practically limited to those configurations for which $x = \tanh(\mu)$, as they overwhelm all others. Hence, in the limit of $N \to \infty$, the μ-Sum is equivalent to the R-Sum defined above with $c = (1 - x)/2$; when $\mu = 0$, the μ-Sum is equivalent to the original U-Sum for N finite but as large as one wishes. Also, for very large N, the U-Sum is equivalent to the R-Sum at $c = 0.5$ (or $\bar\sigma = 0$).

References 10, 25, 29–31 describe the construction of orthonormal basis sets appropriate for the U-Sum, R-Sum and μ-Sum sets. One begins by considering only a single lattice point p at which one defines two linearly independent functions of configuration: 1 and σ_p (for binary systems). Then a *local* orthonormal set may be constructed for example by Gram-Schmidt orthogonalization, yielding the set $[1, \varphi(\sigma_p)]$. For the whole supercell, the required complete orthonormal set (CONS) is obtained by taking the direct product of the local bases at all N points

$$\begin{pmatrix} 1 \\ \varphi(\sigma_1) \end{pmatrix} \otimes \begin{pmatrix} 1 \\ \varphi(\sigma_2) \end{pmatrix} \otimes \cdots \otimes \begin{pmatrix} 1 \\ \varphi(\sigma_N) \end{pmatrix}. \tag{4}$$

There results a set of 2^N *cluster functions* $\Phi_\alpha(\sigma)$ in which the index α represents a *cluster* of lattice sites, those at which the local basis function is chosen to be $\varphi(\sigma)$. All other points, at which the local function is unity, represent the complement $\bar\alpha$ (in the whole supercell) of the cluster α. There are thus as many cluster functions, including the "empty" cluster [from taking all 1's in Expression (4)] and the "full" cluster [from taking all φ's in (4)], as there are configurations, *i.e.* 2^N in the binary case.

In the U-Sum case, the cluster functions are

$$\Phi_\alpha(\sigma) = \prod_{p \in \alpha} \sigma_p, \tag{5}$$

since the set $[1, \sigma_p]$ is already locally orthonormal. In the R-Sum case, the cluster functions are products of functions of the type

$$\varphi(\sigma_p) = \frac{(\sigma_p - \bar\sigma)}{\sqrt{1 - \bar\sigma^2}}, \tag{6}$$

where $\bar\sigma$ may be taken as the average value of σ over the N lattice sites, for example.[29,31] In multicomponent systems, the local orthonormal basis functions are discrete Chebychev polynomials.[10]

With these (or other) choices, the constructed sets have the required *orthonormality* property

$$\rho_0 \sum_\sigma \Phi_\alpha(\sigma)\Phi_\beta(\sigma) \equiv \langle \Phi_\alpha, \Phi_\beta \rangle = \delta_{\alpha\beta}, \tag{7}$$

where $\delta_{\alpha\beta}$ is the Kronecker delta. For finite N, in the U-Sum (more generally, the μ-Sum) case, the equality of the number of clusters and of configurations yields the *completeness* relation

$$\sum_{\alpha} \Phi_\alpha(\sigma)\Phi_\alpha(\sigma') = \delta(\sigma,\sigma'), \tag{8}$$

where $\delta(\sigma,\sigma')$ is the discrete space delta function, equal to ρ_0^{-1} when the configurations σ and σ' are identical, zero otherwise. Note that, for large N, the δ-function diverges rapidly. In the R-Sum case, completeness only holds in the limit $N \to \infty$.

IV. Configurational energy

Cluster expansion methods are most frequently applied to the configurational energy $E(\sigma)$, hence the following discussion will be limited to it, though the formalism is quite general and is valid for any function of configuration. Also, for simplicity we shall present the derivations in the U-Sum framework, which is the most generally used. In the energy case, the cluster expansion coefficients, usually called *Effective Cluster Interactions* (ECI) and denoted by V_α, can be obtained formally from Eq. (2) by breaking up the sum over configurations into two parts: one pertaining to the configuration σ_α of the cluster α itself, and one pertaining to those of its complement $\overline{\alpha}$

$$V_\alpha = 2^{-|\alpha|} \sum_{\sigma_\alpha} \prod_{p \in \alpha} \sigma_p \left[2^{-|\tilde{\alpha}|} \sum_{\sigma_{\tilde{\alpha}}} E(\sigma_\alpha; \sigma_{\tilde{\alpha}}) \right]. \tag{9}$$

In this equation and in following ones, the symbol $|\alpha|$ denotes the number of points in the cluster indicated. The expression enclosed in square brackets (trace over the complement of the cluster) represents the *average of all supercell configurations which contain the specified cluster configuration* σ_α *on cluster* α. This definition is an important one; it follows directly from the defining equation of the inner product, Eq. (2), and is thus completely rigorous. To fully grasp the physical meaning of the ECI definition it is helpful to consider the example of a cluster consisting of a pair of lattice points, p and q, say. Then, by Eq. (9), the corresponding *Effective Pair Interaction* (EPI) is given by

$$V_{pq} = \tfrac{1}{4}\left(\overline{E}_{AA} + \overline{E}_{BB} - \overline{E}_{AB} - \overline{E}_{BA} \right), \tag{10}$$

where $\overline{E}_{IJ}$ is the average energy of all configurations in the system which contain an I-type atom at p and a J-type atom at q.

The present situation is very different from what it was in the past when there were no formal methods of determining interaction parameters "from first principles," *i.e.*, from electronic structure calculations. Previously, *ad hoc* pair potential models were constructed, and interaction parameters were then *fitted* to available thermodynamic data, for example. From the theoretical viewpoint, it appeared curious that alloy energies could be described accurately by just a few near-neighbor "pair potentials." But precisely, the *effective* interactions defined by Eq. (9) or (10) are *not* pair potentials: they are total energy differences, so that even nearest-neighbor (nn) EPI's contain all possible electronic interactions in the solid, so long as the computation of the average energies $\overline{E}_{IJ}$ is performed adequately. It is partly because Eqs. (9) and (10) involve energy *differences* that the magnitude of cluster interactions usually decreases quite rapidly with cluster size (pair spacing or number of points in the cluster).

Cluster expansion methods have significantly clarified the physical meaning of interaction parameters, for example those required by thermodynamic calculations of phase equilibria. From the inner product definition of Eq. (2) applied to energies, it is clear that the ECIs are configuration-independent. However, in the R-sum approach, the ECIs can effectively be made concentration-dependent by relating the parameter $\bar{\sigma}$ of Eq. (6) to the average concentration c, as mentioned above. In that case, we denote these interactions explicitly by the symbol $\tilde{V}_\alpha(c)$, assuming for convenience that they contain the normalization factors previously attached to the cluster functions themselves. The ECIs V_α and $\tilde{V}_\alpha(c)$ will therefore have different values. Likewise, in the μ-Sum case, choices of the arbitrary parameter μ will produce infinitely many basis sets of cluster functions along with their related ECIs. The fact that physical properties can be represented by different sets of cluster expansion parameters has caused consternation in some circles. There is no cause for alarm, however: the only thing that matters is that a given configuration-dependent property be represented by cluster coefficients which correspond to the basis cluster functions chosen, just as an arbitrary function can be represented by Fourier series, Legendre polynomials, whatever, with suitable expansion coefficients.

As a result, ECIs defined in different bases must be related to one another.[29-31] Explicitly, the relationship between concentration independent (V_α, U-Sum) and concentration dependent ($\tilde{V}_\alpha(c)$, R-Sum) interactions is,[26,30] for pair clusters and beyond,

$$\tilde{V}_\alpha(c) = \sum_{\beta \supset \alpha} (\xi_1)^{|\beta|-|\alpha|} V_\beta , \qquad (11a)$$

and

$$V_\alpha = \sum_{\beta \supset \alpha} (-\xi_1)^{|\beta|-|\alpha|} \tilde{V}_\beta(c) , \qquad (11b)$$

where ξ_1 is the "point" correlation, equal to $\bar{\sigma}$, and the summations are over the superclusters β of α, i.e. those which contain the smaller cluster. Remarkably, the supercluster expansions of Eq. (13) were found to converge rapidly for certain prototype binary systems.[29] More recent calculations carried out by Ouannasser on the Ti-Pt system[32] confirmed the earlier results, as seen in Fig. 1a and Fig. 1b. Figure 1a illustrates Eq. (11a): the full heavy line represents the calculated concentration dependence of the nn pair interaction and the thin horizontal line represents the corresponding concentration-independent EPI. V_2 and $\tilde{V}_2(c)$ lines cross at concentration $c = 0.5$ since, for large N (here less than 1000 points), the U-Sum and R-Sum values must coincide at that central composition. When triplet interactions are included in the supercluster expansion of Eq. (11a), one gets the dashed line, and when quadruplets also are included, one gets almost perfect convergence: the resulting dotted line practically coincides with the full line except at the dilute concentration limits where convergence is more sluggish. Excellent convergence is also observed for Eq. (11b): as more terms, triplets, quadruplets are included in the expansion, the concentration dependence of $\tilde{V}_2(c)$ progressively flattens out to approach the concentration-independent V_2 line, as seen in Fig. 1b. It is thus a matter of convenience which of the U-Sum, R-Sum or μ-Sum expansions one wishes to use for a given problem.

As a further illustration of internal consistency, let us demonstrate the fundamental property of complete orthonormal sets for cluster expansions (U-Sum, for simplicity). Consider the energy $E(\sigma_0)$ at some particular configuration σ_0. According to its cluster expansion, from Eq. (1), and the expression of the ECI, from Eq. (2), we have successively,

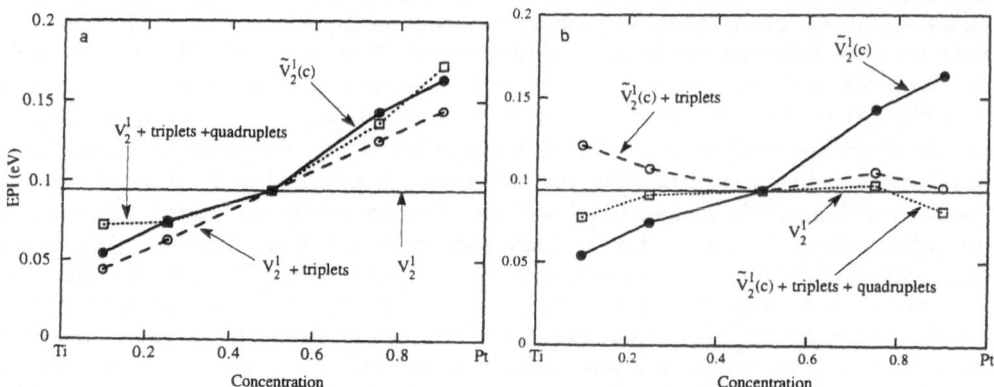

Figure 1. (a) and (b): Graphical representation of Eqs. (11a) and (11b), respectively for the case of nn pair interactions in *fcc* Pt-Ti. The full heavy line represents the calculated concentration dependence of the nn pair interaction and the thin horizontal line represents the corresponding concentration-independent EPI. Dashed and dotted curves show the progressive effects of including superclusters of the pair: triplets and quadruplets.[32]

$$E(\sigma_0) = \sum_\alpha V_\alpha \Phi_\alpha(\sigma_0)$$

$$= \sum_\alpha \left[\rho_0 \sum_\sigma \Phi_\alpha(\sigma) E(\sigma) \right] \Phi_\alpha(\sigma_0)$$

$$= \sum_\sigma \left[\rho_0 \sum_\sigma \Phi_\alpha(\sigma) \Phi_\alpha(\sigma_0) \right] E(\sigma) = E(\sigma_0), \qquad (12)$$

since, by the completeness relation of Eq. (8), the expression in square brackets in the third line is equal to one or zero depending upon whether the two configurations σ and σ_0 are or not identical. Given the complete orthonormal property of cluster functions, Eq. (12) is of course the expected result. It is included here because a similar, though erroneous, derivation was recently presented[33] in a misguided attempt to discredit the general property of equivalence of cluster expansions pertaining to different bases.

The expectation value of a function of configuration, such as the thermodynamic average of the configurational energy, can be expressed easily by cluster expansions:

$$\langle E \rangle = \sum_\alpha V_\alpha \xi_\alpha, \qquad (13)$$

with

$$\xi_\alpha = \langle \Phi_\alpha(\sigma) \rangle. \qquad (14)$$

In the U-Sum case, the parameters ξ_α just introduced are thermodynamic averages (indicated by angle brackets) of products of σ over the cluster sites, and are denoted *multisite correlations*. In the R-Sum case, the ξ_α are averages of products of "local" functions given by Eq. (6), and are thus multisite cumulants. Bilinear form (13) is a very compact expression which shows symbolically how quantum and statistical mechanical computations in some sense can be decoupled in first-principles calculations: the EPI's can be obtained from electronic structure computations at absolute zero of temperature and the correlations (or cumulants) can be obtained by minimization of a CVM free energy functional, for instance. Other thermodynamic properties besides the configurational energy can of course be obtained in a like manner, but not the configurational entropy which is not strictly the average of a configuration-dependent function. That contribution to the configurational free energy is given directly by the CVM, following Kikuchi's original derivation.[2] As shown for example by Ceder,[34] other "excitations," such as those due to vibrational and electronic effects may be incorporated in the cluster expansion formalism, thus leading to temperature-dependent (though configuration-independent) ECIs. Examples will be mentioned below.

V. Computation of ECIs

Successful prediction of thermodynamic properties of alloys depend critically on how well the values of effective cluster interactions can be calculated. Thanks to the cluster expansion formalism, we now have a means of determining these parameters in a very fundamental manner which does not require *ad hoc* fitting procedures. In principle, one must carry out the calculations according Eq. (2) in general, in the expanded form of Eq. (9), or in the more compact form of Eq. (10), valid for pair interactions. Note that the method is *exact*, though the various computational procedures used in practice will differ according to the approximations used.

The most straightforward approach in numerical computations is to start from the inner product definition, Eq. (9), and to perform the indicated sum over configurations in either U-Sum, R-Sum or μ-Sum fashion. In practice, that distinction is rather unimportant: for a number of atoms N equal to a few hundred, results U-Sum of computations are indistinguishable from those of R-Sum computations at $c = 0.5$ ($\bar{\sigma} = 0$) or μ-Sum computations at $\mu = 0$.

The limit $N \to \infty$ at which U- and R-Sums become identical must be understood as follows: with fixed-concentration summations, it can be shown[29,31] that Eqs. (7) and (8) are satisfied only to order $1/N$. The physical meaning of this restriction has eluded some investigators who claim that, physically, the evaluation of ECIs for an alloy of average configuration $\bar{\sigma}'$, say, cannot be carried out by summing over configurations in a "medium" of different average configuration $\bar{\sigma}''$, since certain cluster configurations of $\bar{\sigma}'$ will not be represented in those of $\bar{\sigma}''$. That may indeed be true for clusters of "size" close to that of the whole computational region itself, but in practice this leads to no difficulty: usually, ECI clusters rarely contain more than about $n = 10$ lattice points. For such modest sizes, all possible cluster configurations are easily included in embedding media of arbitrary average configuration provided that n be large with respect to N. That is the only practical limitation that needs to be taken into account to satisfy completeness relations.

This direct application of the inner product formula for cluster interactions has led to the method of *Direct Configurational Averaging* (DCA).[29,35] The problem with that rigorous approach is that it is highly computationally intensive: electronic structure calculations must be carried out for N of the order of a few hundred and about 50 random configurations (typically) for each cluster interaction. For that reason, the DCA has thus far been implemented only in the Tight Binding framework.[35] Hence, what is gained by performing a calculation which is theoretically unassailable is partially lost by performing the electronic structure calculations with a simplified, non-electronically self-consistent Hamiltonian. Still, the TB-DCA has proved to be a very useful tool, particularly in its computationally efficient recursion and orbital peeling formalisms: larger number of ECI calculations could be performed than by other methods, in both binary and ternary systems; in particular, the cluster interactions shown in Fig. 1 were obtained by DCA methods.[32] Further progress along those lines would require extension towards electronic self-consistency, to *total* rather than to *band structure* energy only, and to contributions due to atomic displacements.

At present, one cannot have it both ways: either one emphasizes good configurational statistics, via the DCA, for example, or one refines the energy computations by using advanced electronic structure techniques. In the latter alternative, one must necessarily simplify the sampling of configurations to whatever is deemed computationally feasible. What is usually done is to select a modest number (about 10 or 20) of small-unit-cell structures ($N < 16$, typically), compared to the DCA numbers of about 50 and a few hundred, respectively, as mentioned above. With such small samplings, one can afford to perform self-consistent, full-potential, total energy electronic structure calculations (or any subset of these) on fully relaxed ordered structures, if so desired. However, since the summation in Eq. (7) is now over a small set of selected configurations, orthogonality no longer holds.

Instead of using Eq. (2) or (9) to compute the ECIs, it is then necessary to proceed as follows: decide on a plausible number n of ECIs required to represent energies (or some other configuration-dependent function), select m ($\geq n$) small-unit-cell structures, and calculate the energies of those structures as accurately as deemed necessary. For those m structures, write Eq. (1) in the following "symmetry-adapted" form

$$E(\sigma_s) = \sum_{r=1}^{n} m_r V_r \bar{\bar{\Phi}}_r(\sigma_s), \tag{15}$$

where the energy of structure (configuration) s is normalized to a single lattice point (of the disordered state), the index r represents the *orbit* of the cluster of given type generated by the symmetry operations of the (disordered) lattice, m_r is the number of r-clusters per lattice point, and $\bar{\bar{\Phi}}_r(\sigma_s)$ are orbit-averaged cluster functions. The latter are known for each configuration s, so are the multiplicities m_r, so that Eqs. (15) may be regarded as a system or m linear equations in n unknowns, V_r. If one takes more structures than clusters, system (15) can be solved by least-squares, or singular-value decomposition, with weighting factors possibly introduced to provide faster convergence of the cluster expansion. In its original form, this method was proposed by Connolly and Williams;[36] it is now generally known as the Structure Inversion Method (SIM).

Because of its flexibility and of the possibility of using very accurate electronic structure total energy calculations, the SIM had generally provided the best numerical results to date. Some have criticized the method of being short on physics and long on "fitting." This is an incorrect view: the SIM is based squarely on the cluster expansion method, which is *exact*. Merely, the inversion method consists of inverting a linear system involving a small set of special ("ordered") configurations, without knowing *a priori* whether the expansion has or not converged. The convergence test must be performed after solution of system (15); one must then check whether the energies $E(\tau)$ of structures *not* used in the solution are adequately approximated by the truncated structure expansion. If that is not the case, the calculation must be repeated with a larger set of structures σ_s (or of better-selected ones) until cluster-expanded and directly computed energies agree to desired accuracy.[37]

It is instructive to redo the derivation of Eq. (12) but in the present case of non-orthogonal cluster functions. For an arbitrary structure τ, one finds, leaving out the multiplicity m_τ for simplicity (or incorporating it in the ECIs themselves):

$$E(\tau) = \sum_{s=1}^{m} C_s(\tau)E(\sigma_s), \qquad (16)$$

with

$$C_s(\tau) = \sum_{r=1}^{n} \bar{\bar{\Phi}}_r^{-1}(\sigma_s)\bar{\bar{\Phi}}_r(\tau), \qquad (17)$$

the latter summation *not* being equal to a delta function because of absence of orthogonality. In Eq. (17), $\bar{\bar{\Phi}}_r^{-1}(\sigma_s)$ may be regarded as a pseudo-inverse obtained from the singular value decomposition. Thus, the energy $E(\tau)$ of an arbitrary configuration may be approximated to required accuracy by a linear combination (16) of energies of standard structures, $E(\sigma_s)$. This is an important result: it means that it may not always be necessary to perform costly *ab initio* calculations for complicated structures; a cluster expansion performed with simpler ones may do just as well. Knowing that electronic structure calculations scale approximately as N^3, where N is the number of atoms in the unit cell, it becomes far more advantageous to carry out a number m of calculations (computational time scaling linearly with m) on small-N structures rather than one calculation on a large-N structure.

There exists a third method, historically the first one proposed. Instead of sampling a number of atomic configurations, this method neglects all local fluctuations in order to create an artificial mean-field medium which in fact restores periodicity and hence allows electronic structure calculations to proceed as in a monatomic crystal. The most sophisticated of these averaging methods is the so-called coherent potential approximation (CPA), which, as the name implies, generates an electronically self-consistent potential at all lattice sites. Initially, the CPA appeared to be *the* (electronic) "alloy theory," since, in much or the Physics literature of the time, the word "alloy" was practically synonymous with "complete (configurational) disorder."[38] Of course, "alloy theory" is really concerned not so much with randomness as with *partial order*. It was therefore necessary to perturb the CPA medium, *i.e.* to introduce compositional inhomogeneities. Several more or less equivalent methods were suggested for doing this: the Generalized Perturbation Method of Gautier and Ducastelle (GPM),[39] the concentration wave method of Györffy and Stocks ($S^{(2)}$),[40] and the Embedded Cluster Method (ECM) of Gonis *et al.*[41] Advantages and disadvantages of these methods will not be discussed here. For a detailed treatment of the GPM, for example, see the book by Ducastelle.[22]

VI. Other energy contributions

Electronic structure contributions are surely not the only ones to consider, even at absolute zero of temperature: there are magnetic effects, electronic and vibrational excitations, and in alloys, static displacements of atoms away from their ideal lattice positions. The associated energies, actually free energies, are obviously functions of configuration and can thus be cluster-expanded as described above. Then, as shown by Ceder[34] and by Chiolero,[42] for example, it becomes possible to incorporate these contributions in generalized temperature-dependent ECIs, thereby accounting as well for vibrational and electronic entropy.

For alloys, it is preferable to work with energies ΔE (or entropies, or free energies, or any other extensive quantity) *of mixing*, *i.e.* the energy of a given configuration (σ) referred to that of the concentration-weighted average of the pure components $(E_A$ and $E_B)$

$$\Delta E(\sigma) = E(\sigma) - (c_A E_A + c_B E_B) , \tag{18}$$

where c_A and c_B are the atomic concentrations $1 - c$ and c respectively. The quantity ΔE can then serve as starting point for ECI computations, provided that this energy of mixing be defined in an appropriate manner. Indeed it matters greatly whether the elements are taken in their equilibrium crystalline ground states, or calculated on the same lattice as that of the configuration (σ) in question, if different. In the latter case, should one take as reference state those at which A and B have their optimal atomic volumes Ω_I $(I = A, B)$ on the lattice considered, or should one take the same lattice parameters for both pure elements and for the alloy? These questions are just those that classical thermodynamics also faces: the choice of standard states. Depending on that choice, the energy ΔE can take on widely different values.

Unlike classical thermodynamics, cluster methods explicitly consider atomic configurations described with respect to a reference lattice. Then equilibria, stable, metastable or unstable on one lattice (*fcc*, say) must be compared to those on all possible others (*bcc*, ...). This is done by determining *structural energy* differences,

$$\Delta E_I = E_I^{fcc} - E_I^{bcc}, \qquad I = A \text{ or } B , \tag{19}$$

which may be calculated quite accurately by appropriate electronic structure techniques. A given lattice, however, or lattice framework, may have to accommodate global or local distortions because different atoms have different "sizes" Ω_A and Ω_B. The associated problem of elastic relaxations is far from solved, but is discussed briefly in the author's review article[26] and elsewhere, in particular in publications of Zunger and collaborators.[37,43] The latter author considers (a) volume deformations, (b) cell-external deformations, and (c) cell-internal deformations. The first one (a) pertains to change in unit cell volume, the second one (b) to changes of cell shape at constant volume, such as the c/a ratio in tetragonal structures, or b/a in orthorhombic ones, and the third one (c) to local atomic displacements, δ, of atoms from the sites of a reference (fixed) lattice.

As a result, the energy, or other configuration-dependent function, may depend on many parameters, symbolically: $E(\sigma, \Omega, c/a, \ldots, \delta, \ldots)$. When performing a full energy calculation one must relax the structure with respect to these or other parameters, which can be a time consuming job. An example of such computations was presented by Amador at this symposium for the case of trialuminides. It was observed that the correct structure is predicted only when these sometimes subtle relaxation effects $(\Omega, c/a, \delta)$ are correctly taken into account.[44] In another approach, the ECIs themselves may be considered as inheriting these deformation dependencies from the energies appearing in Eq. (18). It is then possible to minimize the CVM free energy functional, $F(\xi_\alpha, \Omega, c/a, \ldots)$, with respect to these variational parameters, ξ_α being the cluster correlation functions defined in Eq. (14). An example of this procedure, applied to volume deformation only, was proposed by Sluiter *et al.* in their study of phase equilibria in the Al-Li system.[18,45] In this symposium, Sanchez extended the treatment to c/a relaxation as well. Although the calculations are of course much more elaborate, a wealth of information can be obtained through such *free* energy minimizations, for example molar volumes and cell distortions as a function of types of ordered structures, concentration and temperature, the temperature dependence being the one due to configurational effects, not vibrational.

The minimization procedure just described does not solve the problem of incorporating elastic energy corrections into the total energy (or free energy) calculations. The optimal volume is determined adequately, but local displacements are not determined at all, in fact they are assumed not to take place, and so the energy is systematically overestimated. To see that, let us split the complete energy expression into two parts

$$E(\sigma) = E_{\text{constr}}(\sigma) - E_{\text{relax}}(\sigma), \qquad (20)$$

where $E_{\text{constr}}(\sigma)$ represents the energy of the system with all atoms "constrained" to lie precisely at the lattice sites, and $E_{\text{relax}}(\sigma)$ represents the relaxation energy which is *gained* when the atoms are allowed to move to their equilibrium positions. In the notation of Eq. (20), the relaxation energy is always positive, and taking it into account always lowers the energy.

Thus the method of optimizing globally with respect to overall atomic volume is quite unforgiving: the atoms of pure A are compressed (expanded) and the atoms of pure B expanded (compressed) to fit on the lattice with average atomic volume Ω. A method which requires less distortions was suggested by Amador et al.:[46] in this formalism, nn atomic tetrahedra $AAAA$, $AAAB$, $AABB$ are allowed to relax partially, their individual optimal volumes being determined by the average atomic volumes of the corresponding ordered structures: *fcc*, $L1_2$, $L1_0$. This method is well adapted to *fcc* lattices in the nn tetrahedron approximation, and has given excellent results for phase equilibria in Ni-Pt[46] and Cu-Au.[47]

The most general way of solving the full, local relaxation problem would be to consider configurations of interest on large supercells, to calculate these electronic structure of the configuration first with all atoms at their ideal lattice sites, calculate forces on all atoms, then allow relaxation to take place, recalculating forces at each incremental displacement step. The difference between the energy at the zeroth (constrained state) step and at the last (equilibrium) is the relaxation energy, as defined in Eq. (20). Such a procedure is generally not feasible, so simplifications must be sought; for example, one can give up on the idea of determining actual atomic *displacements* on each atom, but instead obtain only average relaxation *energies*. The cluster expansion method can help in implementing this more modest program. As before, we have the choice between "large sampling" (DCA method) and "small sampling" (SIM). In the DCA approach, we can define relaxation ECIs in a manner completely analogous to that leading to Eq. (9) and (10), where now the energies $\overline{E}_{IJ}$ of Eq. (10) must be interpreted as differences between the average of the "constrained" and "relaxed" energies of all configurations having I-type atom at lattice site p and J-type at q. The "constrained" energies may be calculated by some suitable electronic structure method, with all atoms at their ideal lattice sites, then the "relaxation" energies can be obtained by using an approximate force model to relax the atoms to their equilibrium positions. Now that we have ECIs made up of two parts, the constrained and relaxation contributions, the evaluation of relaxed energies of arbitrarily complicated structures can be calculated by cluster expansion in the usual manner. In preliminary calculations, Xie et al. found surprisingly rapid convergence of relaxation EPIs with pair spacing and much smaller values (in magnitude) of multiplet interactions.[48]

In the SIM approach, we calculate (fully) *relaxed* energies of a number of simple ordered structures, before inversion is carried out. If enough structures of low symmetry are considered, where relaxations are expected to be important, we can expect that the resulting ECIs will contain the relaxation contributions as well. An interesting application of these ideas was given by Chiolero.[49] In this study, this author used the embedded atom method (EAM) to calculated relaxed energies of about 200 Cu-Au configurations in supercells (of orthorhombic symmetry and higher) containing about 20 atoms each. All three lattice parameters, a, b, c, were optimized and all internal atomic positions were relaxed. Least squares analysis then produced the necessary ECIs. This was perhaps the most complete calculation of its kind, unfortunately it could be implemented only in the EAM context. If long-range elastic interactions are present, it is best to carry out calculations in k-space, as proposed by Zunger and his collaborators. In a recent study, Wolverton *et al.*[50] have obtained excellent agreement between k-space SIM and direct simulation.

Such calculations illustrate the point made earlier: if cluster expansions are performed with ECIs incorporating relaxation effects, then energy computations of configurations of arbitrary complexity automatically include all relaxation effects, without the need to calculate displacements. It follows that free energies calculated with these "renormalized" ECIs will likewise include all relevant relaxation effects. For example, one could perform CVM or even Monte Carlo simulation computations and obtain the desired thermodynamics, elastic distortions included, without having to "move the atoms" explicitly, *i.e.* without having to perform the difficult and time-consuming task of calculating equilibrium displacements. Needless to say, such cluster expansion procedures result in huge savings of computational effort. The difficulty of course is to calculate the "relaxation ECIs." Once again we face the problem of improving sampling or of improving electronic structure accuracy. The two are pretty much mutually exclusive: good sampling may be attained with simple electronic structure methods such as Tight Binding or EAM, but total energy LDA methods such as FLAPW or FLMTO can only be implemented on small-supercell configurations. Let us hope that the search for "order-N" methods will eventually produce algorithms capable of combining adequate sampling and accurate "first-principles" electronic structure calculations.

Not only *static* displacements, due to internal relaxations, should be taken into account, but also *dynamical* ones (atomic vibrations), leading to temperature dependent ECIs. Various treatments have been proposed, for example one based on electronic structure calculations of Debye-model parameters.[51] At this conference, both Ceder and Asta presented interesting vibrational free energy calculations, requiring the determination of dynamical matrix force constants by approximate electronic structure methods. In addition to his rather elaborate EAM-SIM calculations of static displacement corrections in Cu_3Au, Chiolero has extended the method as well to dynamical displacements.[42,49] In a rather different approach, Finel has proposed a "Gaussian CVM" model[52] which he recently applied to vibrational entropy effects in pure elements, the force-constant matrix being determined by a second-moment tight-binding approximation.[53] Extension of the Gaussian CVM to alloys is in preparation. Whatever precise technique is to be used, since the vibrational energy and entropy are configuration-dependent, these functions can be cluster expanded, so that only dynamical matrix calculations of simple ordered structures must be carried out in detail.

Other thermal excitations which may contribute to the temperature-dependence of the ECIs are *electronic ones*. Using the Sommerfeld model, Wolverton *et al.* were able to show that at least some of the discrepancy observed between and calculated ECIs in Ni_3V could be attributed to the neglect of the electronic free energy contribution.[54] Usually, the electronic excitation contribution is not expected to be a large one, however.

To illustrate the effect of various contributions to alloy free energies, let us now briefly describe the work of Asta *et al.*[55] on the Cd-Mg system.

VII. Phase diagrams

Although there is much recent work to choose from, as an example of phase diagram calculations, let us examine the Cd-Mg system which is of interest for the following reasons: although several studies of the ground states of order and prototype phase diagrams of the *hcp* Ising model have been performed,[16] very few *ab initio* calculations of phase diagrams for alloy systems containing *hcp* structures have been undertaken. In Fig. 2a, the solid state portion of the assessed experimental Cd-Mg phase diagram[55] has been re-drawn. At high temperatures (below the melt), continuous solid solutions between *hcp* Cd and Mg phases are found. As the temperature is lowered, three *hcp*-based ordered phases are stabilized in the composition ranges around stoichiometries Cd_3Mg, CdMg, $CdMg_3$. The equiatomic compounds crystallize in the B19 structure, and the two others in the DO_{19} structure, which are the respective analogs of the CuAu ($L1_0$) and Cu_3Au ($L1_2$) structures which have the same atomic arrangements on the close-packed planes. Therefore, the Cd-Mg system may be described as the *hcp* counterpart of the well-known *fcc*-based Cu-Au system, which, as mentioned in Sect. II, was the one which, historically, motivated the cluster formalism in the first place. Another reason for performing thermodynamic calculations in the Cd-Mg system is that experimental data for ordered and disordered phases are readily available for comparison. Also, it is anticipated that non-configurational effects, such as discussed in the previous section, may play a significant role for the following reasons: the Debye temperature of Cd is only one-half that of Mg so that the differences between the vibrational free energy contributions of Cd- and Mg-rich alloys should be sizable. In addition, the c/a ratio in elemental Mg is close to ideal, whereas it is far from ideal in Cd.

The computational method adopted by Asta[55] combines *ab initio* electronic structure calculations augmented by some experimental input. The SIM was used with 9 starting structures, including the known equilibrium ones. The formation energies of these structures along with the zero-K equilibrium atomic volumes and bulk moduli were calculated by the LMTO-ASA method. The energy of each structure was optimized with respect to volume only since the ASA was considered to be insufficiently reliable to handle the large Cd-rich deviations of c/a from ideally. The non-ideality in question was corrected by standard continuum elasticity methods making use of the solid solution c/a data obtained many years ago by Hume-Rothery and Raynor.[56] The SIM produced 9 ECIs including in- and out-of-basal-plane nn pairs, nnn pair, three distinct nn triplets, and the nn tetrahedron. The CVM in the *hcp* Tetrahedron-Octahedron approximation was used to calculate the phase diagram shown in Fig. 2b. Comparison with the experimentally determined diagram of Fig. 2a reveals that the overall features are fairly well reproduced by this purely "first-principles" calculation, in particular the congruent ordering of CdMg and the peritectoid reaction for $CdMg_3$. However, the Cd_3Mg phase is incorrectly predicted to appear congruently at high temperature instead of by peritectoid reaction. The cause of this discrepancy lies in the neglect of the c/a relaxation of Cd and of Cd-rich solid solutions. When this

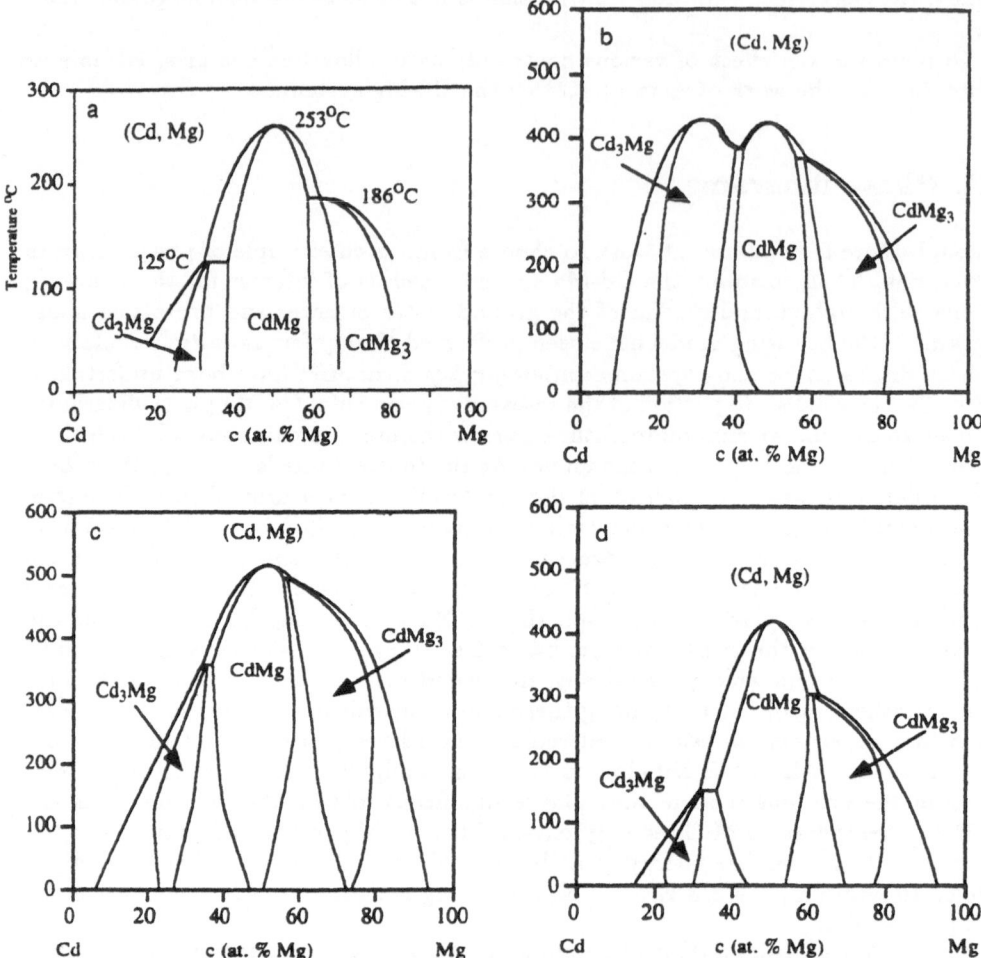

Figure 2. Experimentally determined (a) and successive approximations (b,c,d) of calculated solid-phase portion of Cd-Mg phase diagram.[55]

relaxation is taken into account, one obtains the much improved diagram of Fig. 2c; Cd_3Mg is now correctly predicted to take place by a peritectoid reaction, but the calculated ordering temperatures are still too high.

Agreement with experiment can be improved further by including both electronic and vibrational excitations, as discussed in the previous section. Here, the vibrational free energy contribution was obtained by cluster-expanding the Debye temperature itself rather than by attempting to calculate the dynamical matrix of the structures used in the SIM. Thus, the vibrational correction was performed semi-empirically, for convenience. The electronic free energy contribution was obtained from standard approximations featuring the electronic density of states at the Fermi level, $Ne(\sigma)$ obtained as a by-product of the electronic structure calculations for the 9 structures involved in the SIM. Configuration and temperature dependence of the electronic free energy was obtained by cluster expansion of $Ne(\sigma)$. When both vibrational and electronic excitations are taken into account, the phase diagram of Fig. 2d results. Agreement with the experimental diagram (Fig. 2a) is now striking. Predicted ordering temperatures are still a bit too high, but the overall shape, crystallography of ordered phases and types of transitions are reproduced remarkably well. By comparing the successive phase diagrams of Fig. 2, from (a) to (d), it is also gratifying to note that, as more corrections are included, the agreement with the experimental phase diagram improves at each step.

VIII. Conclusions

If reproducing a known phase diagram were the only objective of these calculations, the justification for such an effort might well be questioned. Actually, much more information can be extracted from the calculations than can be gotten from the experimentally-determined diagram. It is true that the parameters of a model free energy can be backed out of empirical phase diagram information by fitting, but the fit (when it works at all!) will never be better than the model chosen; compensating errors can be committed, and the inaccuracies absorbed in the empirical ECIs, often called incorrectly "pair potentials." With *ab initio* (or partially so) calculations, not only can one determine what is the approximately correct model to use, but one can *understand* the physics behind the calculations. For instance, in the Cd-Mg calculation presented, it was possible to break up the total formation energy (or enthalpy), entropy, free energy into separate contributions: configurational, elastic, electronic, vibrational. *Ab initio* calculations also help avoid mistakes committed in the past when properties of certain alloy phases were extrapolated from experimental data into regions where they were in fact unstable, not merely metastable. Only electronic structure "computer experiments" were able to detect the error, as shown by Craievich, Sanchez and co-workers.[57]

In summary, what precipitated the remarkable recent advances in Alloy Theory was the concept of *cluster*, an idea which was introduced by Ryo Kikuchi when he first proposed his elegant Cluster Variation Method. Perhaps he himself did not realize, at the time, how the cluster concept was going to revolutionize not only the statistical mechanics of phase transformations, but the whole field of thermodynamics by introducing in a very natural way the formalism of orthonormal expansions in cluster functions. It is this formalism which then bridges the gap between statistical and quantum mechanics, two disciplines which had generally gone their separate ways, but which can now combine in a coherent whole.

Let us reiterate: the cluster notion would perhaps not have imposed itself so readily if Ising model thermodynamics had been confined to exercises in Monte Carlo simulation or abstruse mathematical developments of exactly solvable but often unrealistic model systems. From the start, the CVM talked the language of clusters, as the name implies, and solved practical problems numerically but in a way which made the thermodynamics quite transparent, which provided precise enthalpy, entropy, free energy functions. It has been said that the Monte Carlo method gives results which are accurate but imprecise whereas the Cluster Variation method is precise but inaccurate. However, with the right cluster approximations, transition temperatures can be predicted to within a few percent of the best-known approximations. What presently limits the application of cluster-type alloy theory methods to phase equilibria is not the inaccuracies of the CVM, but those of the electronic structure calculations of formation energies: consider, a relatively modest error of 0.01 eV corresponds on a phase diagram to an error of nearly 100 K in temperature, which is unacceptable in practical applications. Even if electronic structure accuracy cannot be improved by one or two orders of magnitude in future, still the theoretical approach has merit, providing an understanding of phase equilibria which was simply not available until quite recently.

The CVM and its offspring, the cluster expansion method, has ushered in a veritable "first principles alloy thermodynamics" which shows great formal and practical promise. As for the future of the CVM itself as a practical means for calculating transition temperatures, its use is still limited because of some nagging problems: for the uninitiated, the formalism may seem a bit heavy and not easy to implement, as opposed to the Monte Carlo method whose basic formalism is very simple to comprehend and to code. However, the CVM is typically 100 times more computationally efficient than the Monte Carlo. Moreover, once a general CVM code has been developed, it can be used as a standard tool, easily adapted to a variety of situations provided that the determination of cluster parameters be suitably automated. Several of the conference participants have developed the required codes based on group theory: J. M. Sanchez, A. Finel, G. Ceder, perhaps others. Presently, Marcel Sluiter is developing a very general and user-friendly CVM code which incorporates, among other features, a method of cluster selection based on an algorithm derived by Vul and the author.[58] Also, the SIM is virtually built into this set of computer programs, making the package particularly attractive to general users. Still, one problem refuses to go away: that of treating long-range interactions. In principle, the maximal cluster used in the CVM free energy must at least contain the largest cluster required for representing the energy (enthalpy). For long-range effective interactions, one is quickly confronted with the "combinatorial explosion" which makes the problem untractable, even with "automated" cluster codes. Several approximate methods of handling distant interaction without increasing the maximal cluster size have been proposed,[59] but these methods still need to be tested critically. Finally, as hybrid method may be of interest, that of combining the CVM and Monte Carlo methods, as originally proposed by Schlijper,[60] and as mentioned at this Conference by Ceder and by Anthony.

Acknowledgements

In closing I would like to show my appreciation to the many collaborators who have worked with me at UCLA then at Berkeley and who have made my venture in this field so exciting and rewarding. First of all my thanks go to Ryo Kikuchi who taught me the CVM and thereby completely changed the orientation of my research. Together with Juan Miguel Sanchez, we learned from the Master, and Juan went on to develop almost *all* of the tools of the cluster formalism which today researchers throughout the world are taking for granted, often ignoring where the original derivations came from. I cannot mention by name all other collaborators, students, post doctoral fellows, visiting scientists who partook in this wonderful adventure, but I take great pride in the observation that most of them today are continuing with great success and originality some aspects of the work which we worked on over the years. A sort of "cluster power" tends to keep people clustered together and maintains them in the field! Actually, about ten of these close associates are attending this conference held in honor of the one who made it all possible, Professor Ryo Kikuchi.

Early work was funded at UCLA by the Army Research Office, then at Berkeley by the National Science Foundation and by the Department of Energy through the Lawrence Berkeley Laboratory.

References

1. R. Kikuchi, *Phys. Rev.* **79**, 718 (1950).
2. R. Kikuchi, *Phys. Rev.* **81**, 988 (1951).
3. R. Kikuchi, *Prog. Theor. Phys.* Suppl. **35**, 1 (1966).
4. C. M. van Baal, *Physica (Utrecht)* **64**, 571 (1973).
5. D. de Fontaine and R. Kikuchi, in *Application of Phase Diagrams in Metallurgy and Ceramics*, N.B.S. Special publication 496, edited by G. C. Carter (1978), p. 967.
6. W. Shockley, *J. Chem. Phys.* **6**, 130 (1938).
7. J. Kanamori, *Prog. Theor. Phys.* **35**, 66 (1966).
8. M. J. Richards and J. W. Cahn, *Acta Metall.* **19**, (1971); S. M. Allen and J. W. Cahn, *Scripta. Metall.* **7**, 1261 (1973).
9. J. M. Sanchez and D. de Fontaine, *Phys. Rev. B* **17**, 2926 (1978).
10. J. M. Sanchez, F. Ducastelle, and D. Gratias, *Physica* **128A**, 334 (1984).
11. T. Morita, *J. Phys. Soc. Jap.* **12**, 753 (1957); *J. Math. Phys.* **13**, 115 (1972).
12. J. M. Sanchez and D. de Fontaine, *Phys. Rev. B* **21**, 216 (1980).
13. J. M. Sanchez and D. de Fontaine, *Phys. Rev. B* **25**, 1759 (1982).
14. T. Mohri, J.M. Sanchez, and D. de Fontaine, *Acta Metall.* **33**, 1171 (1985).
15. M. Sluiter, P. Turchi, F. Zezhong, and D. de Fontaine, *Physica A* **148**, 61 (1988).
16. R. McCormack, M. Asta, D. de Fontaine, and G. Ceder, *Phys. Rev. B* **48**, 6767 (1993).
17. C. Sigli and J. M. Sanchez, *Acta Metall.* **36**, 367 (1988).
18. M. H. F. Sluiter, D. de Fontaine, X.Q. Guo, R. Podloucky, and A. J. Freeman, *Phys. Rev. B* **42**, 10460 (1990).
19. J. M. Sanchez, *Physica* **111A**, 200 (1982).
20. G. Ceder, *Alloy Theory and its Application to Long-Period Superstructure Ordering in Metallic Alloys and High-Temperature Superconductors*. Doctoral Dissertation, University of California, Berkeley (1991); University Microfilms Int'l. Ann Arbor, MI (# 9203517).
21. J. M. Sanchez and D. de Fontaine, *Structure and Bonding in Crystals* (Academic Press, New York, 1981), Vol. II, p. 117.
22. F. Ducastelle, *Order and Phase Stability in Alloys*, (North-Holland, New York, 1991).
23. G. Inden and W. Pitsch, "Atomic Ordering," in *Phase Transformations in Materials*, edited by P. Haasen, (VCH Press, Weinheim and New York, 1991), p. 497.
24. J. M. Sanchez, V. Pierron-Bohnes, and F. Mejía-Lira, *Phys. Rev. B* **51**, 3429 (1995).
25. D. de Fontaine, in *Solid State Physics*, edited by H. Ehrenreich, F. Seitz, and D. Turnbull, (Academic Press, 1979), Vol. 34, p. 74.
26. D. de Fontaine, *Solid State Physics*, edited by H. Ehrenreich and D. Turnbull, (Academic Press, 1994), Vol. 47, p. 33
27. A. Finel, *Contribution à l'Etude des Effets d'Ordre dans le Cadre du Modèle d'Ising: Etats de Base et Diagrammes de Phase*, doctoral dissertation, Université Pierre et Marie Curie, Paris, 1987 (unpublished).
28. K. Kawasaki, in *Phase Transitions and Critical Phenomena*, Vol. 2, Ed. C. Domb and M. S. Green, (Academic Press, New York, 1973), p. 465.
29. M. Asta, C. Wolverton, D. de Fontaine, and H. Dreyssé, *Phys. Rev. B* **44**, 4907 (1991); C. Wolverton, M. Asta, H. Dreyssé, and D. de Fontaine, *Phys. Rev. B* **44**, 4914 (1991).

30. D. de Fontaine, A. Finel, H. Dreyssé, M. Asta, R. McCormack, and C. Wolverton, in *Proc. NATO Advanced Research Workshop on Metallic Alloys: Experimental and Theoretical Perspectives*, Boca Raton, FL, July 16-21, 1993, (1994).

31. J. M. Sanchez, *Phys. Rev. B* **48**, 14013 (1993).

32. S. Ouannasser and H. Dreyssé, in preparation.

33. A. Gonis, P. P. Singh, P. E. A. Turchi, and X. -G. Zhang, *Phys. Rev. B* **51**, 2122 (1995).

34. G. Ceder, *Compt. Mater. Sci.* **1**, 144 (1993).

35. H. Dreyssé, A. Berera, L. T. Wille, and D. de Fontaine, *Phys. Rev. B* **39**, 2442 (1989); C. Wolverton, H. Dreyssé, and D. de Fontaine, *Mat. Res. Soc. Symp. Proc.* **193**, 183 (1991); C. Wolverton, G. Ceder, D. de Fontaine, and H. Dreyssé, *Phys. Rev. B* **48**, 726 (1993); C. Wolverton and D. de Fontaine, *Phys. Rev. B* (1994); C. Wolverton and A. Zunger, *Phys. Rev. B* **50**, 10548 (1994).

36. J. W. D. Connolly and A. R. Williams, *Phys. Rev. B* **27**, 5169 (1983).

37. Z. W. Lu, S. -H. Wei, A. Zunger, S. Frotta-Pessoa, and L.G. Ferreira, *Phys. Rev. B* **44**, 512 (1991); A. Zunger, in *Statics and Dynamics of Alloy Phase Transformations*, edited by P. E. A. Turchi and A. Gonis, (NATO ASI Series, Kluwer, Dordrecht, 1994), p. 361 and references cited therein.

38. J. S. Faulkner, *Prog. Mater. Sci.* **27**, 1 (1982).

39. F. Ducastelle and F. Gautier, *J. Phys.* **F6**, 2039 (1976).

40. B. L. Györffy and G. M. Stocks, *Phys. Rev. Lett.* **50**, 374 (1983).

41. A. Gonis and J. W. Garland, *Phys. Rev. B* **16**, 1495, 2424 (1977); **18**, 3999 (1978).

42. A. Chiolero, *Doctoral Dissertation*, University of Fribourg, Switzerland (1995).

43. D. B. Laks, L. G. Ferreira, S. Froyen, and A. Zunger, *Phys. Rev. B* **46**, 12587 (1992).

44. C. Amador, J.J. Hoyt, B.C. Chakoumakos, and D. de Fontaine, *Phys. Rev. Lett.* **74**, 4955 (1994).

45. M. H. F. Sluiter, Y. Watanabe, D. de Fontaine, and Y. Kawazoe, submitted for publication.

46. C. Amador, W. R. L. Lambrecht, M. van Schilfgaarde, and B. Segall, *Phys. Rev. B* **47**, 15276 (1993).

47. C. Amador and G. Bozzolo, *Phys. Rev. B* **49**, 956 (1994).

48. P. Xie and D. de Fontaine, unpublished work at UC Berkeley.

49. A. Chiolero and D. Baeriswyl, *J. Stat. Phys.* **76**, 347 (1994).

50. C. Wolverton and A. Zunger, *Phys. Rev. Lett.*, in press.

51. V. L. Moruzzi, J. F. Janak, and K. Schwarz, *Phys. Rev. B* **37**, 790 (1988); J. M. Sanchez, J. P. Stark, and V. L. Moruzzi, *Phys. Rev. B* **44**, 5411 (1991).

52. A. Finel, *Progr. Theor. Phys.* **115**, 59 (1994).

53. A. Finel and R. Tétot, preprint.

54. C. Wolverton and A. Zunger, *Phys. Rev. B* **49**, 16058 (1994) and *Phys. Rev. B*, in press.

55. M. Asta, R. McCormack, and D. de Fontaine, *Phys. Rev. B* **48**, 748 (1993).

56. W. Hume-Rothery and G. V. Raynor, *Proc. Roy. Soc.* **A174**, 471 (1940).

57. P. J. Craievich, M. Weinert, J. M. Sanchez, and R. E. Watson, *Phys. Rev. Lett.* **72**, 3076 (1994).

58. D. A. Vul and D. de Fontaine, *Mat. Res. Soc. Symp. Proc.* **291**, 401 (1993).

59. M. Sluiter, *Comp. Mat, Sc.* **2**, 293 (1994).

60. A. G. Schlijper, A. R. D. Vandebergen, and B. Smit, *Phys. Rev. A* **41**, 1175 (1990).

Cluster Variation Method, Effective Field Method, and the Method of Integral Equation on the Regular and Random Ising Models

Shigetoshi Katsura

Tohoku College of Engineering and Information Sciences
6-45-16 Kunimi, Aobaku
Sendai 981
JAPAN

Abstract

The effective field method for the Ising model is closely related to the cluster variation method. In the first part we consider an effective field theory of the simple cubic lattice. Density matrices of a vertex, $\rho^{(1)}$, a pair, $\rho^{(2)}$, a square, $\rho^{(4)}$, and a cube, $\rho^{(8)}$, are written in terms of a single bond effective field λ, an effective field due to a square, λ', that due to a cube, λ'', an effective interaction due to a square, ν, and that due to a cube ν'. The reducibilities of $\rho^{(2)}$ to $\rho^{(1)}$, $\rho^{(4)}$ to $\rho^{(2)}$, and $\rho^{(8)}$ to $\rho^{(4)}$ lead to relations between λ, λ', λ'', ν, and ν'. These relations give the critical temperature of the Ising model on the simple cubic lattice as $kT/J = 4.60402$.

In the second part the $\pm J$ random Ising model is considered. The distribution of the single bond effective field h is denoted by $g(h)$. The consistency among the effective fields leads to an integral equation. We can derive phase diagrams of the system on the square and the simple cubic lattices, the triangular and the face-centered cubic lattices. The shape of the phase diagram of the former is symmetric with respect to $+J$ and $-J$ sides, while that of the latter is asymmetric in reflection of the crystal structure. Our method also gives a qualitatively well behaved phase diagram of $Eu_p Sr_{1-p} S$. The integral equation can be exactly solved at $T = 0$ and discrete solutions and continuous solutions are obtained. The latter give the phase boundary between spin glass, MSG, and ferromagnetic phases.

Theory and Applications of the Cluster Variation and Path Probability Methods
Edited by J.L. Morán-López and J.M. Sanchez, Plenum Press, New York, 1996

I. Cumulant Expansion of the Free Energy and the Consistency of the Reduced Density Matrices

Now we start the cumulant expansion of the free energy. Though the terminology of cumulant expansion and Möbius inversion fornula was not used, its use was begun by Domb[1] and Baker.[2] The cumulant free energies of a vertex $F^{(c)}(1)$, a pair $F^{(c)}(2)$, a square $F^{(c)}(4)$, and a cube $F^{(c)}(8)$, are defined by

$$\begin{pmatrix} F(1) \\ F(2) \\ F(4) \\ F(8) \end{pmatrix} = \begin{pmatrix} 1 & 0 & 0 & 0 \\ 2 & 1 & 0 & 0 \\ 4 & 4 & 1 & 0 \\ 8 & 12 & 6 & 1 \end{pmatrix} \begin{pmatrix} F^{(c)}_{(1)} \\ F^{(c)}_{(2)} \\ F^{(c)}_{(4)} \\ F^{(c)}_{(8)} \end{pmatrix}, \tag{1}$$

where $F(1)$, $F(2)$, $F(4)$ and $F(8)$ are cluster free energies of a vertex, a pair, a square and a cube, respectively. The matrix element M_{ij} is the number of the appearance of the j cluster in the i cluster. The inversion of Eq. (1) is given by

$$\begin{pmatrix} F^{(c)}(1) \\ F^{(c)}(2) \\ F^{(c)}(4) \\ F^{(c)}(8) \end{pmatrix} = \begin{pmatrix} 1 & 0 & 0 & 0 \\ -2 & 1 & 0 & 0 \\ 4 & -4 & 1 & 0 \\ -8 & 12 & -6 & 1 \end{pmatrix} \begin{pmatrix} F(1) \\ F(2) \\ F(4) \\ F(8) \end{pmatrix}. \tag{2}$$

The total free energy per site in several approximations are given by $F = \sum a_i F_i^{(c)}$ where a_i is the number of the appearence of i cluster in the total lattice per site. For example,

1) Pair Bethe approximation

$$F = F^{(c)}(1) + \frac{z}{2}F^{(c)}(2)$$
$$= (1 - z)F(1) + \frac{z}{2}F(2), \tag{3}$$

where z is the coordination number, e.g., for the simple cubic lattice, $z = 6$.

2) Square approximation for the square lattice

$$F = F^{(c)}(1) + 2F^{(c)}(2) + F^{(c)}(4)$$
$$= F(1) - 2F(2) + F(4). \tag{4}$$

3) Square approximation for the cubic lattice

$$F = F^{(c)}(1) + 3F^{(c)}(2) + 3F^{(c)}(4)$$
$$= 7F(1) - 9F(2) + 3F(4). \tag{5}$$

4) Cube approximation for the cubic lattice

$$F = F^{(c)}(1) + 3F^{(c)}(2) + 3F^{(c)}(4) + F^{(c)}(8)$$
$$= -F(1) + 3F(2) - 3F(4) + F(8). \tag{6}$$

We consider the simple cubic lattice. Density matrices of a vertex, a pair, a square and a cube cluster are defined by

$$\rho^{(1)} = \exp\left(L_1 s_1\right),$$

$$\rho^{(2)} = \exp\left[L_2(s_1 + s_2) + K_2 s_1 s_2\right],$$

$$\rho^{(4)} = \exp\left[L_4(s_1 + s_2 + s_3 + s_4) + K_4(s_1 s_2 + s_2 s_3 + s_3 s_4 + s_4 s_1)\right],$$

$$\rho^{(8)} = \exp\left[L_8(s_1 + s_2 + s_3 + s_4 + s_5 + s_6 + s_7 + s_8)\right.$$
$$+ K_8(s_1 s_2 + s_2 s_3 + s_3 s_4 + s_4 s_1 + s_5 s_6 + s_6 s_7$$
$$\left. + s_7 s_8 + s_8 s_5 + s_1 s_5 + s_2 s_6 + s_3 s_7 + s_4 s_8)\right]. \tag{7}$$

The free energy $F(j)$ is given by $F(j) = \log \mathrm{Tr}\, \rho^{(j)}$, for $j = 1, 2, 4, 8$. The stationarity conditions $\delta F / \delta L_i = 0$ lead to self-consistent relations between the effective fields L_i. In Eq. (7) L_1, L_2, L_4, L_8 are the effective fields times β at a spin of a vertex, a pair, a square, and a cube, respectively. They are expressed by

$$L_1 = 6\lambda_1,$$
$$L_2 = 5\lambda_1 + 4\lambda_2,$$
$$L_4 = 4\lambda_1 + 6\lambda_2 + 2\lambda_3,$$
$$L_8 = 3\lambda_1 + 6\lambda_2 + 3\lambda_3, \tag{8}$$

where λ_1 is a single bond (outside the cluster) effective field, λ_2 is an effective field due to a square which includes a pair of the cluster, λ_3 is an effective field due to a cube which includes a square of the cluster. K_2, K_4, K_8 are the effective interactions for a bond in a pair, a square and a cube respectively. They are expressed by

$$K_2 = K + 4\nu_1 + 4\nu_2,$$
$$K_4 = K + 3\nu_1 + 2\nu_2 + 2\nu_3,$$
$$K_8 = K + 2\nu_1 + \nu_2 + 2\nu_3, \tag{9}$$

where ν_1 is an effective field due to a square which includes the pair, ν_2 is an effective field due to a cube which includes the pair of the cluster, and ν_3 is an effective field due to a cube which includes the square of the cluster. K is the original bare interaction. Figure 1 explains the meaning of L_i, K_i, λ_i and ν_i.

It is noted that

$$K = K_2 - 4K_4 + 4K_8.$$

Up to $\mathcal{O}(L_i)$, they are written as (excluding the constant term)

$$\rho^{(1)} = (1 + l_1 s_1),$$

$$\rho^{(2)} = (1 + l_2 s_1)(1 + l_2 s_2)(1 + t_2 s_1 s_2),$$

$$\rho^{(4)} = (1 + l_4 s_1)(1 + l_4 s_2)(1 + l_4 s_3)(1 + l_4 s_4)$$
$$\times (1 + t_4 s_1 s_2)(1 + t_4 s_2 s_3)(1 + t_4 s_3 s_4)(1 + t_4 s_4 s_1),$$

$$\rho^{(8)} = (1 + l_8 s_1)(1 + l_8 s_2)(1 + l_8 s_3)(1 + l_8 s_4)$$
$$\times (1 + l_8 s_5)(1 + l_8 s_6)(1 + l_8 s_7)(1 + l_8 s_8)$$
$$\times (1 + t_8 s_1 s_2)(1 + t_8 s_2 s_3)(1 + t_8 s_3 s_4)(1 + t_8 s_4 s_1)$$
$$\times (1 + t_8 s_5 s_6)(1 + t_8 s_6 s_7)(1 + t_8 s_7 s_8)(1 + t_8 s_8 s_5)$$
$$\times (1 + t_8 s_1 s_5)(1 + t_8 s_2 s_6)(1 + t_8 s_3 s_7)(1 + t_8 s_4 s_8), \tag{10}$$

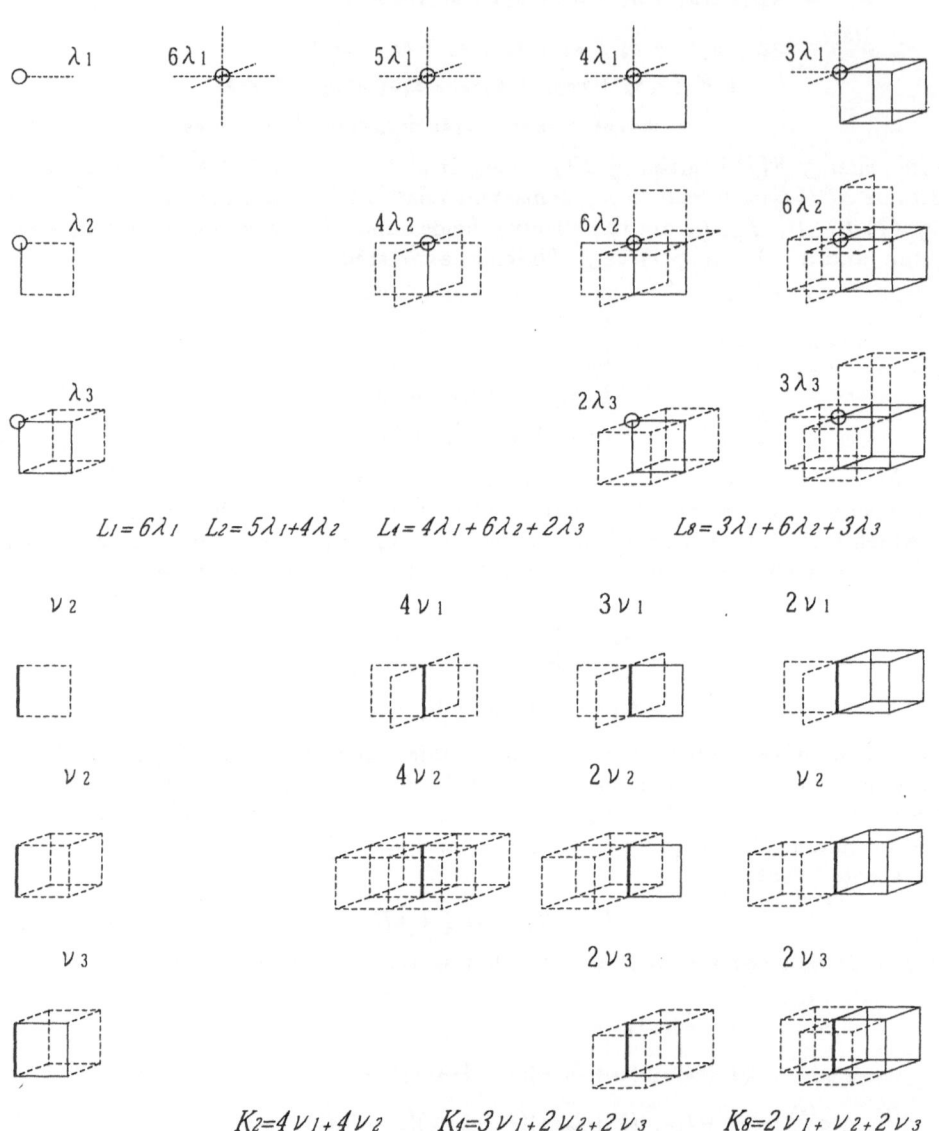

Figure 1. Effective field and effective interaction for a vertex, a pair, a square, and a cube. λ_1: effective field due to a bond, λ_2: effective field due to a square, λ_3: effective field due to a cube. ν_1: effective interaction due to a square, ν_2: effective interaction due to a cube which include a respective bond, ν_3: effective interaction due to a cube which includes a respective bond and a respective square.

where $l_i = \tanh L_i$ for $i = 1, 2, 4, 8$, and $t_i = \tanh K_i$ for $i = 2, 4, 8$. The partial traces of $\rho^{(1)}$, $\rho^{(2)}$, $\rho^{(4)}$, and $\rho^{(8)}$ are given by (the hat denotes the normalization),

$$\hat{\rho}^{(1)} = 1 + \langle s_1 \rangle_{(1)},$$

$$\hat{\rho}^{(2)} = 1 + \langle s_1 \rangle_{(2)}(s_1 + s_2) + \langle s_1 s_2 \rangle_{(2)} s_1 s_2,$$

$$\begin{aligned}
\hat{\rho}^{(4)} = &\, 1 + \langle s_1 \rangle_{(4)}(s_1 + s_2 + s_3 + s_4) \\
&+ \langle s_1 s_2 \rangle_{(4)}(s_1 s_2 + s_2 s_3 + s_3 s_4 + s_4 s_1) \\
&+ \langle s_1 s_3 \rangle_{(4)}(s_1 s_3 + s_2 s_4) \\
&+ \langle s_1 s_2 s_3 \rangle_{(4)}(s_1 s_2 s_3 + s_2 s_3 s_4 + s_3 s_4 s_1 + s_4 s_1 s_2) \\
&+ \langle s_1 s_2 s_3 s_4 \rangle_{(4)} s_1 s_2 s_3 s_4,
\end{aligned}$$

(11)

where

$$\langle s_1 \rangle_{(1)} = l_1, \qquad \langle s_1 \rangle_{(2)} = (1 + t_2) l_2, \qquad \langle s_1 \rangle_{(4)} = \left(1 + 2\frac{t_4 + t_4^2 + t_4^3}{1 + t_4^4}\right) l_4,$$

$$\langle s_1 s_2 \rangle_{(2)} = t_2, \qquad \langle s_1 s_2 \rangle_{(4)} = \frac{t_4 + t_4^3}{1 + t_4^4}, \qquad \langle s_1 s_3 \rangle_{(4)} = \frac{2t_4^2}{1 + t_4^4},$$

$$\langle s_1 s_2 s_3 \rangle_{(4)} = \frac{2(1 + 2t_4 + t_4^2) t_4}{1 + t_4^4} l_4, \qquad \langle s_1 s_2 s_3 s_4 \rangle_{(4)} = \frac{2t_4^2}{1 + t_4^4},$$

and

$$\mathrm{Tr}_2\, \hat{\rho}^{(2)} = 1 + \langle s_1 \rangle_{(4)}(s_1 + s_2 + s_3 + s_4) + \langle s_1 s_2 \rangle_{(4)},$$

$$\begin{aligned}
\mathrm{Tr}_{5678}\, \hat{\rho}^{(8)} = &\, 1 + \langle s_1 \rangle_{(8)}(s_1 + s_2 + s_3 + s_4) \\
&+ \langle s_1 s_2 \rangle_{(8)}(s_1 s_2 + s_2 s_3 + s_3 s_4 + s_4 s_1) \\
&+ \langle s_1 s_2 s_3 \rangle_{(8)}(s_1 s_2 s_3 + s_2 s_3 s_4 + s_3 s_4 s_1 + s_4 s_1 s_2), \\
&+ \langle s_1 s_2 s_3 s_4 \rangle_{(8)} s_1 s_2 s_3 s_4,
\end{aligned}$$

with

$$\langle s_1 \rangle_{(8)} = \frac{1 + 3t_8 + 6t_8^2 + 12t_8^3 + 24t_8^4 + 42t_8^5 + 58t_8^6 + 60t_8^7 + 39t_8^8 + 11t_8^9}{1 + 6t_8^4 + 16t_8^6 + 9t_8^8},$$

$$\langle s_1 s_2 \rangle_{(8)} = \frac{t_8(1 + 2t_8^2 + 12t_8^4 + 14t_8^6 + 3t_8^8)}{1 + 6t_8^4 + 16t_8^6 + 9t_8^8},$$

$$\langle s_1 s_3 \rangle_{(8)} = \frac{2t_8^2(1 + t_8^2 + 7t_8^4 + 5t_8^6)}{1 + 6t_8^4 + 16t_8^6 + 9t_8^8},$$

$$\langle s_1 s_2 s_3 \rangle_{(8)} = \frac{2t_8(1 + 3t_8 + 6t_8^2 + 13t_8^3 + 24t_8^4 + 31t_8^5 + 26t_8^6 + 15t_8^7 + 7t_8^8 + 2t_8^9)}{1 + 6t_8^4 + 16t_8^6 + 9t_8^8},$$

$$\langle s_1 s_2 s_3 s_4 \rangle_{(8)} = \frac{2t_8^2(1 + 2t_8^2 + 10t_8^4 + 2t_8^{(6)} + t_8^8)}{1 + 6tr_8^4 + 16t_8^6 + 9t_8^8},$$

1) Pair (Bethe) approximation

We use $\hat{\rho}^{(1)}$ and $\hat{\rho}^{(2)}$ and neglect λ_2 (and λ_3, t_4, t_8), and put $t_2 = t$. From $\langle s_1 \rangle_{(1)} = \langle s_1 \rangle_{(2)}$ and $l_1 = 6\lambda_1$ and $l_2 = 5\lambda_1$, the critical temperature is given by

$$6\lambda = (1 + t_2)5\lambda.$$

Hence

$$t_2 = t = \tfrac{1}{5},$$

this is the critical temperature of the Bethe approximation. It is noted that results for the Bethe approximation is exact for the Bethe lattice.[3,4]

2) Square approximation

We use $\hat{\rho}^{(1)}$, $\hat{\rho}^{(2)}$ and $\hat{\rho}^{(4)}$ and neglect $\hat{\rho}^{(8)}$ and t_8. From $\langle s_1 \rangle_{(1)} = \langle s_1 \rangle_{(2)}$ and $\langle s_1 \rangle_{(1)} = \langle s_1 \rangle_{(4)}$, $l_1 = 6\lambda$, $l_2 = 5\lambda_1 + 4\lambda_2$, we have

$$6\lambda_1 - (1 + t_2)(5\lambda_1 + 4\lambda_2) = 0,$$

$$6\lambda_1 - \left(1 + 2\frac{t_4 + t_4^2 + t_4^3}{1 + t_4^4}\right)(4\lambda_1 + 6\lambda_2) = 0. \tag{12}$$

Using $\langle s_1 s_2 \rangle_{(2)} = \langle s_1 s_2 \rangle_{(4)}$, we have $t_2 = ((t_4 + t_4^3)/(1 + t_4^4))$. Eliminating t_2 and expressing (12) in

$$M\begin{pmatrix}\lambda_1 \\ \lambda_2\end{pmatrix} = 0,$$

where

$$M_{1j} = \text{coeff. of } \lambda_j \text{ in } (\langle s_1 \rangle_{(1)} - \langle s_1 \rangle_{(2)}), \qquad j = 1, 2$$
$$M_{2j} = \text{coeff. of } \lambda_j \text{ in } (\langle s_1 \rangle_{(1)} - \langle s_1 \rangle_{(4)}), \tag{13}$$

the critical temperature is given by $\det M = 0$, which results in a polynomial of 8th order of t_4. Among the 8 values of t_4, we choose $t_4 = 2 - \sqrt{3}$. Then $t_2 = 2/7$, $\nu = \tfrac{3}{4}\log 3 - \tfrac{1}{2}\log 5$, hence[5,6]

$$\tanh K = \frac{22}{103}, \qquad \frac{kT}{J} = \left[\log\left(\frac{5^{3/2}}{3^2}\right)\right]^{-1} = 4.609733.$$

3) A modified Cube approximation (new result)

We use $\rho^{(1)}$, $\rho^{(2)}$, $\rho^{(4)}$, $\rho^{(8)}$, λ_1, λ_2, λ_3, ν_1, ν_2, and ν_3. Using the relations (at the first order, $L_i = l_i$),

$$l_1 = 6\lambda_1,$$
$$l_2 = 5\lambda_1 + 4\lambda_2,$$
$$l_4 = 4\lambda_1 + 6\lambda_2 + 2\lambda_3,$$
$$l_8 = 3\lambda_1 + 6\lambda_2 + 3\lambda_3, \tag{14}$$

which hold at the zero external field limit, the self consistent relations

$$\langle s_1 \rangle_{(1)} - \langle s_1 \rangle_{(2)} = 0, \quad \langle s_1 \rangle_{(1)} - \langle s_1 \rangle_{(4)} = 0, \quad \langle s_1 \rangle_{(1)} - \langle s_1 \rangle_{(8)} = 0,$$

give

$$M(\lambda_1, \lambda_2, \lambda_3)^T = 0. \tag{15}$$

Table I
Critical temperature in several methods.*

Method	kT_c/J
MFA	6.0
Bethe	4.93621
Cactus-sq[6]	4.89275
Cactus-cube[7]	4.83951
Yvon-square[8]	4.76107
Yvon-cube[8]	4.70604
Kikuchi(2)[5]	4.60973
Present result	4.60402
Kikuchi(3)[5]	4.58099
High Temp. Series	4.5103

where M is the 3×3 matrix for which M_{ij} is a function of t_2, t_4 and t_8. Here, superscript T denotes the transpose. They are given by

$$M_{1j} = \text{coeff. of } \lambda_j \text{ in } (\langle s_1 \rangle_{(1)} - \langle s_1 \rangle_{(2)}),$$
$$M_{2j} = \text{coeff. of } \lambda_j \text{ in } (\langle s_1 \rangle_{(1)} - \langle s_1 \rangle_{(4)}), \qquad \text{for } j = 1, 2, 3,$$
$$M_{3j} = \text{coeff. of } \lambda_j \text{ in } (\langle s_1 \rangle_{(1)} - \langle s_1 \rangle_{(8)}), \tag{16}$$

the critical temperature is given by $\det M = 0$.

As the relation between t_2, t_4 and t_8 we use

$$\langle s_1 s_2 \rangle_{(2)} - \langle s_1 s_2 \rangle_{(4)} = 0, \quad \text{and} \quad \langle s_1 s_2 \rangle_{(2)} - \langle s_1 s_2 \rangle_{(8)} = 0.$$

Eliminating t_2 and t_4 from (15) and (16), we have the equations for t_8 at the critical point

$$-9,801\, t_8^{45} - 352,836\, t_8^{44} + \cdots$$
$$\cdots + 228\, t_8^6 - 231\, t_8^5 - 108\, t_8^4 - 74\, t_8^3 - 36\, t_8^2 - 7\, t_8 + 6 = 0. \tag{17}$$

Among 45 values of t_8, the ones which are real and satisfie $0 < t_8 < 1$, are $t_8 = 0.850431$ and $t_8 = 0.251669$. The former is discarded and the latter gives

$$t_2 = 0.288416, \qquad t_4 = 0.270222$$

Hence $K_2 = 0.296837$, $K_4 = 0.277103$, $K_8 = 0.257194$, $K = 0.217201$, and $kT_c/J = 1/K = 4.60402$. The value of the critical point is compared with other cases in Table I.

---

* It is noted that Fujiki et al.[9] and Katori and Suzuki,[10] use further the consistency of $\langle s_1 s_3 \rangle$, $\langle s_1 s_2 s_3 \rangle$, and $\langle s_1 s_2 s_3 s_4 \rangle$ and got Kikuchi value.[5]

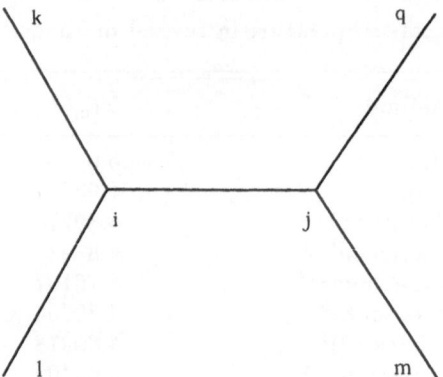

Figure 2. Bethe lattice ($z=3$).

II. Effective Field Theory of the Random Ising Models

II.1 Bethe Approximation and the Integral Equation for the Distribution Function

Now we consider the random Ising model by the method of effective field theory. Generalizing the basic equation of section I to the case where the effective field L is different from site to site, and the external field H may exist, we denote the effective field at the site i from the site k by h_{ik}. First we consider the Bethe lattice, *i.e.*, the Bethe approximation[4,11] (Fig. 2). Then the one-body effective field at the site i, $H_i^{(1)}$, and the two-body effective field at the site i in the pair ij, $H_{i-ij}^{(2)}$, are given by

$$L_i^{(1)} \equiv \beta H_i^{(1)} = \sum_{k=1}^{z} \beta h_{ik} + C \,,$$

$$L_{i-ij}^{(2)} \equiv \beta H_{i-ij}^{(2)} = \sum_{\substack{k=1 \\ k \neq j}}^{z} \beta h_{ik} + C \,, \qquad (18)$$

(simply denoted by $L_i^{(2)}$) where

$$C \equiv \beta \mu H,$$

with H the external field. The partial trace $\mathrm{Tr}_j \, \rho^{(2)}(s_i, s_j)$ is given by

$$\mathrm{Tr}_j \, \widehat{\rho}_{ij}^{(2)} = \frac{\exp\left[\left(L_i^{(2)} + \mathrm{arctanh}(\tanh K_{ij} \tanh L_j^{(2)})\right)s_i\right]}{2 \cosh\left[L_i^{(2)} + \mathrm{arctanh}(\tanh K_{ij} \tanh L_j^{(2)})\right]} \,. \qquad (19)$$

The reducibility

$$\mathrm{Tr}_j \, \widehat{\rho}_{ij}^{(2)} = \widehat{\rho}_i^{(1)} \,,$$

where

$$\hat{\rho}_i^{(1)} = \frac{\exp\left(L_i^{(1)} s_i\right)}{2\cosh L_i^{(1)}},\qquad(20)$$

leads the recurrence relation

$$L_i^{(1)} = L_i^{(2)} + \operatorname{arctanh}\left(\tanh K_{ij}\,\tanh L_j^{(2)}\right).\qquad(21)$$

Eliminating $L_i^{(1)}$ with use of Eq. (21), we have a set of N equations with N unknowns. Instead of solving the N equations, we consider a distribution function $G^{(2)}(H_i^{(2)})$. In the paramagnetic, ferromagnetic, and spin glass states, we regard the distribution of the effective field to be independent of the site. Then we have an integral equation for $G^{(2)}(H_i^{(2)})$,

$$G^{(2)}(H_i^{(2)}) = \int \delta\left(H_i^{(2)} - \frac{1}{\beta}\sum_{k\neq j}\operatorname{arctanh}\left(\tanh\beta J_{ik}\,\tanh\beta H_k^{(2)}\right)\right)$$

$$\times \prod_{k\neq l} P(J_{ik})\,G^{(2)}(H_k^{(2)})\,dJ_{ik}\,dH_k^{(2)},\qquad(22)$$

here $\int_{-\infty}^{\infty}\cdots\int_{-\infty}^{\infty}$, is simply denoted by $\int$. The distribution fuction of the exchange energy is denoted by $P(J)$. For the $\pm J$ model, $P(J) = p\delta(J + J_0) + (1 - p)\delta(J - J_0)$, where p is the concentration of the ferromagnetic bonds.

Fourier transform of Eq. (22) is carried out[12] and

$$G^{(2)}(H^{(2)}) = \frac{1}{2\pi}\int \exp\left(ix(H^{(2)} - H)\right)(S(x))^{z-1}\,dx$$

with

$$S(x) = \int G^{(2)}(H^{(2)})\exp\left(-\frac{x}{\beta}\operatorname{arctanh}(\tanh\beta H^{(2)}\,\tanh\beta J)\right)dH^{(2)}\,dJ,\qquad(23)$$

is obtained. Thus (22) is transformed into

$$S(x) = \frac{1}{2\pi}\int K(x, y)(S(x))^{z-1}\,dx,\qquad(24)$$

with a kernel

$$K(x, y) = \int P(J)\exp\left(iy(H^{(2)} - H) - i\frac{x}{\beta}\operatorname{arctanh}(\tanh\beta J\,\tanh\beta H^{(2)})\right)dJ\,dH^{(2)},\qquad(25)$$

$S(x)$ is the Fourier transform of the single bond distribution function $g(h)$ of the single bond effective field h,

$$g(h) = \frac{1}{2\pi}\int \exp(ix(h - H))\,S(x)\,dx.\qquad(26)$$

The magnetization σ and the spin glass order parameter q are given by

$$\sigma = \int \tanh \beta H^{(1)} G^{(1)}(H^{(1)}) \, dH^{(1)},$$

$$q = \int \tanh^2 \beta H^{(1)} G^{(1)}(H^{(1)}) \, dH^{(1)}. \tag{27}$$

The distribution function $g(h)$ of the single bond effective field h satisfies the integral equation

$$g(h) = \int \delta\left(h - \frac{1}{\beta} \text{arctanh}(\tanh \beta J \tanh \beta H')\right)$$

$$\times \delta(H' - h' - h'') g(h') g(h'') P(J) \, dH' \, dh' \, dh'' \, dJ. \tag{28}$$

When the external field is zero, the paramagnetic state is characterized by $\sigma = 0$ and $q = 0$, the ferromagnetic state by $\sigma \neq 0$ and $q \neq 0$, and the spin glass state by $\sigma = 0$ and $q \neq 0$. From (27) we see at $H = 0$ the uniform and the spin glass susceptibilities, $\chi_u = \partial \sigma / \partial H$ and $\chi_g = \partial q / \partial (H^2)$ are given by

$$\frac{\chi_u}{\beta} = \frac{1 + \overline{\tanh \beta J}}{1 - (z-1)\overline{\tanh \beta J}}, \tag{29}$$

$$\frac{\chi_g}{\beta^2} = \frac{1 + \overline{\tanh^2 \beta J}}{1 - (z-1)\overline{\tanh^2 \beta J}}, \tag{30}$$

For the $\pm J$ model, $\overline{\tanh \beta J} = (2p - 1) \tanh \beta J_0$, $\overline{\tanh^2 \beta J} = \tanh^2 \beta J_0$. The transition temperature between the paramagnetic and the ferromagnetic states are given by

$$t_c \equiv \tanh \beta_c J = \frac{1}{(2p-1)(z-1)} \tag{31}$$

and that between the paramagnetic and the spin glass states are given by

$$t_g^2 \equiv \tanh^2 \beta_g J = \frac{1}{z-1}. \tag{32}$$

Figure 3 is the phase diagram in the Bethe approximation, $i.e.$, the exact result for the Bethe lattice. Equations (31) and (32) were obtained by Matsubara and Sakata[13] (and by Carlson $et~al.$[14]).

A state expressed by $g(h) = \frac{1}{3}[\delta(h + J_0) + \delta(h) + \delta(h - J)]$ is an exact solution of (28) at $T = 0$ and represents a spin glass state. The energy E and the entropy S of that state were calculated to be $E = \frac{23}{27} E_F$ (with E_F the energy of the ferromagnetic state) and $S/k = -\frac{11}{54} \log 2 + \frac{8}{54} \log 3 = 0.2156$, and the latter is positive[15] in contrast to the infinitely long-ranged model.

II.2 Face-Centered Cubic Lattice

The pair approximation for the random Ising model is generalized to more complex systems. Here we consider the face centered cubic lattice (Fig. 4). We denote the effective field at site i in the vertex, in the pair, and in the tetrahedron cluster by $l_i^{(1)}$,

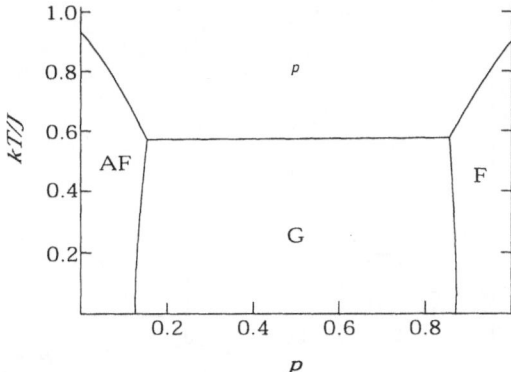

Figure 3. Phase diagram of $\pm J$ model on the Bethe lattice for $z = 3$. P: paramagnetic state, F: ferromagnetic state, AF: antiferromagnetic state, GLP: spin-glass state. See also Fig. 12.

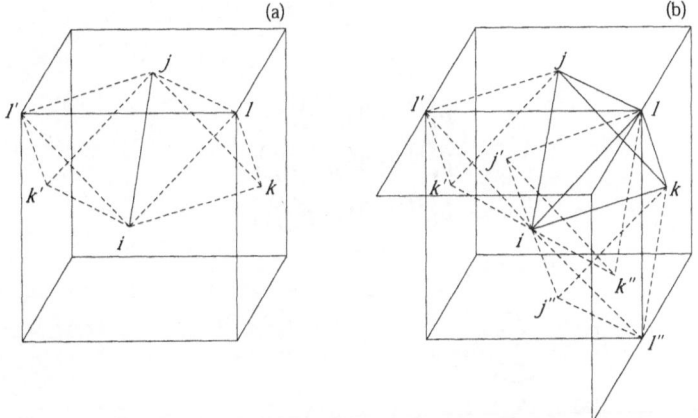

Figure 4. Tetrahedron in the *fcc* lattice. (a) $l^{(2)}$, $2\lambda'$, (b) $l^{(4)}$, $3\lambda'$.

$l_i^{(2)}$, $l_i^{(4)}$, respectively. We also denote the effective interaction in the pair, and in the tetrahedron by $t_{ij}^{(2)}$, $t_{ij}^{(4)}$, respectively. Then the density matrices of the vertex, the pair, and the tetrahedron are given by

$$\rho_i^{(1)} = 1 + l_i^{(1)} s_i,$$

$$\rho_{ij}^{(2)} = \left(1 + l_i^{(2)} s_i\right)\left(1 + l_j^{(2)} s_j\right)\left(1 + t_{ij}^{(2)} s_i s_j\right)$$

$$\simeq \left(1 + l_i^{(2)} l_j^{(2)} t_{ij}^{(2)}\right) + \left(l_i^{(2)} + l_j^{(2)} t_{ij}^{(2)}\right) s_i + \left(l_j^{(2)} + l_i^{(2)} t_{ij}^{(2)}\right) s_j,$$

$$\rho_{ijkl}^{(4)} = \prod_{\mu\nu}\left(1 + t_{\mu\nu}^{(4)} s_\mu s_\nu\right)\prod_\mu\left(1 + l_\mu^{(4)} s_\mu\right), \tag{33}$$

$\mu\nu$ in $\prod_{\mu\nu}$ runs over all the connected bonds of the cluster $ijkl$, and μ in $\prod$ runs over all vertices of the cluster $ijkl$. We postulate that the physical quantities obtained from $\rho_i^{(1)}$, $\rho_{ij}^{(2)}$, and $\rho_{ijkl}^{(4)}$ are equal. Then the following reducibility conditions result

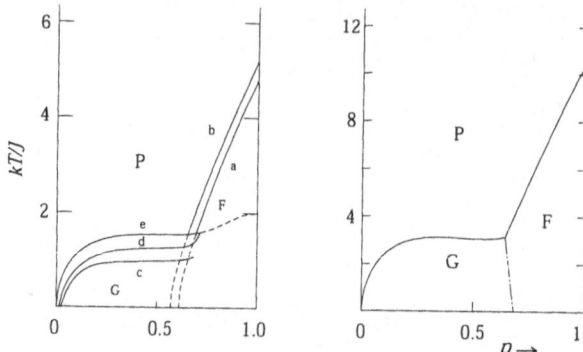

Figure 5. Left: phase diagram of the binary mixture in the *fcc* lattice. (b), (e) tetrahedron cactus approximation, (a), (d) Full tetrahedron approximation. P: paramagnetic state, F: ferromagnetic state, G: spin glass state. Right: tetrahedron cactus approximation, the boundary between ferromagnetic and spin-glass states is shown.

$$\widehat{\rho}_i^{(1)} = \mathrm{Tr}_j\, \widehat{\rho}_{ij}^{(2)},$$
$$\widehat{\rho}_i^{(1)} = \mathrm{Tr}_{jkl}\, \widehat{\rho}_{ijkl}^{(4)},$$
$$\widehat{\rho}_{ij}^{(2)} = \mathrm{Tr}_{kl}\, \widehat{\rho}_{ijkl}^{(4)}, \tag{34}$$

with $l_i^{(1)}$, $l_i^{(2)}$, and $l_i^{(4)}$ related by

$$l_i^{(1)} = \sum_{m=1}^{12} \lambda_{im},$$
$$l_i^{(2)} = \sum_{\substack{m=1 \\ m\neq j}} \lambda_{im} + \lambda'_{ijkl} + \lambda'_{ijk'l'},$$
$$l_i^{(4)} = \sum_{\substack{m=1 \\ m\neq j,k,l}} \lambda_{im} + \lambda'_{ijk'l'} + \lambda'_{ikjl''} + \lambda'_{ilk''j'}, \tag{35}$$

and $t_{ij}^{(2)}$, and $t_{ij}^{(4)}$ by

$$t_{ij}^{(2)} = \tanh K_{ij}^{(2)}, \qquad K_{ij}^{(2)} = K_{ij} + \nu_{ijkl} + \nu_{ijk'l'},$$
$$t_{ij}^{(4)} = \tanh K_{ij}^{(4)}, \qquad K_{ij}^{(4)} = K_{ij} + \nu_{ijk'l''}, \tag{36}$$

where λ_{im} is the effective field at the site i due to the bond im, and λ'_{ijkl} is the effective field at the site i due to the tetrahedron $ijkl$. ν_{ijkl} is the effective interaction at the bond ij due to the tetrahedron $ijkl$. The solution of (34)–(36) in the regular Ising model reproduces the critical temperature of Kikuchi's[16,17,18] value of the *fcc* lattice, $kT_c/J = 10.025$.

The system of (34-1) and (34-2) (neglect of (34-3)) refers to the exact solution of the tetrahedron cactus lattice. Figure 5 shows the phase diagram of the *fcc* lattice in the full tetrahedron approximation and in the tetrahedron cactus approximation. For details of the calculation see Ref. (17).

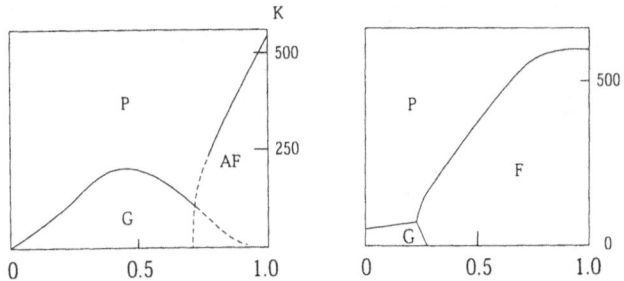

Figure 6. Transition temperature of the random $(\pm J, 0)$ model in the triangular cactus lattice $(z = 6)$.

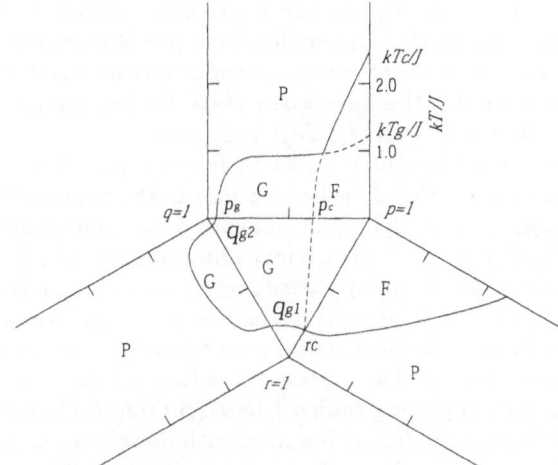

Figure 7. Experimental phase diagrams of $(Cr_p V_{1-p})_{1-\delta}$ (left) and $Mn_p Cu_{1-p}$ (right).

The cluster approximation gives qualitatively different results for lattices which can and cannot be divided into two equivalent lattices. Examples of the latter are, kagome, triangular, and face-centered cubic lattices. Fig. 6 shows the phase diagram of a model with ferromagnetic, antiferromagnetic, and nonmagnetic bonds $(P(J) = p\delta(J - J_0) + q\delta(J + J_0) + r\delta(J), p + q + r = 1)$ in the triangular lattice.[19] The results in Fig. 5 resemble, qualitatively, the experiments for $(Cr_p V_{1-p})_{1-\delta}$Te and $Mn_p Cu_{1-p}$ and others (Fig. 7).

II.3 The Site-Random Ising Model, a Model for $Eu_p Sr_{1-p}S$, and Other Models

Hitherto we treated the spin glass in the bond-random Ising model. Our method can be extended to the site-random Ising model of A and B, of which the exchange integrals are J_{AA}, J_{BB}, and J_{AB}. The uniform, staggered and spin glass susceptibilities are obtained.[20] The transition temperatures are given by

$$1 \mp (z - 1)(p_A t_{AA}^n + p_B t_{BB}^n) + (z - 1)^2 p_A p_B (t_{AA}^n t_{BB}^n - t_{AB}^n t_{BA}^n) = 0, \qquad (37)$$

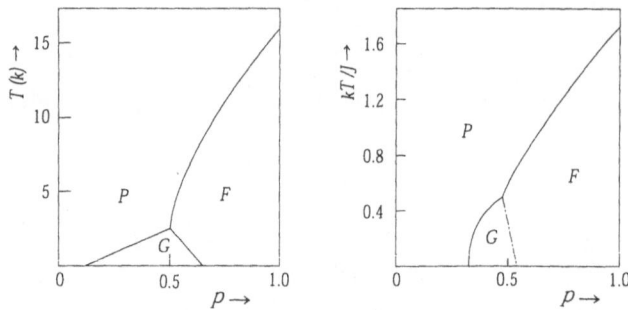

Figure 8. Experimental (left) and theoretical (right) phase diagrams of $Eu_pSr_{1-p}S$.

where t_{AA} is $\tanh\beta J_{AA}$, etc. Upper and lower signs with $n=1$ give the Curie and Neel temperatures, respectively. Upper sign with $n=2$ gives the spin glass transition temperatures. It was shown that when a proper second neighbor interaction which enhances frustration exists, the spin glass state becomes easy to appear. Related experiments are those for $Eu_{1-p}Gd_p$, Fe_pCr_{1-p}, etc.

$Eu_pSr_{1-p}S$ is a crystal in which Eu or Sr forms a *fcc* lattice and J_{Eu-Eu} is ferromagnetic or antiferromagnetic when Eu–Eu pair is the first or the second neighbors, respectively, and J_{Eu-Sr} and J_{Sr-Sr} are always zero. The experimental phase diagrams are shown in Fig. 8. The theoretical phase diagram in a site model with first and second neighbor interactions by the square approximation is in qualitative agreement with experiment.[21,22] The method in the pair approximation was applied to the classical planar model, classical Heisenberg model, and quantum Heisenberg-Ising spin glass. The Curie, Néel and spin glass transition temperatures for the former two systems are obtained by replacing $\tanh\beta J$ by $I_1(\beta J)/I_0(\beta J)$ and $L(\beta J)$, respectively. Here $I_n(x)$[23] and $L(x)$ are the Bessel function with imaginary argument, and Langevin function, respectively. It was shown for quantum system that, when the anisotropic parameter η ($=0$: Ising, $=1$: isotropic Heisenberg) exceeds a critical value η_0, the spin glass disappears (Fig. 9).[24]

It was shown that the Sherrington-Kirkpatrick result for the infinitely long ranged model is reproduced by taking the limit N in a pair approximation with a distribution $P(J)$ such that

$$\int J\,P(J)\,dJ = J_0/N\,,$$

$$\left(\int J^2\,P(J)\,dJ\right)^{\frac{1}{2}} = J_\Delta/N^{1/2}\,. \tag{38}$$

II.4 Discrete and the Continuous Solution of the Integral Equation at $T=0$

The integral equation (22), that is (24) with the kernel (25) for $z=3$, has the following exact discrete solutions,[25]

$$g(h) = \sum_{n=-N}^{N} u_n\delta(h - n/N)\,, \qquad N = 1,2,3,\ldots \tag{39}$$

where the coefficient u_n satisfies

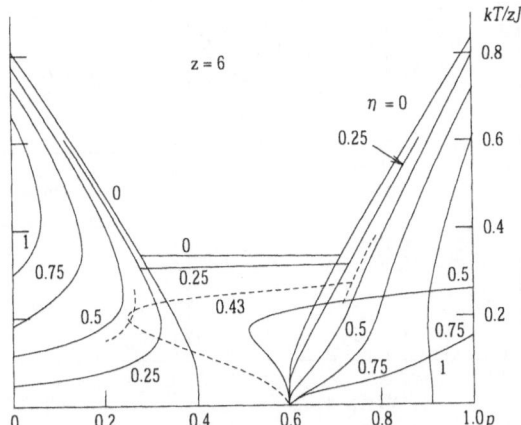

Figure 9. Phase diagram of the random Heisenberg-Ising model. When $\eta < \eta_c$ the spin glass does not appear.

$$\sum_{n=-N}^{N} u_n u_{l-n} = u_l, \qquad -N+1 \le l \le N-1, \qquad (40)$$

$$u_l = 0, \qquad N+1 \le |l|,$$

$$\sum_{l=-N}^{N} u_l = 1.$$

The symmetric case of (40) ($u_l = u_{-l}$) for a given N is called the KFIFG (Katsura N) equation.[25,26] The number of solutions of (40) for a given N is 2^N. Among them, physically meaningful solutions are restricted to those for which u_n is real and $0 \le u_n \le 1$ (u_n is a probability).

The physically acceptable solutions include:
1. $u_0 = 1$ and all other $u_n = 0$.
2. $u_{-N} = u_0 = u_N = 1/3$, and all other $u_n = 0$.
3. When N is not a prime and M is a divisor of N, then the u_n for such N include the u_n for M. For example, when N is even, a solution $u_{\pm N} = (2\sqrt{2}+1)/14$, $u_{\pm N/2} = (3-\sqrt{2})/14$, $u_0 = (3-\sqrt{2})/7$ and all other $u_n = 0$, exists. That is, the number of physically acceptable solutions is $\omega(N) + 1$, where ω is the number of divisors of N. The Gröbner basis of (40) for given N leads

$$F_N(u_0) = 0,$$
$$f_{Nl}(u_0) = u_l, \qquad (41)$$

where $F_N(u_0)$ and $f_{Nl}(u_0)$ are polynomials of u_0 of degree 2^N and $(2^N - 1)$, respectively.

From numerical calculation of Eq. (41), as shown in Fig. 10, we found that as N tends to infinity u_l, becomes roughly continuous except for u_{-N}, u_0, u_N. This fact suggests that in the limit of $N \to \infty$, the solution tends to three delta functions at

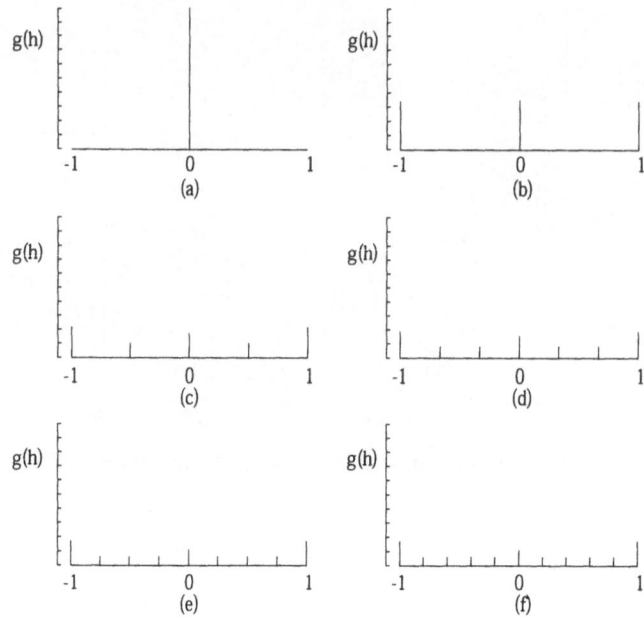

Figure 10. Discrete distribution function of the effective field. (a) $N = 1, 2, 3, 4, 5, \ldots$ (b) $N = 1, 2, 3, 4, 5 \ldots$ (c) $N = 2, 4, \ldots$ (d)$N = 3$, (e) $N = 4$, (f) $N = 5$.

$h = \pm 1$ and at $h = 0$ and a continous function in $-1 < h < 1$, which is now to be seeked. At $T = 0$, and $p = 1/2$, the kernel $K(x, y)$ in (25) is calculated to be

$$K(x, y) = 2\pi \delta(y) \cos x - \frac{\sin(y - x)}{y} - \frac{\sin(y + x)}{y} + \frac{\sin(y - x)}{y - x} + \frac{\sin(y + x)}{y + x}. \quad (42)$$

Here we use the addition theorem of spherical Bessel functions

$$\frac{\sin(y \pm x)}{y \pm x} = \sum_{n=0}^{\infty} (\mp 1)^n (2n + 1) j_n(x) j_n(y). \quad (43)$$

Let

$$S(x) = a + b \cos x + \sum_{n=0}^{\infty} d_n j_n(x). \quad (44)$$

Inserting (43) and (44) into (24), we get algebraic equations for unknowns a, b, d_l

$$a = a^2 + \frac{b^2}{2},$$

$$b = 2ab + \frac{b^2}{2} + bd_0 - \sum_l \sum_m (-1)^{l+m} d_{2l} d_{2m} I_{0,2l,2m},$$

$$d_{2n} = 2a d_{2n} + 2b(1 + 4n) \sum_l (-1)^l d_{2l} I_{2l,2n}$$

$$+ (1 + 4n) \sum_l \sum_m (-1)^{l+m} d_{2l} d_{2m} I_{2l,2m,2n}, \quad (45)$$

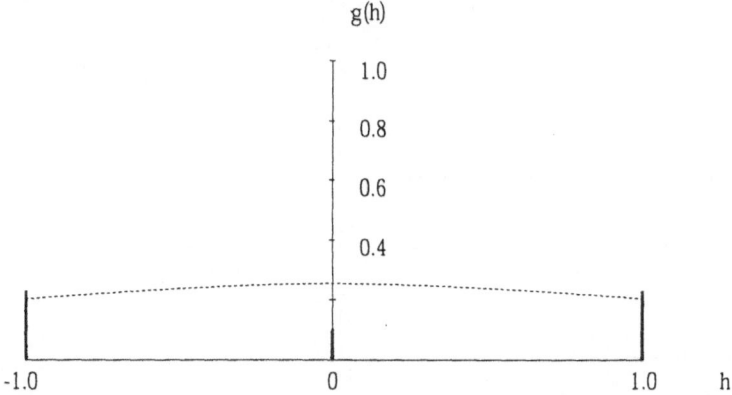

Figure 11. Continuous distribution function of the effective field.

where

$$I_{2l,2m} = \frac{1}{\pi} \int_{-\infty}^{\infty} \cos x \, j_{2l}(x) \, j_{2m}(x) \, dx \, ,$$

$$I_{2l,2m,2n} = \frac{1}{\pi} \int_{-\infty}^{\infty} j_{2l}(x) \, j_{2m}(x) \, j_{2n}(x) \, dx \, . \tag{46}$$

The integral (46) can be obtained by residue calculation. We truncate d_l beyond d_4; then we get
  1) $a = 1$, $b = 0$, $d_{2n} = 0$
  2) $a = 1/3$, $b = 2/3$, $d_{2n} = 0$
  3) $a = 0.10683$, $b/2 = 0.21843$, $d_0 = 0.45631$, $d_2 = 0.05759$
as physical solutions.

The solution 1) represents the paramagnetic solution. The solution 2) represents the spin glass solution previously mentioned. The solution 3) represents another spin glass solution with a continuous distribution. That is,

$$g(h) = \begin{cases} a\delta(h) + \dfrac{b[\delta(h+1) + \delta(h-1)]}{2} + \dfrac{d_0}{2} - \dfrac{d_2(3h^2 - 4)}{4}, & \text{if } |h| \le 1; \\ 0, & \text{if } |h| > 1. \end{cases} \tag{47}$$

Equation (47) is shown in Fig. 11.

In the case where the concentration p of the ferromagnetic bonds varies from 0 to 1, several other solutions than those stated above appear.[27] We determine the unknown coefficients a, b, d_l, (l up to 3). Then the obtained solutions as functions of p are shown in Fig. 12. We have three critical concentrations, $p_2 = 7/8 = 0.875$, $p_1 = 0.869427$, and $p_3 = 11/12$. p_2 is the point where the discrete symmetric solution disappears and the discrete asymmetric solution appears. p_1 is the point where the asymmetric continuous solution disappears and connects to the symmetric continuous solution. p_3 is the point where the asymmetric continuous solution disappears. p_3 was obtained also by Kwon and Thouless.[28] Solving the integral equation at finite temperature is being attempted and the integral kernel is evaluated.[29]

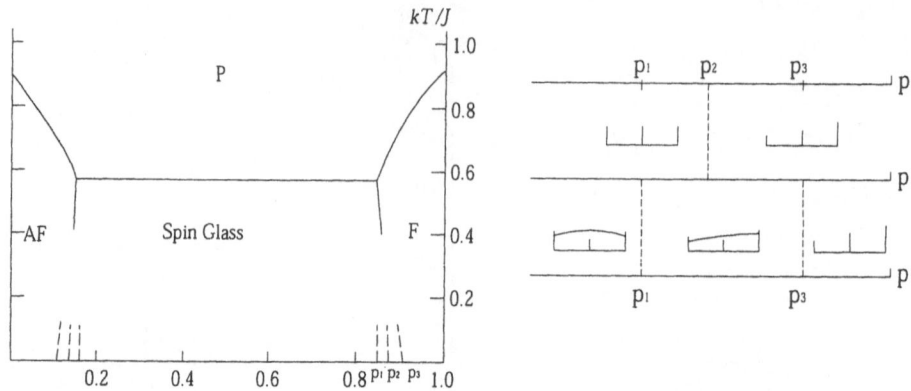

Figure 12. Phase diagram of the random Ising model in the Bethe lattice. $p_1 = 0.869427$, $p_2 = 0.875$, and $p_3 = 0.91667$ are the critical concentrations at $T = 0$. Types of the distribution function in the repective range are shown.

III. Conclusion

In this paper we describe the effective field theory of the regular and random Ising models. The method consists of reducibility of the density matrices of a vertex, a pair, and a cluster. The one which uses the consistency of a vertex and a pair is equivalent to the Bethe approximation. The one which uses a consistency of a vertex and a cluster is the exact solution of the cactus tree lattice. The one which uses the consistency of a vertex, a pair, and a cluster, is equivalent to Kikuchi's CVM in the regular system. Our method was extended to random systems and we treated the spin glass. Phase diagrams of several systems which approximate the experiments are obtained. In random system the distribution function of the effective field is considered and an integral eqution is formulated. The integral equation in the Bethe lattice was solved exactly at $T = 0$ and approximately at $T \neq 0$, and the phase diagram of the spin glass was derived.

Acknowledgements

The author expreses sincere thanks to Professor Kikuchi and the organising committee who gave the author a chance of this talk.

References

1. C. Domb, *Adv. Phys.* **9**, 149 (1960).
2. G. A. Baker, Jr., H. E. Gilbert, J. Eve, and G. S. Rushbrooke, *A Data Compendium of Linear Graph*, BNL 50053 T-46B (Brookhaven National Laboratory 1967) and references therein. See also T. Morita, *J. Math. Phys.* **13**, 115 (1972).
3. M. Kurata, R. Kikuchi and T. Watari, *J. Chem. Phys.* **21**, 435 (1953).
4. S. Katsura and M. Takizawa, *Prog. Theor. Phys.* **51**, 82 (1974).
5. R. Kikuchi, *Phys. Rev.* **81** (1951) 988.
6. S. Katsura and S. Fujiki, *J. Phys. C* **13**, 4711; *J. Phys. C* **13**, 4723 (1980).

7. S. Fujiki, Y. Abe, and S. Katsura, *Comp. Phys. Comm.* **25**, 119 (1982).
8. T. Morita, *Physica A* **98**, 566 (1980).
9. S. Fujiki, M. Katori, and M. Suzuki, *J. Phys. Soc. Jpn.* **59**, 2681 (1990) and private communication.
10. M. Katori and M. Suzuki, *Prog. Theor. Phys.* Suppl. **115**, 83 (1994).
11. S. Katsura and S. Fujiki, *J. Phys. C* **12**, 1087 (1979).
12. S. Katsura, *Prog. Theor. Phys.* Suppl. **87**, 139 (1986).
13. F. Matsubara and M. Sakata, *Prog. Theor. Phys.* **55**, 672 (1976).
14. J. S. M. Carlson, J. T. Chayes, L. Chayes, J. P. Sethna, and D. H. Thouless, *Europhys. Lett.* **5**, 355 (1988).
15. S. Katsura, *Physica A* **104**, 333 (1980).
16. R. Kikuchi and H. Sato, *Acta Metall.* **22**, 1089 (1974).
17. S. Katsura and I. Nagahara, *J. Phys. C* **13**, 4995 (1980).
18. S. Katsura and R. Kikuchi, *Physica* **123** A, 595 (1984).
19. I. Nagahara, S. Fujiki and S. Katsura, *J. Phys. C* **14**, 3781 (1981).
20. S. Katsura, S. Fujiki and S. Inawashiro, *J. Phys. C* **12**, 2839 (1979).
21. S. Katsura and I. Nagahara, *Z.* Phys. **41**, 349 (1981).
22. S. Katsura and A. Matsuno, *phys. status solidi B* **119**, 73 (1983).
23. S. Katsura and T. Shirakura, *phys. status solidi B* **113**, 327 (1982).
24. S. Katsura and K. Shimada, *phys. status solidi B* **97**, 663 (1980).
25. S. Katsura, W. Fukuda, S. Inawashiro, N. M. Fujiki, and R. Gebauer, *Cell. Biophys.* **11**, 309 (1987).
26. W. Boege, R. Gebauer, and H. Kredel, *J. Symbolic Compt.* **1**, 83 (1986).
27. M. Sasaki and S. Katsura, *Physica A* **155**, 206 (1989); *Physica A* **157**, 1195 (1989).
28. C. Kwon and D. J. Thouless, *Phys. Rev. B* **37**, 7649 (1988).
29. M. Seino and S. Katsura, *Prog. Theor. Phys.* Suppl. **115**, 237 (1994).

The Cluster Variation Method, The Cluster Consistency Method, and The Quantum Cluster Variation Method

Tohru Morita

Department of Computer Science
College of Engineering
Nihon University
Koriyama 963
JAPAN

Abstract

A review is given of my work on the Cluster Variation Method (CVM). The Cluster Consistency Method (CCM), which is the consistency approach equivalent to the Cluster Variation Method, is first presented and its equivalence to the variational approach is shown. We then consider the Quantum Cluster Variation Method (QCVM), which was developed to discuss high-temperature as well as low-temperature properties of quantum systems within an approximation.

I. Introduction

In a recent review of my papers on the Cluster Variation Method,[1] I recalled the textbook of Fowler and Guggenheim[2] where approximate methods of statistical mechanics are classified into two approaches, variational approach and consistency approach. The Cluster Variation Method proposed by Kikuchi, presented in Refs. 3, 4 and also in Ref. 5, was based on the maximum of a term of the partition function and its reformulations, *e.g.* my papers Refs. 6 and 7, are based on the variational principle of the free energy. The variation of the free energy was taken formally in Ref. 8 and then it was found that the result of the variation provides a method which is equivalent to the original variational approach and should be classified in the consistency approach. The formal structure of both approaches in the cluster variation method became elegantly given in terms of a Möbius function.[9-11]

In the present paper, the two approaches in the Cluster Variation Method will be cited by CVM and CCM (Cluster Consistency Method). They give equivalent results.

Theory and Applications of the Cluster Variation and Path Probability Methods
Edited by J.L. Morán-López and J.M. Sanchez, Plenum Press, New York, 1996

Here we shall explain the CCM first and then show that the distribution functions obtained in the CCM satisfy the variational principle of the CVM. We next proceed to the QCVM, which is an extension of the CVM.[12] The extension intends to discuss the low temperature as well as high temperature properties of a quantum system within an approximation. QCVM was developed for studying the Hubbard model, the $t-J$ model,[13] and the Heisenberg model. The QCVM is applicable to quantum lattice gases. It was applied to a lattice gas equivalent to the Heisenberg model in Ref. 14. This application is now discussed including some new results.

The presentations in sections II and III follow another review of my papers given in Ref. 15.

II. Set of Clusters Q and Sets q_α

In the Cluster Variation Method, we usually consider a lattice, and a set of lattice sites is called a cluster. An approximation of the method is characterized by a set of clusters, which we denote by Q. In the set Q, we include **0** and **1**, where **0** is the cluster of no lattice sites and **1** is the cluster of all the lattice sites in the system. For each cluster α in Q except **0**, we consider a distribution function ϱ_α, which satisfies the normalization condition

$$\mathrm{Tr}_\alpha \, \varrho_\alpha = 1, \tag{1}$$

where Tr_α denotes the summation over all the configurations of the cluster α. The word "distribution function" is used. But it is appropriate to use "density matrix" for ϱ_1 and "reduced density matrix" for ϱ_α for α not equal to **1** for quantum systems.

Greek characters α, β and γ are used to denote clusters in Q below. An order relation $\beta \leq \alpha$ is introduced to show that β is a subcluster of α, and then ϱ_β and ϱ_α for $\beta < \alpha$, that is $\beta \leq \alpha$ and $\beta \neq \alpha$, are related by the condition

$$\varrho_\beta = \mathrm{Tr}_{\alpha \backslash \beta} \, \varrho_\alpha, \qquad \beta < \alpha. \tag{2}$$

This condition is called a consistency condition or a reducibility condition. The clusters in Q form a partially ordered set by this order relation. As an illustration how Q is constructed, an example given in Ref. 12 is shown in Fig. 1.

Here we introduce the Möbius function $\mu(\beta, \alpha)$ for β and α in the partially ordered set Q.[16] It is defined such that $\mu(\beta, \alpha)$ can take nonzero value only if $\beta \leq \alpha$ and they satisfy

$$\sum_{\substack{\beta \\ \gamma \leq \beta \leq \alpha}} \mu(\beta, \alpha) = \delta_{\gamma \alpha}. \tag{3}$$

In order to express an operator for a cluster β in Q excluding **1**, we introduce a complete set of linearly independent operators q_β for each β in $Q \backslash \{1\}$, so that any operator for the cluster β can be expressed as a linear combination of the operators in q_β.[11,17] We choose q_β such that it consists of unity and those denoted by $s_{\gamma \nu}$ for $\gamma \in Q$ and $\gamma \leq \beta$ where ν labels different operators for the same cluster γ. If $\gamma < \beta$, q_γ is a subset of q_β. Now equality (2) can be expressed as the consistency of all the averages of the operators in the set q_β, which gives

$$\langle s_{\gamma \nu} \rangle_\beta = \langle s_{\gamma \nu} \rangle_\alpha, \qquad \gamma \leq \beta, \quad \beta < \alpha, \tag{4}$$

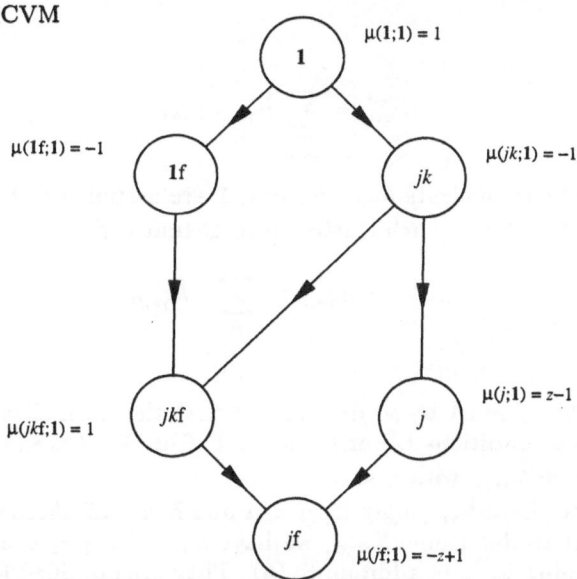

Figure 1. Set Q for the pair approximation in the QCVM and the Möbius function $\mu(\alpha, 1)$ for α in Q, taken from Ref. 12.

where $\langle a \rangle_\alpha$ for an operator a of a cluster α denotes

$$\langle a \rangle_\alpha = \mathrm{Tr}_\alpha \, a \varrho_\alpha. \tag{5}$$

This set (4) is seen to be equivalent to its subset

$$\langle s_{\gamma\nu} \rangle_\gamma = \langle s_{\gamma\nu} \rangle_\alpha, \qquad \gamma < \alpha. \tag{6}$$

This set (6) gives an enumeration of the number of conditions in the consistency conditions without repetition.[18]

In CVM, the Hamiltonian H_1 of the system we consider is assumed to be expressed as follows

$$H_1 = \sum_{\substack{\gamma\nu \\ \gamma<1}} h_{\gamma\nu,1} \, s_{\gamma\nu}, \tag{7}$$

where $h_{\gamma\nu,1}$ are coefficients.

III. The Cluster Consistency Method

In this section, α, β and γ are clusters in $Q\backslash\{0, 1\}$, except when otherwise is explicitly stated.

In the CCM, the distribution functions ϱ_α are expressed as

$$\varrho_\alpha = \exp\{(F_\alpha - H_\alpha^{\mathrm{eff}})/(k_B T)\}, \tag{8}$$

where k_B is the Boltzmann constant, T is the temperature and H_α^{eff} is expanded in the "Fourier series" in the set of operators q_α as

$$H_\alpha^{\text{eff}} = \sum_{\substack{\gamma\nu \\ \gamma \leq \alpha}} h_{\gamma\nu,\alpha}\, s_{\gamma\nu}. \tag{9}$$

In the CCM, the coefficients $h_{\gamma\nu,\alpha}$ for $\alpha < 1$ are assumed to be the sum of $h_{\gamma\nu,1}$ and contributions $\bar{h}_{\gamma\nu,\beta}$ from such clusters β in Q that $\beta > \gamma$ and $\beta \not\leq \alpha$

$$h_{\gamma\nu,\alpha} = h_{\gamma\nu,1} + \sum_{\substack{\beta \\ \gamma<\beta<1 \\ \beta\not\leq\alpha}} \bar{h}_{\gamma\nu,\beta}. \tag{10}$$

Now F_α and $\bar{h}_{\gamma\nu,\alpha}$ must be so determined that the normalization condition (1) and the consistency conditions (2) or (6) are satisfied. We have (1) to determine F_α and (6) to determine $\bar{h}_{\gamma\nu,\alpha}$ with $\gamma < \alpha$.

In the above, we have $\bar{h}_{\gamma\nu,\alpha}$ only for $\gamma < \alpha$ and hence all these can be determined by (6). If we want to determine $h_{\gamma\nu,\alpha}$, we have $h_{\gamma\nu,\alpha}$ for $\gamma \leq \alpha$ and hence we need equations determining $h_{\gamma\nu,\gamma}$ in addition to (6). They are obtained from (10). In order to get them, we write (10) for $\alpha = \gamma$

$$h_{\gamma\nu,\gamma} = h_{\gamma\nu,1} + \sum_{\substack{\beta \\ \gamma<\beta<1}} \bar{h}_{\gamma\nu,\beta}. \tag{11}$$

We define $\bar{h}_{\gamma\nu,\gamma}$, which do not appear in (10), by $\bar{h}_{\gamma\nu,\gamma} = -h_{\gamma\nu,\gamma}$ and then we have

$$h_{\gamma\nu,1} = -\sum_{\substack{\beta \\ \gamma\leq\beta<1}} \bar{h}_{\gamma\nu,\beta}. \tag{12}$$

Substituting this into (10), we obtain

$$h_{\gamma\nu,\alpha} = -\sum_{\substack{\beta \\ \gamma\leq\beta\leq\alpha}} \bar{h}_{\gamma\nu,\beta}, \tag{13}$$

which is consistent with (12) if we put $\bar{h}_{\gamma\nu,1} = 0$. The Möbius inversion formula[16] applied to (13) with $\gamma \leq \alpha \leq 1$ states that

$$-\bar{h}_{\gamma\nu,\alpha} = \sum_{\substack{\beta \\ \gamma\leq\beta\leq\alpha}} h_{\gamma\nu,\beta}\, \mu(\beta,\alpha). \tag{14}$$

Putting $\alpha = 1$ in this equation and using $\bar{h}_{\gamma\nu,1} = 0$ and $\mu(1,1) = 1$, we obtain

$$h_{\gamma\nu,1} = -\sum_{\substack{\beta \\ \gamma\leq\beta<1}} h_{\gamma\nu,\beta}\, \mu(\beta,1). \tag{15}$$

These are the equations determining $h_{\gamma\nu,\gamma}$, when we desire to determine $h_{\gamma\nu,\alpha}$.

If γ is such that α in Q satisfying $\gamma \leq \alpha < 1$ is nothing but γ, γ is called a basic cluster and then $\mu(\gamma,1) = -1$ by (3) and $h_{\gamma\nu,\gamma} = h_{\gamma\nu,1}$ by (15).

The form (8), (9) and (10) with (6) was obtained in Refs. 8 and 18, by the variational principle of the CVM. The form (8) and (9) with (15) and (6) was obtained in Ref. 10.

IV. The Cluster Variation Method

In order to show that the set of ϱ_α given by CCM in the preceding section satisfy the variational principle of CVM, we define S_α for α in Q by

$$S_\alpha = -k_B \, \text{Tr}_\alpha \, \varrho_\alpha \ln \varrho_\alpha, \tag{16}$$

and we start from the variational principle of the free energy that

$$F = \min_{\varrho_1} (E_1 - TS_1), \tag{17}$$

where S_1 is given by (16) for $\alpha = 1$ and

$$E_1 = \langle H_1 \rangle_1 = \sum_{\substack{\gamma\nu \\ \gamma < 1}} h_{\gamma\nu,1} \, \langle s_{\gamma\nu} \rangle_1. \tag{18}$$

Here the normalization condition (1) for $\alpha = 1$ is imposed in the variations. We now define $\bar{S}_\beta$ for β in Q such that

$$S_\alpha = \sum_{\substack{\beta \\ \beta \leq \alpha}} \bar{S}_\beta. \tag{19}$$

The Möbius inversion formula then states that

$$\bar{S}_\beta = \sum_{\gamma \leq \beta} S_\gamma \, \mu(\gamma,\beta), \tag{20}$$

which is confirmed to be the solution of (19) simply by substituting this into the righthand side of (19) and using (3).

In the CVM, we assume that $\bar{S}_1 = 0$ and then (20) for $\beta = 1$ gives

$$S_1 = -\sum_{\gamma < 1} S_\gamma \, \mu(\gamma,1). \tag{21}$$

By using (18), (6) and (21), we can express the variational function $E_1 - TS_1$ in (17) in terms of ϱ_β with $\beta < 1$. In the CVM, we determine the free energy as the minimum value of this expression with respect to the variations of ϱ_β with $\beta < 1$, under the normalization conditions (1) and the reducibility conditions (2) or (6) for $\alpha < 1$. We write this variational function by using (15), (6) and (21) as

$$F_1\{\varrho_\beta, \beta \in Q\backslash\{0,1\}\} = -\sum_{\substack{\beta \\ \beta < 1}} F_\beta\{\varrho_\beta\} \, \mu(\beta,1), \tag{22}$$

where

$$F_\beta\{\varrho_\beta\} = \text{Tr}_\beta \, \varrho_\beta(H_\beta^{\text{eff}} + k_B T \ln \varrho_\beta). \tag{23}$$

The variations of this expression (22) with respect to ϱ_β for $\beta < 1$, considering (1) and ignoring the constraints (6), give the ϱ_α given by (8). The obtained ϱ_α satisfy (1) and (6) and hence the variations considering (1) and (6) give the same result.

V. The Quantum Cluster Variation Method

The set Q in the CVM is defined as a set of clusters of lattice sites. But the properties used for Q are that Q is a partially ordered set, the distribution function ϱ_α is defined for every α in Q, and ϱ_β is calculated if ϱ_α with $\alpha > \beta$ is known. The QCVM[12] is introduced for Bose or Fermi lattice gas. In QCVM, we consider a set Q_c of clusters of lattice sites and then consider the Q such that it involves two elements α and αf for an element α in Q_c. $q_{\alpha f}$ is a subset of q_α and consists of linearly independent operators which are 1, linear or quadratic forms of operators creating or annihilating a particle on a site in the cluster α. In $\varrho_{\alpha f}$ is a quadratic form of creation and annihilation operators. $\varrho_{\alpha f}$ is determined from ϱ_α by the conditions that

$$\langle s_{\gamma\nu}\rangle_{\alpha f} = \langle s_{\gamma\nu}\rangle_\alpha \qquad \text{for } \gamma\nu \in q_{\alpha f}. \tag{24}$$

We assume the order that $\alpha f < \alpha$ and $\beta f < \alpha f$ if $\beta < \alpha$. We include the cluster 1f in Q and H_{1f}^{eff} is the Hamiltonian of an ideal gas.

VI. The Heisenberg Model

In reference 14, the pair approximations in QCVM is applied to the Heisenberg model, whose Hamiltonian is given by

$$H_{\mathbf{H}} = -J \sum_{(j,k)} s_j \cdot s_k - h \sum_j s_j^z, \tag{25}$$

where s_j is the spin operator on the jth lattice site and s_j^z is its z component, J and h are the exchange integral and the external field, and (j,k) denotes that the sum is taken over pairs of nearest-neighbour lattice sites j and k.

We study the system described by the Hamiltonian H_1 that is expressed as follows

$$H_1 = \mu \sum_j \sum_\sigma n_{j\sigma} - h \sum_j (n_{j+} - n_{j-}) - \frac{J}{2} \sum_{(j,k)} \sum_\sigma (a_{j\sigma}^+ a_{k\sigma} + a_{k\sigma}^+ a_{j\sigma})$$

$$- J \sum_{(j,k)} \sum_\sigma n_{j\sigma} n_{k\sigma} + U \sum_j \sum_\sigma n_{j\sigma}(n_{j\sigma} - 1), \tag{26}$$

where $\sigma = \pm$, $n_{j\sigma} = a_{j\sigma}^+ a_{j\sigma}$, and $a_{j\sigma}^+$ and $a_{j\sigma}$ are Bose operators. U is tended to infinity to prohibit double occupancy of particles of the same $j\sigma$. If we denote the sums of terms with $\sigma = +$ and $-$, respectively, in (26) as H_+ and H_- and set $\mu = zJ/2$, we can show that each of H_+ and H_- is equivalent to $H_{\mathbf{H}}$, where z is the coordination number of the lattice. In this identification, we have to restrict to

$$\langle n_{j+}\rangle + \langle n_{j-}\rangle = 1. \tag{27}$$

The magnetization of a lattice site j is given by

$$\langle s_j^z\rangle = \tfrac{1}{2}(\langle n_{j+}\rangle - \langle n_{j-}\rangle). \tag{28}$$

Equation (26) represents a gas of Bose particles with spin $\frac{1}{2}$. Spin waves at low temperatures are expressed by Bose particles of long wave-length. The gas keeps the

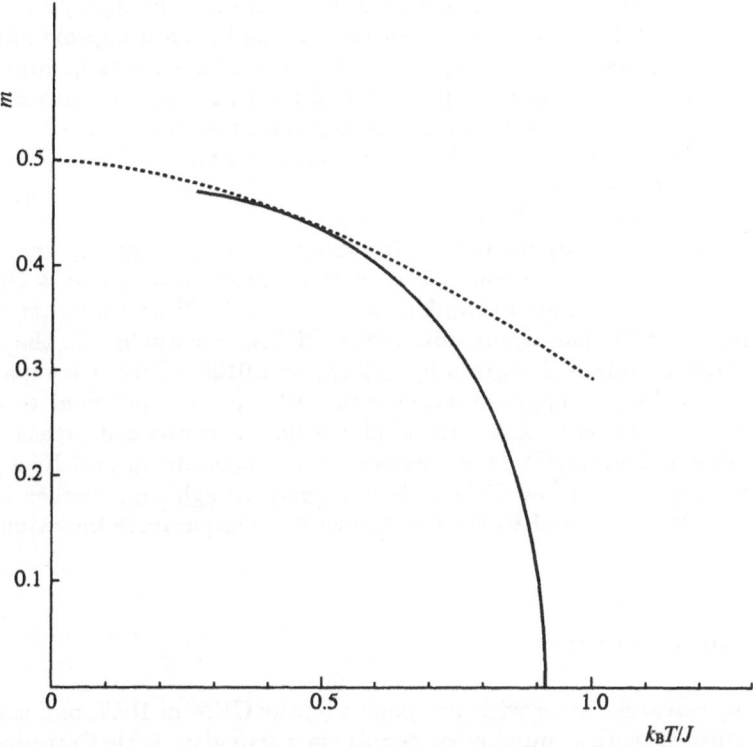

Figure 2. The spontaneous magnetization as a function of the temperature T in the pair approximation of QCVM for the Heisenberg model on the *sc* lattice. The dashed line is the spin wave theory without interactions.

symmetry of up and down so that we can discuss the phase transition between the ferro- and paramagnetic phases.

In the paramagnetic state at zero external field, we get $\langle n_{j+} \rangle = \langle n_{j-} \rangle = \frac{1}{2}$ by (27) and (28). In a study of the 2-dimensional Heisenberg model, Takahashi[19] obtained low-temperature properties by studying the Hamiltonian H_- which is described in terms of Bose operators a_{j-}^+ and a_{j-} and imposing the condition $\langle n_{j-} \rangle = \frac{1}{2}$, which is expected at zero external field. If we use (26), we get the equivalent result by his method but also can extend the argument to non-zero external field or non-zero magnetization.

We applied the pair approximation of the QCVM in Ref. 14, where the set of clusters Q as shown in Fig. 1 consists of j and jf for each lattice site j, jk and jkf for each pair of nearest-neighbour lattice sites j and k, and 1f for the whole system 1. The reduced density matrices ϱ_α for α in the set Q are expressed as (8) and H_α^{ef} given by (9) involve 14 parameters. The parameters were determined to satisfy the consistency conditions provided by the CCM. The curve in solid line in Fig. 2 shows the spontaneous magnetization of the system on the simple cubic lattice, obtained by numerical calculation. In the figure, dashed line shows the curve when we consider the spin wave excitation without interaction. The figure indicates that the curve in the present approximation is good down to one half of the critical temperature T_c but is not good below it. The critical temperature T_c is given by $k_B T_c/J = 0.915$, which is compared with 0.91024 and 0.9107 in the pair and square approximation[20] in the CVM. The pair approximation is equivalent to the constant coupling approximation.[21] The value by high-temperature expansion and the Padé approximant is 0.8387.[22] At low temperatures, we have a spontaneous magnetization where $\langle n_{j-} \rangle$ is very small if the external field h is positive and $\langle n_{j+} \rangle$ is almost unity, which occurs in such a way that the excitation energy contributing $\langle n_{j+} \rangle$ becomes very small proportionally to the temperature, without a Bose condensation.

A corresponding study for the antiferromagnet was attempted, where the Hamiltonian is the sum of two Hamiltonians of Bose lattice gases, one is the one based on a Néel state as was done by Anderson and Kubo,[23−25] and another based on the other Néel state. The pair approximation in QCVM was applied to the system. The Néel temperature obtained is given by $k_B T_N/J = 1.0139$, which is compared with the value 1.0155 in the pair approximation of in CVM that is equivalent to the constant coupling approximation.[26] The numerical calculation shows a decrease of sublattice magnetization as lowering the temperature, which suggests an anti-Néel point.

It is my regret that the QCVM is not good enough and further improvement over the QCVM is required to get the correct low-temperature behaviours near zero temperature.

Acknowledgements

I started my research career with two papers on the CVM in 1957, and is still working on it. Discussion with a number of people, in particular, with Professor Tomoyasu Tanaka in 1966 and Professor Kazuyuki Tanaka in recent years, suggested developments stated in this report. Numerical calculation including Fig. 2 was done by Midori Takahashi in her graduate study in Tohoku University. I appreciate their collaboration.

References

1. T. Morita, *Prog. Theor. Phys.* Suppl. No. 115, 27 (1994).
2. R. Fowler and E. A. Guggenheim, *Statistical Thermodynamics*, 2nd Ed. (Cambridge U. P., 1952) Chap. XIII.
3. R. Kikuchi, *Phys. Rev.* **81**, 988 (1951).
4. M. Kurata, R. Kikuchi and T. Watari, *J. Chem. Phys.* **21**, 434 (1953).
5. R. Kikuchi, *Prog. Theor. Phys.* Suppl. No. 115, 1 (1994).
6. T. Morita, *J. Phys. Soc. Jpn.* **12**, 753 (1957).
7. T. Morita, *J. Phys. Soc. Jpn.* **12**, 1060 (1957).
8. T. Morita, *J. Math. Phys.* **13**, 115 (1972).
9. T. Morita, *J. Stat. Phys.* **59**, 819 (1990).
10. T. Morita, *Prog. Theor. Phys.* **85**, 243 (1991).
11. T. Morita, *Phys. Lett. A* **161**, 140 (1991).
12. T. Morita, *Prog. Theor. Phys.* **92**, 1081 (1994).
13. K. Tanaka, H. Ebisawa and T. Morita, *Physica C* **235-240**, 2187 (1994).
14. T. Morita, *J. Phys. Soc. Jpn.* **64**, 1211 (1995).
15. T. Morita, proc. 1994 Korea-Japan Joint Workshop on Statistical Mechanics, J. Korean Phys. Soc. **28** S333 (1995).

16. G.-C. Rota, *Z. Wahrsch.* **2**, 340 (1964).
17. T. Morita and T. Tanaka, *Phys. Rev.* **145**, 288 (1966).
18. T. Morita, *J. Stat. Phys.* **34**, 319 (1984).
19. M. Takahashi, *Prog. Theor. Phys.* Suppl. No. 87, 233 (1986).
20. T. Tanaka, K. Hirose and K. Kurati, *Prog. Theor. Phys.* Suppl. No. 115, 41 (1994).
21. P. W. Kasteleijn and J. Van Kranendonk, *Physica* **22**, 317 (1956).
22. G. A. Baker, H. E. Gilbert, J. Eve and G. S. Rushbrooke, *Phys. Rev.* **162**, 800 (1967).
23. P. W. Anderson, *Phys. Rev.* **86**, 694 (1952).
24. R. Kubo, *Phys. Rev.* **87**, 568 (1952).
25. T. Morita, *Prog. Theor. Phys.* **20**, 614 (1958).
26. P. W. Kasteleijn and J. Van Kranendonk, *Physica* **22**, 367 (1956).

Note added in proof: Stimulated by discussion with Professor Hiroshi Sato in the occasion of the Workshop on the Cluster Variation Method and the Path Probability Method in Teotihuacan, Mexico, I wrote a paper entitled as Time-Dependent Cluster Consistency Method and includes the Path Probabiluty Method as a special case. That paper was published in *Prog. Theor. Phys.* **94**, *761 (1995)*.

The Cluster Expansion Method

J. M. Sanchez

Center for Materials Science and Engineering
The University of Texas at Austin
Austin, Texas 78712
U.S.A.

Abstract

The Cluster Expansion Method leading to the formal description of configurational functions in alloys is reviewed in the context of several recent developments, modifications and numerous applications. A general and rigorous formulation of the method is shown to unify the approaches proposed by the author and collaborators in 1984 with the concentration restricted sum method of Asta and co-workers.

I. Introduction

The Cluster Variation Method (CVM), developed by Professor Ryoichi Kikuchi[1] over forty five years ago to describe the configurational entropy of Ising lattices, is widely recognized as a major contribution to theoretical statistical physics. At present, the CVM has clearly transcended its intended application to Ising lattices and is now routinely applied to the study of numerous other disordered or partially ordered systems. Thus, while the interest in Ising lattices has declined over the years, the CVM has emerged as one of the main tools for the study of the subtle effects of SRO present in real alloys and other disordered systems. The diverse and numerous applications found in this volume are excellent examples of the widespread acceptance currently enjoyed by the CVM.

In the most general sense, the CVM extends the intuitive and appealing concepts of mean-field theory into the realm of systems with strong statistical correlations. Although the CVM requires that we abandon the simplicity of the single-site approximation, higher level approximations are essential for the correct theoretical description of alloys. In principle, the methodology implicit in the CVM can also be used to describe a wide range of configurational dependent problems that are commonly encountered in condensed matter physics. In practice, however, and with the notable exception of the kinetic theory known as the Path Probability Method, also developed by Professor Kikuchi,[2] the idea of using clusters to describe strong correlations has

Theory and Applications of the Cluster Variation and Path Probability Methods
Edited by J.L. Morán-López and J.M. Sanchez, Plenum Press, New York, 1996

been primarily restricted to the calculation of configurational entropies in statistical mechanics problems.

In dealing with generalizations and applications of the CVM, a formal mathematical framework is essential. In particular, the formalism should allow us to deal systematically with the complexities encountered outside the secure and well understood domain of the single site-approximation. All the elements of such a "cluster algebra" were introduced in early CVM studies of Ising lattices[3-5] and then formalized by Sanchez et al.[6] several years ago. The essence of the method is to describe the configurational energy of alloys and, in general, any function of configuration, in terms of orthogonal discrete Chebyshev's polynomials. The configurational average of these basis functions are the well known and widely used multisite correlation functions.[3-5,7] For example, in this representation, the average configurational energy takes the form of a bilinear expression in the effective cluster interactions (ECI) and the corresponding multisite correlation functions. This form of the energy is referred in the literature as the Cluster Expansion (CE). It should be noted that some elements of the cluster algebra, such as the multisite correlation functions, are standard items in statistical physics and, as such, have been in use for many years, although in a different context.[3-5,7] The *important* aspect of the method described in Ref. 6 is that *any* function of configuration can be cluster expanded in terms of basis functions that have a simple and intuitive meaning, namely that their configurational averages are the familiar multisite correlation functions.

The CE method has found widespread use in the *ab initio* computation of phase diagrams. Thus, the methodology and basic ideas of the cluster algebra have been described extensively in numerous publications and excellent reviews articles with, in most cases, only minor changes in the original notation.[8-10] More recently, variations of the original method have also been proposed.[11,12] For the purpose of this article, we will distinguish between two formulations of the CE method which are based on different definitions of the scalar product. The restricted sum method of Asta et al.[11] defines the scalar product in a subspace of configurational space with fixed concentrations, whereas in the unrestricted sum method, developed in Refs. 6 and 12, the scalar product is defined in the full configuration space. As we shall see, these two schemes are closely related although there are some limitations on the range of applicability of the restricted sum method.

Although the most common application of the CE method is the expansion of the configurational energy, the method can be used quite generally to study other configuration dependent quantities. For example, the CE method has been applied to describe the vibrational free energy by Sanchez et al.[13] using a Debye-Grüneisen model and subsequently by Garbulski and Ceder[14] in the harmonic approximation. The cluster expansion of the vibrational free energy results in a straightforward renormalization of the interactions giving rise to temperature dependent ECIs.

The CE method brings both conceptual simplicity and a rigorous foundation to the first principles computation of thermodynamic properties in disordered systems. In practice, however, cluster expansions are difficult to carry out and are generally unavailable for realistic cases of the total energy or, for that matter, even selected components of the total energy of alloys. Thus, the method has been used primarily as a guiding principle to fit the energy of disorder systems in terms of an *a priori* set of ECI's. This procedure has led to some criticism and misunderstanding of the approach. In particular, the Ising-like form of the bilinear energy expression resulting from the CE has been misinterpreted as an approximation when, in fact, the expression is exact. The deeply rooted bias against the Ising model as a realistic model of alloys has resulted in extraordinary efforts to show the shortcomings of the CE. Among the most recent efforts is that of Gonis et al.[15] who have developed arguments leading to the claim that the cluster expansion is valid for finite systems (of any size) but breaks down in the thermodynamic limit. As we shall see, the issue of the thermodynamic limit is quite irrelevant to the difficulties pointed out in Ref. 15, which can be traced to the particular definition of the scalar product used in the restricted sum formulation.[11] The concerns raised by Gonis et al. are easily shown to be groundless if one uses a complete basis set in configurational space, as done in the unrestricted sum method.[6,12]

The objective of this paper is to review the CE, as proposed by the author and collaborators in 1984,[6] in the context of several recent developments, modifications and numerous applications. The next section gives an overview of the cluster algebra commonly used in the cluster expansion method.[12] The emphasis is placed in providing a rigorous foundation for the method as well as to clearly establish the limitations of the restricted sum approach[11] for the treatment of arbitrary functions of configurations.[15]

II. Cluster Expansion

The original development of the cluster algebra was aimed at constructing a basis in configurational space with four attributes.[6,12] In essence, these properties were that the basis be *complete* and *orthogonal* both in the configurational space of *finite clusters* as well as in the *thermodynamic limit*. Orthogonality is, of course, a matter of convenience whereas completeness is an absolute necessity in order to be able to describe *any* function of configuration. The somewhat obvious requirement that the method be applicable to finite cluster is important for actual implementations. Finally, the thermodynamic limit is invoked for mathematical convenience and, clearly, it is not a critical issue in the validity of the cluster expansion since all cases of physical interest correspond to finite systems.

The *unrestricted sum* formulation has these four attributes; *i.e.* the basis sets are orthogonal and complete in finite clusters as well as in the thermodynamic limit. In contrast, the *restricted sum* method is not applicable to small finite clusters since the configuration space is restricted to alloy configurations with a given concentration. Thus, the restricted sum method is, by construction, only applicable to large crystals or, in a stricter sense, only in the thermodynamic limit. Furthermore, in the restricted sum method, the basis functions and the scalar product between them, are defined *only* in a subspace of constant concentration in configurational space.

As shown later in this section, the restricted sum method cannot be used to describe arbitrary function of configuration. The break down of the method is seen to occur when attempting to expand functions that depend on multisite correlations of order-N, where N is the number of sites in the crystal. However, the method is valid (in the thermodynamic limit or for large crystals) when applied to the description of the configurational energy of physical systems for which the *order* of the multisite cluster interactions is *finite*. This limitations of the restricted sum scheme has caused some undue confusion leading Gonis et al.[15] to devise a proof that the cluster expansion of Asta et al.[11] implies the vanishing of the energy for all configurations, an obvious physical and mathematical absurdity. Such difficulties disappear, of course, with complete basis sets which, of necessity, requires the use of unrestricted sums.[6,12] We will return to discuss the validity and possible limitations of the different cluster expansions after introducing the basic mathematical elements of the formalism.

II.1 Basis functions

As it is customary, the configurational variable in a binary system takes values 1 and -1 for each type of atoms. If we consider only a point "cluster," the configurational space is discrete and one-dimensional. In this space, the scalar product between two functions $f(\sigma)$ and $g(\sigma)$, is defined as[12]

$$\langle f(\sigma), g(\sigma) \rangle_\mu = \frac{1}{2\cosh(\mu)} \sum_{\sigma=\pm 1} e^{-\mu\sigma} f(\sigma) g(\sigma), \tag{1}$$

where $e^{-\mu\sigma}$ is a weight function with μ an arbitrary constant. Clearly μ plays the role of a chemical potential and, in the thermodynamic limit and assuming that we are only interested in describing *extensive* functions, it determines the concentrations of those alloy configurations that participate in the scalar product.

For each value of μ, a complete and orthonormal basis in the one-dimensional space spanned by the discrete variable σ is given by a set of two polynomials, $\phi_0^\mu(\sigma)$ and $\phi_1^\mu(\sigma)$ of order 0 and 1, respectively.[12] These are given by

$$\phi_0^\mu(\sigma) = 1, \tag{2}$$

$$\phi_1^\mu(\sigma) = \cosh(\mu)\left[\sigma - \tanh(\mu)\right] = \frac{(\sigma - \bar{\sigma})}{\sqrt{1 - \bar{\sigma}^2}}, \tag{3}$$

with $\bar{\sigma} = \tanh(\mu)$.

Given the polynomials $\{\phi_n^\mu(\sigma)\}$ for the point cluster, an orthogonal and complete basis for any finite cluster, including the whole crystal with N lattice sites, can be easily constructed. Following Ref. 6, we note that the configurational space of a cluster α consisting of n_α lattice sites $\{p_1, p_2, \ldots, p_{n_\alpha}\}$ is obtained by constructing the direct product of the subspaces associated to each site.

Thus, an orthonormal basis in the space spanned by the n_α-dimensional discrete vector $\vec{\sigma}_\alpha = \{\sigma_1, \sigma_2, \ldots, \sigma_{n_\alpha}\}$, with σ_i the occupation operator at site p_i, is given by characteristic functions $\Phi_\alpha^\mu(\vec{\sigma}_\alpha)$ defined by all possible products of the form

$$\Phi_\alpha^\mu(\vec{\sigma}_\alpha) = \prod_{i=1}^{n_\alpha} \phi_1^\mu(\sigma_i). \tag{4}$$

The corresponding scalar product between two functions $f(\vec{\sigma}_\gamma)$ and $g(\vec{\sigma}_\gamma)$ in the configurational space of cluster γ is defined by[12]

$$\langle f, g \rangle_\mu = \frac{1}{\left[2\cosh(\mu)\right]^{n_\gamma}} \sum_{\sigma_1=\pm 1} \sum_{\sigma_2=\pm 1} \cdots \sum_{\sigma_{n_\gamma}=\pm 1} e^{n_\gamma \mu x} f(\vec{\sigma}_\gamma) g(\vec{\sigma}_\gamma), \tag{5}$$

where, as mentioned, n_γ is the number of sites in γ and

$$x = \frac{1}{n_\gamma} \sum_{i=1}^{n_\gamma} \sigma_i. \tag{6}$$

Note that in the thermodynamic limit x is related to the average concentration c_A of "A" atoms (for which $\sigma = +1$) by $x = 1 + 2c_A$.

The orthogonality and completeness of the characteristic functions for any finite cluster follows trivially from the properties of the polynomials $\phi_n^\mu(\sigma)$. For a crystal of N lattice points, orthogonality and completeness are given, respectively, by

$$\langle \Phi_\alpha^\mu, \Phi_\beta^\mu \rangle_\mu = \frac{1}{[2\cosh(\mu)]^N} \sum_{\vec{\sigma}_N} e^{N\mu x}\, \Phi_\alpha^\mu(\vec{\sigma}_\alpha)\, \Phi_\beta^\mu(\vec{\sigma}_\beta) = \delta_{\alpha,\beta}\,, \qquad (7)$$

and

$$\sum_\alpha \Phi_\alpha^\mu(\vec{\sigma}_\alpha)\, \Phi_\alpha^\mu(\vec{\sigma}_\alpha') = \frac{2^N}{(1+\overline{\sigma})^{N_A}(1-\overline{\sigma})^{N_B}}\, \delta_{\vec{\sigma}_\alpha,\vec{\sigma}_\alpha'}\,, \qquad (8)$$

where α and β are two arbitrary clusters, $\vec{\sigma}_\alpha$ and $\vec{\sigma}_\alpha'$ two arbitrary configurations of cluster α, and $N_A = N(1+x)/2$ and $N_B = N(1-x)/2$ are the numbers of A and B atoms, respectively. The δ function equals 1 when both arguments are the same and zero otherwise. The sums in Eqs. (7) and (8) are of course unrestricted.

A point of clarification with regard to the quantities x and $\overline{\sigma} \equiv \tanh(\mu)$ is in order. The variable $\overline{\sigma}$ (or μ) is used here strictly to label the basis and is to be kept fixed for a given choice of basis. The variable x, on the other hand, corresponds to the average value of the occupation number σ_i and it varies, independently of the choice of basis, with the atomic concentration of the different configurations. It should be noted that the basis used in the original derivation of the cluster expansion of Sanchez et al.[6] corresponds to the especial case of $\mu = 0$ in the present formalism. As we shall see later in this section, the restricted sum method of Asta et al.[11] is also a special case that follows from taking the thermodynamic limit with the choice of $\overline{\sigma} \equiv \tanh(\mu) = x$.

II.2 Expansions

An immediate and straightforward application of the basis defined by Eqs. (4) is the expansion of functions of configuration. For example if we consider a function $F(\vec{\sigma}_\alpha)$ defined in the configuration space of cluster α, the expansion takes the form[6,12]

$$F(\vec{\sigma}_\alpha) = \sum_{\beta \subseteq \alpha} F_\beta^\mu\, \Phi_\beta^\mu(\vec{\sigma}_\beta)\,, \qquad (9)$$

where the expansion coefficients F_β^μ are given by the projection of the function $F(\vec{\sigma}_\alpha)$ on the basis function $\Phi_\beta^\mu(\vec{\sigma}_\beta)$ (see Eq. (5))

$$F_\beta^\mu = \langle F(\vec{\sigma}_\alpha), \Phi_\beta^\mu(\vec{\sigma}_\beta) \rangle_\mu\,. \qquad (10)$$

The representation of any function of configuration in terms of the characteristic functions $\{\Phi_\beta^\mu(\vec{\sigma}_\beta)\}$, Eq. (9), is valid for any cluster α, including the whole crystal with N sites. In particular, the formalism remains valid for $N \gg 1$ or, equivalently, in the thermodynamic limit.

A few observations are in order. First, one should note that by changing the value of μ, the same function $F(\vec{\sigma}_\alpha)$ can be described using an infinity of basis sets. Thus, for a given function $F(\vec{\sigma}_\alpha)$ the expansion coefficients F_β^μ are not unique, obviously depending on the choice of basis. Furthermore, a simple relation can be derived for the expansion coefficients F_β^μ and $F_\beta^{\mu'}$ for two different basis.[11,12] This relationship follows trivially from the fact that the basis are related by a linear transformation in configuration space.

Consider two basis sets $\{\Phi_\alpha^\mu\}$ and $\{\Phi_\alpha^{\mu'}\}$ corresponding to two choices of μ and μ', with $\mu \neq \mu'$. The linear transformation relating the two basis is[12]

$$\Phi_\alpha^\mu(\vec{\sigma}_\alpha) = \sum_{\beta \subseteq \alpha} a_{\alpha\beta}^{\mu\mu'} \, \Phi_\beta^{\mu'}(\vec{\sigma}_\beta), \tag{11}$$

with the coefficients $a_{\alpha\beta}^{\mu\mu'}$ given, for $\mu \neq \mu'$ and $\alpha \supseteq \beta$, by[12]

$$a_{\alpha\beta}^{\mu\mu'} = \frac{\left(1 - \overline{\sigma}^2\right)^{n_\alpha/2}}{\left(1 - \overline{\sigma}'^2\right)^{n_\beta/2}} \left(\overline{\sigma}' - \overline{\sigma}\right)^{n_\alpha - n_\beta}. \tag{12}$$

The linear transformation given by Eq. (12) has been a source of some confusion in the literature. Gonis *et al.*[15] questioned the validity of Eq. (12) for the trivial case of $\mu = \mu'$. In order to avoid similar misunderstandings, it may be appropriate to re-emphasize here that Eq. (12) is valid for $\mu \neq \mu'$, whereas for the trivial case of $\mu = \mu'$, the obvious result is $a_{\alpha\beta}^{\mu\mu} = \delta_{\alpha,\beta}$.

There, however, important and interesting results that follow from the linear transformation, Eq. (12). For example, expanding the function $F(\vec{\sigma}_\alpha)$ in two different basis μ and μ' gives[12]

$$F(\vec{\sigma}_\alpha) = \sum_{\beta \subseteq \alpha} F_\beta^\mu \, \Phi_\beta^\mu(\vec{\sigma}_\beta) = \sum_{\beta \subseteq \alpha} F_\beta^{\mu'} \, \Phi_\beta^{\mu'}(\vec{\sigma}_\beta), \tag{13}$$

where the expansion coefficients in the two different basis are clearly related. Asta *et al.*[11] derived the relationship between the effective cluster interactions defined in the restricted sum scheme and the original unrestricted cluster expansion[6] which, as mentioned, corresponds to the case $\mu = 0$. The general relation between the expansion coefficients in any two basis follows directly from Eq. (12)[12]

$$F_\alpha^\mu = \sum_{\beta \supseteq \alpha} a_{\beta\alpha}^{\mu'\mu} \, F_\beta^{\mu'}. \tag{14}$$

As we shall see, Eq. (14) implies that the energy for any configuration with a fixed average concentration c_A' can be obtained from the knowledge of the energies at a different concentration c_A, the only restriction being that the energy be an extensive quantity. This observation was first made by Asta *et al.*[11] who established a relation between the ECI at an arbitrary concentration c_A' with those obtained at concentration $c_A = c_B = 0.5$.

II.3 Restricted Sum Method

In order to make contact with the restricted sum method of Asta *et al.*,[11] we note that the scalar product defined in Eq. (5) between two functions $f(\vec{\sigma})$ and $g(\vec{\sigma})$ can also be written as[12]

$$\langle f, g \rangle_\mu = \sum_{n=-N}^{+N} \left[G(x, \overline{\sigma})\right]^N \left(\frac{1}{W_{N,x}} \sum_{\vec{\sigma}_N}' f(\vec{\sigma}_N) g(\vec{\sigma}_N)\right), \tag{15}$$

where the index n in the first sum is $n = xN = N_A - N_B$, with N the total number of atoms, and N_A and N_B the number of A and B atoms, respectively. The second sum is restricted to configurations $\vec{\sigma}_N$ with fixed concentrations given by x and N; *i.e.* configurations with N_A and N_B atoms. The total number of such configurations, $W_{N,x}$, is

$$W_{N,x} = \frac{N!}{[N(1+x)/2]! \, [N(1-x)/2]!} = \frac{N!}{N_A! \, N_B!} . \tag{16}$$

The factor $G(x, \bar{\sigma})$ in Eq. (15) is given by

$$G(x, \bar{\sigma}) = \left(\frac{e^{\mu x}}{2 \cosh(\mu)} \right) (W_{N,x})^{1/N} . \tag{17}$$

We note that for $N \gg 1$, the factor $[G(x, \bar{\sigma})]^N$ approaches the Kronecker's delta, $\delta_{x,\bar{\sigma}}$, which equals 1 for $x = \bar{\sigma}$ and zero otherwise. Although the equality between $[G(x, \bar{\sigma})]^N$ and $\delta_{x,\bar{\sigma}}$ is only strictly obtained in the thermodynamic limit $(N \to \infty)$, the result is essentially valid for sufficiently large values of N.

The restricted sum method now follows naturally from Eq. (15) if we assume that, for all cases of interest, the factor in the curly bracket in Eq. (15) has a weaker divergence than $[G(x, \bar{\sigma})]^{-N}$ *for all values of* x. With this limitation in mind, the scalar product defined by Eq. (5), or by Eq. (15), can be written in the thermodynamic limit in the form of a restricted sum

$$(f, g)_\mu \equiv \frac{1}{W_{N,\bar{\sigma}}} {\sum_{\vec{\sigma}_N}}' f(\vec{\sigma}_N) g(\vec{\sigma}_N), \qquad N \gg 1, \tag{18}$$

where, as before, the sum is restricted to configurations with constant average concentration, and where we use the notation $(f, g)_\mu$ to denote the scalar product in the concentration restricted space. Note that the average concentration of the configurations in the restricted sum is now determined by the choice of basis functions (*i.e.* by μ) through the condition $x = \bar{\sigma} \equiv \tanh(\mu)$.

The cluster expansion of the configurational energy $E(\vec{\sigma}_N)$ in the restricted sum method takes the form

$$E(\vec{\sigma}_N) = \sum_\alpha \tilde{V}_\alpha^\mu \, \Phi_\alpha^\mu(\vec{\sigma}_\alpha), \qquad N \gg 1, \tag{19}$$

where the effective cluster interactions $\tilde{V}_\alpha^\mu$ are given by

$$\tilde{V}_\alpha^\mu = \left(E(\vec{\sigma}_N), \Phi_\alpha^\mu(\vec{\sigma}_\alpha) \right)_\mu = \frac{1}{W_{N,\bar{\sigma}}} {\sum_{\vec{\sigma}_N}}' E(\vec{\sigma}_N) \Phi_\alpha^\mu(\vec{\sigma}_\alpha). \tag{20}$$

It should be clear from the discussion leading to Eq. (18), that the restricted sum approach is not expected to be valid for arbitrary functions of configurations. The breakdown of the method can be seen with a simple example. Let us consider a crystal with N sites and a function of configuration, $F(\vec{\sigma}_N)$, that we choose equal to the characteristic function of order N, i.e. $F(\vec{\sigma}_N) = \Phi_N^\mu(\vec{\sigma}_N)$. The cluster expansion of this function is trivial, with a single non-vanishing coefficient $F_N^\mu = \langle F, \Phi_N^\mu \rangle_\mu = 1$. As we show below, the restricted sum method fails to reproduce this result.

To proceed, we note that since the relevant concentration is determined by $x = \bar{\sigma} \equiv \tanh(\mu)$, the characteristic function of order N may be written as

$$\Phi_N^\mu(\vec{\sigma}_N) = (-1)^{N_B} \left(\frac{c_A^{c_B} c_B^{c_A}}{\sqrt{c_A c_B}} \right)^N , \tag{21}$$

with $c_A = 1 - c_B = (1 + \bar{\sigma})/2$. The projection of $F(\vec{\sigma}_N)$ onto $\Phi_N^\mu(\vec{\sigma}_N)$ according to Eq. (18) can be calculated exactly for any value of N. After some straightforward algebra we obtain

$$(F(\vec{\sigma}_N), \Phi_N^\mu(\vec{\sigma}_N))_\mu = \frac{1}{W_{N,\vec{\sigma}}} {\sum_{\vec{\sigma}_N}}' \Phi_N^\mu(\vec{\sigma}_N) \, \Phi_N^\mu(\vec{\sigma}_N) = \left[\left(\frac{c_B}{c_A} \right)^{c_A - c_B} \right]^N . \qquad (22)$$

Except for the case $c_A = c_B = 0.5$ which, as mentioned, corresponds to the unrestricted sum cases developed in Refs. 6 and 12, the right-hand side of Eq. (22) vanishes in the limit of large N. This simple example shows that the orthogonality and completeness relations given by Eqs. (7) and (8) are not obeyed in a concentration restricted configurational space. We note that in their recent work, Gonis *et al.*[15] have explicitly assumed the orthogonality and completeness of the basis set in the restricted sum approach. This assumption has unfortunately led these authors to a series of mathematical errors in their derivations and, understandably, to several wrong conclusions with regard to the validity of the cluster expansion.

The root of the difficulty in the example given above is, of course, the fact that for $c_A \neq c_B$ the norm of the characteristic function $(\Phi_N^\mu(\vec{\sigma}_N), \Phi_N^\mu(\vec{\sigma}_N))_\mu$ vanishes (or approaches zero) for sufficiently large values of N. The same result applies to the norm of any characteristic function $\Phi_N^\mu(\vec{\sigma}_\alpha)$ of order m_α, for which m_α is of the order of N, *i.e.* the ratio m_α/N remains non-zero when $N \to \infty$.

For a cluster of m points, the norm of the characteristic function $\Phi_m^\mu(\vec{\sigma}_m)$ is given by

$$(\Phi_m^\mu, \Phi_m^\mu)_\mu = \frac{N_A! \, N_B!}{N!} \sum_{m_A = m_1}^{m_2} \frac{(N-m)!}{(N_A - m_A)! \, (N_B - m_B)! \, m_A! \, m_B!} \frac{m!}{(c_A c_B)^m} \, , \qquad (23)$$

where the running indices m_A and $m_B = m - m_A$ are the number of A and B atoms in the cluster, respectively, and where the lower (m_1) and upper (m_2) limits in the sum depend on the value of m. For $m < N_A < N_B$ the limits are $m_1 = 0$ and $m_2 = m$ and the norm is

$$(\Phi_m^\mu, \Phi_m^\mu)_\mu = \left(\frac{c_A}{c_B} \right)^m \frac{N_B! \, (N-m)!}{N! \, (N_B - m)!} \, {}_2F_1(-m, -N_A; N_B - m + 1, c_B^2/c_A^2), \qquad (24)$$

$$\text{for } m < N_A < N_B \text{ and } N_A < m < N_B,$$

where ${}_2F_1(a, b; c, z)$ is Gauss' hypergeometric function. As indicated, the result of Eq. (24) is valid for $m < N_A < N_B$ as well as $N_A < m < N_B$. For the remaining case of $m > N_A > N_B$, the limits in Eq. (23) are $m_1 = m - N_B$ and $m_2 = N_A$ and the norm of the characteristic function is given by

$$(\Phi_m^\mu, \Phi_m^\mu)_\mu = \left(\frac{c_B}{c_A} \right)^{m - 2N_B} \frac{N_A! \, m!}{N! \, (m - N_B)!} \, {}_2F_1(m - N, -N_B; m - N_B + 1, c_B^2/c_A^2), \quad (25)$$

$$\text{for } m > N_B > N_A.$$

We note that the exact results given by Eqs. (24) and (25) are valid for *all* values of N, small or large.

For the sake of completeness we investigate next the asymptotic behavior of the norms of the characteristic functions for $N \gg 1$. We consider only the case of fixed concentrations c_A and c_B with $c_A \neq c_B$ since, for $c_A = c_B$, the norms equal 1 exactly for all values of N and of the order m of the characteristic function. For a fixed value of the ratio $z = m/N$ between the order of the characteristic function and the number of atoms in the crystal, the asymptotic behavior for large N is given by the largest term in Eq. (23), which occurs for $m_A = \lambda N$, where λ is given by

$$\lambda = \frac{z}{2} - \frac{c_A c_B}{2(c_A - c_B)} + \frac{1}{2(c_A - c_B)} \sqrt{4 c_A c_B^2 z(c_A - c_B) + [c_A c_B - z(c_A - c_B)]^2}. \quad (26)$$

Using the asymptotic form of the factorial function, we obtain

$$(\Phi_m^\mu, \Phi_m^\mu)_\mu \rightarrow \left[\frac{z^z (1-z)^{1-z} \, c_A^{c_A+z-2\lambda} \, c_B^{c_B-z+2\lambda}}{\lambda^\lambda (z-\lambda)^{z-\lambda} (c_A - \lambda)^{c_A-\lambda} (c_B - z + \lambda)^{c_B-z+\lambda}} \right]^N, \quad \text{for } N \gg 1. \quad (27)$$

It is straightforward to see that the right hand side of Eq. (27) tends to zero for $z \neq 0$, indicating the breakdown of the orthogonality relations in the restricted sum method for basis functions of order N. We note that the asymptotic form of Eq. (27) is also valid for $\{z, \lambda\} \rightarrow 0$, in which case the norm of the characteristic functions tend to 1.

III. Validity of the Restricted Sum Method

As seen in the previous section, the orthogonality relations between the characteristic functions, which provide of course a complete set in the full configurational space, break down in restricted sum method. The origin of the breakdown is the definition of the scalar product. The failure of the restricted sum method in the general case is, however, irrelevant for all cases of physical interest. In particular, all *extensive* functions will contain at most many-body interactions of finite order. Such functions can be expanded fully in a concentration restricted subspace for *any* concentration c_A. For example, the cluster expansion of the energy takes the form

$$E(\vec{\sigma}) = \sum_\alpha{}'' \tilde{V}_\alpha^\mu \, \Phi_\alpha^\mu(\vec{\sigma}_\alpha), \quad (28)$$

where the double prime in the sum indicates that only clusters with a finite number of points, *i.e.* such that $n_\alpha/N \rightarrow 0$, are present in the expansion. The expansion coefficients $\tilde{V}_\alpha^\mu$ are then given exactly by the restricted sum projection $(E(\vec{\sigma}), \Phi_\alpha^\mu(\vec{\sigma}))_\mu$.

We now turn our attention to the problem of describing the energies $E(\vec{\sigma}_N')$ for configurations $\vec{\sigma}_N'$ with fixed concentrations c_A', in terms of ECI's, $\tilde{V}_\alpha^\mu$, obtained using energies $E(\vec{\sigma}_N)$ with concentration $c_A \neq c_A'$. According to Eq. (28), we have

$$E(\vec{\sigma}_N') = \sum_\alpha{}'' \tilde{V}_\alpha^\mu \, \Phi_\alpha^\mu(\vec{\sigma}_\alpha'). \quad (29)$$

We stress once again that the sum in Eq. (29) *must* be constrained to clusters of finite order since, otherwise, the concentration restricted expansion of $E(\vec{\sigma}_N')$ would not be possible. We can now combine Eqs. (28) and (29) to write the energy of configuration $\vec{\sigma}_N'$ as

$$E(\vec{\sigma}_N') = \frac{1}{W_{N,\sigma}} \sum_{\vec{\sigma}_N}{}' E(\vec{\sigma}_N) \sum_\alpha{}'' \Phi_\alpha^\mu(\vec{\sigma}_\alpha') \Phi_\alpha^\mu(\vec{\sigma}_\alpha), \quad N \gg 1. \quad (30)$$

In Eq. (30) the configurations $\vec{\sigma}_N$ and $\vec{\sigma}'_N$ are different since we have assumed that $c_A \neq c'_A$. Gonis et al.[15] have then invoked the completeness relations of the characteristic functions, Eq. (8), and the fact that $\vec{\sigma}_N \neq \vec{\sigma}'_N$, to conclude that the right hand side of Eq. (30) vanishes. These authors thus arrived at the result that $E(\vec{\sigma}'_N)$ vanishes for all configurations which, as mentioned is a mathematical and physical absurdity. The flaw in the proof of Gonis et al.[15] rests in their assumption that the second sum in Eq. (30) vanishes for $\vec{\sigma}_N \neq \vec{\sigma}'_N$. Clearly, since the sum in Eq. (30) is restricted to clusters of finite order we have

$$\sum_{\alpha}{}'' \Phi^\mu_\alpha(\vec{\sigma}'_\alpha)\, \Phi^\mu_\alpha(\vec{\sigma}_\alpha) \neq 0, \tag{31}$$

for $\vec{\sigma}_N \neq \vec{\sigma}'_N$. Thus, Eq. (30) remains valid, simply expressing the linear relationship that exist between ECI's (of finite order) calculated using different concentration restricted subspaces of the full configurational space.

IV. Concluding remarks

We conclude by stressing that, despite some unfounded concerns that have been raised in the literature, the cluster expansion method rest on solid and rigorous grounds. As shown here, a general formulation of the method requires the use of the correct definition of the scalar product which, in order to ensure orthogonality and completeness of the basis functions, must include the appropriate weight factors as defined by Eq. (5).

The definition of ECI's is clearly not unique, since we may choose an infinity of basis functions by changing the weight factors in the scalar product.[12] However, the formalism of the cluster expansion suggests a natural and systematic approach to the definition these interactions. In particular, the choice of basis functions allows us to introduce a natural concentration dependence in the interactions. In turn, the ability to describe the configurational energy using different basis functions may prove valuable for the optimization of the convergence of the cluster expansion.

References

1. R. Kikuchi, *Phys. Rev.* **81**, 988 (1951).
2. R. Kikuchi, *Prog. Theor. Phys.* Suppl. **35**, 1 (1966).
3. J. M. Sanchez, and D. de Fontaine *Phys. Rev. B* **17**, 2926 (1978).
4. J. M. Sanchez, and D. de Fontaine *Phys. Rev. B* **21**, 216 (1980).
5. J.M. Sanchez, and D. de Fontaine *Phys. Rev. B* **25**, 1759 (1982).
6. J. M. Sanchez, F. Ducastelle and D. Gratias, *Physica A* **128**, 334 (1984).
7. K. Kawasaki, in *Phase Transitions and Critical Phenomena*, edited by C. Domb and M. S. Green, (Academic Press, New York, 1973), Vol. 2, p. 465.
8. F. Ducastelle, *Order and Phase Stability in Alloys*, (North Holland, Amsterdam, 1991).
9. D. de Fontaine, in *Solid State Physics*, edited by H. Ehrenreich and D. Turnbull, (Academic Press, New York, 1994), Vol. 47, p. 33.
10. A. Zunger, in *Static and Dinamics of Alloy Phase Transformations*, edited by P. E. A. Turchi and A. Gonis, NATO ASI Series. Series B, Physics. (Plenum Press, New York, 1994), Vol. 319, p. 361.

11. M. Asta, C. Wolverton, D. de Fontaine, and H. Dreyssé *Phys. Rev. B* **44**, 4907 (1991); C. Wolverton, M. Asta, H. Dreyssé, and D. de Fontaine *Phys. Rev. B* **44**, 4914 (1991)
12. J. M. Sanchez *Phys. Rev. B* **48**, 14013 (1993).
13. J. M. Sanchez, J. P Stark and V. L. Moruzzi *Phys. Rev. B* **44**, 5411 (1991).
14. G. D. Garbulski and G. Ceder *Phys. Rev. B* **49**, 6327 (1994).
15. A. Gonis, P. P. Singh, P. E. A. Turchi and X.-G. Zhang *Phys. Rev. B* **51**, 2122 (1995).

Lattice Models and Cluster Expansions for the Prediction of Oxide Phase Diagrams and Defect Arrangements

G. Ceder, P. D. Tepesch, G. D. Garbulsky, and A. F. Kohan

Department of Materials Science and Engineering,
Massachusetts Institute of Technology,
Cambridge, MA 02139,
U.S.A.

Abstract

First-principles quantum mechanics can be combined with a lattice model cluster expansion to predict the arrangement and configurational entropy of defects in concentrated solid solutions. We demonstrate that for the CaO-MgO system, the effective interactions in the cluster expansion can be computed accurately from first-principles. In systems in which substitutional disorder on the cation and anion sublattices is coupled through charge compensation, two coupled binary expansions can be used. This is applied to the Gd_2O_3-ZrO_2 system. Even with the strong electrostatic contribution to the configurational energy, a short-range cluster expansion accurately models the energetics of these ionic materials.

I. Introduction

The properties of many important oxides are intimately related to the material's point defect content and configuration. There are many sources of point defects in oxides, from off-stoichiometry to impurities or dopants. In many cases, the concentration of defects and their influence on transport properties can be qualitatively understood with point defect models. The basic assumption in these models is that defects or defect complexes are randomly distributed and do not interact. It is clear that this approximation will break down at high enough concentration. Given the strong interaction between charged defects, this breakdown could occur already at much lower defect concentrations than in metals.

Theory and Applications of the Cluster Variation and Path Probability Methods
Edited by J.L. Morán-López and J.M. Sanchez, Plenum Press, New York, 1996

As an alternative to point defect models, one can use lattice models to study the energetics and configurations of defects. The use of a lattice model with an Ising-like Hamiltonian to compute the thermodynamics of materials with substitutional disorder has now been well justified.[1] Even more, the interpretation of the lattice model Hamiltonian as the result of successive coarse-graining of the full system Hamiltonian[2] combined with the configurational or cluster expansion has led to a well-determined procedure to obtain thermodynamic properties of a material from first-principles. This approach is not limited to a static array of ions and defects on a rigid lattice but is capable of fully incorporating the effect of ionic displacements and lattice vibrations.

In this paper we address some of the potential problems for applying lattice models to oxides, such as the large number of microscopic species, the electrostatic interaction between defects and the computation of the effective lattice model interactions from first-principles.

I. Lattice Models and the Cluster Expansion

A lattice model for a material can be defined by taking all the sites on which atomic disorder occurs. This disorder can be caused by vacancies, interstitials or substitutional defects. For example, in $YBa_2Cu_3O_{7-\delta}$, oxygen deficiency creates vacancies on the oxygen sublattice, causing an order-disorder transformation with the oxygen ions on that same sublattice.[3] The lattice model for this system contains all the sites on which the oxygen occupation is variable. If only two possible species can sit on a given site (such as in the case of oxygen and oxygen vacancies on the oxygen sublattice) a two-state spin variable, σ_i ($\sigma_i = \pm1$), can represent the occupation of site i. The formalism can be extended to ternary systems by using a three-state variable ($\sigma_i = 1, 0, -1$). A lattice configuration is indicated with $\{\sigma\} = (\sigma_1, \sigma_2, \sigma_3, \ldots, \sigma_N)$ and its energy is given by the value of the lattice model Hamiltonian $H(\{\sigma\})$.

Very few approximations are made in modeling substitutional disorder with a lattice model. The only inherent assumption in such an approach is that the phase space of the system can be exactly partitioned in a set of ensembles, $C(\{\sigma\})$ each characteristic of a specific substitutional state of atoms. All states in such an ensemble are represented by the same configuration $\{\sigma\}$ on the lattice model but differ by their state of other excitations (vibrational, electronic, etc.). Hence, each lattice model configuration $\{\sigma\}$ is merely a label for a (infinite) collection of states of the system. This partitioning of states allows one to write the partition function of the system exactly as the partition function of a lattice model

$$Z(T) = \sum_{\{\sigma\}} \exp\left\{\frac{-F(\{\sigma\}, T)}{k_B T}\right\},\tag{1}$$

where the effective lattice Hamiltonian contains the effect of all excitations that do not modify the substitutional state

$$F(\{\sigma\}, T) = -k_B T \ln\left(\sum_{\substack{\text{Excitations} \\ \text{in } C(\{\sigma\})}} \exp\left\{\frac{-E}{k_B T}\right\}\right).\tag{2}$$

One expects this coarse-graining procedure to be valid for systems in which the vibrational and electronic excitations (and other excitations if present) occur on a much faster time scale than the substitutional exchanges. Only in that case can the faster excitations be considered ergodic on the substitutional time scale and can an effective Hamiltonian such as Eq. (2) be defined. Although this condition is surely satisfied in most metallic and oxide systems there are some well documented materials[4,5] such as AgI and CsC_{24} where this hypothesis may be challenged due to the almost liquid-like nature of the disorder.

The challenge in first-principles calculations is of course to compute $F(\{\sigma\}, T)$. Most often $F(\{\sigma\}, T)$ is approximated by the groundstate energy in the ensemble $C\{\sigma\}$, neglecting the effect of vibrations and electronic excitations. In that case, the configurational Hamiltonian also becomes temperature independent. In most of this paper, we will make this approximation, but investigate the effect of lattice vibrations on the first-principles prediction of the CaO-MgO phase diagram (section V.3).

Although the previous deduction justifies the use of a lattice Hamiltonian it does not describe the dependence of $F(\{\sigma\}, T)$ on the configuration $\{\sigma\}$. This dependence can be explicitly obtained by expanding $F(\{\sigma\}, T)$ in an orthogonal set of basis functions.[6] For a binary system this expansion is given by

$$F(\{\sigma\}, T) = \sum_\alpha V_\alpha(T)\sigma_\alpha, \tag{3}$$

where α is a cluster of points on the lattice and the summation is over all possible clusters. The functions σ_α are defined as the product of all occupation variables σ_i in cluster α and form a complete set in the space of all configurations, making Eq. (3) formally exact. The expansion coefficients $V_\alpha(T)$ are called effective cluster interactions (ECI) and their value can be determined by fitting Eq. (3) to empirical or first-principles calculations of the energy for some representative configurations. Note that if one wants to obtain the temperature dependence of the ECI it is necessary to fit to free energies (of systems with fixed $\{\sigma\}$) rather than ground state energies. These effective interactions have to be distinguished from potentials. The ECI are numbers (for a given temperature) and have no distance-dependence. They parameterize the dependence of the energy on the configurational variables and incorporate changes in lattice parameter and ionic relaxations implicitly.

The importance of the expansion in Eq. (3) lies in its convergence. For most systems, it has been found that the ECI rapidly approach zero as the distance between the points in the cluster α increases, or as the number of points in α increases. As a result of this convergence the dependence of the energy on configuration can be described with just a few effective cluster interactions. If the value of these ECI can be computed from first-principles, Eq. (3) presents a non-empirical energy model for the system with none of the traditional assumptions for the configuration of defects.

The cluster expansion (CE) described above has been used extensively to predict the phase diagrams of systems with binary disorder.[7-16] Although the formal extension to ternary systems is straightforward and can be achieved in several ways[17,18] its applications have been limited to prototype systems due to the increasing numerical complexity of the cluster expansion formalism. Although oxides typically contain a large number of microscopic species we show in the next section how, in most cases, these systems can be described with binary or ternary expansions.

III. Alio-Valently Doped Oxides: the Coupled Cluster Expansion

Oxides present a complexity not often found in metallic systems. Due to the creation of charge compensation defects, the number of microscopic species that participate in the disorder can be considerably larger than the number of chemical constituents. In CaO-stabilized ZrO_2, for example, the introduction of CaO introduces an equal number of substitutional Ca ions on the cation sublattice and oxygen vacancies on the anion sublattice. The oxygen vacancy is created to compensate for the different valence of the Ca and Zr cations. These substitutional defects can be randomly distributed, associated (randomly distributed in pairs) or, at high defect concentrations, they can order into superstructure arrangements. A microscopic model for ionic systems, such as CaO-ZrO_2, with binary disorder on cation and anion sublattice thus needs to account for at least four components. Although a cluster expansion for a true quaternary system is prohibitively complex these alio-valently doped oxides can formally be treated with a binary cluster expansion.[19] The simplification to a binary problem depends on the following assumption: Two distinct sublattices, L_1 and L_2, each contain two species, A and B on L_1, C and D on L_2, that do not occupy the other sublattice. This will be valid for most doped systems where one sublattice contains cations and the other anions. We will refer to these systems as coupled-sublattice systems. In the example of CaO-doped ZrO_2 the cation sublattice contains Ca and Zr and the anion sublattice contains O and vacancies. The configurational state on each individual sublattice can be described by a set of occupation variables and a basis of cluster functions, similarly to the binary expansion described in the introduction. For sublattice L_1, the occupation variable, σ_i, is equal to 1 (-1) if an A (B) species sits at the site i ($i = 1, 2, \ldots, N_1$). Similarly, on sublattice L_2, δ_j is equal to 1 (-1) if a C (D) species sits at the site j ($j = 1, 2, \ldots, N_2$). N_1 and N_2 are the number of sites on sublattice L_1 and L_2 respectively. For each lattice we define cluster functions in terms of these occupation variables

$$\Phi_\alpha = \prod_{i \in \alpha} \sigma_i, \quad \text{and} \quad \Psi_\beta = \prod_{j \in \beta} \delta_j, \tag{4}$$

where α is any cluster of points in sublattice L_1, and β is any cluster of points in sublattice L_2. These basis sets can describe the configurational state on each individual sublattice, but not the correlation between the states on the different sublattices. In order to introduce the interaction between sublattices, it is necessary to work in the tensor product space of configurations on both sublattices, with dimension $2^{N_1+N_2}$. A convenient basis for this space is the set of tensor products of the basis functions in the separate subspaces

$$\Theta_{\alpha\beta} = \Phi_\alpha \Psi_\beta \quad \text{for all } \alpha \text{ and } \beta. \tag{5}$$

Since the set of $\Theta_{\alpha\beta}$ forms a basis in the space of configurations for these systems, the configurational Hamiltonian can be exactly written as

$$H(\{\sigma\}, \{\delta\}) = \sum_{\alpha\beta} V_{\alpha\beta} \Theta_{\alpha\beta}. \tag{6}$$

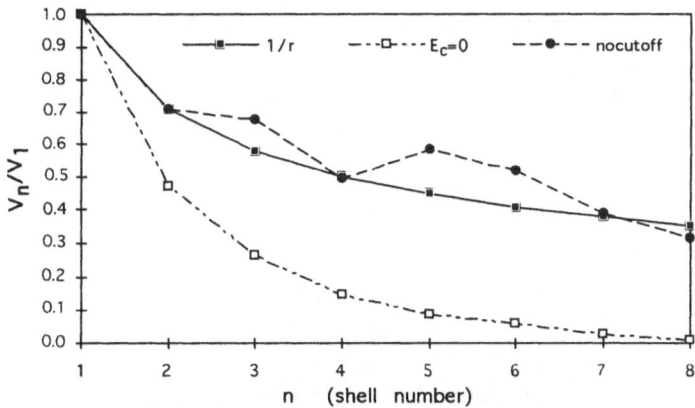

Figure 1. Comparison between the exact ECI (filled squares with solid line) and the ECI obtained with two different fits (broken lines) for a model with fixed point charges on an *fcc* lattice. The ECI obtained by fitting a Hamiltonian with first to eighth nearest-neighbor interactions to the energy of 291 structures are shown in solid circles. The ECI indicated with open squares were obtained by only fitting to those structures with energy below zero.

The ECI ($V_{\alpha\beta}$) contain both inter- and intra-sublattice coupling terms. The model presented here can, in principle, be applied to study any configurational problem in ionic systems. It represents a major departure from the traditional point defect models, prevalent in studies on oxides. Because the basis set contains functions that describe both the state of order of a given sublattice, and the association between species on different sublattices, the full range of configurational behavior from isolated defects to long-range ordering, and the transition between the different regimes can be described within a single model. Some applications of the method will be investigated in section V.

IV. Convergence of the Effective Cluster Interactions

The practical usefulness of both the regular cluster expansion and the coupled cluster expansion depends on the rapid convergence of the ECI with distance. Although this convergence has been well established in metallic and semiconductor systems,[20-22] the transferability of these conclusions to ionic systems, may seem questionable, given the strong electrostatic interactions in these materials. This fear may be exacerbated by the fact that for a simple model of unscreened fixed point charges the exact ECI only decay with distance as $1/r$. (The analytical calculation of the ECI is trivial in the case of a pairwise interacting system in which the ions are not allowed to relax). However, even for this worst-possible system (no screening, fixed point charges) a rapidly convergent cluster expansion can be constructed that reproduces the energy of the important configurations very well.[23] The only configurations for which the $1/r$ decay is important are the ones with very high energy. This is demonstrated in Figs. 1 and 2. Figure 1 shows the first through eighth nearest-neighbor effective pair interactions for a system of point charges on a fixed *fcc* lattice. The exact ECI are given by the solid line in Fig. 1. The $1/r$ dependence of the ECI does not lead to a rapidly convergent expansion and therefore prohibits the use of a real-space cluster

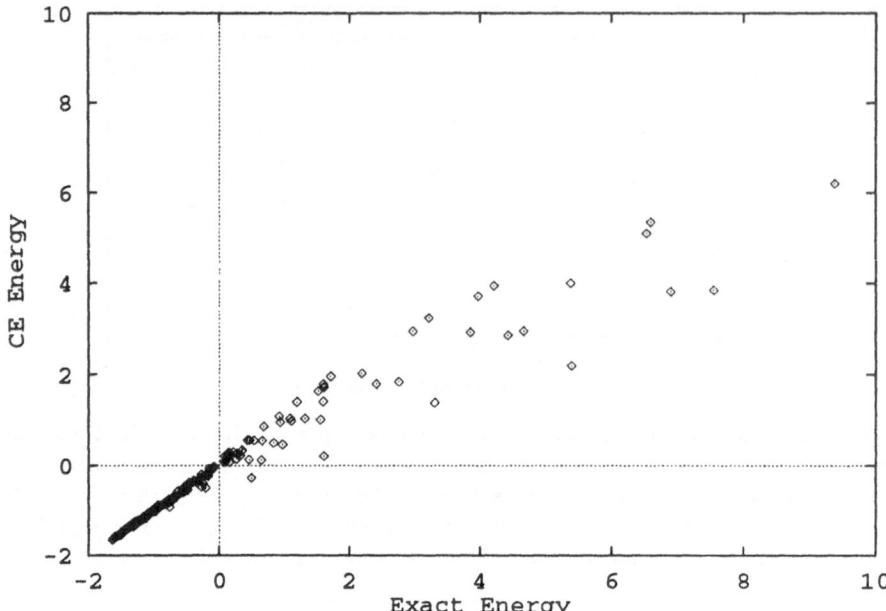

Figure 2. Energies as reproduced by the rapidly converging cluster expansion (open squares in Fig. 1) *versus* exact electrostatic energies. The energies are in units of $q^2/(8\pi\epsilon_0 d)$ where d is the nearest-neighbor distance.

expansion. This slow decay of the exact ECI is the result of the high energy regions in configuration space. It is possible to obtain a much faster converging expansion by not requiring that the cluster expansion reproduce these regions with high energy. The open squares in Fig. 1 are obtained by computing the electrostatic energy of a large number of configurations (291) and fitting the CE (with 8 ECI) only to those energies that are below some cutoff. The resulting ECI clearly converge much faster than $1/r$. If no cutoff is applied and the expansion is fitted to the energy of all configurations the slow convergence of the exact ECI is recovered as evidenced by the solid circles in Fig. 1. These results dramatically illustrate that the convergence of the cluster expansion can be accelerated by limiting its application to the part of configuration space with low energy.

The quality of the fit for the rapidly converging expansion is extremely good as evidenced by Fig. 2 which shows the energy computed from the CE *versus* the exact electrostatic energy. These energies are in dimensionless units (Madelung constants). The error is largest for the configuration with high energy, but these states are not important for ground state and phase diagram calculations.

The rapid decay of the electrostatic ECI fitted to the low energy part of config-uration space can be understood from the requirement of local charge neutrality in structures with low energy. Low energy structures do not have long-ranged charge imbalance, implying that, *on the average*, ions do not effectively interact with their environment outside some radius of charge neutrality. Of course, ions interact indi-vidually over much larger distances, but on average, the electrostatic field from the region far away does not depend on the details of the arrangement in that region, pro-vided the charge is constant there. Only in structures in which there are large charged

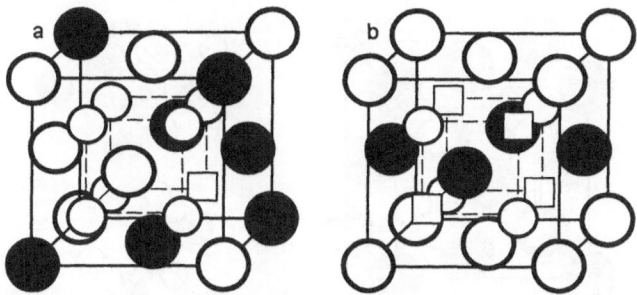

Figure 3. (a) One eighth of the pyrochlore unit cell. Large circles are cation sites. Small circles are oxygen sites. The square represents the vacancy. (b) The $L1_0$-B1 ordered state. The cations are ordered in the $L1_0$ structure and the anions in the B1 structure.

regions will long-ranged effective interaction be necessary. When this is the case, the short-range correlation functions are not enough to describe the charge imbalance, hence a short range cluster expansion can not reproduce the electrostatic energy of the system.

The above computations demonstrate that *even in a rigid electrostatic system that undergoes no screening or charge transfer, a rapidly converging cluster expansion can be constructed that accurately reproduces the low-energy configurations.* The convergence rate of the expansions can be systematically varied by changing the cutoff energy in the fit used to determine the ECI. In real ionic materials one can expect the convergence to be even better due to energy lowering mechanisms such as charge transfer and screening. One can thus be confident that the phase diagram and order-disorder transitions in ionic systems can be modeled well with relatively short-ranged cluster expansions.

V. Applications

V.1 Coupling Between Cation and Anion Ordering

In highly doped oxides, both the cation and anion sublattice can undergo an order-disorder transition. These transitions can occur at distinct temperatures or couple into a single order-disorder transition. The possibility to create separate order-disorder transitions is important in pyrochlore forming oxides such as $Gd_2Zr_2O_7$. Figure 3a shows the pyrochlore structure: At low temperature the cations are ordered on an *fcc* lattice with the oxygen ions ordered over the tetrahedral interstices. One out of every eight oxygen sites is vacant. This structural vacancy plays a key role in the application of these materials as fast oxygen conductors. At elevated temperature the oxygen-vacancy arrangement partially disorders providing mobile vacancies for oxygen diffusion. Unfortunately this oxygen disorder is accompanied by increasing cation disorder which is believed to suppress oxygen conductivity through an increase in the oxygen migration enthalpy. An understanding of the coupling between the state of order on cation and anion sublattices is therefore crucial to the further optimization of the properties of these materials.

The limited case of coupling between two order parameters to a common strain order parameter has been studied already using a Landau expansion.[24] Here we will look

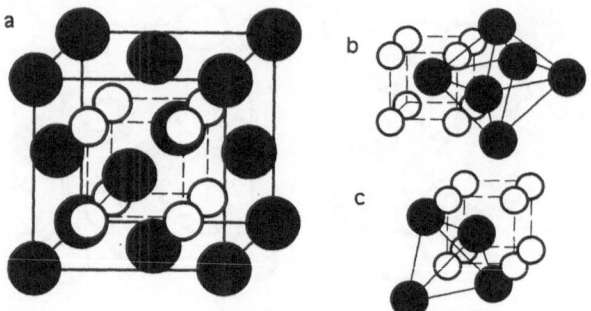

Figure 4. (a) The fluorite structure. The large dark circles are cation sites and the small empty circles are the anion sites. (b) The octahedron of cations and one of the two anion cube clusters used in the Cluster Variation Method. (c) The second anion cube cluster and the filled-tetrahedron cluster which consists of the nearest-neighbor cation tetrahedra and the anion in its center.

at the direct coupling between the two order parameters by computing order-disorder transitions for two systems with distinctly different symmetry relations between the two sublattices.

The coupling behavior naturally depends on the symmetry breaking of the ordered state. If only pairwise coupling ECI exist between L_1 and L_2, the ions on L_1 can be thought of producing a chemical field on the sites of L_2. If ordering on sublattice L_1 causes this field to be of lower translational symmetry than that of sublattice L_2, inequivalent sites will be produced on L_2 upon ordering of L_1. In this case, at the order-disorder transition of L_1, the order parameter on L_2 will necessarily change. We refer to this case as *site-symmetry breaking*. If no such symmetry breaking takes place on one sublattice due to ordering on the other sublattice we call the transition *site-symmetry preserving*. For the latter case there will be no coupling between the order parameters on L_1 and L_2 within a mean field approach, whatever the coupling interactions.

The argument for the site-symmetry preserving transitions is strictly true if only pair coupling interactions are considered. With multiplet coupling ECI spanning two or more points on each sublattice, ordering on L_1 can create effective intra-sublattice interactions on L_2, thereby changing the ordering behavior of the species on L_2. The results presented below will show that non-mean field behavior can also cause some coupling between the state of order on each sublattice, even in the site-symmetry preserving case. To illustrate the differences between the two cases, we study two representative types of ordering on the sites of the fluorite structure. In the completely disordered state, both systems have the fluorite symmetry (space group Fm3m) with two distinct sublattices: an *fcc* cation sublattice and a simple cubic anion sublattice formed by the tetrahedral interstitial positions of the cation lattice (Fig. 4a).

The first type of ordering is that of the pyrochlore structure (Fig. 3a) and is site-symmetry breaking. An example of a site-symmetry preserving case on both sublattices is given in Fig. 3b, where the cations are ordered in the $L1_0$ ordered state on the fcc sublattice and anions in the rocksalt structure (B1) on the simple cubic sublattice. We refer to this arrangement of species as the $L1_0$-B1 structure. The structure merely serves to illustrate site-symmetry preserving behavior and, to our knowledge, has not been observed in nature.

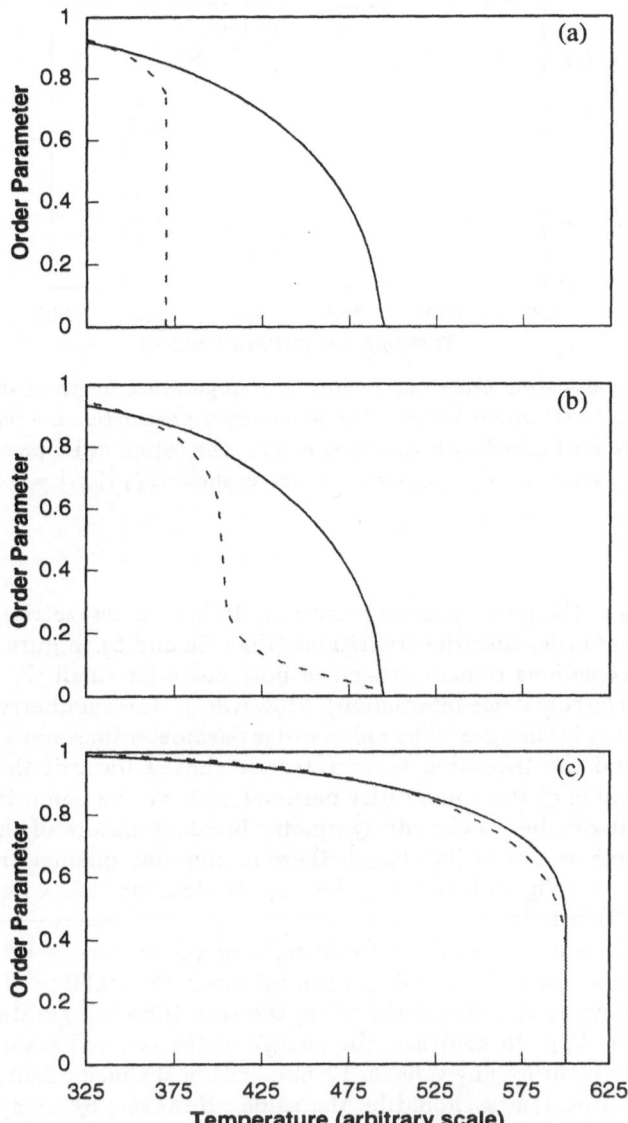

Figure 5. Anion (dashed lines) and cation (solid lines) order parameters *vs.* temperature for three different values of $|V_1^c/V_{max}|$ for the pyrochlore ordered state. (a) $|V_1^c/V_{max}| = 0$. (b) $|V_1^c/V_{max}| = 0.1$. (c) $|V_1^c/V_{max}| = 1.25$.

We have studied the qualitative behavior of these systems by computing the cation and anion order parameters with a set of ECI that stabilize the respective ground states. Coupling between the sublattices was achieved with a single nearest-neighbor cation-anion ECI (V_1^c). The order parameters were computed with the Cluster Variation Method (CVM)[25] in a 5-6-8-8 point approximation. The maximal clusters in the CVM are shown in Fig. 4b. The qualitative difference between the two model systems is in the different types of symmetry breaking, and is reflected in the behavior of the

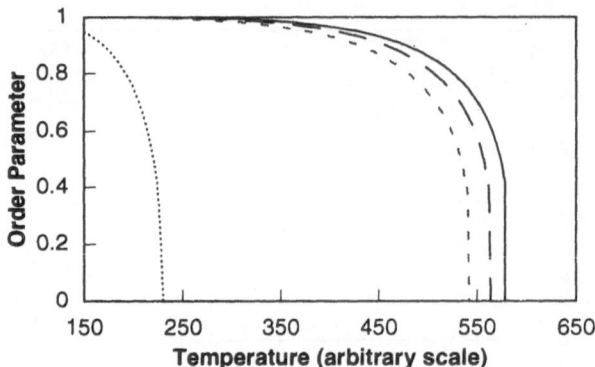

Figure 6. Anion and cation order parameters *vs.* temperature for three different values of $|V_1^c/V_{max}|$ for the $L1_0$-$B1$ ordered state. The anion order parameters are represented by the dotted line and are unchanged with the value of V_1^c. The cation order parameter is slightly shifted with V_1^c. Solid line: $|V_1^c/V_{max}| = 0$. Large dashes: $|V_1^c/V_{max}| = 0.5$. Small dashes: $|V_1^c/V_{max}| = 0.75$.

order parameters at the transition temperatures. In both cases, setting $V_1^c = 0$ results in two independent order-disorder transitions (Figs. 5a and 6). Figures 5b and 6 show that these two transitions remain present in both cases for small $|V_1^c/V_{max}|$ (V_{max} is the maximum intra-sublattice interaction). However, in the symmetry-breaking case, the anion transition is changed: The anion order parameter does not go to zero above the transition, and the transition temperature is shifted towards the cation transition. The finite value of the anion order parameter above the anion transition in the pyrochlore ordering is due to the site-symmetry breaking nature of the cation ordering Finally, at high values of $|V_1^c/V_{max}|$, there is only one distinct transition in the pyrochlore structure (Fig. 5c), while in the $L1_0$-$B1$ structure two completely distinct transitions (Fig. 6) remain.

It is clear that, in both systems, introducing a non-zero value of V_1^c shifts the transition temperatures. Since the coupling term increases the stability of the pyrochlore ground-state relative to the disordered state, the transition temperatures in this system increase with $|V_1^c|$. In contrast, the energy of the ordered state in the $L1_0$-$B1$ ground state is the same for any value of V_1^c and, within the mean-field approximation, the transitions temperatures should be the same. However, by analyzing the CVM correlation functions, it can be shown that the small shift in the cation transition in Fig. 6 is due to non-mean-field effects (short-range order).

V.2 Ionic Conductivity in ZrO_2-Gd_2O_3

The coupled cluster expansion can also be used to study configurations of defects in a dilute concentration. This is illustrated below for the case of Gd_2O_3-doped ZrO_2. ZrO_2 materials doped with subvalent oxides such as Gd_2O_3, Y_2O_3 or CaO are used as fast oxygen ion conductors in applications ranging from oxygen sensors in cars to solid electrolytes for fuel cells. The subvalent dopant cations are introduced in ZrO_2 to create charge compensating vacancies on the oxygen sublattice, thereby increasing the anionic conductivity by several orders of magnitude for a few percent of dopant. At higher concentrations the conductivity goes through a maximum and begins to

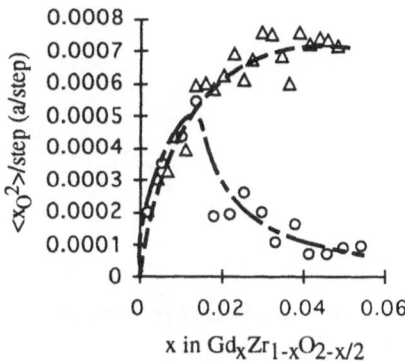

Figure 7. Mean squared oxygen displacement as a function of composition at 2000 K for two different cation conditions. Open circles are for cations fixed in equilibrium. Triangles are computed with the cations randomly distributed. In each case the cations are not allowed to move when sampling the mean-squared displacement for the oxygen.

decrease. Although this behavior has been attributed to the association and ordering of defects[26] there is little experimental or theoretical information to clarify the nature of the defect rearrangements.

We model the system with a cubic fluorite lattice which is valid except for compositions very near pure ZrO_2 where the real system undergoes symmetry breaking transitions upon cooling. To determine the values of the ECI in the coupled cluster expansion we computed the energy of 188 structures with distinct cation and anion arrangements. The energy of each structure was minimized with respect to volume and ion positions, keeping cubic relationships between all parameters. Energies were computed with a simple empirical pair potential model. In this model the ions interact electrostatically and through a short range Buckingham potential to ensure the repulsion between ions at short distances. The parameters for the potentials were taken from the literature.[27] Although one cannot expect quantitatively accurate results from this empirical potential model we expect the qualitative conclusion to be valid. Quantum mechanical techniques could be used to obtain more accurate results (see section V.3).

The ECI were determined from the structural energies by fitting the cluster expansion to the 118 structures with excitation energy less than 0.65 eV/cation above the groundstate line. By not using the structures with higher energy in the fit, the convergence of the cluster expansion is improved (see section IV).

These ECI were used in Monte Carlo simulations at different Gd_2O_3 concentrations and temperatures. Because of the large number of pair and multiple interactions the Monte Carlo simulation could be significantly accelerated by using a cluster sampling technique.[28] In this technique, the local environment around a lattice point is tabulated with a series of characteristic integers each representing a unique part of the environment. For each environment the Monte Carlo exchange energy can be tabulated at the start of the program using the v-matrix construction that is often used in the Cluster Variation Method.[29] We also kept track of the mean squared displacement of oxygen ions in the simulation. This quantity is proportional to the oxygen diffusivity if we neglect the change in migration barrier with local environment. Figure 7 shows the oxygen displacement as a function of Gd_2O_3 concentration for two different cation arrangements. For the open circles, cations were equilibrated

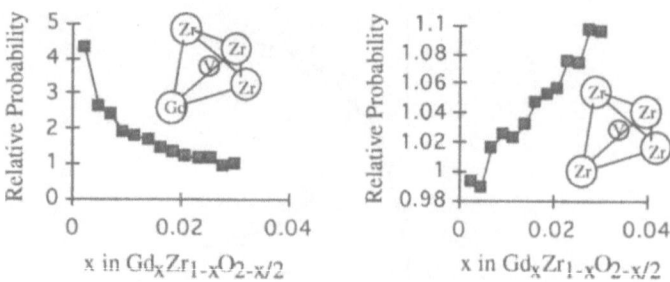

Figure 8. Relative probabilities as a function of concentration at 2000 K for two configurations of cations and a vacancy.

at 2000 K but then fixed during the sampling of the oxygen displacement. Keeping the cations fixed to sample the oxygen displacement is realistic due to the much higher migration enthalpy of the cations. Similar to what can be observed from the experimental diffusion data for this material the oxygen diffusivity first increases with Gd_2O_3 concentration but then sharply drops at about $X_{Gd_2O_3} \approx 0.015$. The second curve (open triangles) dramatically illustrates the effect of cation disorder. In this simulation, the cations were randomly distributed over their lattice sites, again keeping them fixed while sampling the oxygen displacement. It is evident from these computational experiments that cation disorder can significantly enhance the oxygen diffusivity. When the cations are allowed to equilibrate, a strong association occurs between oxygen vacancies and cation arrangements, limiting the oxygen conductivity. This phenomena has been observed experimentally: While quenched samples tend to have the highest oxygen diffusivity, aging tends to reduce the diffusivity.[30]

Figure 8 illustrates that the character of the defect association is changing as the amount of Gd_2O_3 is increased. These graphs show the relative probability for the oxygen vacancy to sit in two specific cation tetrahedra. The probabilities are relative to those for a completely random distribution of cations and anions. Whereas the electrostatic interaction between the Gd cation and the oxygen vacancy causes them to associate in a nearest-neighbor position at low Gd_2O_3 concentration (Fig. 8a), this association weakens at higher Gd_2O_3 content where the vacancy is more likely to be surrounded by four Zr cations.

Although a more detailed study is necessary to accurately isolate the configurational changes that are responsible for the maximum in the oxygen conductivity, this example serves to illustrate the usefulness of the coupled cluster expansion. Once the ECI are computed the statistical mechanics of the defect arrangements can be treated without the need to resort to simplifying assumptions.

V.3 Solubility Limits in the CaO-MgO System

In the previous example, the ECI were computed from an empirical potential model. One can expect a much more accurate temperature scale in the result if quantum mechanical methods are used. However, given the large number of structural energies that were used to fit ECI for the ZrO_2-Gd_2O_3 system such an effort will be computationally very costly. To establish the accuracy of first-principles methods for the computation of oxide phase diagrams, we performed an *ab initio* study of a simple system: CaO-MgO. Both CaO and MgO form the rocksalt structure and have substantial solubility in each other. Because of the identical valence of Ca and Mg, the oxygen

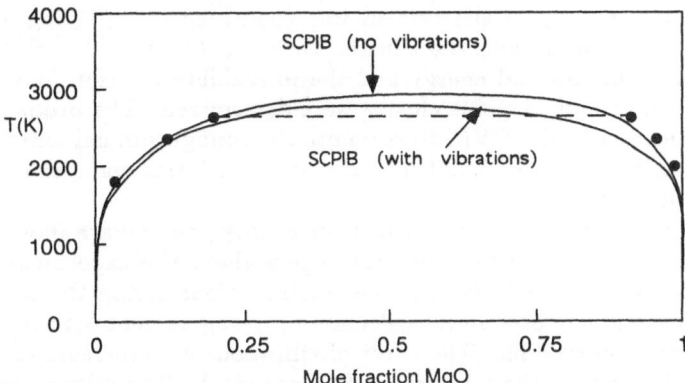

Figure 9. Solid part of the CaO-MgO phase diagram computed with and without vibrational effects. A few experimental data points[37] are represented with filled circles.

sublattice remains fully occupied in CaO-MgO mixtures so that disorder only takes place on the cation sublattice. Hence, this system can be modeled with a standard binary cluster expansion. In this case, to obtain as high an accuracy as possible, we included the effect of lattice vibrations on the free energy of the system. The effect of vibrations can be incorporated by expanding, in Eq. (2), the free energy of the system with fixed cation configuration instead of the static ground state energy. The vibrational effects make the ECI temperature dependent. In the harmonic approximation the temperature-dependence can be linearized above the Debye temperature.[31] We will therefore write the ECI as a chemical part and a vibrational part

$$V_\alpha(T) = V_\alpha^{\text{chem}} + k_B T V_\alpha^{\text{vib}} . \tag{7}$$

The configuration-independent term in the vibrational free energy is excluded from Eq. 7. More details of this procedure can be found in Ref. 31. To determine the values of the chemical ECI we have computed the zero-temperature energy of structures with a number of different cation arrangements and fitted the temperature-independent part of the ECI. The energies were computed with the Self-Consistent Potential Induced Breathing (SCPIB) method,[32,33] also referred to as the Spherical Self-Consistent Atomic Deformation (SSCAD) method. The SCPIB is a first-principles method in which the total energy of the crystal is computed as the sum of the self energy of each ion plus the interatomic interaction energy. The interaction term is computed from the overlap of the self-consistent ionic charge densities. The energies of all structures are minimized with respect to all lattice parameters and internal degrees of freedom. The cluster expansion used in the fit was well converged with pair interaction up to the sixth cation–cation nearest-neighbor, three three-body terms and two four-body terms.[34]

The vibrational components to the ECI, V_α^{vib}, were determined in a similar procedure, by fitting them to the vibrational free energy of structures with a number of different cation arrangements. The vibrational free energy was computed by diagonalizing the dynamical matrix for each structure at a large number of k-points. The elements of the dynamical matrix were obtained indirectly from the SCPIB wave functions.

The total free energy of the system was computed by applying a combination of Monte Carlo simulation and the Cluster Variation Method.[1,6,11,25] From a Monte Carlo simulation the internal energy and the probability distributions for cation arrangements on a 13 and 14 point cluster were determined. The probability distributions were then used in the CVM to compute the configurational entropy. With this procedure, described in more details in Refs. 35 and 36, free energies can be obtained in a single Monte Carlo run.

The resulting phase diagrams display a miscibility gap and are shown in Fig. 9. As only the solid state part is computed, the region above the experimentally observed eutectic temperature (dashed line) is metastable. Considering the fact that no adjustable parameters were used in the calculation, the agreement with the experimental solubility limits is remarkable. The effect of vibrations is to increase solubility limits. In this particular system the vibrational ECI are small. The maximum temperature of the miscibility gap is lowered by about 10%, reducing the agreement with the experimental data. The correction due to the vibrations is asymmetric: The increase in solubility of CaO in MgO is much greater than the increase in solubility of MgO in CaO.

VI. Conclusions

The technique of cluster expanding the configurational energy dependence is a versatile tool to predict the defect arrangement and phase diagrams of oxides. Very good agreement between experimental data and first-principles prediction could be obtained for the CaO-MgO system. In systems in which the electrostatic ion-ion interaction contributes directly to the ECI, a rapidly convergent cluster expansion can still be constructed by limiting the range in configuration space over which it is valid.

For systems in which cation substitution is accompanied by the introduction of a charge compensating defect, disorder on both the cation and anion sublattices has to be accounted for. By appropriately coupling cluster expansions that describe the configuration of each sublattice, the association and ordering of defects in these multicomponent systems can be studied with no *a priori* assumptions regarding their configurations.

Acknowledgements

G. C. gratefully acknowledges travel support from the National Science Foundation to participate in this workshop. P. D. T. acknowledges support from the Petroleum Research Fund, grant ACS-PRF29133-AC5, and a fellowship from the Department of Defense. This work was supported in part by the MRSEC Program of the National Science Foundation under Award Number DMR-9400334.

References

1. D. de Fontaine, in *Solid State Physics*, edited by H. Ehrenreich and D. Turnbull (Academic Press, 1994), Vol. 47, p. 33.
2. G. Ceder, *Computational Materials Science* 1, 144 (1993).
3. G. Ceder, *et al.*, *Phys. Rev. B* 41, 8698 (1990).

4. B. J. Wuensch, *Materials Science and Engineering B* **18**, 186 (1992).
5. R. Clarke, N. Caswell, S. A. Solin, and P. M. Horn, *Phys. Rev. Lett.* **43**, 2018 (1979).
6. J. M. Sanchez, F. Ducastelle, and D. Gratias, *Physica* **128A**, 334 (1984).
7. M. Asta, R. McCormack, and D. de Fontaine, *Phys. Rev. B* **48**, 748 (1993).
8. M. Asta, D. de Fontaine, M. Van Schilfgaarde, and M. Sluiter, *Phys. Rev. B* **46**, 5055 (1992).
9. G. Ceder, *et al.*, *Acta Metall. Mater.* **38**, 2299 (1990).
10. G. Ceder, M. Asta, and D. de Fontaine, *Mat. Res. Soc. Symp. Proc.*, **169**, 189 (1990).
11. F. Ducastelle, *Order and Phase Stability in Alloys*. Eds. F. R. de Boer and D. G. Pettifor, Cohesion and Structure (North-Holland, Amsterdam, 1991), Vol. 3.
12. L. G. Ferreira, S. Wei, and A. Zunger, *Phys. Rev. B* **40**, 3197 (1989).
13. J. M. Sanchez, J. P. Stark, and V. L. Moruzzi, *Phys. Rev. B* **44**, 5411 (1991).
14. M. Sluiter, P. Turchi, F. Zezhong, and D. de Fontaine, *Phys. Rev. Lett.* **60**, 716 (1988).
15. M. Sluiter, P. E. A. Turchi, F. J. Pinski, and G. M. Stocks, *J. Phase Equilibria* **13**, 605 (1992).
16. P. E. A. Turchi, *et al.*, *Phys. Rev. Lett.* **67**, 1779 (1991).
17. R. McCormack, Ph.D. thesis, U.C. Berkeley (1995).
18. G. Ceder, G. D. Garbulsky, D. Avis, and K. Fukuda, *Phys. Rev. B* **49**, 1 (1994).
19. P. D. Tepesch, G. D. Garbulsky, and G. Ceder, *Phys. Rev. Lett.* **74**, 2272 (1995).
20. C. Wolverton, G. Ceder, D. de Fontaine, and H. Dreyssé, *Phys. Rev. B* **48**, 726 (1993).
21. A. Zunger, in *Statics and Dynamics of Alloy Phase Transformations*, edited by P. E. A. Turchi and A. Gonis, p. 361 (1994).
22. D. B. Laks, L. G. Ferreira, S. Froyen, and A. Zunger, *Phys. Rev. B* **46**, 12587 (1992).
23. G. Ceder, G. D. Garbulsky, and P. D. Tepesch, *Phys. Rev. B* **51**, 11257 (1995).
24. E. Salje and V. Devarajan, *Phase Transitions* **6**, 235 (1986).
25. R. Kikuchi, *Phys. Rev.* **81**, 988 (1951).
26. A. Nakamura and J. B. J. Wagner, *J. Electrochemical Soc.* **133**, 1542 (1986).
27. M. P. van Dijk, A. J. Burggraaf, A. N. Cormack, and C. R. A. Catlow, *Solid State Ionics*, **17** 159 (1985); A. Dwivedi and A. N. Cormack, *Journal of Solid State Chemistry*, **79**, 218 (1989).
28. G. Ceder, P. D. Tepesch, C. Wolverton, and D. de Fontaine, in *Statics and Dynamics of Alloy Phase Transformations*, edited by P. E. A. Turchi and A. Gonis, p. 571 (1994).
29. G. Ceder, Ph.D. thesis, U.C. Berkeley (1991).
30. T. Takahashi and Y. Suzuki, *Proc. J. Intl. Etude Piles Combust.* **2**, 378 (1967); W. Baukal, *Electrochim. Acta* **14**, 1071 (1969).
31. G. D. Garbulsky and G. Ceder, *Phys. Rev. B* **49**, 6327 (1994).
32. L. L. Boyer and M. J. Mehl, *Ferroelectrics* **150**, 13-24 (1993).
33. L. L. Boyer, M. J. Mehl, and M. R. Peterson, *Ferroelectrics* **151**, 7 (1994).
34. P. D. Tepesch, *et al.*, *J. Am. Ceram. Soc.*, in press (1996).
35. A. G. Schlijper, *Phys. Rev. A* **41**, 1175 (1990).
36. A. G. Schlijper and B. Smit, *J. Stat. Phys.* **56**, 247 (1989).
37. Y. Yin and B. B. Argent, *Journal of Phase Equilibria* **14** 588 (1993).

Diffuse Scattering of Neutrons in Ni_3V and Pt_3V: Test of the Gamma Expansion Method Approximation in a Degenerate Case

R. Caudron,[1,2] D. Le Bolloc'h,[1,2] A Finel,[1] and M. Barrachin[1]

[1] *Direction des Matériaux (OM)*
Office National d'Etudes et de Recherches Aérospatiales (ONERA)
BP 72, 92322 Châtillon Cedex
FRANCE

[2] *Laboratoire Leon Brillouin*
CEN Saclay, 91191 Gif sur Yvette Cedex
FRANCE

Abstract

Our results of *in situ* diffuse scattering of neutrons in Ni_3V and Pt_3V are presented. In the Ising model framework, effective pair interactions (EPI) up to the 9th neighbor were extracted from the data. These EPI were used to explain successfully the transition temperature of both compounds, the core structure of the dislocations in Ni_3V, and the occurrence of long periods in Pt_3V. The need of approximations yielding long-range EPI is stressed. We give a description of the Gamma Expansion Method (GEM). Because of its mean field nature, we expected the GEM to fail in the highly degenerate case of Pt_3V. We tested it in that case, and found it completely successful. This gives more confidence in the mean field approximations.

I. Introduction

In the case of ordering or clustering on an underlying lattice, the ordering energy of a substitutional binary alloy A_cB_{1-c} can be expressed as a rapidly convergent sum of pair and higher order multiplet interactions between the atomic species:

$$H = \tfrac{1}{2}\sum_{mn} V_{mn}(p_m - c)(p_n - c) + \tfrac{1}{3!}\sum_{lmn} V_{lmn}(p_l - c)(p_m - c)(p_n - c) + \cdots , \quad (1)$$

Theory and Applications of the Cluster Variation and Path Probability Methods
Edited by J.L. Morán-López and J.M. Sanchez, Plenum Press, New York, 1996

where c is the concentration and the p_n are occupation numbers, taking on the values 0 or 1 depending on the species sitting at site n. In principle, the V_{mn} could be deduced from electronic structure calculations, but the smallest ones amount only to fractions of meV, whereas the calculations deal with binding energies, *i.e.* with a few eV: it can be understood why the calculation techniques are not yet sufficiently accurate to compute detailed interactions, and why we find it better, until now, to extract them from experimental data.

However, the General Perturbation Method[1] leads to qualitative arguments enabling, in particular for transition metals alloys, to select the most relevant interactions. This method is founded on a perturbation development of the order energy, the reference state, namely the random alloy, being calculated within the Coherent Potential Approximation (CPA). For transition alloys, this procedure, within the Tight Binding Approximation, leads to simple and general results:

- The pair interactions are dominant *versus* the other multiplet interactions, *i.e.* the order energy can be written:

$$H = \tfrac{1}{2} \sum_{n,m} J_{nm} \sigma_n \sigma_m - \sum_n h_n \sigma_n, \qquad (2)$$

  the σ's, related to the p's by $p = \tfrac{1}{2}(1 - \sigma)$, are spin-like operators, taking on -1 or 1 values, the J's are the corresponding effective pairwise interactions (EPI) and h is the chemical potential difference. This is the reason why we take into account only the pair interactions, though, generally, multiplets cannot be excluded, in which case, through renormalization effects, they would induce temperature and concentration variations of the EPI,[2] which would superimpose on the intrinsic variations explicitly predicted on grounds of band structure calculations.[1]

- The interactions between the second, third and fourth neighbors are of the same order of magnitude, and generally small compared to the first neighbor interaction. Further interactions are still smaller. This hierarchy is governed by the number of first neighbor (110) jumps needed to connect the origin to the neighbor under consideration, with an advantage to the straight paths (220) fourth neighbors for instance).

The first purpose of the present paper is to show, through two experimental examples, how the EPI can be extracted from diffuse scattering of neutrons or x-rays, and how these energetic parameters can be used to predict order-disorder transition temperatures, together with ground state structures and important properties of the ordered state. The quoted examples will make evident the need of obtaining *long-range* EPI.

To obtain the EPI, the Ising model must be solved, which cannot be made exactly, so that either costly Monte Carlo simulations must be used, or approximations must be devised, which allow to save computing time. Among these approximations, the CVM is very useful but, unless further approximations are used,[3,4] the number of EPI's available through this approximation is limited to at most 4 or 5. We intend to focus, in this paper, on the Gamma Expansion (GEM) approximation, which was first devised by Tokar,[5] and already tested in various cases by Masanskii and Tokar[6] and Reinhard and Moss.[7] The second purpose of our paper is to give a complementary evaluation of this approximation, in a degenerate case where we expected it to fail.

So, in the first part, we will present the experimental procedure we currently use to extract EPI from *in situ* diffuse scattering of neutron (and which is also used for x-rays). Then, we will explain how we have applied this procedure to the experimental cases of Ni$_3$V and Pt$_3$V, and how we have used the EPI to explain properties of the ordered state. At this level, we will draw some preliminary conclusions, which point out the need of long ranged potentials to account for the whole physics of some specific order-disorder problems. We will follow by a short description of the GEM approximation, and by the method we used to test it in a case close to the experimental situation of Pt$_3$V.

II. Experimental Procedure

II.1 Experimental Setup

The experiments of diffuse scattering of neutrons by single crystal samples were performed on the dedicated diffuse scattering spectrometer G44 at the Laboratoire Leon Brillouin, CEN Saclay, France. The spectrometer is equipped with a vacuum chamber (10^{-6} torr) and a furnace which surround the sample and enable us to reach 1300°C. At temperatures corresponding to the disordered state of our samples, strong phonon annihilation processes occur, and an energy analysis is necessary to reject the corresponding intensity. For this purpose, the spectrometer is equipped with a chopper and a time of flight analysis, which allow us to reject inelastic scattering processes with an energy resolution of 5 and 3 meV for phonon annihilation and creation, respectively.[8] In these conditions, we explored two planes of high symmetry, the (001) and (011) planes, by tilting the samples so as to set (001) or (011) axes vertical. The incident wave length was $\lambda = 2.59$ Å.

II.2 Data Reduction

The general expression for the intensity scattered by a binary alloy at the scattering vector $\vec{q}$ is :

$$I(\vec{q}) = \sum_i \sum_j b_i b_j \exp\left[i\vec{q}\cdot(\vec{R}_i - \vec{R}_j + \vec{u}_i - \vec{u}_j)\right]. \tag{3}$$

Where $\vec{R}_i$ is the ith lattice position, b_i and b_j are the scattering factors for the atom sitting at i and j, $\vec{u}_i$ and $\vec{u}_i$ are the displacement vectors from the lattice position to the true atom position. The sums run on sites of the *fcc* lattice (structure of the two alloys studied, Ni$_3$V and Pt$_3$V). Up to now, our statistical methods do not enable us to deal with combined effects of chemical order and displacements. Thus, it is necessary to single out the contribution of the local order only: for this purpose, we used a method inspired from Ref. 9. If we consider displacements $\vec{u}_{ij} = \vec{u}_i - \vec{u}_j$ small compared to the vectors $\vec{R}_{ji} = \vec{R}_i - \vec{R}_j$, then the exponential can be expanded, and the intensity can be written as a sum of two terms: the first one is the SRO contribution, periodic in the reciprocal lattice. The second one is the contribution of the static atomic displacements, whose effects are more important far away from the origin of the reciprocal lattice.

After separation of the Bragg scattering and thermodynamic averaging, the contribution due solely to the SRO can be written:

$$I(\vec{q}) = I_{\text{Laue}} \cdot \alpha(\vec{q}), \tag{4}$$

$$\alpha(\vec{q}) = \sum \alpha(\vec{R}) \exp(i\vec{q}\cdot\vec{R}), \tag{5}$$

where the sum runs on the lattice vectors $\vec{R}$ with respect to some origin O. The definition of $\alpha(\vec{R})$ is based on the correlation functions of the occupation operators:

$$4c(1-c)\alpha(\vec{R}) = \langle \sigma(\vec{R})\sigma(0)\rangle - \langle \sigma(0)\rangle^2 . \tag{6}$$

This SRO contribution is just a modulation of the Laue intensity:

$$I_{\text{Laue}} = c(1-c)(b_A - b_B)^2 , \tag{7}$$

and is linear in the $\alpha(\vec{R})$'s. From the expression of the intensity, the different parameters characterising the contributions of the local order and distortions can be fitted using a multilinear least squares routine based on the singular value decomposition.[10]

II.3 Determination of the Effective Potentials

Generally, statistical physics methods are devised in order to solve approximately the Ising model [Eq. (2)] with a given set of potentials J_{mn}, and to compute the properties of the system, including the SRO parameters α_{mn} [$= \alpha(\vec{R})$]. Such a procedure can be termed *direct*. Deducing potentials from SRO requires to solve the *inverse* problem, and we use a trial and error method based on two distinct statistical physics approaches.

II.3.1 Cluster Variation Method

In a first investigation, an inverse Cluster Variation Method (CVM) has been used. The CVM is an analytic method which consists in approximating the probability to obtain a given configuration Ω by a probability $\rho_{\text{CVM}}(\beta_M)$ associated to a basic cluster β_M. Actually, this approximation consists in replacing the exact entropy by a linear combination of entropies of finite clusters included in the basic cluster. Because of the limited memory size of the computers available nowadays, the method is limited to the 4th neighbors in the *fcc* case. The smallest cluster we must use in such a case is the so called 13-14 point approximation.[11]

II.3.2 Monte Carlo

In order to investigate longer-ranged interactions, we had to employ an inverse Monte Carlo (MC) code. Fundamentally different from the CVM, the MC approach is a numerical simulation, free from approximation but subjected to fluctuation and very expensive in terms of computation time.

Now, we will, for two typical cases, give the EPI we obtained by this method in the *disordered* state, and show what can be made of them, in order to predict physical properties of the *ordered* state.

III. Applications

III.1 Ni_3V

For this alloy, which orders at 840°C with the $L1_2$ structure, the experiments were performed at 1100°C.[12] The experimental maps with the SRO contribution only are shown on Fig. 1 left.

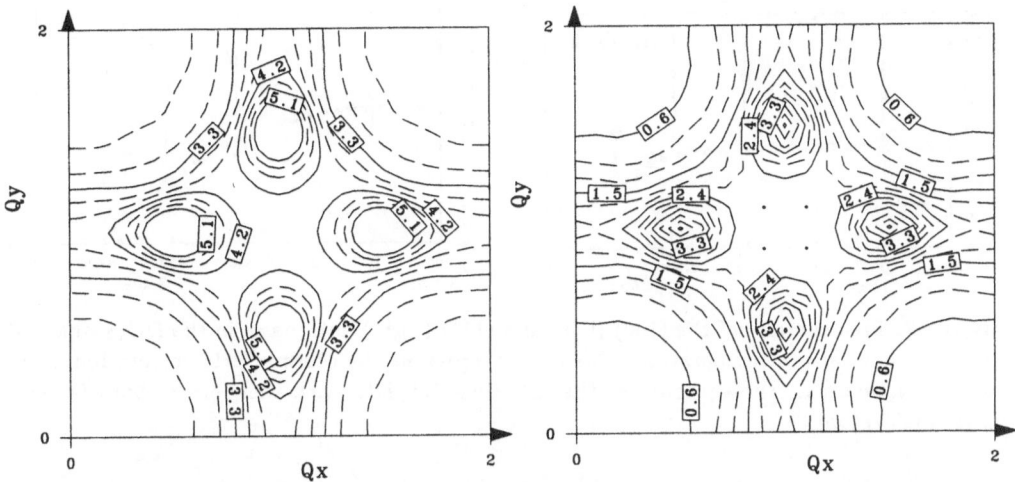

Figure 1. diffuse intensity $\alpha(\vec{q})$ in Ni₃V at $T = 1100°C$ in the (100) plane (Laue units); left: experimental; right: Monte Carlo simulation, with the optimal interaction set of Table I.

Table I

Effective pair interactions in Ni₃V (meV)

J_1	J_2	J_3	J_4	J_7	J_8	J_9
36.2	−7.8	−0.5	3.5	−0.5	−2.4	−1.9

The inverse MC yielded potentials up to 9th neighbors, but we kept only the best set displayed in Table I.[13] As a check, these interactions were introduced in a MC simulation and reproduced closely the experimental map, especially the amplitudes and the shape of the intensity maxima (Fig. 1 right).

These effective potentials can be used to predict physical properties: first, by Monte Carlo simulations, we have calculated the transition temperature T_c between the disordered phase and the DO₂₂ compound, which is the stable low temperature phase of Ni₃V. They are in reasonable agreement with the experimental T_c (experimental: 1045°C; calculated: 840°C). The energy difference between the two phases susceptible of being stabilized at low temperature for this stoichiometry (the DO₂₂ and the L1₂ phases) has been calculated:

$$\Delta E = e(\text{DO}_{22}) - e(\text{L1}_2) = -J_2 + 4J_3 - 4J_4 - 4J_6 + 8J_7 = -12 \text{ meV}. \qquad (8)$$

This result is in agreement with the observed stability of DO₂₂ and interesting comparisons have been made with electronic structure calculations:[13,14] the energy difference between these two phases is eight times lower than the one found by *ab initio* calculations at 0 K. This discrepancy is partially explained by the electronic excitations.[13,14]

A stringent test of the transferability of the set of potentials is to use them to compute antiphase free energies in the DO₂₂ ordered state and to compare the computed values to those obtained experimentally through a study of the dissociation widths

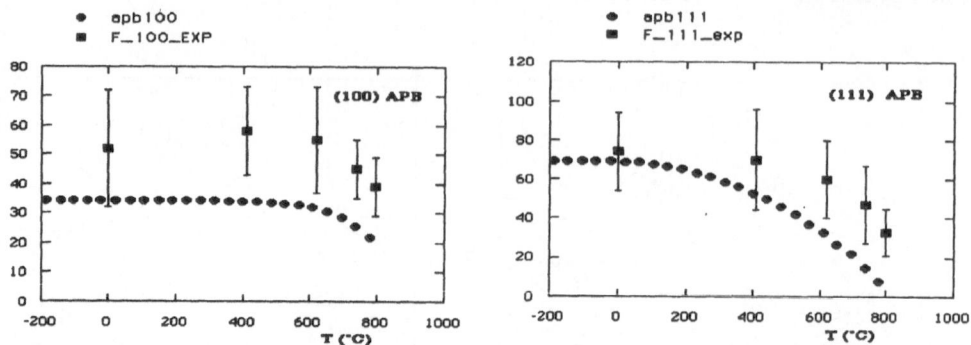

Figure 2. Free energies of the (100) (left) and (111) (right) antiphases in the DO$_{22}$ phase of Ni$_3$V, as functions of temperature. The circles represent the Monte Carlo calculation using our interaction set. the squares are the experimental data with their errors bars (energy units: meV/atom)

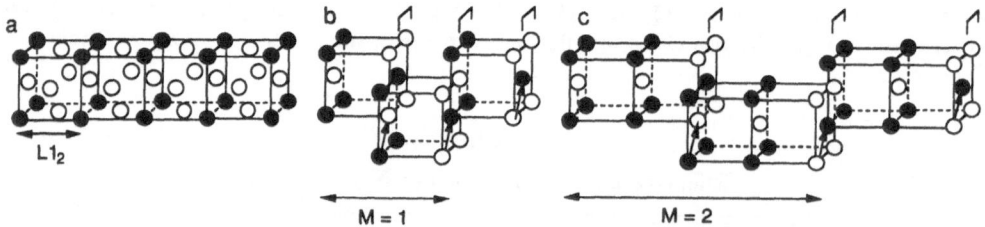

Figure 3. Structures L1$_2$ (a), DO$_{22}$ (M = 1) (b), and M = 2 (c). The two last structures (DO$_{22}$ and M = 2) can be described from the basic structure L1$_2$ by introducing a modulation of antiphase boundaries (APB).

of dislocations in the DO$_{22}$ phase.[15] The results are shown in Fig. 2, along with the experimental data. The agreement is very good.

III.2 Pt$_3$V

This alloy, which, like Ni$_3$V, orders at low temperature on the *fcc* lattice, has the peculiarity of exhibiting different types of order in the ordered state. Electron microscopy observations[16] have shown the presence of long period structures (Fig. 3) derived from the L1$_2$ phase and characterized by the mean distance M between antiphase boundary along the (100) direction: the DO$_{22}$ (M = 1) phase between 0 and 900°C and the M = 4/3 phase (three antiphases for four cubes) up to the transition temperature $T_c = 1040$°C.

These configurations can also be modified by varying slightly the electronic structure of the alloy.[16] Such a situation is typical of a quasi-degenerate case, where the energy differences between each phase in competition in the ordered state are very low: we will try to explain it with the help of the EPI extracted from our measurements of diffuse scattering.

The measurements were performed at 1120°C. The local order parameters have been deduced from the experiment, using the method outlined in subsection II.2. The (100) map of the measured SRO is displayed in Fig. 4 left.

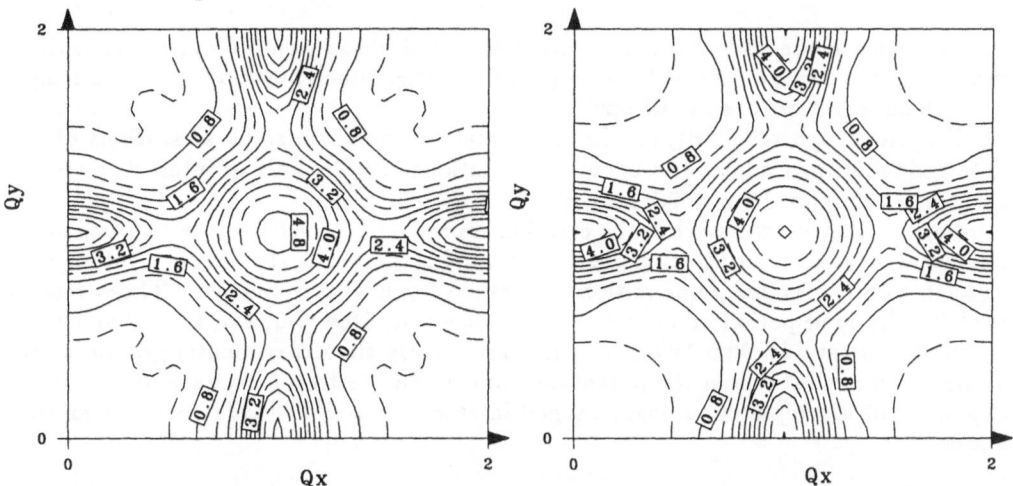

Figure 4. Diffuse intensity $\alpha(\vec{q})$ in the (100) plane of Pt$_3$V, at 1120°C measured (left), and simulated by Monte Carlo (right), using the nine first interactions of Table II. (Laue units)

Table II

Effective pair interactions in Pt$_3$V (meV)

	J_1	J_2	J_3	J_4	J_5	J_6	J_7	J_8	J_9
CVM	46.1	−9.4	53.8	5.2	—	—	—	—	—
MC	44.5	−7.0	6.3	5.6	4.9	0.19	−0.88	2.3	−4.0

We obtained the EPI by inverse CVM and inverse MC simulation. The two sets of potentials we have obtained are displayed in Table II.

With the set of 9 potentials, we can reproduce accurately the experimental map (Fig. 4, right) and evaluate a transition temperature by Monte Carlo simulation (experimental: 1310 K; simulated: 1050 K).

We can also use these potentials to distinguish between the long periods which can be stabilized as ground states, but, as we show below, with potentials up to the ninth neighbor, we can only make the difference between DO$_{22}$ (M = 1), M = 2 and the L1$_2$ phase. For that purpose, starting from the L1$_2$ phase, two parameters are needed: the creation energy per site of a single APB, ξ, and the interaction between two APB when they are first neighbors, J_{APB}. If ξ and J_{APB} are both negative, the density of APB's is maximal, and the long period DO$_{22}$ will be the ground state. If ξ is negative and J_{APB} positive, the APB's will be as numerous as possible, but the repulsion between neighboring APB's will favor the structure M = 2.

Within the Ising Model up to the 9th interactions, the two relevant parameters of our problem are written as linear combinations of pair interactions:

$$\xi = -2J_2 + 8J_3 - 8J_4 - 8J_6 + 16J_7 - 4J_8 \,, \tag{9}$$

$$J_{\text{APB}} = 4J_8 \,, \tag{10}$$

To evaluate the stability of the actual $M = 4/3$ phase would require interactions between more distant APB's, hence a potential range much beyond the eighth neighbor, which is out of reach until now.

Despite a high sensitivity of the individual potentials to small alterations of the data reduction procedure, the linear combinations of potentials of interest is stable: $\xi = -4.8$ meV and $J_{APB} = 9.2$ meV.

So, our analysis shows that the creation energy ξ of the APB's is only a few meV, but negative: this is in favor of their creation. Furthermore, the effective interaction J_{APB} between first neighbor antiphases, also very low, is repulsive. These results explain why long period structures can be observed. However, the $M = 2$ structure is found more stable than DO_{22}, which is apparently the experimental ground state. Temperature variations of the potentials, induced by electronic excitations[13,14] could explain that discrepancy. Longer ranged interactions could also favor other ground states.

III.3 Preliminary Conclusion

In this paper, we explain how we have deduced effective pair interactions of a Ising Model from diffuse scattering of neutrons in the disordered state of two binary alloys (Pt_3V and Ni_3V). With these potentials, we have described accurately not only the disordered state, but also properties of the ground state. In Ni_3V, a pair Ising Hamiltonian expanded up to the 9th pair interaction enables us to describe faithfully the experimental map and to simulate accurately the order-disorder transition temperature, the energy difference between DO_{22} and $L1_2$ phases, and the evolution of free energies of the (100) and (111) antiphases in the DO_{22} phase as a function of temperature. In order to describe a quasi-degenerate case as Pt_3V, we have shown that the difference between various configurations is very small. This result is effectively in favor of the stabilization of long period structures in the ground state, but it shows that very long range potentials are needed in order to explain thoroughly the situation, and why we are interested by approximate methods which could yield longer ranged potentials with sensible computing times and/or computer core sizes.

IV. Approximations

The preceding section has shown the need for approximate methods. The CVM approximation, which is celebrated in this conference, can only yield 4 or 5 potentials, because the computer core needed increases exponentially with the size of the basic cluster. However, the CVM can be extended, through a further approximation: the interactions extending beyond the size of the basic cluster are treated in a mean field like way, without modifying the CVM prescriptions for the entropy.[3,17] Our purpose is to describe our contribution to the evaluation of another method:

IV.1 The Gamma Expansion Method Approximation

This approximation is a quite new extension of the widely used mean field *point* approximation (MFA)[18-21] which, in its very principle, neglects the fluctuations, and should yield only informations about the *points* (local concentrations), and nothing about the *pair* correlations of the fluctuations. Nevertheless, they can be obtained, somewhat artificially, by taking into account the spatial variation of the local concentration, either directly or through the assumption that the phases of the Fourier components of the fluctuations are random [Random Phase Approximation (RPA)].[20] Equivalently, the fluctuation-dissipation theorem can be applied to the mean field solution, though it is exact only for the true equilibrium state.[19] Whatever the method, the result is:

$$\alpha(\vec{q}) = \frac{1}{1 + 4c(1 - c)\beta J(\vec{q})} \, . \tag{11}$$

The $\alpha(\vec{R})$ and the $J(\vec{R})$ are expressed through their Fourier transforms $\alpha(\vec{q})$ and $J(\vec{q})$ [Eq. (5)]. As the $\alpha(\vec{q})$ are obtained directly in the reciprocal space, it is very natural to compute the $J(\vec{q})$ and to submit them to an inverse Fourier transform (IFT), in order to obtain the $J(\vec{R})$. The simplicity of this procedure explains its popularity. Unfortunately, this simplicity has a price, which is a consequence of the approximations used to obtain (11): this equation does not possess enough degrees of freedom to ensure its compatibility with the requirement that, in the *inverse* problem, an atom should not, in principle, interact with itself, *i.e.* $J(\vec{R} = 0) = 0$ or, in the *direct* problem, the application of Eq. (6) to the $\vec{R} = 0$ case leads to:

$$\alpha(\vec{R} = 0) = 1 = \Omega^{-1} \int \alpha(\vec{q}) \, d^3\vec{q} \, . \tag{12}$$

In order to satisfy one or another of those requirements, without much fundamental justification, people were obliged to *multiply* Eq. (11) by a *factor D*.[22,23]

Other drawbacks of this approximation are inconsistencies in the determination of the long-range order parameter in the ordered state, and an overestimate of the transition temperature, as the correlations in the disordered state, which lower its energy, are not taken into account.[20]

These problems can be overcome by noticing that the RPA violates the perquisite that the sum of the intensities of the fluctuating modes should be equal to the number of sites. To meet this condition, a Lagrange multiplier method can be applied: this introduces a kind of chemical potential, which looks like an interaction of an atom with itself, *i.e.* an *additive* constant $4c(1 - c)\beta J(\vec{R} = 0)$ to the *denominator* of Eq. (11).† This approximation, compared to the RPA, generally yields a better estimate of the transition temperature and, for the purpose of fulfilling the requirement (12), provides an adjustable parameter which, because it has a physical meaning, is much more natural than the factor proposed previously.[22,23] This conclusion, which has also been reached by Hoffman,[25] should already have been more widespread: a deeper insight in the very nature of the MFA is indicated by the structure of (11) which suggest that, somehow, a summation of a geometrical series has been implicitly performed. For this reason, a correction for the inconsistencies of (11) should rather have been sought in the denominator.

These remarks explain the unexpected success of the Mean Field Method [Eq. (11)] *with no D correcting factor in the numerator*, applied to experimental cases[26,27] when compared to *Reverse* Monte Carlo, as termed by Ref. 26 (it uses the Gerold and Kern method,[28] which is distinct from the Inverse Monte Carlo we currently use). Indeed, the spherical model formula can be written as follows, solved for the inverse problem, *i.e.* with the EPI given explicitly as functions of the short-range order parameters:

$$J(\vec{q}) = \frac{1}{4c(1 - c)\beta\alpha(\vec{q})} - \frac{1}{4c(1 - c)\beta} + J(\vec{R} = 0). \tag{13}$$

---

† Exactly the same formula has been used by Ref. 24 through a method termed "Onsager cavity field."

Clearly, if no multiplying factor was used in (11), and if nobody cared about $J(\vec{R} = 0)$ which is supposed to be adjusted to satisfy (12), the Inverse Fourier Transform of $J(\vec{q})$ corresponded to the spherical model rather than the less realistic Bragg-Williams model. The term $-1/[4c(1 - c)\beta] + J(\vec{R} = 0)$, which is constant in q-space, yielded only a component for $(\vec{R} = 0)$ which, because its physical meaning was not clear, arose little interest.

To summarize this historical overview, the way Ref. 26 and Ref. 27 have used Eq. (13) correspond in fact to the *spherical* model rather than the Mean Field approximation.

This revival of the point mean field approximation, improved by the spherical model (SM), was initiated by Masanskii and Tokar,[6] because the SM is the zero order approximation of the theory developed by Tokar[5] under the name of Gamma Expansion Method (GEM), which is a new method of solving the field-theoretic lattice models. It is based on a series expansion which is valid if the pair correlation functions decay exponentially with distance. Its main result is a set of corrections to the spherical model results J^{sph} for the three first shells in the *direct* space:†

$$J_1 = J_1^{\text{sph}} + \tfrac{1}{4}kT\left(A\alpha_1^2 + B\alpha_1^3\right),$$

$$J_2 = J_2^{\text{sph}} + \tfrac{1}{4}kTA\alpha_2^2,$$

$$J_3 = J_3^{\text{sph}} + \tfrac{1}{4}kTA\alpha_3^2,$$

with

$$A = \frac{1}{2}\left[\frac{1 - 2c}{c(1 - c)}\right]^2, \quad \text{and} \quad B = \frac{1}{6}\frac{[1 - 6(1 - c)]^2 - 3(1 - 2c)^4}{[c(1 - c)]^4},$$

and the other EPI are unchanged. These formula can easily be applied to the experimental situation: $1/\alpha(\vec{q})$ is readily computed from the experimental data, and the IFT of $J(\vec{q})$ yields the $J(\vec{R})$. For the Tokar corrections, an estimate of $\alpha(\vec{R})$ which is not too affected by truncation is obtained by an IFT of the data with a limited number of $\vec{R}$.

IV.2 Test Procedure of the GEM Approximation

The GEM approximation being close to the Mean Field, its pitfalls should become apparent in situations where the Mean Field theory fails. As we have shown in a previous paper,[29] in presence of degeneracy, which is a consequence of frustration, the Mean Field approximation predicts an incorrect location of the diffuse intensity maximum, which is a severe *qualitative* error.

From this standpoint, the tests or Reinhard and Moss[7] were performed on rather favorable situations: cubic-centered alloys are free from frustration, and, for the face-centered examples chosen by Reinhard and Moss, the ratio $\Delta E/J_1$ of the energy difference between the DO_{22} and $L1_2$ ground states to the interaction between first neighbors is not les than 0.25. This parameter is a measure of the degeneracy: in the

---

† As the Hamiltonian implicitly used by Ref. 6 and Ref. 7 is $H = \tfrac{1}{2}\sum'_{nm} V_{nm}\sigma_n\sigma_m - \sum_n h_n\sigma_n,$[26] with the summation restricted to $n > m$ i.e. $H = \tfrac{1}{4}\sum_{n,m} V_{nm}\sigma_n\sigma_m - \sum_n h_n\sigma_n$, whereas ours is $H = \tfrac{1}{2}\sum_{n,m} J_{nm}\sigma_n\sigma_m - \sum_n h_n\sigma_n$ (without restricted summations), their EPI are twice ours. As a consequence, the denominator of Eq. (11) contains $4c(1 - c)\beta J(\vec{q})$ instead of $2c(1 - c)\beta V(\vec{q})$ in Refs. 6 and 7, and the Tokar corrections to the V are twice our corrections to the J.

case of J_1 alone and positive, the face-centered lattice is strongly frustrated, and it is very hard for the system to find its ground state.

With $\Delta E/J_1$ around 0.05, the Pt₃V case can be considered as much closer to degeneracy than all the examples treated by Reinhard and Moss. It is then worthwhile to test the GEM approximation in such a case. However, we used a slightly different procedure than Reinhard and Moss: instead of comparing the potentials obtained by the GEM with those of inverse Monte Carlo when both applied to the same experimental distribution of diffuse intensity, we submitted to a GEM algorithm the $3D$ distribution obtained by a Monte Carlo (MC) simulation and compared the resulting EPI to the input values of the MC. This approach tests more efficiently the approximation itself, and difficulties linked with the sampling in the reciprocal space are avoided.

So, the EPI set displayed in Table II, which is valid for Pt₃V, was introduced in a MC simulation, which was run for four different temperatures: 1393, 1453, 1693 and 1993 K. The cubic box of the simulation contained 37,000 (4×21^3) atoms and 1000 MC steps per spin were performed for each temperature. The short-range order parameters $\alpha(\vec{R})$ up to the 16th neighbors were computed by averaging the successive configurations. These parameters were Fourier transformed on a $3D$ Cartesian mesh including 1,321 points, spaced every 0.1 RLU,† and contained in a $1 \times 1 \times 1$ RLU cube, including its borders, edges and vertices.

Then, taking the formula (13), we computed $J(\vec{q})$, and we performed its Inverse Fourier transform (IFT). Two methods were used for the IFT: on one hand, a linear regression of $J(\vec{q})$ by $\sum J(\vec{R}) \exp(i\vec{q} \cdot \vec{R})$, taking into account the symmetries of the lattice points $\vec{R}$; on the other hand, a direct summation of the data $\sum J(\vec{q}) \exp(-i\vec{q} \cdot \vec{R})$ was performed in the cube of edge (001) in which it had been computed, with a $\frac{1}{2}$ weight for the faces of the cube, $\frac{1}{4}$ for the edges and $\frac{1}{8}$ for the vertices. This type of summation, which concerns two Brillouin zones, is strictly equivalent to the integration in the true Brillouin zone, but the problem of the weight at the borders is solved in a much easier way. The results of the two IFT methods are the same, even for the lowest temperature, which is the most unfavorable case.

IV.3 Results

These results, which are valid for the bare spherical model (SM), without the Tokar correction,[5] are shown in Figs. 5, 6, and 7 for the five first EPI, as a function of the temperature for which both the Monte Carlo and the SM procedure were run. The input values and the results corrected, according to the method of Tokar[5] are also shown. As is expected from the Mean Field Approximation, the higher the temperature, the better the agreement; at the lowest temperature, the J_1 value is only 4% too strong, and the Tokar correction some bit degrades the agreement, which, however, remains good (about 10%); for J_2, the SM result is already very close to the input value and, with the Tokar correction, the agreement is almost perfect. The correction is negligible for J_3 and zero for the longer ranged neighbors, and the input values are well reproduced. The overall corrected results are shown in Fig. 8.

Up to the ninth shell, the output of the GEM is very close to the input of the MC simulation. The latter being zero from the tenth neighbor onwards, the results of the GEM should also be zero. As explained above, this tends to be true at the

---

† Relative Lattice Units: units of the reciprocal space, such that the *fcc* Bragg peaks show up at 200, 110, ...

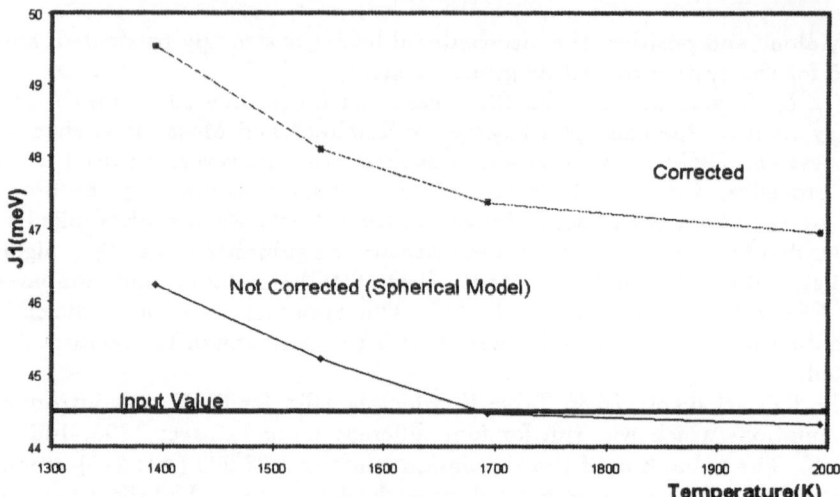

Figure 5. First effective pair interaction(EPI) J_1, obtained by the spherical model (SM), applied to the $3D$ distribution of the diffuse intensity issued from a Monte Carlo (MC) simulation whose input was the EPI shown in Table II. The values are plotted as a function of the temperature for which the MC simulation *and* the SM were run. Also shown are the input value J_1 of the MC and the results, corrected along the GEM,[5,6] procedure.

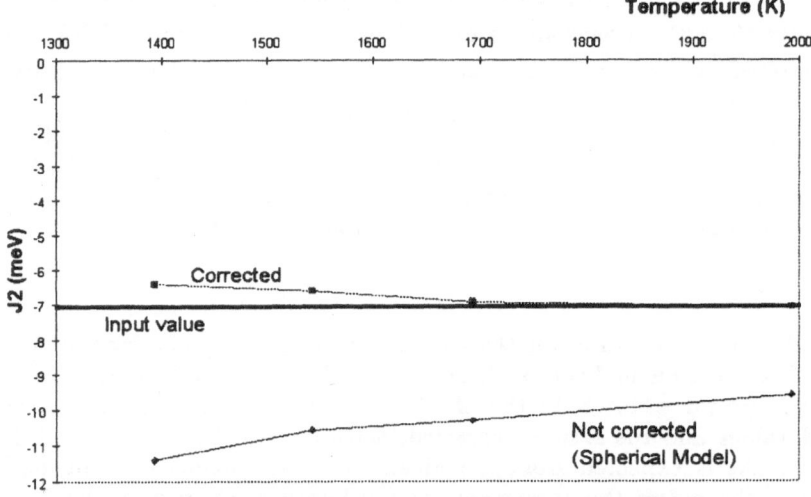

Figure 6. Same plot as Fig. 5, but for J_2 instead of J_1.

highest temperature tested but, as the critical temperature is approached, stronger and stronger humps show up around the 15th, the 21th and the 23th neighbor. These anomalies can be attributed to the limited range (16th neighbor) of the short-range order parameters, which induce cut-off effects, *i.e.* spurious oscillations in q-space. The overall shape of $\alpha(\vec{q})$ is not too sensitive to these oscillations but, around the origin, $\alpha(\vec{q})$ is very low,† and its inverse $1/\alpha(\vec{q})$ is strong, so that the small oscillations

---

† Close to the transition temperature, the short-range order intensity is strong, and con-

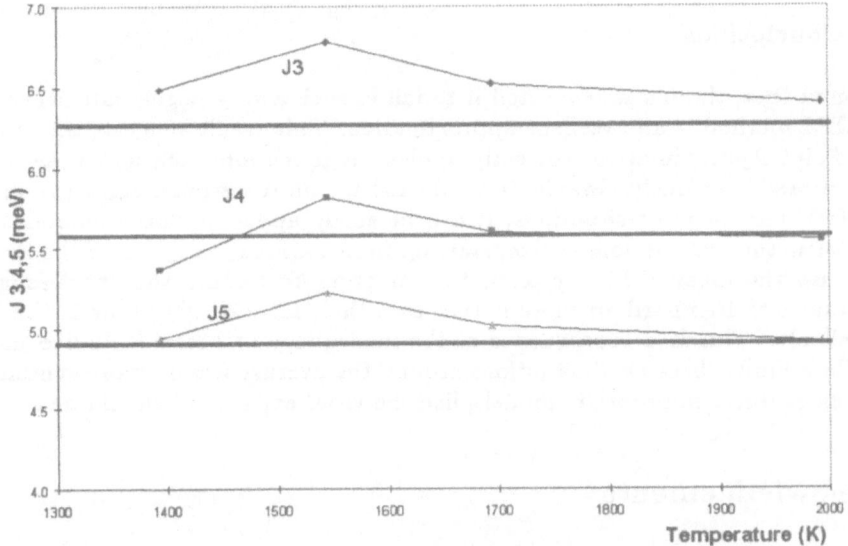

Figure 7. Same plot as Fig. 5, but for J_3, J_4 and J_5 instead of J_1.

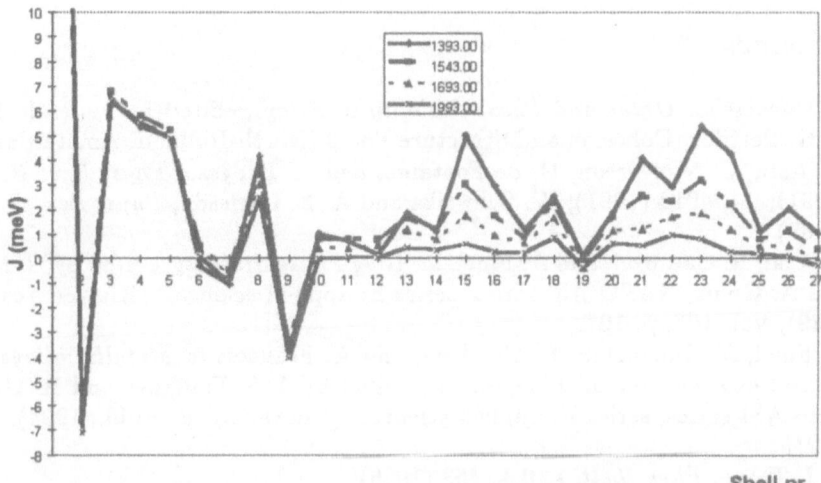

Figure 8. Effective pair interactions(EPI), except J_1, obtained by the GEM approximation, including corrections to the the spherical model up to the third shell. The GEM approximation was applied to the $3D$ distribution of the diffuse intensity issued from a Monte Carlo (MC) simulation whose input was the EPI shown in Table II. The temperature for which the MC simulation *and* the GEM were run are displayed in the inset.

are amplified. Anyway, these effects, which cannot be attributed to the GEM itself, occur outside the relevant region, and should be overcome by appropriate numerical methods.

centrated around (100). As (12) must be fulfilled, only a small amount of intensity remains around the origin.

IV.4 Conclusions

We found that, though we expected it to fail in such a very degenerate case as Pt_3V, the GEM method is an excellent approximation. This result suggests that the Mean Field Point Approximation, correctly applied, is much more efficient than we would have guessed previously. Maybe it would fail for more extreme cases (J_1 alone, for instance), but, in most situations, it can be safely applied. This validates the electronic structure calculations which, starting from a CPA description of the disordered state, use the mean field approximation in order to deduce the short-range order correlations.[24] Reinhard an Moss have shown that, for c close to 0 or 1, the GEM is less reliable. This is a consequence of the inadequacy of Mean Field-like models in the dilute limits, because fluctuations around the average are no more symmetric. In those cases, more appropriate models, like the virial expansion, should be considered.

Acknowledgements

We thank F. Ducastelle for fruitful discussion, and D. Regen for the elaboration of high quality single crystals.

References

1. F. Ducastelle, *Order and Phase Stability in Alloys*, edited by F. R. de Boer and D. G. Pettifor, Cohesion and Structure Vol. 3 (North-Holland, Amsterdam, 1991).
2. M. Asta, C. Wolverton, D. de Fontaine, and H. Dreysse, *Phys. Rev. B* **44**, 4907 (1991); **44**, 4914 (1991); W. Schweika and A. E. Carlsson, *Phys. Rev. B* **40**, 4990 (1989).
3. F. Solal, R. Caudron, and A. Finel, in *Alloy Phase Stability*, edited by G. M. Stocks and A. Gonis, NATO ASI series, series E: applied sciences, (Kluwer Acad. Publ., 1989), Vol. 163, p. 107.
4. A. Finel, M. Barrachin, R. Caudron, and A. François in *Metallic alloys, Experimental and Theoretical Perspectives*, edited by J. S. Faulkner and R. G. Jordan, Nato ASI series, series E: applied sciences, (Kluwer Acad. Publ., 1994), Vol. 256, p. 215.
5. V. I. Tokar, *Phys. Lett.* **110A**, 453 (1985).
6. I. V. Masanskii, V. I. Tokar, and T. A. Grishchenko, *Phys. Rev. B* **44**, 4647 (1991).
7. L. Reinhard and S. C. Moss, *Ultramicroscopy* **52**, 232 (1993).
8. R. Caudron, M. Sarfati, M. Barrachin, A. Finel, F. Ducastelle, and F. Solal, *J. Phys.* **I2**, 1145 (1992).
9. J. C. Sparks and B. Borie, in *Local Atomic Arrangements studied by X-Ray Diffraction*, edited by J. B. Cohen and J. E. Hilliard (Gordon and Breach, 1966).
10. C. L. Lawson and R. J. Hanson, *Solving Linear Least Squares Problem*, (Prentice-Hall, Englewood-Cliffs, New-Jersey, 1974).
11. A. Finel, Ph.D. thesis, Paris VI, 1987.
12. M. Barrachin, Ph.D. thesis, Paris XI-Orsay, 1993.
13. M. Barrachin, A. Finel, R.Caudron, A. Pasturel, and A. François, *Phys. Rev. B* **50**, 12980 (1994).
14. C. Wolverton and A. Zunger, *Phys. Rev. B* **52**, 8813 (1995).
15. A. François, Ph.D. thesis, Paris VI, 1992.

16. E. Cabet and A. Loiseau, *J. Phys.* (France) IV colloque C7, Supp. III **3**, 2051 (1993).
17. A. Finel, in *Statics and Dynamics of Alloy Phase Transformations*, edited by A. Gonis and P. E. A. Turchi, NATO ASI series B: Physics, (Plenum Press, New York, 1994), Vol. 319, p. 495.
18. W. L. Bragg and E. J. Williams, *Proc. Roy. Soc. London, Ser. A* **145**, 699 (1934); **152**, 231 (1935).
19. M. A. Krivoglaz, *Theory of X-Rays and Thermal Neutron scattering by Real Crystals*, (Plenum, New York, 1969).
20. R. H. Brout, *Phase Transitions*, (Benjamin, New York, 1965).
21. P. G. de Gennes and J. Friedel, *J. Phys. Chem. Solids*, **4**, 71 (1958).
22. P. C. Clapp and S. C. Moss, *Phys. Rev.* **142**, 418 (1966); **171**, 754 (1968); S. C. Moss and P. C. Clapp, *Phys. Rev.* **171**, 764 (1968).
23. S. Lefebvre, F. Bley, M. Fayard, and M. Roth, *Acta Metall.* **29**, 749 (1981).
24. J. B. Staunton, D. D. Johnson, and F. J. Pinski, *Phys. Rev. B* **50**, 1450 (1994).
25. D. W. Hoffman, *Met. Trans.* **3**, 3231 (1972).
26. W. Schweika and H. G. Haubold, *Phys. Rev. B* **37**, 9240 (1988).
27. B. Schönfeld, l. Reinhard, G. Kostorz, and W.Bührer, *Phys. Status. Solidi B* **148**, 457 (1988).
28. V. Gerold and G. Kern, *Acta Metall.* **35**, 393 (1987).
29. F. Solal, R. Caudron, F. Ducastelle, A. Finel, and A. Loiseau, *Phys. Rev. Lett.* **58**, 2245 (1987).

Cluster Variation Method Applications to Large Ising Aggregates

F. Aguilera-Granja and J.L. Morán-López

Instituto de Física "Manuel Sandoval Vallarta"
Universidad Autónoma de San Luis Potosí
San Luis Potosí, S.L.P., 78000
MEXICO

Abstract

The magnetic properties of very large Ising aggregates with various shapes are studied using the pair, triangle and square approximations of the Cluster Variation Method (CVM). To derive analytic expressions for the Curie temperature of clusters with an arbitrary number of sites we assume that the probability distributions do not depend on the position of the spins. Our results indicate that for very large clusters the critical temperature depends more on the geometrical characteristics (number of sites, number of pairs, number of planes, etc.) than on the position dependence of the cluster probabilities. We found that for a fixed number of atoms, the highest critical temperature corresponds to the more spherical aggregates. We compare our results with Monte Carlo simulations and with other CVM calculations in which the position dependence of the pair probabilities were taken into account. In the case of the high order approximations (triangle and square), we found that for aggregates with a number of atoms smaller than n^*, the phase transition disappears. The number n^* depends on the geometrical characteristics of the system.

I. Introduction

The magnetic properties of atomic aggregates has attracted renewed interest due to the impresive development of experimental techniques. Now, aggregates with a particular size can be grown in a controlled way. Furthermore, the small clusters of transition and rare-earth metals promise interesting technological applications.[1−4] Recent experiments show how the magnetic properties of clusters, in the range of a few to hundred atoms, depend on the size.[5−10] In addition, from the fundamental point of view, the understanding of the dependence of the magnetic properties of clusters on the number of atoms and geometrical structure is by itself a very interesting problem.

Theory and Applications of the Cluster Variation and Path Probability Methods
Edited by J.L. Morán-López and J.M. Sanchez, Plenum Press, New York, 1996

Here, we are interested in the magnetic properties of Ising aggregates with a large number of atoms. We assume an effective magnetic moment at each site of the atomic aggregate and that all the spins are described by an average probability distribution independent of the spin position. This probability distribution is calculated by means of a simplified version of the Cluster Variational Method.[11] The model presented here is similar to the one published by Hellenthal.[12] However, in our case we consider correlations between first and second-nearest neighbors, neglected in previous contribution.

It is known that well defined phase transitions exist only in infinite systems. In the case of finite systems, the spin fluctuations produce a remanent magnetization that prevails at high temperatures. The value of the magnetization tail depends on the number of atoms.[7,13−15] However, one can still define a pseudo-phase transition. In Monte Carlo simulations (MCS), for example, one identifies a critical temperature with the point at which the specific heat takes its maximum value.[13−15]

In Sec. II we present the model and the results obtained by means of various approximations. Our conclusions and summary are contained in Sec. III.

II. Model

To describe the magnetic properties of the finite size systems we adopt the Ising model with spin $\frac{1}{2}$. We consider only ferromagnetic interactions between nearest neighbors and assume that all the spins in the aggregate are described by an average probability distribution, which is the same for all the spins regardless of the position. This approximation allow us to handle very large atomic aggregates and to deduce analytical expressions for the critical temperature and the values of the short-range order parameters at that temperature.

The energy parameters ε_{ij} that describe the ferromagnetic interactions are

$$\varepsilon_{ij} = \begin{cases} -J & \text{for parallel pair of spins,} \\ +J & \text{for anti-parallel pair of spins,} \end{cases} \tag{1}$$

where $J > 0$. The energy for an ensemble of M copies of the system is written as a sum of the energy of the nearest-neighbor pairs times the number of copies of the system in the ensemble. Then, for the entire ensemble

$$E = M\gamma \sum_{ij} \varepsilon_{ij} y_{ij} = MJ\gamma(4y_{12} - 1) = -MJ\gamma\sigma, \tag{2}$$

where γ is the number of pairs within every one of the copies of the ensemble, y_{ij} is the probability of finding a pair of nearest-neighbors spins i–j, and σ a short-range order parameter defined in terms of the pair probabilities. The parameter σ has the property that takes the value one for $T = 0$ and vanishes as T goes to infinity.

II.1 Pair Approximation

The single site probabilites are denoted by x_i with $i = 1$ (2) for spin up (down). The single site probabilites, in terms of the pair probabilities, are given by $x_i = \sum_j y_{ij}$, and the pair probabilities satisfy the normalization condition $\sum_{ij} y_{ij} = 1$.

The magnetic behavior of the system is described in terms of a long-range order (LRO) parameter $\xi = x_1 - x_2 = [(y_{11} + y_{12}) - (y_{22} + y_{21})]$ and of the pair probability y_{12}, which can be interpreted as a short-range order (SRO) parameter. Taking into account the normalization constraint, and the fact that $y_{12} = y_{21}$, these two order parameters are the only independent variables.

Defining $\{\bullet\}_M = \prod_i (Mx_i)!$ and $\{\bullet\!-\!\!\bullet\}_M = \prod_{ij}(My_{ij})!$ one can write the entropy for an ensemble of M samples in the pair approximation as

$$\exp\left(\frac{S}{k_{\mathrm{B}}}\right) = \left(\frac{M!}{\{\bullet\}_M}\right)^\alpha \times \left(\frac{(\{\bullet\}_M)^2}{\{\bullet\!-\!\!\bullet\}_M\, M!}\right)^\gamma. \tag{3}$$

where α and γ are the number spins and the number of spin pairs for any of the M copies of the ensemble, respectively. The entropy of one of the samples of the ensemble can be rewritten as follows

$$\frac{S}{k_{\mathrm{B}}} = (2\gamma - \alpha)\sum_i \mathcal{L}(x_i) - \gamma\sum_{i,j}\mathcal{L}(y_{ij}) + (\gamma - \alpha), \tag{4}$$

where $\mathcal{L}(v) = v\ln v - v$. The free energy $(\mathcal{F} = E - TS)$ for one of the M copies of the ensemble is given by

$$\Phi \equiv \frac{\beta\mathcal{F}}{M} = \gamma\beta J(4y_{12} - 1) - (2\gamma - \alpha)\sum_i \mathcal{L}(x_i) + \gamma\sum_{i,j}\mathcal{L}(y_{ij}) - (\gamma - \alpha). \tag{5}$$

Here, $\beta = 1/k_{\mathrm{B}}T$. The minimization of Φ with respect to the order parameters leads to the coupled equations

$$\frac{\partial\Phi}{\partial\xi} = -(2\gamma - \alpha)\ln(x_1/x_2) + \gamma\ln(y_{11}/y_{22}) = 0, \tag{6a}$$

$$\frac{\partial\Phi}{\partial y_{12}} = 4\beta J - \ln\left(y_{11}y_{22}/y_{12}^2\right) = 0. \tag{6b}$$

These equations can rewritten as

$$(x_1/x_2)^{2\gamma-\alpha} = (y_{11}/y_{22})^\gamma, \tag{7a}$$

$$y_{11}y_{22} = y_{12}^2\exp(4\beta J). \tag{7b}$$

In order to simplify Eqs. (7) and their solution, we introduce a variable Θ, defined as follows

$$\frac{x_1}{x_2} \equiv \exp(6\Theta\gamma) = \frac{1+\xi}{1-\xi} = \frac{\exp(+A)}{\exp(-A)}, \tag{8}$$

with $A = 3\Theta\gamma$. Notice that at and above the Curie temperature Θ vanishes, then $x_1 = x_2$. In terms of Θ the long-range order parameter is given by $\xi = \tanh A$. Now, Eq. (7a) can be rewritten as

$$\frac{y_{11}}{y_{22}} \equiv \exp[6\Theta(2\gamma - \alpha)] = \frac{x_1 - y_{12}}{x_2 - y_{12}} = \frac{\exp(+B)}{\exp(-B)}, \tag{9}$$

where $B = 3\Theta(2\gamma - \alpha)$. From Eq. (9) one can get an expression for y_{12} in terms of A and B

$$y_{12} = \frac{1}{2}\left(1 - \frac{\tanh A}{\tanh B}\right). \tag{10}$$

The other two pair probabilities are given by

$$y_{11} = \frac{1}{2}\left(1 + \frac{1}{\tanh B}\right)\tanh A, \tag{11a}$$

$$y_{22} = \frac{1}{2}\left(\frac{1}{\tanh B} - 1\right)\tanh A. \tag{11b}$$

By substituting Eqs. (10) and (11) in Eq. (7b) one obtains a general expression for the temperature ($\beta = 1/k_B T$) in term of A and B

$$\left[\frac{1}{2}\left(1 - \frac{\tanh A}{\tanh B}\right)\right]^2 \exp(4\beta J) = \frac{1}{4}\left(1 - \tanh^2 B\right)\left(\frac{\tanh A}{\tanh B}\right)^2, \tag{12}$$

One can simplify Eq. (12). This leads to the following expression for the temperature in terms of Θ

$$k_B T = \frac{2J}{\ln\left[\sinh 3\Theta\gamma / \sinh 3\Theta(\gamma - \alpha)\right]}, \tag{13}$$

By using Eq. (8) and taking into account the normalization constraint one can write x_1 and x_2 as a function of Θ

$$x_1 = \frac{\exp(3\Theta\gamma)}{2\cosh(3\Theta\gamma)}, \quad \text{and} \quad x_2 = \frac{\exp(-3\Theta\gamma)}{2\cosh(3\Theta\gamma)}. \tag{14}$$

Equations (13) and (14) represent the general solution of the problem in terms of Θ. To calculate the transition temperature, one takes in Eq. (13) the limit as Θ goes to zero

$$k_B T_c = \frac{2J}{\ln\left[\gamma/(\gamma - \alpha)\right]}. \tag{15}$$

It is worth to notice that this is a very general expression and it holds for any kind of structure regardless of the size. Furthermore, it is very easy to prove that the short-range order parameter σ is given by

$$\sigma = \frac{2\tanh 3\Theta\gamma - \tanh 3\Theta(2\gamma - \alpha)}{\tanh 3\Theta(2\gamma - \alpha)}. \tag{16}$$

At T_c the SRO is given by

$$\sigma_c = \frac{\alpha}{2\gamma - \alpha}. \tag{17}$$

The behavior of the parameters ξ and σ close to T_c is also interesting. From $\xi = \tanh 3\Theta\gamma$, and expanding Eq. (13) in Θ one obtains

$$\xi \approx 2\left(\frac{\gamma}{\alpha}\right)\sqrt{\frac{3J\sigma_c}{k_B T_c}}\left(1 - \frac{T}{T_c}\right)^{1/2}. \tag{18}$$

The behavior of σ close to T_c (for $T < T_c$) can be obtained from Eq. (13) together with a series expansion of Eq. (16). The shift in the SRO is

$$\frac{\sigma(T)}{\sigma_c} - 1 \approx 8\left(\frac{\gamma}{\alpha^3}\right)\left(\alpha^2 - 4\alpha\gamma + 3\gamma^2\right)\left(\frac{J\sigma_c}{k_B T_c}\right)\left(1 - \frac{T}{T_c}\right). \tag{19}$$

For $T > T_c$ the behavior is simply given by $\sigma(T) = \tanh(J/k_B T)$.

On the other hand, in the low temperature limit and by using a similar procedure, one gets $\xi \approx 1 - 2\exp\{-4(\gamma/\alpha)J/k_B T\}$ and $\sigma \approx 1 - 4\exp\{-4(\gamma/\alpha)J/k_B\}$. Notice that σ decays faster that ξ. Next, we apply the theory to various systems.

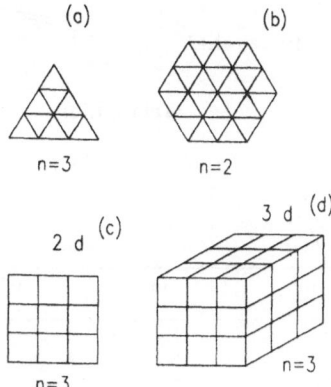

Figure 1. The various types of aggregates studied in this contribution. (a) Triangular, (b) hexagonal, (c) square and (d) simple cubic.

II.1.1 Bulk Properties

As a first application we consider the case of a ferromagnetic system with N atoms and z nearest-neighbors. In this case $\alpha = N$ and $\gamma = \frac{1}{2}Nz$. In the thermodynamic limit, the critical value of T and the SRO σ at that temperature are given by

$$\frac{k_B T_C^B}{J} = \frac{2}{\ln[z/(z-2)]}, \qquad \text{and} \qquad \sigma_C = \frac{1}{z-1}, \tag{20}$$

respectively, where B in T_C^B stands for bulk. This equation is the Bethe's expression for the critical temperature in infinite systems.[16]

II.1.2 Two Dimensional Triangular Lattice

In the case of the triangular lattice we consider two kinds of aggregates, triangular and hexagonal arrangements as shown in Figs. 1a and 1b, respectively. For the sake of simplicity we call them TRI and HEXA, respectively. The parameter used to describe the size of the system is the length of the side in units of lattice constants. In Figs. 1a and 1b we show a triangular aggregate with $n = 3$ and an hexagonal aggregate with $n = 2$. In the TRI case, the number of sites and pairs are given by $\alpha = N = \frac{1}{2}(n+1)(n+2)$ and $\gamma = \frac{3}{2}n(n+1)$, and for the HEXA case $\alpha = N = 3n^2 + 3n + 1$ and $\gamma = 3n(3n+1)$. T_C and σ at the critical temperature for the TRI clusters are given by

$$\frac{k_B T_C}{J} = \frac{2}{\ln\{3n/[2(n-1)]\}}, \qquad \text{and} \qquad \sigma_C = \frac{n+2}{5n-2}. \tag{21}$$

On the other hand for the HEXA clusters one obtains

$$\frac{k_B T_C}{J} = \frac{2}{\ln\{[3n(3n+1)]/[6n^2-1]\}}, \qquad \text{and} \qquad \sigma_C = \frac{3n^2+3n+1}{15n^2+3n-1}. \tag{22}$$

In Fig. 2a we show the results for $T_C(N)$ as a function of the total number of atoms in the aggregate. One observes that HEXA aggregates disorder at higher T_C than the TRI clusters. Similar results have been observed in the case of three dimensional

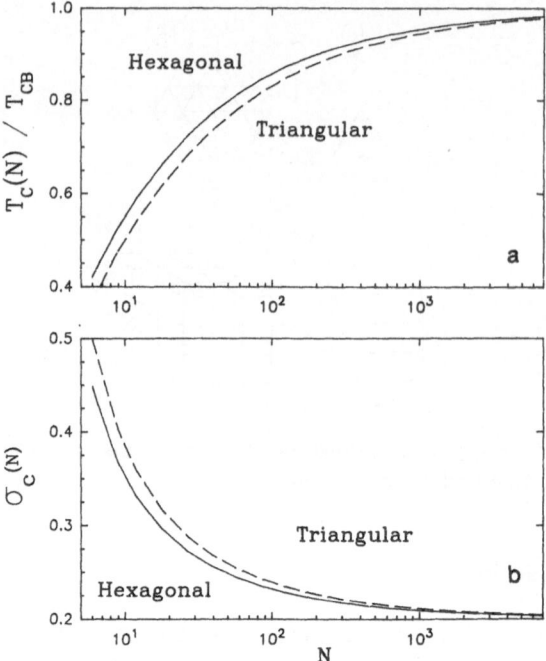

Figure 2. (a) Normalized critical temperature for triangular and hexagonal aggregates as a function of the number of spins. (b) Short-range order at the critical temperature for the same cases as (a).

systems like cubo-octahedral and icosahedral aggregates;[8] *rounded* shape aggregates disorder at higher T_C than *sharp* shape aggregates. In the case of the short-range order parameter σ_C, the *rounded* clusters present lower values than the *sharp* ones, This is illustrated in Fig. 2b. Notice that the difference in the temperature and in the short-range order parameter can persist up to more than 2000 spins.

The asymptotic behavior for the SRO as a function of the number of atoms at T_C is $\sigma_C(N) \approx \sigma_C(\infty) + C_1/\sqrt{N}$, with $C_1 = 6\sqrt{2}/25$ and $4\sqrt{3}/25$, for the TRI and HEXA clusters, respectively. The shift in the temperature is $1 - T_C(N)/T_C^B \approx C_2/\sqrt{N}$, with $C_2 = 1/[\sqrt{2}\ln(3/2)]$ and $1/[\sqrt{3}\ln(3/2)]$ for TRI and HEXA systems, respectively.

II.1.3 Symmetric Finite Cubic System

The symmetric finite size cluster with a cubic structure (square, simple cubic, ..., etc.) are characterized by edges with n cells. Figures 1c and 1d illustrate aggregates in two and three dimensional systems, respectively. The number of spins in a cubic aggregate of d-dimensions is $\alpha = N = (n+1)^d$ and the number of pairs is $\gamma = dn(n+1)^{(d-1)}$. The transition temperature T_C and the short-range order σ_C are given by

$$\frac{k_B T_C}{J} = \frac{2}{\ln\{dn/[(d-1)n-1]\}}, \quad \text{and} \quad \sigma_C = \frac{n+1}{(2d-1)n-1}. \tag{23}$$

Equation (23) holds for any dimensionality and for any size of the atomic aggregates with cubic structure. This equation reduces to the bulk expression [Eq. (20)] as n goes to infinity, since $d = z/2$ in the cubic crystals. The behavior of T_C as a function of

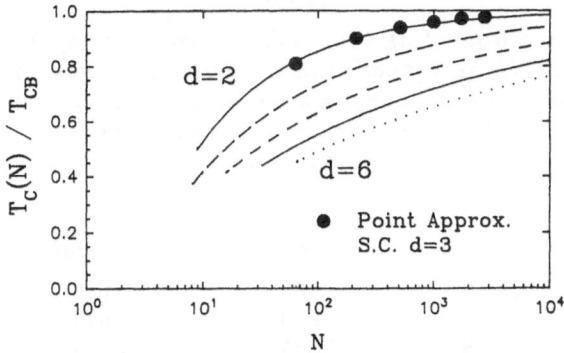

Figure 3. Normalized critical temperature for aggregates with square symmetry as a function of the number of spins for dimensions 2 to 6 (square lattice, simple cubic lattice, ...). The circles correspond to the Bragg-Williams approximation for a $3-d$ simple cubic system with position dependence of the spins.[9]

the number of atoms in the system is presented in Fig. 3 for $d = 2, 3, \ldots, 6$. The first feature to notice is that as the dimensionality of the system increases the normalized critical temperature $(T_C(N)/T_C^B)$ decreases. This behavior is understandable due to the fact that the higher the dimension the larger the number of atoms necessary to reach the *bulk* limit. In the $2d$ case the smallest aggregate that present spontaneous order is the one with $n = 2$ (3×3 spins). There is no order for $n = 1$ since in this approximation the aggregate is equivalent to one dimensional system (infinite one dimensional chain with $z = 2$). In the case of d larger than two, there is spontaneous order for $n = 1$. In this case the number of atoms of the smallest aggregate with spontaneous order is 2^d.

In order to have an idea about the level of approximation of the calculations presented here, we include in Fig. 3 the results obtained in the point approximation (Bragg-Williams) for aggregates with simple cubic structure (three dimensions). In these results the position dependence in the probabilites were taken into account. The comparison indicates a qualitative agreement between the location dependent calculation and the approximation presented here. Based on the knowledge of the different approximations as applied to bulk systems,[17] one can expect that a calculation within the pair approximation in which the position dependence of the spins is considered, will differ only few per cent.

A comparison of our results with the Monte Carlo Simulations (MCS)[15] in the two dimensional case for the normalized $T_C(n)/T_C^B$ as a function of the number of cells in an edge is shown in Table I. We can see that our calculations yield always slightly smaller values than the MCS results. The T_C in the MCS results correspond to the maximum value of the specific heat. In the case of higher dimension the agreement is better due to the fact that fluctuations neglected in our approximation becomes less important.[18] The normalized critical temperature levels off faster in the case of low dimensional systems than in the case of the high dimensional ones. The asymptotic limit for T_C is given by

$$1 - \frac{T_C(N)}{T_C^B} \approx \left\{ \frac{1}{(d-1)\ln[d/(d-1)]} \right\} \frac{1}{\sqrt[d]{N}}, \tag{24}$$

Table I

Comparison of the normalized critical temperature $T_C(n)/T_C^B$ as a function of the linear dimension of the system for a two dimensional square lattice with results obtained by Monte Carlo simulations.[15]

n	$T_C(n)/T_C^B$	
	MCS	This Work
4	0.7384	0.7066
6	0.8103	0.7917
8	0.8538	0.8384
10	0.8794	0.8680
14	0.9110	0.9034
20	0.9435	0.9311
30	0.9594	0.9533
40	0.9705	0.9647
60	0.9820	0.9763
100	0.9884	0.9857

where $T_C^B = T_C(\infty)$ is the Curie temperature of the bulk or infinite system. The asymptotic behavior of the SRO at the critical temperature is $\sigma_c(N)/\sigma_c(\infty) - 1 \approx 2d/[(2d-1)\sqrt[d]{N}]$.

II.2 Triangular Approximation

The single site and the pair probabilities are denoted as before by x_i and y_{ij}, respectively. The probability to find a triplet formed by the spins i–j–k is denoted by t_{ijk}. In terms of the pair probabilities, the single site probabilites are given by $x_i = \sum_j y_{ij}$, and the pair probabilities are given in terms of the triplets by $y_{ij} = \sum_k t_{ijk}$. As usual, the triplet probabilities satisfy the normalization constraint $\sum_{ijk} t_{ijk} = 1$. Furthermore, the triplet probabilites satisfy also $t_{ijk} = t_{pq\ell}$, where $\{p, q, \ell\}$ is a permutation of $\{i, j, k\}$. A similar relationship is satisfied by the pair probabilities.

In this case we describe the magnetic behavior of the system in terms of two long-range order parameters and one short-range order parameter. These three order parameters are the only independent variables and are defined as follows

$$\xi = x_1 - x_2, \tag{25a}$$

$$\eta = t_{112} - t_{122}, \tag{25b}$$

$$\sigma = y_{11} + y_{22} - 2y_{12}. \tag{25c}$$

The first two parameters are LRO and vanish at T_C. The last one is the SRO and remains finite at T_C. Notice that all the order parametrs are defined between zero and one. By using Eqs. (25) one can write all the probabilities in terms of the order parameters. Those expressions are given in Table II. From that Table one can read $t_{111} = 1/8 + \xi/2 + 3\sigma/8 - \eta/2$.

The internal energy is the same as in Eq. (2). To calculate the entropy we use the concept of the Correlation Correction Factor.[19] In triangular aggregates, the number

Table II

The single, pair, and triplet probabilities in terms of
the long- and short-range order parameters.

	1	ξ	σ	η
x_1	1/2	1/2	–	–
x_2	1/2	–1/2	–	–
y_{11}	1/4	1/2	1/4	–
y_{22}	1/4	–1/2	1/4	–
y_{12}	1/4	–	–1/4	–
t_{111}	1/8	1/2	3/8	–1/2
t_{222}	1/8	–1/2	3/8	1/2
t_{112}	1/8	–	–1/8	1/2
t_{122}	1/8	–	–1/8	–1/2

of sites, pairs and triangles in terms of the length size (n) are: $\alpha = N = \frac{1}{2}(n+1)(n+2)$, $\gamma = \frac{3}{2}n(n+1)$ and $\delta = n^2$, respectively. Notice that $(n+1)$ is the number of atoms on the edges. Taking into acoount these geometrical factors one can write for the entropy the expression[19]

$$\frac{S}{k_B} = (2\gamma - \alpha - 3\delta)\sum_i \mathcal{L}(x_i) - (\gamma - 3\delta)\sum_{i,j}\mathcal{L}(y_{ij}) - \delta\sum_{i,j,k}\mathcal{L}(t_{ijk}) + (\gamma - \alpha - \delta). \quad (26)$$

The minimization of $\Phi = \beta F$ with respect to the independent variables leads us to the following set of equations

$$\frac{\partial \Phi}{\partial \xi} = \Phi_\xi = \frac{1}{2}(n-2)(n-1)\left[\frac{1}{2}\ln\left(\frac{1+\xi}{1-\xi}\right)\right] - \frac{3}{2}n(n-1)\left[\frac{1}{2}\ln\left(\frac{1+\sigma+2\xi}{1+\sigma-2\xi}\right)\right]$$
$$+ n^2\left[\frac{1}{2}\ln\left(\frac{1+3\sigma+4(\xi-\eta)}{1+3\sigma-4(\xi-\eta)}\right)\right] = 0, \quad (27a)$$

$$\frac{\partial \Phi}{\partial \sigma} = \Phi_\sigma = -\frac{3}{2}n(n+1)\beta J - \frac{3}{2}n(n-1)\left\{\frac{1}{4}\ln\left[\frac{(1+\sigma+2\xi)(1+\sigma-2\xi)}{(1-\sigma)^2}\right]\right\}$$
$$+ n^2\left\{\frac{3}{8}\ln\left(\frac{[1+3\sigma+4(\xi-\eta)][1+3\sigma-4(\xi-\eta)]}{[(1-\sigma)+4\eta][(1-\sigma)-4\eta]}\right)\right\} = 0, \quad (27b)$$

$$\frac{\partial \Phi}{\partial \eta} = \Phi_\eta = n^2\left\{\frac{1}{2}\ln\left[\frac{1+3\sigma-4(\xi-\eta)}{1+3\sigma+4(\xi-\eta)}\right] + \frac{3}{2}\ln\left[\frac{(1-\sigma)+4\eta}{(1-\sigma)-4\eta}\right]\right\} = 0. \quad (27c)$$

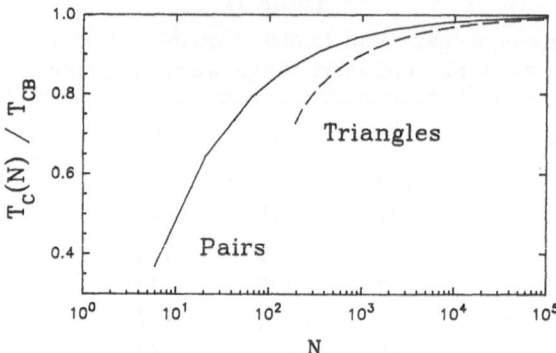

Figure 4. Normalized critical temperature for triangular lattice as a function of the number of spins for the pair and triangle approximation without considering the position dependence of the spins.

In order to find the order parameters as a function of the temperature and size aggregate we solve the set of Eqs. (27) numerically. However, the critical temperature of the system (T_C) is obtained by evaluating the Hessian determinat, which is composed by the second order derivaties of Φ with respect to the LRO parameters in the paramagnetic state. In this case the Hessian is given by

$$\left(\Phi_{\xi\xi}\,\Phi_{\eta\eta} - \Phi_{\eta\xi}\,\Phi_{\xi\eta}\right)\big|_{T=T_C} = 0\,, \tag{28}$$

where Φ_{vu} means the second order derivaties of Φ with respect to v and u, respectively. From the evaluation of the Hessian we get the following polynomial equation

$$(4 - 6n + 2n^2)\sigma_c^2 + (6 + 3n - 3n^2)\sigma_c + (2 + 3n + n^2) = 0\,, \tag{29}$$

where σ_c correspond to the value of σ at T_C. The solution of this equation is given by

$$\sigma_c = \frac{1}{4}\left(3 - \frac{\sqrt{(n^2 - 17n - 2)(n+1)}}{(n-1)\sqrt{n-2}} + \frac{6n-12}{n^2 - 3n + 2}\right)\,, \tag{30}$$

The transition temperature T_C can be calculated analyticaly by a simple substitution of σ_c in Eq. (27b) in the paramagnetic state

$$\frac{2J}{k_B T_C} = \frac{n}{n+1}\ln\left(\frac{1+3\sigma_c}{1-\sigma_c}\right) - \frac{n-1}{n+1}\ln\left(\frac{1+\sigma_c}{1-\sigma_c}\right)\,. \tag{31}$$

The results for T_C as a function of the total of spins are shown in Fig. 4, where we include the results of the pair approximation for the same system [Eq. (21)]. By looking to the behavior of σ_c, one can notice that there is not real solution for $n \leq 17$, therefore there is not phase transition for aggregates smaller than that size. We interpret this as an improvement of our calculation with respect of the pair approximation that predicts transitions in aggregates of any size. From Fig. 4, one can observe that the normalized T_C has smaller values in comparison with the pair approximation.

In this case, the asymptotic behavior of the Curie temperature is given by

$$1 - \frac{T_C(N)}{T_c^B} \approx \left\{\begin{array}{c} \frac{8}{5} + \ln\frac{9}{5} \\ \sqrt{2}\ln\frac{5}{3} \end{array}\right\} \frac{1}{\sqrt{N}}\,, \tag{32}$$

Table III

Redefinition of the probabilities used in the
square approximation.

Old Variable	New Variable	Weight $w(v_i)$
x_1	x_1	1
x_2	x_2	1
y_{11}	y_1	1
y_{12}	y_2	2
y_{22}	y_3	1
z_{1111}	z_1	1
z_{1112}	z_2	4
z_{1122}	z_3	4
z_{1212}	z_4	2
z_{1222}	z_5	4
z_{2222}	z_6	1

where T_C^B is the critical temperature for an infinite triangular lattice ($k_B T_C^B / J = 2/\ln(5/3)$ in this approximation).

II.3 Square Approximation for Systems with Cubic Structure

II.3.1 Simple cubic system

The single site probabilites and pair probabilities are denoted by x_i and y_{ij}. The square probabilities are denoted by z_{ijkl}, where the subindices take values 1 for spin up and 2 for spin down. The single site probabilites are given in terms of the pair probabilities by $x_i = \sum_j y_{ij}$, and the pair probabilities in terms of the square probabilities by $y_{ij} = \sum_{kl} z_{ijkl}$. As usual all the probabilities satisfy the normalization constraint. The square probabilites have the property that $z_{ijkl} = z_{pqrs}$, where $\{p, q, r, s\}$ is a cyclic permutaion of $\{i, j, k, \ell\}$. The previous relationship reduce the number of independent square probabilites to six. Taking this into account we redefine the variables as shown in Table III. This notation is the same as the one used in the original formulation of the CVM.[11] The statistical weight factors $w(v_i)$ defined in Table III, are the number of different configurations with the same probability. There are three different types of statistical weight factors, corresponding to x, y and z and denoted by $w(x_i)$, $w(y_i)$ and $w(z_i)$, respectively.

In this case the magnetic behavior of the system is described in terms of two long-range order (LRO) parameters and three SRO parameters. The two LRO parameters are defined as follows

$$\xi_1 = x_1 - x_2, \tag{33a}$$
$$\xi_2 = z_2 - z_5. \tag{33b}$$

As SRO parameters we use y_2, z_3 and z_4. In Table IV, we give the dependent probabilities in terms of the order parameters.

Table IV

The single, pair, and square probabilities in terms of
the long- and short-range order parameters.

	1	ξ_1	ξ_2	y_2	z_3	z_4
x_1	1/2	1/2	—	—	—	—
x_2	1/2	−1/2	—	—	—	—
y_1	1/2	1/2	—	−1	—	—
y_3	1/2	−1/2	—	−1	—	—
z_1	1/2	1/2	−1	−2	—	1
z_2	—	—	1/2	1/2	−1/2	−1/2
z_5	—	—	−1/2	1/2	−1/2	−1/2
z_6	1/2	−1/2	1	−2	—	1

The internal energy is given by Eq. (2). The entropy can be written in terms of the probabilities and the number of spins (α), pairs (γ) and squares (δ). In terms of the number of cells along the edges (n), $\alpha = N = (n+1)^3$, $\gamma = 3n(n+1)^2$ and $\delta = 3n^2(n+1)$. The expression for the entropy is

$$\frac{S}{k_B} = -s\sum_i w(x_i)\mathcal{L}(x_i) + p\sum_i w(y_i)\mathcal{L}(y_i) - \delta\sum_i w(z_i)\mathcal{L}(z_i) - \frac{S_0}{k_B}. \tag{34}$$

where $S_0/k_B = (\alpha - \gamma + \delta)$, $p = 4\delta - \gamma$ and $s = \alpha - 2\gamma + 4\delta$. The minimization of the free energy with respect to the independent variables leads to the following set of equations

$$\frac{\partial F}{\partial \xi_1} = F_{\xi_1} = 0 \quad : \quad (z_1/z_6)^\delta (y_3/y_1)^p (x_1/x_2)^s = 1, \tag{35a}$$

$$\frac{\partial F}{\partial \xi_2} = F_{\xi_2} = 0 \quad : \quad z_1^2 z_6 = z_1 z_5^2, \tag{35b}$$

$$\frac{\partial F}{\partial y_2} = F_{y_2} = 0 \quad : \quad H^{4\gamma} = \left(\frac{z_1 z_6}{z_2 z_5}\right)^{2\delta} \left(\frac{y_2^2}{y_1 y_3}\right)^p, \tag{35c}$$

$$\frac{\partial F}{\partial z_3} = F_{z_3} = 0 \quad : \quad z_3^2 = z_2 z_5, \tag{35d}$$

$$\frac{\partial F}{\partial z_4} = F_{z_4} = 0 \quad : \quad z_1 z_4^2 z_6 = (z_2 z_5)^2, \tag{35e}$$

where $H = \exp(J/k_B T)$. In the paramagnetic state ($\xi_1 = \xi_2 = 0$) the set of Eqs. (35) reduces to

$$H^{2\gamma} = \left(\frac{z_1}{z_2}\right)^{2\delta} \left(\frac{y_2}{y_1}\right)^p, \quad z_2 = z_3 \quad \text{and} \quad z_2^2 = z_1 z_4. \tag{36}$$

In this phase the variables satisfy the relation[11]

$$y_2 = 3z_3 + z_4. \tag{37}$$

By using Table III, Eq. (37), and a transfer function for the paramagnetic state defined as $\phi \equiv z_3/z_4$, one can rewrite the variables and H in terms of the transfer function and z_4 as follows

$$y_1 = y_3 = (\phi^2 + 3\phi)z_4, \quad \text{and} \quad y_2 = (3\phi + 1)z_4, \tag{38a}$$

$$z_1 = z_6 = \phi^2 z_4, \quad z_2 = z_3 = z_5 = \phi z_4, \quad \text{and} \quad z_4 = \frac{1}{2(\phi^2 + 6\phi + 1)}, \tag{38b}$$

$$H^2 = \left(\frac{3\phi + 1}{\phi^2 + 3\phi}\right)^{\frac{3-m}{1+m}} \phi^{\frac{2}{1+m}}, \tag{38c}$$

where $m = 1/n$. To evaluate T_C we calculate the Hessian

$$(\mathcal{F}_{\xi_1 \xi_1} \mathcal{F}_{\xi_2 \xi_2} - \mathcal{F}_{\xi_1 \xi_2} \mathcal{F}_{\xi_2 \xi_1})\big|_{T=T_C} = \left(\frac{\delta}{z_1} - \frac{p}{y_1} + 2s\right)\left(\frac{1}{z_1} + \frac{1}{z_2}\right) - \frac{\delta}{z_1^2} = 0, \tag{39}$$

where $\mathcal{F}_{vu}$ denotes the second order derivaties of the free energy with respect to v and u, respectively. In terms of the transfer function the Hessian can be rewritten as

$$(\phi + 1)\left(\frac{1}{\phi^2} - \frac{(p/\delta)}{\phi^2 + 3\phi} + \frac{(s/\delta)}{\phi^2 + 6\phi + 1}\right) - \frac{1}{\phi^2} = 0. \tag{40}$$

This equation leads us to a third order polynomial in the transfer function

$$(1 - a + b)\phi^3 + (9 - 7a + 4b)\phi^2 + (19 - 7a + 3b)\phi + (3 - a) = 0, \tag{41}$$

where $a = p/\delta$ and $b = s/\delta$. For an infinite system $a = 3$ and $b = 7/3$. In this case the polymomial reduces to $\phi(\phi - 3)(\phi - 5) = 0$; the root with physical meaning is $\phi = 3$. The H^2 takes the value $(125/81)$, and $(k_B T_C^B/J) = 2/\ln(125/81)$.

For finite systems, the polymomial can be written in terms of the size aggregate as

$$(1 - m + m^2)\phi^3 - (8 - 5m - 4m^2)\phi^2 + 3(5 + 3m + m^2)\phi + 3m = 0, \tag{42}$$

where $m = 1/n$. We discuss now the asymptotic behavior of the critical temperature when n goes to infinity. The polynomial equation in this case can be approximated by a quadratic equation: $(1 - m)\phi^2 + (5m - 8)\phi + 3(5 + 3m) = 0$. An approximate solution for the transfer function is given by

$$\phi \approx 3 + \tfrac{15}{2}m + \tfrac{195}{8}m^2 + \tfrac{285}{2}m^3 + \tfrac{142305}{128}m^4 + \cdots. \tag{43}$$

Now using this expression for ϕ in the normalized $(H/H(\infty))$ and performing a series expansion for small values of m one obtains

$$\left(\frac{H(m)}{H(\infty)}\right)^2 \approx 1 + \left(\tfrac{1}{2} - 4\ln\tfrac{5}{9} - 2\ln 3\right)m$$

$$+ \left(\tfrac{5}{8} + 2\ln\tfrac{5}{9} + 8\left(\ln\tfrac{5}{9}\right)^2 + \ln 3 + 8\ln\tfrac{5}{9}\ln 3 + 2(\ln 3)^2\right)m^2$$

$$+ \mathcal{O}(m^3) = 1 + a_1 m + a_2 m^2 + \mathcal{O}(m^3). \tag{44}$$

In the limit when m goes to zero

$$\left(\frac{H(m)}{H(\infty)}\right)^2 = \exp\left[2\left(\frac{1}{\tau} - \frac{1}{\tau_\infty}\right)\right] \approx 1 + 2\left(\frac{1}{\tau} - \frac{1}{\tau_\infty}\right), \tag{45}$$

where $\tau = k_B T_C/J$ and $\tau_\infty = k_B T_C^B/J$, respectively. Combining Eqs. (44) and (45) we obtain

$$\frac{\tau}{\tau_\infty} = \left[1 + \tfrac{1}{2}\tau_\infty(a_1 m + a_2 m^2)\right]^{-1} \approx 1 - \tfrac{1}{2}a_1 \tau_\infty m + \left(\tfrac{1}{4}a_1^2 \tau_\infty^2 - \tfrac{1}{2}a_2 \tau_\infty\right)m^2. \tag{46}$$

To the lowest order Eq. (46) leads to $1 - \tau/\tau_\infty = a_1\tau_\infty m/2$, which can be rewritten in terms of the total number of spins in the aggregate as follows

$$1 - \frac{T_C(N)}{T_C^B} \approx \frac{1}{2}a_1\left(\frac{k_B T_C^B}{J}\right)\frac{1}{\sqrt[3]{N}}.\tag{47}$$

Here, $a_1 = \frac{1}{2} - 4\ln\frac{5}{9} - 2\ln 3$ and $k_B T_C^B/J = 2/\ln\frac{125}{81}$.

II.3.2 Square Lattice System

The formulation for the square lattice within the square approximation is a straight forward exercise considering the previous formulation for the simple cubic structure. In this case $\alpha = (n+1)^2$, $\gamma = 2n(n+1)$, and $\delta = n^2$, for the number of spins, pairs and squares, respectively. All the equations from (33) to (41) hold, however the values of the constants are different. In this case the polynomial is given by

$$m^2\phi^3 + \left(-1 + 6m + 4m^2\right)\phi^2 + \left(8 + 8m + 3m^2\right)\phi + \left(1 + 2m\right) = 0,\tag{48}$$

where $m = 1/n$. In the asymptotic case all the terms of order m^2 can be neglected. The approximated solution of Eq. (48) is given by

$$\phi \approx 4 + \sqrt{17} + \left(28 + \frac{116}{\sqrt{17}}\right)m + \left(168 + \frac{11768}{\sqrt{4913}}\right)m^2 + \cdots.\tag{49}$$

The H^2 function in this case is given by

$$H^2 = \left(\frac{3\phi + 1}{\phi^2 + 3\phi}\right)^{\frac{1-m}{1+m}}\phi^{\frac{1}{m+1}},\tag{50}$$

which in the asymptoptic limit can be approximated by

$$\left(\frac{H(m)}{H(\infty)}\right)^2 \approx 1 + \left[\frac{4(85 + 21\sqrt{17})}{221 + 3\sqrt{17^3}}\right]m + \cdots = 1 + a_1 m + \cdots.\tag{51}$$

The lowest order of Eq. (51) leads to

$$1 - \frac{T_C(N)}{T_C^B} \approx \frac{1}{2}a_1\left(\frac{k_B T_C^B}{J}\right)\frac{1}{\sqrt{N}},\tag{52}$$

where $a_1 = 4(85 + 21\sqrt{17})/(221 + 3\sqrt{17^3})$ and $k_B T_C^B/J = 2/\ln\left[(5 + \sqrt{17})/4\right]$ (consistent with the results of Ref. 11).

The critical value of the SRO σ_C depends in general of the transfer function and in both cases (square and cubic lattice) is given by

$$\sigma_C = \frac{\phi^2 - 1}{\phi^2 + 6\phi + 1}.\tag{53}$$

In the asymptotic case, the transfer function can be approximated by $\phi \approx \phi_0 + b_1 m + b_2 m^2 + \cdots$, where ϕ_0 corresponds to the bulk value (m tends to zero). Using this expression in Eq. (53) one obtains that the SRO in the case of large n is given by

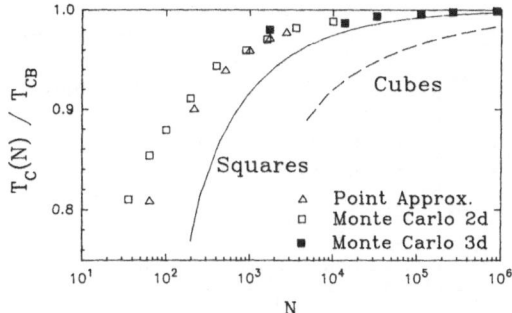

Figure 5. Normalized critical temperature for the square and simple cubic lattice as a function of the number of spins for the square approximation without considering the position dependence of the spins. The square marks correspond to the Monte Carlo Simulation results for square and simple cublic lattice and the triangular marks correspond to the Bragg-Williams approximation; both the MCS and the Bragg-Williams include the position dependence of the spins.

$$\sigma_c = \frac{\phi_0^2 - 1}{\phi_0^2 + 6\phi_0 + 1} + \frac{2b_1\left(3 + 2\phi_0 + 3\phi_0^2\right)}{\left(\phi_0^2 + 6\phi_0 + 1\right)^2} m + \cdots . \tag{54}$$

The first term corresponds to the bulk value $\sigma_c(\infty)$ and the second one is the first correction due to the finite size of the aggregate. In general one can write

$$\sigma_c(N) \approx \sigma_c(\infty) + \frac{A}{\sqrt[d]{N}}, \tag{55}$$

where A is a constant and d is the dimensionality of the system.

The results for the normalized temperatures for the square and simple cubic lattices as a function of the total number of spins are shown in Fig. 5. In the same figure we include the results of Monte Carlo Simulations (MCS) for the square and simple cubic lattice.[13-15] The results of the point (Bragg-Williams) approximation for the simple cubic lattice when the position dependence is considered[9] are also plotted. In the case of MCS the points correspond to the maximum value in the specific heat due to the absence of singularities in the thermodynamic quantities. The first thing to be noticed is that there is a qualitative agreement between the MCS, the point approximation with position dependence and our results, although our approximation predict the lowest temperatures. We also would like to notice that our calculations predict that there is no phase transitions for very small aggregates. In simple cubic aggregates the phase transition appears only at $N = 4913$ spins ($n^* = 16$), and in the case of the square lattice for $N = 196$ spins ($n^* = 13$). The reason is that in both cases there is not a positive real solution for the transfer function for smaller values of n. Although we can proceed as in the case of the MCS and take the maximum in the specific heat we prefer to show only those points for which the transfer function ϕ has a positive real solution.

In the Fig. 6 we present a direct comparison of the results for T_C for the square and cube aggregates with the MCS results for the same type of systems. In the case of the square lattice for small aggregates the MCS results are higher than our calculation and for aggregates larger than 600 atoms they fall below. For infinite systems our

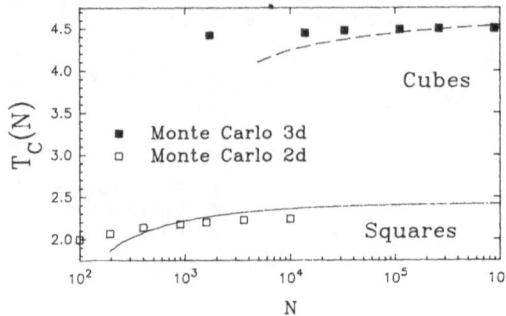

Figure 6. Direct comparison of $T_C(N)$ with the Monte Carlo simulation results. In the MCS results the position dependence of the spins is considered while in our calculation it has been neglected.

calculation gives a value 6.7% larger than the exact value. In the case of the cubic aggregate our calculation for T_C is smaller than the MCS value for aggregates smaller than approximately 300,000 atoms and above the MCS results for larger aggregates. In the limit of infinite systems our calculation for T_C is 2.1% larger than the best value known for the critical temperature.

III. Summary and Conclusions

We applied a simplified version of the CVM approximations to the study of the magnetic behavior of large Ising aggregates. In this model we considered pair, three- and four-body spin correlations. This simplified version assumes that all the probabilities are independent of the position of the spins. The advantage of this approximation is that one can work out analytical expressions as a function of the cluster size and investigate the asymptotic behavior. The comparison of our results with those in which the location dependence of the spins are taken into account shows that our approximation underestimate the critical temperature only a few per cent. We found that the shift of the temperature $(1 - T_C(N)/T_C^B)$ vanishes as $1/\sqrt[4]{N}$. Our calculation suggest that the geometrical characteristics (number of sites, number of pairs, etc.) determine the critical temperature and that location dependence effects play a less important role in large aggregates.

We showed that for a fixed number of atoms the *rounded* aggregates posses higher critical temperatures than the *sharp* ones, and that in the magnetic order shows the opposite behavior.

Finally, aware of the limitations of our model, we consider our results as a first step in a series of approximations in which one can improve the solution of the Hamiltonian by using larger clusters, and by considering the position dependence of the spins.

Acknowledgements

We acknowledge the Consejo Nacional de Ciencia y Tecnología (México) Grants Nos. 1774–E9210 and 485100–5–3883E.

References

1. D. M. Cox, D. J. Trevor, R. L. Whetten, E. A. Rohlfing, and A. Kaldor, *Phys. Rev.* B **32**, 7290 (1985).
2. W. A. de Herr, P. Milani, and A. Chatelain, *Phys. Rev. Lett.* **65**, 488 (1990).
3. J. P. Bucher, D. C. Douglas, P. Xia, B. Haynes, and L. A. Bloomfield, *Phys. Rev. Lett.* **66**, 3052 (1991).
4. J. P. Bucher, in *Physics and Chemistry of Finite Systems: From Clusters to Crystals*, edited by P. Jena, S. N. Khanna, and B. K. Rao (Klewer Academic Publishers, Dodrecht, 1992), p. 799.
5. G. M. Pastor, J. Dorantes-Dávila, and K. H. Bennemann, *Phys. Rev.* B **40**, 7642 (1989).
6. P. J. Jensen and K. H. Bennemann, *Z. Phys.* D **21**, 349 (1991).
7. J. Merikowski, J. Timonen, M. Manninen, and P. Jena, *Phys. Rev. Lett.* **66**, 938 (1991).
8. J. L. Morán-López, K. H. Bennemann, J. M. Montejano-Carrizales and F. Aguilera-Granja, *Solid State Commun.* **88**, 101 (1993); *New Trends in Magnetism, Magnetic Materials, and Theirs Applications*, edited by J. L. Morán-López and J. M. Sanchez, (Plenum, New York, 1994), p. 47.
9. F. Aguilera-Granja, J. L. Morán-López and J. M. Montejano-Carrizales, *Surf. Sci.* **326**, 150 (1995).
10. Per-Anker Lindgård and P. V. Hendriksen, *New Trends in Magnetism, Magnetic Materials, and Theirs Applications*, edited by J. L. Morán-López and J. M. Sanchez, (Plenum, New York, 1994), p. 37.
11. R. Kikuchi, *Phys. Rev.* **81**, 988 (1951).
12. W. Hellenthal, *Z. Phys.* **170**, 303 (1962).
13. K. Binder, H. Rauch, and V. Wildpaner, *J. Phys. Chem. Solids* **31**, 391 (1970).
14. A. M. Ferrenberg and D. P. Landau, *J. Appl. Phys.* **70**, 6215 (1991).
15. D. P. Landau, *Phys. Rev.* B **14**, 255 (1976); *Finite Size Scaling and Numerical Simulation of Statistical Systems*, edited by V. Privman, (World Scientific, Singapore, 1994), p. 225.
16. H. A. Bethe, *Proc. R. Soc. London Ser.* A **150**, 552 (1935).
17. D. de Fontaine, *Solid State Physics*, edited by H. Ehrenreich, F. Seitz, and D. Turnbull, (Academic Press, 1979), Vol. 34, p. 74.
18. K. Huang, *Statistical Mechanics*, (John Wiley and Sons, 1987).
19. R. Kikuchi, *Prog. Theor. Phys.* Suppl. **115**, 1 (1994).

Thermodynamic Properties of Coherent Interphase Boundaries in *fcc* Substitutional Alloys

Mark Asta

Computational Materials Science Department
Sandia National Laboratories
MS 9161, P. O. Box 969, Livermore, CA, 94551
U.S.A.

Abstract

The structural and thermodynamic properties of coherent interphase boundaries (IPB's) have been studied using the Cluster Variation Method. Calculations have been performed for IPB's in model ordering and phase-separating substitutional alloy systems with *fcc*-based crystal structures. Special attention has been devoted in this study to an analysis of the effects which temperature and crystallographic orientation have upon the finite-temperature properties of IPB's. Values of the interfacial excess free energies, *i.e.* the interphase energies, have been calculated as a function of temperature for IPB's with {100}, {110} and {111} crystallographic orientations. Additionally, the dependences of the composition, long-range and short-range order parameters on distance within compositionally diffuse interfacial regions have been computed. It will be demonstrated that both the thermodynamic and structural properties of IPB's can be strongly dependent upon the temperature and the nature of the energetic parameters in the alloy system.

I. Introduction

In multiphase materials, the interfacial layer between two coexisting phases is commonly referred to as an interphase boundary (IPB). In this paper we will be interested in the equilibrium structural and thermodynamic properties of coherent IPB's between two different phases with crystal structures based upon the same parent lattice in a substitutional alloy system. Our interest in the properties of IPB's is motivated by the important role which they play in determining both the shape and the kinetics of formation of precipitates in multiphase alloy microstructures.

Theory and Applications of the Cluster Variation and Path Probability Methods
Edited by J.L. Morán-López and J.M. Sanchez, Plenum Press, New York, 1996

The equilibrium shape of alloy precipitates are determined by two factors: the elastically "soft" directions in the material which can accommodate coherency strain with the lowest cost in energy, and the dependence upon crystallographic orientation of the *interphase energy* which is defined as the excess free energy per unit area associated with an IPB. For very small precipitates, with large surface to volume ratios, the shape is generally that which minimizes the total interfacial free energy.

The interphase energy plays a very important role in traditional kinetic theories of precipitation including both classical nucleation theory[1] and the Lifshitz and Slyozov[2] and Wagner[3] theory of coarsening. As a specific example, in classical nucleation theory the steady-state nucleation rate is proportional to $\exp(-\Delta G^*/k_B T)$, where ΔG^*, the free energy of formation of a critical nucleus, is proportional to the *cube* of the interphase energy.

A structural feature of IPB's which has important consequences for the validity of classical kinetic theories is the "width" of the interface. The IPB width is defined as the distance over which the value of a relevant order parameter and/or concentration variable varies in going from one phase to the other. As was pointed out by Cahn and Hilliard,[4] if the width of the IPB is on the order of the size of the critical nucleus, the thermodynamic formalism used in classical nucleation theory is of questionable validity. The dependences of concentration, long-range and short-range-order parameters upon distance within the IPB layer are also important structural features which should generally be considered in the interpretation of data obtained from transmission-electron-microscopy and small-angle X-ray and neutron scattering experiments performed on multiphase materials.

In this paper results of Cluster Variation Method (CVM)[5] calculations of the structural and thermodynamic properties of coherent IPB's will be presented for both model ordering and phase-separating alloy systems with *fcc*-based crystal structures. We will focus on results relating to the important features of IPB's discussed above: namely the details of the structural variations across IPB's, the values and temperature dependences of interphase energies, and the effects which crystallographic orientation have upon these properties. The next section will be devoted to a brief review of previous theoretical work relating to the properties of IPB's. The computational approach used in the present study will then be outlined in section III, and results for model alloy systems will be presented in section IV and summarized in section V.

I. Previous Studies

Pioneering work on the theory of IPB's was performed by van der Waals over a hundred years ago.[6] The theory of van der Waals was later extended by Cahn and Hilliard.[7] The formalism of the van der Waals/Cahn-Hilliard (vdWCH) theory is based upon an integral equation for the free energy of an inhomogeneous system. As a specific example, consider the case of a flat, coherent IPB between two disordered phases possessing the same crystal structures but different average concentrations. For this one-dimensionally inhomogeneous system, the free energy in the vdWCH theory is written as follows

$$F = A \int_{-\infty}^{\infty} \left[f_0(c) + \kappa (dc/dz)^2 \right] dz, \tag{1}$$

where A is the cross-sectional area, c is the composition (which depends upon the distance z), $f_0(c)$ is the bulk free energy density, and κ is defined in terms of second-order derivatives of f_0 with respect to distance.[7] The equilibrium width and the excess free energy (*i.e.*, the interphase energy) of a flat IPB can be derived once the functional form of the composition profile $c(z)$ which minimizes Eq. (1) is determined. In their 1958 paper, Cahn and Hilliard[7] pointed out that for cubic systems the tensor κ is isotropic and therefore the interphase energy and composition profile must also be when Eq. (1) is valid.

Within the vdWCH theory, it is possible to show that the interphase energy goes to zero and the IPB width diverges with power-law dependences on temperature as a critical point is approached.[7] The values of the critical exponents associated with these power-law temperature dependences have been derived within the vdWCH theory by Widom and Fisk.[8,9]

In addition to the work based upon the vdWCH theory, numerous studies of the properties of coherent IPB's have been performed using Ising (or, equivalently, lattice-gas) models in two and three dimensions. In these studies a wide variety of statistical-mechanical techniques have been utilized including Onsager's exact solution of the two-dimensional Ising model,[10] regular-solution,[11-14] and CVM[15-21] mean-field calculations, and Monte Carlo simulations.[21-26] Following Cenedese,[20] much of the work performed within the context of Ising models can be divided into two classes: those based upon the so-called *sum* method and those which make use of an *inner product* formalism.

The inner product method, originally due to Woodbury and Clayton,[27-29] is a powerful technique which allows one to determine a value of the interphase energy from a knowledge of information relating to bulk phases only. The inner product method was elegantly reformulated in a 1972 paper by Kikuchi[16] and it is reviewed by Cenedese in the present proceedings.[20] With the exception of the original work by Woodbury and Clayton[27-29] and that of Kikuchi[17] and Cenedese,[20] there have been surprisingly few applications of the inner product formalism to the study of IPB properties.

In the sum method, IPB's are studied by considering the thermodynamic properties of large systems containing distinct spatial regions of different types of bulk phases with interfaces between them. The total free energy of such a system is given by the sum of the free energies of the bulk phases combined with the excess free energies associated with the interfaces. If the free energies can be computed for both the composite system (containing interfaces) and the relevant bulk phases (without interfaces), using for example Monte Carlo simulations or mean-field calculations, the excess free energy associated with the interfaces can be determined. From the results of such calculations structural information about the IPB's can also be determined, as shown below.

In many of the Ising-model studies of coherent IPB properties it has been found that non-negligible anisotropy exists at temperatures far from critical points for the values of the interphase energy (see for example the results presented by Lee and Aaronson[14]). Anisotropy is seen even for cubic systems where Eq. (1) predicts complete isotropy. The origin of this discrepancy between the Ising model calculations and the predictions of the vdWCH theory can be attributed to the neglect in Eq. (1) of spatial derivatives of $f_0(c)$ which are higher than second order and which become increasingly important as the width of the IPB decreases.

For the remainder of this paper we will focus on the properties of coherent IPB's in substitutional alloy systems with *fcc*-based crystal structures. Previous studies of coherent IPB's in *fcc phase-separating* alloys have been performed using regular solution models and the sum method approach;[13,14] to the best of the author's knowledge, there have been no studies which use higher order mean-field (CVM) approximations or Monte Carlo simulations for such systems. The first study of both the finite-temperature thermodynamic and structural properties for coherent IPB's in *ordering* fcc alloy systems was performed by Kikuchi and Cahn.[19] In the work of Kikuchi and Cahn, IPB's (as well as antiphase boundaries) in Cu-Au alloys were studied using the sum-method approach combined with the tetrahedron approximation of the CVM.[5] In the CVM calculations the energy was parametrized by effective interaction parameters for the subclusters of the nearest-neighbor tetrahedron; the values of these parameters were obtained from a fit to the experimentally determined Cu-Au phase diagram.[30] Since interaction parameters which span only the range of the nearest-neighbor pair on the *fcc* lattice were used in these calculations, values of the interphase energy for IPB's between disordered and $L1_2$ phases were predicted to vanish at zero tempera-ture. As was noted by Kikuchi and Cahn, interactions with ranges longer than the nearest-neighbor are responsible for nonvanishing values of the interphase energy for IPB's between fcc-disordered and $L1_2$ alloy phases at zero temperature.

In a study of antiphase boundaries in $L1_2$ alloys, Finel[21] calculated the equilibrium width of an IPB between *fcc*-disordered and $L1_2$ phases at the coherent order-disorder temperature using both Monte-Carlo simulations and CVM calculations. It was found that CVM calculations which made use of the tetrahedron approximation gave rise to a calculated IPB width which was significantly smaller than that found by Monte Carlo. By contrast, Finel showed that more accurate CVM calculations based upon the tetrahedron-octahedron (TO)[31,32] approximation were able to reproduce very well the Monte Carlo results for the IPB width. Therefore, TO appears to be the lowest level of approximation within the CVM which can be used to perform quantitatively accurate studies of IPB structural properties in *fcc* alloys. An additional advantage of the TO relative to the tetrahedron approximation of the CVM is that it allows one to include the effect of interactions within the range of the second neighbor pair on the *fcc* lattice. Therefore, the TO approximation can be used to perform more realistic studies of IPB's in alloys where the interphase energy is not required to vanish at zero temperature.

In this paper we present the results of CVM calculations for the structural and thermodynamic properties of coherent IPB's in model *fcc* phase-separating and or-dering alloy systems. In these calculations use was made of the TO approximation of the CVM for the reasons discussed in the previous paragraphs. In particular, the motivation for our study of ordering alloy systems is to explore the properties of IPB's (between disordered *fcc* and ordered $L1_2$) using model Hamiltonians which do not enforce vanishing interphase energies at zero temperature. Additionally, we have reinvestigated the case of IPB's in phase separating *fcc* alloys in part to compare the TO-CVM results with those obtained from regular solution models.[14]

II. Method

Before describing the details of the CVM calculations performed in the present study, it is necessary to give a short review of the thermodynamic formalism relevant for the properties of IPB's. A more complete review of the thermodynamic issues discussed in the proceeding paragraphs is given by Cahn.[33]

A useful thermodynamic potential for the study of alloys is the so-called *grand potential* (Ω) defined as follows for a binary (A-B) alloy

$$\Omega(\Delta\mu, T, N) = G - \Delta\mu(N_A - N_B) = \tfrac{1}{2}(\mu_A + \mu_B)N + \sum_{(hkl)} \gamma_{hkl} A_{hkl}, \qquad (2)$$

where G is the Gibbs free energy $(G = N_A\mu_A + N_B\mu_B + \sum_{(hkl)} \gamma_{hkl} A_{hkl})$, N_A and N_B are the number of atoms of type A and B, respectively, N is the total number of atoms $(N = N_A + N_B)$, and $\Delta\mu$ is the *chemical field* which is defined as half the difference between the chemical potentials for the two atomic species: $\Delta\mu = 1/2(\mu_A - \mu_B)$. In Eq. (2) γ_{hkl} and A_{hkl} are the interphase energy and total cross-sectional area, respectively, for IPB's with crystallographic orientations given by Miller indices (hkl). The last term on the furthest-right-hand side of Eq. (2) represents the total work required to (reversibly) create IPB's.

The condition for equilibrium between two bulk phases α and β in an alloy is the equality of chemical potentials for each atomic species i: $\mu_i^\alpha = \mu_i^\beta \; \forall \, i$. For a binary alloy this condition can be formulated in terms of the values of the grand potentials and chemical fields as follows

$$\Omega^\alpha = \Omega^\beta \equiv \Omega_0, \qquad (3a)$$
$$\Delta\mu^\alpha = \Delta\mu^\beta \equiv \Delta\mu_0, \qquad (3b)$$

where the subscript "0" indicates a quantity corresponding to bulk thermodynamic equilibrium.

In the present work we have made use of the sum-method approach described briefly in the previous section. In this approach IPB properties are studied by considering the thermodynamics of composite systems containing different types of phases and interfaces between them. Specifically, let $\Omega_{hkl}(\Delta\mu_0, T, N)$ denote the value of the grand potential for such a composite system containing regions of bulk α and β phases with IPB's oriented along (hkl). If one can calculate $\Omega_{hkl}(\Delta\mu_0, T, N)$, a value of the interphase energy γ_{hkl} can be derived from the following equation

$$\Omega_{hkl}(\Delta\mu_0, T, N) - \Omega_0(\Delta\mu_0, T, N) = \gamma_{hkl} A_{hkl}. \qquad (4)$$

Equation (4) shows that, in order to calculate γ_{hkl} within the sum-method, a knowledge of the equilibrium values of the grand potential (Ω_0) and chemical field $(\Delta\mu_0)$ is required.

For the purpose of calculating the values of the various thermodynamic potentials entering Eqs. (2)–(4), the CVM has been used in the present study. Specifically, the TO approximation of the CVM for the *fcc* lattice has been employed. In the CVM, a value of the grand potential is determined by minimizing a free energy functional $\widetilde{\Omega}$ which can be written using the formalism of Sanchez and de Fontaine[31,34,35] as follows

$$\widetilde{\Omega}(\Delta\mu, T, N, \{\xi_{n,s}(p)\}) = \sum_p \sum_{n,s} E_{n,s} \xi_{n,s}(p) - \Delta\mu \sum_p \xi_1(p)$$
$$+ k_B T \sum_p \sum_{n,s} \phi_{n,s} \sum_J x_{n,s}(J,p) \ln x_{n,s}(J,p). \qquad (5)$$

In equation (5) the first and third terms on the right-hand side give the CVM expressions for the enthalpy and the negative of the temperature multiplied by the configurational entropy, respectively. The second term on the right-hand side corresponds to the product $\Delta\mu(N_A - N_B)$ in Eq. (2). The symbols p in Eq. (5) denote lattice points, of which there are a total of N. The subscripts n and s define a cluster of atoms; n labels the number of atoms and s specifies the spatial range of the cluster and its geometrical arrangement. In the *fcc* TO approximation of the CVM, the sums in n and s run over each tetrahedron and octahedron associated with a given lattice site, as well as all of the subclusters of these "maximal" clusters.

The enthalpy in Eq. (5) is written as a sum over products of effective cluster interactions (ECI's), $E_{n,s}$, and cluster *correlation functions*, $\xi_{n,s}(p)$. The ECI's, whose formal definitions are discussed in detail by Sanchez *et al.*,[36,37] parametrize the energetics associated with atomic rearrangements. A correlation function for a given cluster (n, s) is defined as the ensemble average of a product of occupation variables (spin variables in the Ising model) associated with each of the n atomic sites.[35] In the entropy expression in Eq. (5) the variables $\phi_{n,s}$ are the Kikuchi-Barker coefficients;[5,38] values of these variables for the clusters in the *fcc* TO approximation are given, for example, by Sanchez and de Fontaine.[35] The third sum in the entropy expression is over all possible arrangements, J, of A and B atoms on the sites of a given cluster; the variables $x_{n,s}(J,p)$ denote the ensemble *cluster probabilities* for the occurrence of an atomic configuration J on cluster (n, s) at site p. As discussed in detail by Sanchez and de Fontaine[31,35] $x_{n,s}(J,p)$ can be written as a linear combination of the correlation functions corresponding to each cluster which is a subcluster of (n, s) at site p. The coefficients in this linear expansion define the so-called[39] *configuration matrix*. The elements of the configuration matrix and the Kikuchi-Barker coefficients depend only upon the geometries of the clusters and the lattice and they can all be calculated using group-theoretical methods.[40,41]

The correlation functions form a linearly independent set of variables with respect to which $\tilde{\Omega}$ must be minimized in order to calculate the grand potential by the CVM. In the present study the minimizations were performed numerically using the Newton-Raphson technique, as discussed in detail by Sluiter.[42] Once a sequence of such minimizations have been performed for two phases α and β as a function of chemical field at a given temperature, the values $\Delta\mu_0$ and Ω_0 corresponding to bulk thermodynamic equilibrium between α and β can be determined according to Eq. (3).

In order to compute a value of the grand potential $\Omega_{hkl}(\Delta\mu_0, T, N)$ for the type of composite system described in the discussion of the sum method above, CVM free energy calculations have been performed in the present study for large supercells. The geometry of a typical supercell which has been used to study IPB's with (100) orientations in an *fcc* lattice is shown schematically in Fig. 1a. The long direction (z-direction) is perpendicular to the IPB planes for the orientation of interest, and periodic boundary conditions are imposed on all of the boundaries of the supercell.

A value of $\Omega_{hkl}(\Delta\mu_0, T, N)$ is calculated by the CVM using the supercell geometry as follows: first, the values of the point correlation variables from the bulk thermodynamic calculations for phases α and β are used to formulate a starting point for the Newton-Raphson minimization assuming an idealized composition (order parameter) profile indicated schematically by the dashed lines in Fig. 1b. This starting point is appropriate for a supercell containing spatial regions of each phase with abrupt interfaces between them. The CVM grand potential functional (Eq. (5)) is next minimized with respect to the independent correlation functions in the supercell.[43] Provided the supercell has been chosen large enough to accomodate the equilibrium width of the

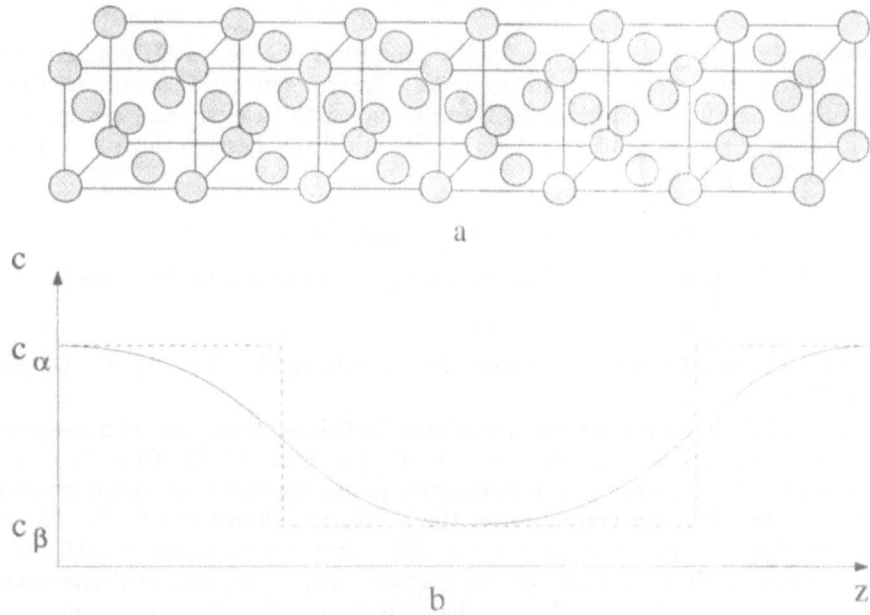

Figure 1. Composition-temperature phase diagrams for *fcc*-based alloy systems calculated with the tetrahedron-octahedron approximation of the CVM. The phase diagrams in (a) and (b) correspond to ordering alloy systems with energetics described by Ising-model Hamiltonians with $\epsilon = -0.1$ and $\epsilon = -1.0$, respectively. The phase diagram in (c) corresponds to a phase-separating alloy system with $\epsilon = 0$.

IPB's, the Newton-Raphson minimization converges to a local minimum characterized by a set of correlation functions corresponding to an inhomogeneous state in the supercell consisting of spatial regions of bulk phases α and β with compositionally diffuse IPB's between them. A typical composition profile obtained after such a CVM minimization is indicated schematically by the solid line in Fig. 1b. It should be emphasized that the Newton-Raphson minimization will only converge to the local minimum for the inhomogeneous state if the length of the supercell perpendicular to the IPB is chosen large enough to accomodate the equilibrium widths of the IPB's. The widths of the IPB's for the types of alloy systems and temperatures considered in the present study required that supercells containing as many as 100 planes (for {110} orientations) be used.

The procedure discussed in the previous two paragraphs yields the value of the grand potential Ω_{hkl} needed to calculate the interphase energy γ_{hkl} according to Eq. (4). In addition, an analysis of the values of the correlation functions which correspond to the inhomogeneous state in the supercell (corresponding to a supercell containing regions of bulk phases and IPB's) provides interesting structural information about the IPB's. For example, the point correlation functions contain information about the concentration and order parameter profiles across the IPB's. Additionally, from values of the pair and point correlation functions the distance dependence of the Warren-Cowley[44] short-range order (SRO) parameters across an IPB can be determined.

IV. Results for Model *fcc* Alloy Systems

In this section results are presented for the properties of coherent IPB's in model ordering and phase-separating *fcc* alloy systems. The energetics of these model alloy systems are described by an Ising model Hamiltonian with nearest- and second-nearest-neighbor effective pair interactions (EPI's). Specifically, we assign the following values to the ECI's in Eq. (5)

$$E_{n,s} = \begin{cases} V_1, & \text{if } (n,s) \text{ is a nearest-neighbor pair} \\ V_2 \equiv \epsilon V_1, & \text{if } (n,s) \text{ is a second-nearest-neighbor pair} \\ 0, & \text{otherwise,} \end{cases} \tag{6}$$

where $\epsilon \equiv V_2/V_1$ is defined as the ratio of the second-neighbor EPI, V_2, to the nearest-neighbor EPI, V_1.

Three model alloy systems are considered in this section: one phase-separating system ($V_1 < 0$) with $\epsilon = 0$, and two ordering systems ($V_1 > 0$) with $\epsilon = -0.1$ and $\epsilon = -1.0$. The composition-temperature phase diagrams for these systems, as calculated by the TO approximation of the CVM, are shown in Fig. 2. The phase diagrams in Figs. 2a and 2b correspond to the ordering alloy systems with $\epsilon = -0.1$ and $\epsilon = -1.0$, respectively. These phase diagrams display regions of thermodynamic stability for a disordered solid-solution phase (*fcc*), as well as for ordered phases with $L1_0$ and $L1_2$ crystal structures. All transitions in Fig. 2a are of first order and three distinct congruent order-disorder points are present at $c = 0.5$ and near compositions $c = 0.25$ and $c = 0.75$. By contrast, in the ordering phase diagram for $\epsilon = -1.0$ (Fig. 2b) the tops of each of the $L1_2$ and $L1_0$ phase fields coincide at a critical point located at $c = 0.5$. The phase diagram in Fig. 2c corresponds to the model phase-separating alloy system and it displays a symmetric miscibility gap with a critical point located at $c = 0.5$.

IV.1 Interphase Energies

In figure 3 results are plotted showing the temperature dependence of calculated interphase energies (γ) for {100}, {110} and {111} IPB orientations. The results in Figs. 3a and 3b correspond to IPB's between disordered *fcc* and ordered $L1_2$ phases in the model ordering alloy systems with $\epsilon = -0.1$ and $\epsilon = -1.0$, respectively. In Fig. 3c calculated values of γ are plotted for IPB's between the two disordered alloy phases located on either side of the miscibility gap in Fig. 2c. Finite-temperature results plotted in Fig. 3 were obtained from CVM-TO calculations using the method described in the previous section. Zero-temperature results were computed analytically from an analysis of the distributions of different bond types in both bulk alloy compounds and across abrupt IPB's. At zero temperature the values of γ for the ordering systems considered here are as follows: $\gamma = -2V_2/a^2$, $-2\sqrt{2}V_2/a^2$, and $-2\sqrt{3}V_2/a^2$ for {100}, {110} and {111} orientations, respectively, where a denotes the *fcc* lattice parameter. The corresponding values of γ for the phase-separating alloy system are as follows: $\gamma = (-16V_1 - 8V_2)/a^2$ for {100}, $\gamma = (-12\sqrt{2}V_1 - 8\sqrt{2}V_2)/a^2$ for {110}, and $\gamma = (-8\sqrt{3}V_1 - 8\sqrt{3}V_2)/a^2$ for {111}.

The temperature scales in Fig. 3 are plotted in units of T/T_0, where T_0 represents the temperature at the top of the $L1_2$ phase field and the miscibility gap for ordering and phase-separating cases, respectively. For the alloy system with the phase diagram shown in Fig. 2a two-phase fields between ordered and disordered alloy phases are

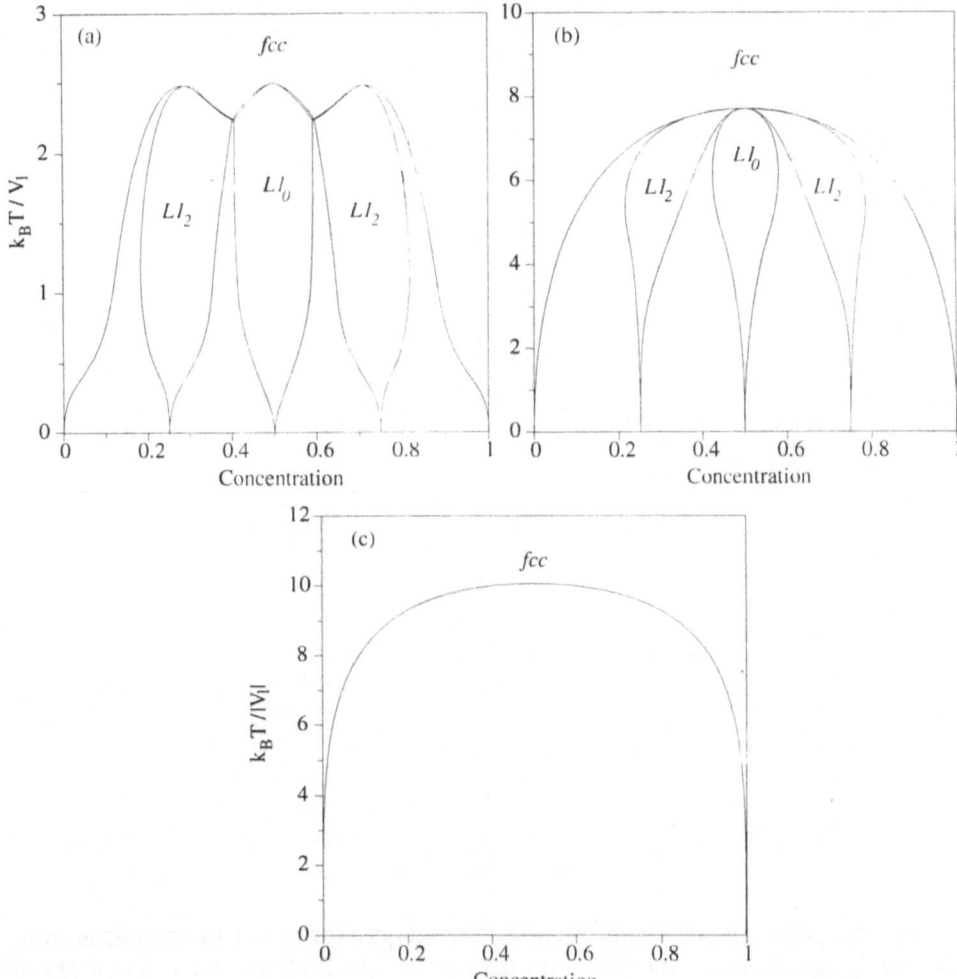

Figure 2. Composition-temperature phase diagrams for *fcc*-based alloy systems calculated with the tetrahedron-octahedron approximation of the CVM. The phase diagrams in (a) and (b) correspond to ordering alloy systems with energetics described by Ising-model Hamiltonians with $\epsilon = -0.1$ and $\epsilon = -1.0$, respectively. The phase diagram in (c) corresponds to a phase-separating alloy system with $\epsilon = 0$.

present on either side of each of the congruent order-disorder transition points; the results plotted for this alloy system in Fig. 3a correspond to IPB's between $L1_2$ and disordered (*fcc*) phases with compositions located to the left (right) of the congruent order-disorder transition point for $L1_2$ alloys with stoichiometry $c = 0.25$ ($c = 0.75$).

For each of the three model alloy systems considered in this study the temperature dependence of the interphase energy is significant. In particular, in going from zero temperature to $T/T_0 = 0.5$, the values of γ plotted for the lowest-energy orientations in Figs. 3a, 3b and 3c decrease by approximately 70%, 45% and 20%, respectively. The strongest and weakest temperature dependences of γ for low values of T/T_0 are found for the ordering case with $\epsilon = -0.1$ and for the phase-separating system,

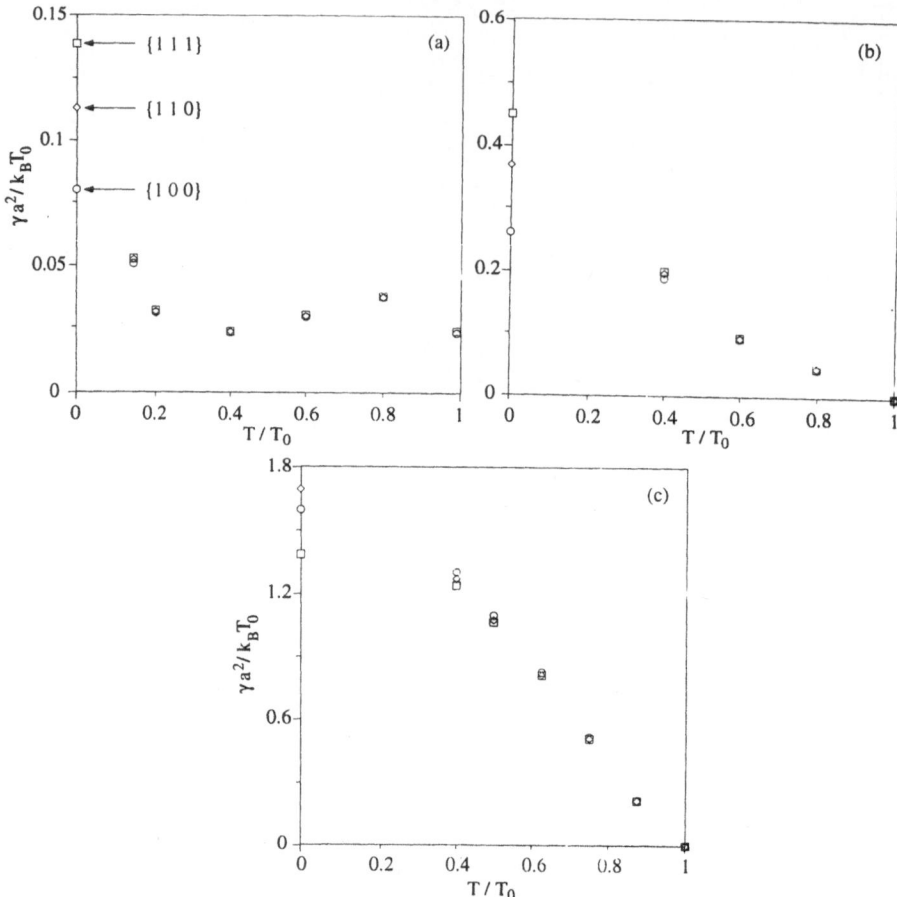

Figure 3. Temperature dependence of interphase energies calculated by the tetrahedron-octahedron approximation of the CVM. On the vertical axis is plotted the ratio $\gamma a^2/k_B T_0$, where γ, a, k_B and T_0 denote the interphase energy, lattice parameter, Boltzmann's constant, and phase-transition temperature, respectively. Results are plotted for model ordering alloy systems with $\epsilon = -0.1$ and $\epsilon = -1.0$ in (a) and (b), respectively. Interphase energies for the phase-separating system are shown in (c). Values of $\gamma a^2/k_B T_0$ are plotted in (a), (b) and (c) for IPB's with {100} (open circles), {110} (open diamonds) and {111} (open squares) crystallographic orientations.

respectively. From a comparative analysis of the results plotted in Fig. 3 and the phase diagrams shown in Fig. 2, it can be seen that generally γ is strongly T-dependent over temperature ranges where large changes occur in the width of the equilibrium two-phase fields. For the ordering alloy system with $\epsilon = -1.0$ and for the phase-separating system, the interphase energies vanish as the critical temperature $(T/T_0 = 1)$ is approached. The interphase energy for the ordering system with $\epsilon = -0.1$ is finite at $T/T_0 = 1$ since the order-disorder transition in this case is of first order. The results plotted in Fig. 3a show an oscillation in the values of γ for each crystallographic orientation as a function of temperature; this non-monotonic temperature-dependence of γ, which was also found by Kikuchi and Cahn[19] in their study of IPB's in Cu-Au, correlates with a qualitatively similar dependence on temperature displayed by the value of the order parameter for the L1$_2$ phase in equilibrium with the disordered (*fcc*) solid solution.

The smallest values of the interphase energies (except very near T_0) are found to be those corresponding to the model ordering alloy system with $\epsilon = -0.1$. By increasing the magnitude of V_2 larger values of γ result for the ordering case (compare Figs. 3a and 3b). The largest values of the interphase energy are obtained for the model phase-separating alloy system in this study. An important difference between the ordering and phase-separating alloy systems is the lowest-energy orientation for IPB's: we find (in agreement with the results of previous calculations[14,19]) that out of the three orientations considered in this study the lowest energy IPB's at all temperatures are parallel to {100} and {111} planes for ordering and phase-separating alloy systems, respectively. In the case of the phase-separating system, the results shown in Fig. 3c are consistent with experimental observations that Guinier-Preston-zone precipitates in the Ag-Al system facet primarily along {111} for temperatures significantly below the top of the metastable miscibility gap for *fcc*-based alloys.[45]

A striking feature of the results plotted in Fig. 3 is the strong effect which temperature has on the degree of crystallographic anisotropy displayed by the values of γ at low T/T_0. While very strong anisotropy is present for all three alloy systems at zero temperature, the difference between the values of γ for each orientation decreases sharply and becomes negligible at temperatures above $T/T_0 \approx 0.2$ for the ordering system with $\epsilon = -0.1$, and above $T/T_0 \approx 0.6$ for both the phase-separating case and the ordering system with $\epsilon = -1.0$. In order to further explore the effect which temperature has on the anisotropy displayed by the values of γ, low-temperature expansions (LTE) (see, for example, Parisi[46]) have been used to calculate values of interphase energies for low values of T/T_0. The results of LTE calculations, carried to second-order in the spin-flip energies, are plotted with open symbols in Fig. 4 for the ordering system with $\epsilon = -0.1$. Also plotted in Fig. 4 with filled symbols are the lowest-temperature results which could be obtained with the CVM. The dotted, dashed and solid lines in Fig. 4, illustrate that the extrapolated LTE results agree very well with the CVM-calculated values of γ at $T/T_0 = 0.14$. The LTE results show that the interphase energy for the {111} orientation decreases initially most rapidly. For {110} orientations, the value of γ remains nearly constant up to $T/T_0 \approx 0.025$ at which point the magnitude of the slope of γ vs. temperature increases quickly. Already at $T/T_0 \approx 0.075$, γ has essentially the same value for {111} and {110} orientations. The weakest temperature dependence of γ at low values of T/T_0 is found for the {100} orientation; the value of the interphase energy for this orientation stays roughly constant until $T/T_0 \approx 0.075$.

For the phase-separating alloy system considered here, several studies[11−14] of the finite-temperature properties of coherent IPB's have been performed using the regular-solution model (RSM) which is equivalent to the point approximation of the CVM. Since it is known that the RSM gives rise to significant errors in its predictions of the thermodynamic properties of bulk alloys,[47] it is interesting to compare the results of previous studies with those presented in this paper for coherent IPB's. In Fig. 5 the values of the interphase energy for {100} and {111} orientations calculated by both the CVM-TO method and the RSM are compared. The RSM numbers have been taken from the paper by Lee and Aaronson.[14] In order to properly interpret the results shown in Fig. 5 it should be noted that on the vertical axis is plotted a *ratio* of the interphase energy to the critical temperature T_0. Therefore, even though the values of γ at zero-temperature do not depend upon the level of approximation

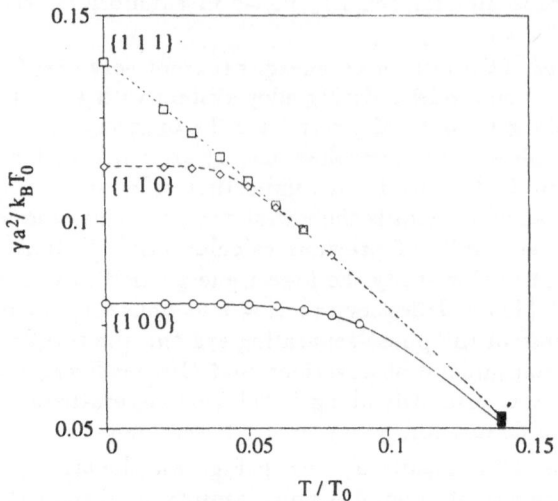

Figure 4. Low-temperature-expansion and CVM results for the interphase energy as a function of temperature for the model ordering alloy system with $\epsilon = -0.1$ are plotted with open and filled symbols, respectively. Results are shown for IPB's with {100} (open circles), {110} (open diamonds) and {111} (open squares) crystallographic orientations. The dashed and solid lines through the data in this figure are drawn as a guide to the eye and are meant to illustrate that the extrapolated low-temperature-expansion results agree well with the CVM-calculated interphase energies at $T/T_0 = 0.14$.

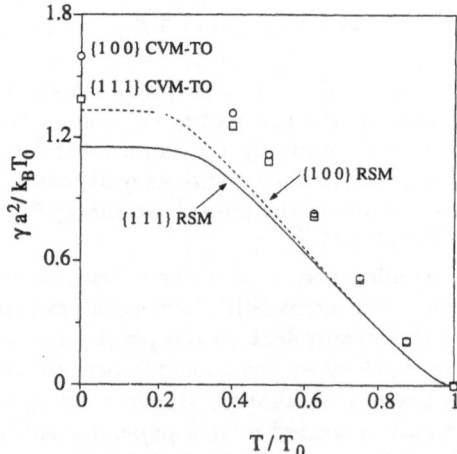

Figure 5. CVM-Tetrahedron-Octahedron (CVM-TO) and Regular-Solution-Model (RSM) calculated values of the interphase energy for the model phase separating alloy system with $\epsilon = 0$. RSM results are plotted for {100} and {111} orientations with dashed and solid lines, respectively. CVM-TO values are indicated by open squares and circles.

used in the finite-temperature calculations, the ratio $\gamma a^2/k_B T_0$ does through T_0. In Fig. 5 it is shown that, compared with the CVM-TO results, the RSM underestimates $\gamma a^2/k_B T_0$ by as much as 25%. This result has important consequences since it is common practice to fit a RSM to experimentally-measured values of T_0 and to then use the fitted parameters to calculate values of the interphase energy. Based on the comparison discussed in this paragraph it is expected that such a practice will lead to underestimated values of the interphase energy.

IV.2 Interphase Boundary Structure

Order-parameter and concentration profiles calculated by the CVM are plotted in Fig. 6 for {100}, {110} and {111} crystallographic orientations. The order-parameter profiles in Figs. 6a and 6b correspond to $L1_2$: *fcc* IPB's in the ordering alloy systems with $\epsilon = -0.1$ and $\epsilon = -1.0$, respectively, at a temperature of $T/T_0 = 0.6$. The concentration profiles in Fig. 6c were calculated for the phase-separating alloy system at $T/T_0 = 0.5$. The order parameter plotted in Figs. 6a and 6b is defined as the difference between the average concentrations for points on each of the two types of sublattices in the $L1_2$ phase; these concentrations are averaged over sublattice points on three, two and four consecutive planes parallel to IPB's with {111}, {100} and {110} orientations, respectively. The number of planes over which the point concentration variables are averaged for the purpose of computing the order parameter is dictated by the repeat distance of the $L1_2$ structure in the $\langle 111 \rangle$, $\langle 100 \rangle$ and $\langle 110 \rangle$ directions.

In figure 6c it can be seen that, to within the accuracy of the calculations, the computed concentration profiles lie along the same curve for each crystallographic orientation. In other words, the CVM results show isotropic concentration profiles for the phase-separating case, in agreement with the predictions of the vdWCH theory (see section II). It is also interesting to note that for the ordering alloy systems, the order-parameter profiles plotted in Figs. 6a and 6b are found to be isotropic as well at $T/T_0 = 0.6$. In the lowest-temperature CVM calculations performed for model ordering alloys ($T/T_0 = 0.14$ and 0.4 for $\epsilon = -0.1$ and -1.0, respectively) a small degree of anisotropy in the order-parameter profiles could be detected: it was found that the width of the {100} IPB order-parameter profile was slightly smaller than those for {110} and {111} orientations.

The results plotted in Fig. 6 show clearly that the equilibrium widths of coherent IPB's can be quite large. In Fig. 7 the IPB widths calculated by the CVM are plotted as a function of temperature for each of the model ordering and phase-separating alloy systems. The widths plotted in Fig. 7 are defined as the distance over which 99% of the change in the order parameter or concentration occurs in going from one phase to the other across a {100} IPB. The filled diamond symbols in Fig. 7 correspond to widths of IPB's in the phase-separating alloy system; results for the ordering alloy systems with $\epsilon = -0.1$ and -1.0 are plotted with open circles and squares, respectively. It is interesting to point out that for the phase-separating alloy system, the CVM-calculated widths plotted in Fig. 7 are found to be roughly 15% larger than those predicted by the RSM.[14]

For each alloy system the calculated widths plotted in Fig. 7 increase monotonically as a function of temperature. For both the phase-separating case and the ordering alloy system with $\epsilon = -1.0$ the width of the IPB diverges as the critical temperature is approached. For the other ordering alloy system ($\epsilon = -0.1$) the width is finite at T_0 since the order-disorder transition at this temperature is first order in this case. Away

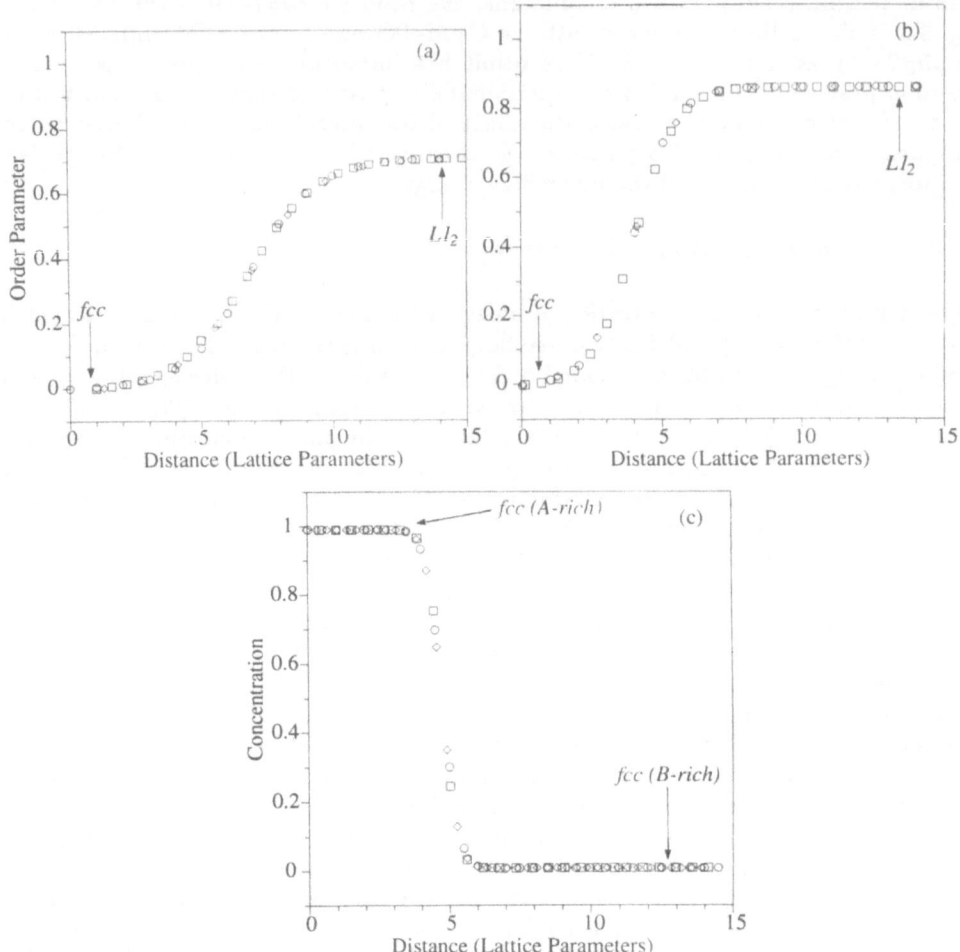

Figure 6. CVM-calculated composition and order-parameter profiles for IPB's in ordering and phase-separating alloy systems. In (a) and (b) the order parameter is plotted as a function of distance across *fcc*: L1$_2$ IPB's for the model ordering systems with $\epsilon = -0.1$ and $\epsilon = -1.0$, respectively. In (c) the distance-dependence of the composition across IPB's between disordered phases with *A*-rich and *B*-rich compositions in the model phase-separating alloy system is shown. In each figure results are plotted for IPB's with {100}, {110} and {111} crystallographic orientations using open circles, diamonds and squares, respectively.

from T_0 the largest widths are found for the ordering system with $\epsilon = -0.1$. For this alloy system the equilibrium widths are found to be larger than 10 lattice spacings over a wide temperature range. The smallest calculated IPB widths are found for the phase-separating system; in this case the widths range between two and four lattice spacings until very near the critical temperature.

In figure 8 CVM-calculated values of the Warren-Cowley SRO parameter (α_1) for nearest-neighbor pairs located *within* planes parallel to IPB's are plotted as a function of the distance across interfacial layers in the model *phase-separating* alloy system. Results in Figs. 8a and 8b correspond to temperatures of $T/T_0 = 0.5$ and

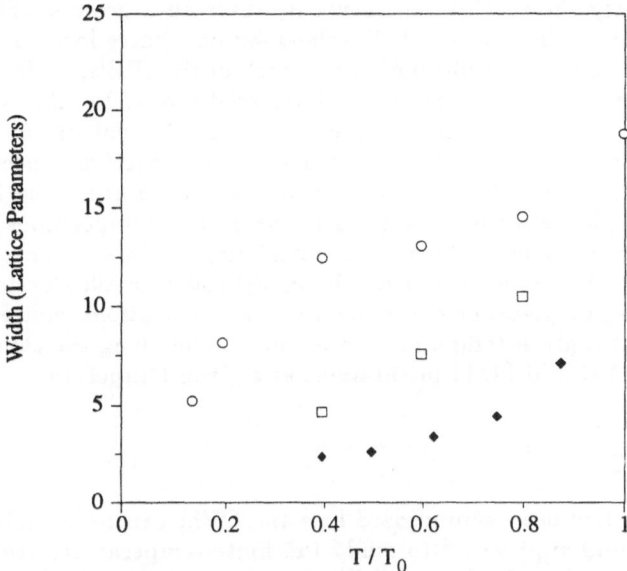

Figure 7. CVM-calculated widths of IPB's are plotted as a function of temperature for the model ordering alloy systems with $\epsilon = -0.1$ (open circles) and $\epsilon = -1.0$ (open squares) and for the phase-separating system with $\epsilon = 0$ (filled diamonds). The widths plotted in this figure are defined as the distance over which 99% of the change in order parameter or concentration occurs in going from one phase to the other across IPB's with {100} orientations.

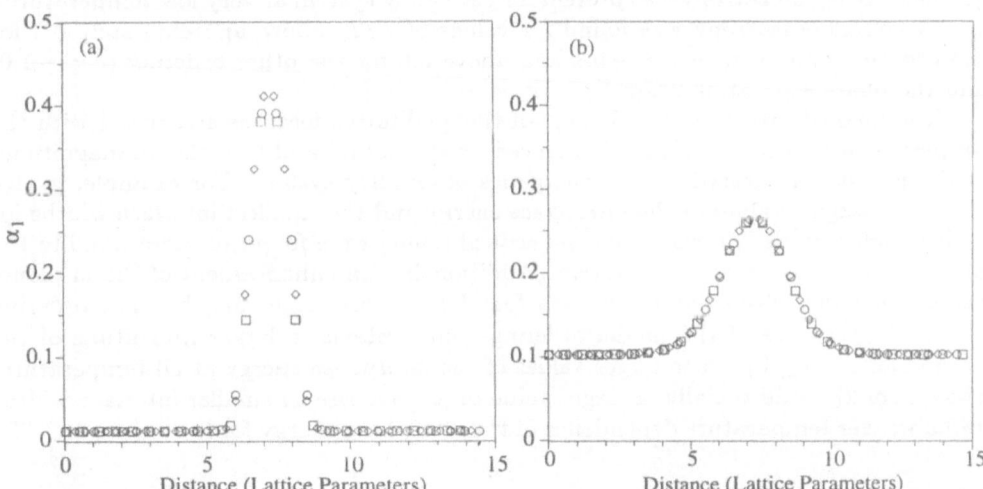

Figure 8. The SRO parameter α_1 corresponding to nearest-neighbor pairs located within planes parallel to IPB's in the model phase-separating alloy system are plotted as a function of distance within the interfacial layer at two temperatures: (a) $T/T_0 = 0.5$ and (b) $T/T_0 = 0.875$. Results are plotted for IPB's with {100}, {110} and {111} crystallographic orientations using open circles, diamonds and squares, respectively.

0.875, respectively. For each temperature α_1 takes on a positive value, indicative of a clustering type of SRO, in the bulk solid-solution phases located on either side of the IPB's. As a function of distance across each of the IPB's, α_1 increases and takes on a maximum value half way across the interfacial layers. In other words, it is found that within the IPB layer the degree of clustering SRO is enhanced. A comparison of the results in Figs. 8a and 8b shows that the degree of SRO enhancement is larger at lower temperatures. The fact that the clustering SRO is enhanced in an IPB is not surprising since planes within an interfacial layer have compositions which lie within the miscibility gap for bulk alloys. It is interesting to note, however, that a similar enhancement of SRO is not found for the model ordering alloy systems. As was the case for the long-range-order parameters and concentrations plotted in Fig. 6, the profiles for α_1 at a given temperature are found to lie along the same curve for each of the {100}, {110} and {111} orientations at a given temperature.

V. Summary

In this paper it has been summarized how the CVM can be combined with the so-called sum-method approach[20] to study the finite-temperature structural and thermodynamic properties of coherent IPB's in substitutional alloys. Results of such a study were presented in the previous section for model phase-separating and ordering alloy systems with *fcc*-based crystal structures.

It was found that calculated values of the interphase energy and interfacial width were strongly temperature dependent in ranges of T where large changes occurred in the values of the concentration (or order parameter) for one or both of the phases located on either side of an IPB. Additionally, the temperature was found to have a rather dramatic effect on the degree of anisotropy displayed by the calculated values of γ: while strong anisotropy was present for each alloy system at very low temperatures, nearly complete isotropy was found for values of T/T_0 above approximately 0.2 for the ordering system with $\epsilon = -0.1$ and above 0.6 for the other ordering ($\epsilon = -1.0$) and the phase-separating cases.

The quantitative values and some of the qualitative features associated with the properties of IPB's are strongly influenced by the nature and the relative magnitudes of the parameters describing the energetics of an alloy system. For example, in this study the largest values of the interphase energy and the smallest interface widths for a given value of T/T_0 (away from the critical point) were found to correspond to the model phase-separating alloy system. Additionally, an enhancement of the in-plane-nearest-neighbor SRO parameter was found for phase-separating but not ordering alloys. In the case of the model ordering alloy systems, a larger magnitude of the parameter $\epsilon = V_2/V_1$ led to larger values of the interphase energy at all temperatures away from T_0. Additionally, a larger value of $|\epsilon|$ gave rise to smaller interface widths and a weaker temperature dependence of the interphase energy for low values of T/T_0.

Acknowledgements

Helpful discussions with Drs. Duane Johnson and Stephen Foiles are gratefully acknowledged. The author also sincerely thanks Professor Gerbrand Ceder for providing his configuration-matrix computer code. This research was supported by the U.S. Department of Energy, Office of Basic Energy Sciences, Materials Science Division, under contract number DE-AC04-94AL85000.

References

1. K. C. Russell, in *Phase Transformations* (ASM International, Metals Park, Ohio, 1970).
2. I. M. Lifshitz and V. V. Slyozov, *J. Phys. Chem. Solids* **19**, 35 (1961).
3. C. Wagner, *Z. Elektrochem.*, Ber. Bunsenges. Phys. Chem. **65**, 581 (1961).
4. J. W. Cahn and J. E. Hilliard, *J. Chem. Phys.* **31**, 688 (1959).
5. R. Kikuchi, *Phys. Rev.* **81**, 998 (1951).
6. J. D. van der Waals, *Z. Physik. Chem.* **13**, 657 (1894).
7. J. W. Cahn and J. E. Hilliard, *J. Chem. Phys.* **28**, 258 (1958).
8. S. Fisk and B. Widom, *J. Chem. Phys.* **50**, 3219 (1969).
9. B. Widom, in *Phase Transitions and Critical Phenomena*, edited by C. Domb and M. S. Green (Academic, New York, 1972).
10. L. Onsager, *Phys. Rev.* **65**, 117 (1944).
11. S. Ono, *Mem. Fac. Engng.*, Kyushu Univ. **10**, 195 (1947).
12. M. Hillert, *Acta Metall.* **9**, 525 (1961).
13. J. L. Meijering, *Acta Metall.* **14**, 251 (1966).
14. Y. W. Lee and J. I. Aaronson, *Acta. Metall.* **28**, 539 (1980).
15. J. Y. Parlange, *J. Chem. Phys.* **48**, 169 (1968).
16. R. Kikuchi, *J. Chem. Phys.* **57**, 777 (1972).
17. R. Kikuchi, *J. Chem. Phys.* **57**, 783 (1972).
18. R. Kikuchi, *J. Chem. Phys.* **57**, 787 (1972).
19. R. Kikuchi and J. W. Cahn, *Acta. Metall.* **27**, 1337 (1979).
20. P. Cenedese, in *Theory and Application of the Cluster Variation and Path Probability Methods*, edited by J. L. Morán-López and J. M. Sanchez (Plenum, New York, 1996), p. 247.
21. A. Finel, in *Statics and Dynamics of Alloy Phase Transformations*, edited by P. E. A. Turchi and A. Gonis (Plenum, New York, 1994).
22. H. J. Leamy and K. A. Jackson, *J. Appl. Phys.* **42**, 2121 (1971).
23. H. J. Leamy, G. H. Gilmer, K. A. Jackson, and P. Bennema, *Phys. Rev. Lett.* **30**, 601 (1973).
24. *Monte Carlo Methods in Statistical Physics*, edited by K. Binder (Springer, New York, 1979), and references cited therein.
25. K. Binder, *Phys. Rev. A* **25**, 1699 (1982).
26. *Applications of the Monte Carlo Method in Statistical Physics*, edited by K. Binder (Springer, New York, 1984), and references cited therein.
27. G. W. Woodbury, Jr., *J. Chem. Phys.* **51**, 1231 (1969).
28. G. W. Woodbury, Jr., *J. Chem. Phys.* **53**, 3728 (1970).
29. D. B. Clayton and G. W. Woodbury, Jr., *J. Chem. Phys.* **55**, 3895 (1971).
30. D. de Fontaine and R. Kikuchi, in *Proceedings of NBS Workshop Applications of Phase Diagrams in Metallurgy and Ceramics*, edited by G. C. Carter (National Bureau Standards Special Publication No. 496, Vol. 2, U. S. Department of Commerce, 1978).
31. J. M. Sanchez and D. de Fontaine, *Phys. Rev. B* **17**, 2926 (1978).
32. S. K. Aggarwal and T. Tanaka, *Phys. Rev. B* **16**, 3936 (1977).
33. J. W. Cahn, in *Interfacial Segregation*, edited by W. C. Johnson and J. M. Blakely (ASM, Metals Park, Ohio, 1977).
34. J. M. Sanchez and D. de Fontaine, *Phys. Rev. B* **21**, 216 (1980).
35. J. M. Sanchez and D. de Fontaine, *Phys. Rev. B* **25**, 1759 (1982).
36. J. M. Sanchez, F. Ducastelle and D. Gratias, *Physica* **128A**, 334 (1984).

37. J. M. Sanchez, in *Theory and Application of the Cluster Variation and Path Probability Methods*, edited by J. L. Morán-López and J. M. Sanchez (Plenum, New York, 1996), p. 167.
38. J. A. Barker, *Proc. R. Soc.* **A216**, 45 (1953).
39. D. de Fontaine, in *Solid State Physics*, edited by H. Ehrenreich and D. Turnbull, (Academic Press, 1994), Vol. 47, p. 33.
40. D. Gratias, J. M. Sanchez and D. de Fontaine, *Physica* **113A**, 315 (1982).
41. G. Ceder, Ph. D. Thesis, University of California at Berkeley, 1991.
42. M. Sluiter, Ph. D. Thesis, University of California at Berkeley, 1988.
43. In order to help ensure that the Newton-Raphson converged to a local minimum corresponding to an inhomogeneous state, rather than one of the global minima for either of the two homogeneous states (corresponding to homogeneous α or β phase in the supercell), the point correlation variables for the first plane in the supercell were held fixed during the minimizations. In order to check that this constraint did not affect the final results for IPB's, additional Newton-Raphson minimizations were performed in several cases using as a starting point the set of correlation functions obtained from the constrained minimization. In every case it was found that the subsequent unconstrained minimization converged to the same local minimum as was obtained from the constrained minimization.
44. B. E. Warren, *X-Ray Diffraction* (Addison-Wesley, Reading, MA, 1969).
45. K. B. Alexander, F. K. Legoues, H. I. Aaronson and D. E. Laughlin, *Acta Metall.* **32**, 2241 (1984).
46. G. Parisi, *Statistical Field Theory* (Addison-Wesley, Menlo Park, CA, 1988).
47. D. de Fontaine, in *Solid State Physics*, edited by H. Ehrenreich, F. Seitz, and D. Turnbull, (Academic Press, 1979), Vol. 34, p. 74.

Efficient CVM Methods
to Study Interphase Boundary Properties

Pierre Cenedese[1] and Ryoichi Kikuchi[2]

[1] CECM/CNRS
  15 rue G. URBAIN
  Vitry-sur-Seine, F94407
  FRANCE

[2] Department of Materials Science and Engineering
  University of California, Los Angeles, Ca 90095-1595
  U.S.A

Abstract

The surface tension σ due to an interphase boundary (IPB) in the $2D$ square lattice and in the $3D$ cubic lattice is computed using the scalar product (SP) and the sum (S) CVM methods. It is shown that the resulting values of σ bracket the exact solution and extrapolation methods to the rigorous limit are outlined.

I. Introduction

Since its discovery by R. Kikuchi,[1] the Cluster Variation Method (CVM) approximation has been applied to various statistical mechanics problems. The CVM is now a useful technique for phase diagram computations and other bulk phase related problems.

The application of the CVM approximation to the thermodynamics of boundaries between phases was first considered by Kikuchi and Cahn.[2] This method, named the sum method (S), is now almost an standard. However, in the early 70's, Kikuchi[3−5] had proposed another CVM scheme to estimated the surface tension σ of extended defects, called the scalar product (SP) CVM approximation which has been dormant for many years.

In the S-CVM method, the CVM approximation is introduced at the earliest stage of formulation by considering a lattice with the translational symmetry broken. On the other hand, in the SP-CVM method, first a formal and rigorous transfer matrix method is used to lead to an exact expression for the surface tension that depends

Theory and Applications of the Cluster Variation and Path Probability Methods
Edited by J.L. Morán-López and J.M. Sanchez, Plenum Press, New York, 1996

explicitly upon two vectors to represent the bulk phases meeting at the boundary, and then the CVM approximation is introduced.

A substantial advantage of the SP-CVM scheme is that only a reduced set of variables is needed. This allows applications to multicomponent systems and leads to a series of hierarchical CVM approximations which converge towards exact solutions.

Another very important feature of the SP formalism is that close to the transition temperature T_c the surface tension tends to zero with a mean field approximation (MFA)* critical exponent $\mu = 1$, while in the S-CVM, which is of the gradient square approximation type, the surface tension behaves with a MFA critical exponent $\mu = 1.5$. This difference in the MFA critical behavior can be used, in conjunction with Suzuki's Coherent Anomaly Method,[6] to show that the S-CVM method surface tension σ^S and the SP-CVM surface tension σ^{SP} bracket the exact solution.

The paper is organized as follows; in sections II and III, we recall the basis of the S-CVM and SP-CVM. Section IV, which is devoted to applications, is divided into two parts. First we summarize some CAM results which are then used to extrapolate series of CVM approximations to the rigorous limit.

II. The Sum (S) Method

The basis of the sum method is to introduce the CVM approximation at the beginning of the formulation. In other words, the system density matrix is factorized in terms of local density matrices by taking into account the spatial inhomogeneity perpendicular to the planar defect.

Quite generally, let the subscripts I and II refer to the two bulk phases meeting at the boundary. For example, in the case of the Ising IPB, phases I and II have magnetization of opposite sign, while for an APB, phases I and II correspond to different variants of a given ordered phase. To discuss the CVM sum method, we now focus on the nearest-neighbor (nn) Ising IPB problem; other cases could be formulated along on similar line.

A lattice plane parallel to the boundary (a plane for short), is indexed by the integer index n, counted from the boundary. Two consecutive planes at positions n and $n + 1$ are indexed by the half integer parameter $\nu = n + \frac{1}{2}$.

The aim of the CVM sum method is to write the total free energy of the system as a sum of local free energies that only depend on a small subset of the variables. For example, in the CVM approximation based on maximal clusters that extend to two consecutive planes, the free energy can be written as

$$\beta \mathcal{F} = \lim_{N \to \infty} \sum_{\nu = -N+\frac{1}{2}}^{N-\frac{1}{2}} \beta f_\nu \,, \tag{1}$$

$$\beta f_\nu = \beta f_\nu \left(\{X_n\}, \{X_\nu\}, \{X_{n+1}\} \right) \,, \tag{2}$$

where $\{X_n\}$ represents the set of correlation functions needed to describe the planar clusters involved in the CVM entropy expression, while $\{X_\nu\}$ is the additional set of correlation functions involved by clusters encompassing two consecutive planes. The total set of correlation functions is simply the union of $\{X_n\}$ and $\{X_\nu\}$, that is

---

* By MFA we understand any cluster approximation.

$$\{X\} = \{X_{-N}\} \cup \{X_{-N+\frac{1}{2}}\} \cup \{X_{-N+1}\} \cup \cdots \cup \{X_{N-1}\} \cup \{X_{N-\frac{1}{2}}\} \cup \{X_N\}.$$

The element functions βf_ν, have the same structure as the bulk free energy; *i.e.* they depend on a small set of cluster probabilities that are linearly linked to the local subset of correlation functions. Provided the local entropy is symmetrically written in terms of the planar clusters, the rectangular matrix linking the probabilities to the local correlation functions is spatially invariant. Owing to the simple form of Eq. (1) and of Eq. (2), the free energy can be minimized using the standard Newton Raphson (NR) algorithm taking advantage of the sparsity pattern of the Hessian matrix. In summary, the CVM sum method works as follows:

- Compute the bulk phases using standard bulk CVM and fix the far field conditions. Minimize the system free energy of Eq. (1), using NR algorithm together with sparse matrix algebra.

As an example, let us consider the Ising IPB, using the point CVM approximation. When x denotes the magnetization, the zero field bulk free energy is written as

$$\beta \mathcal{F}(x) = -\beta J \frac{q}{2} x^2 + \frac{1}{2} \left[(1+x)\ln(1+x) + (1-x)\ln(1-x)\right], \qquad (3)$$

where J is the exchange coupling constant and q the lattice coordination number. At thermal equilibrium, the magnetization x is found to be the solution of the self-consistent equation, $x = \tanh(\beta q J x)$, which has a non-zero solution only for temperature T below the critical temperature $T_c = qJ/k$. For temperatures close to T_c, we then have

$$\lim_{t \to 0^+} \beta \mathcal{F} \approx -\frac{t}{2}x^2 + \frac{x^4}{12},$$

$$t = \frac{T_c - T}{T_c} \geq 0.$$

From this relation, we recover the usual mean field critical behavior for the bulk magnetization x_B, that is,

$$x_B \approx t^{\frac{1}{2}}.$$

Having solved the bulk problem x_B at a given temperature, we form the corresponding Ising IPB. In terms of the local magnetization x_n the free energy is written as follows

$$\beta \mathcal{F}(\{x\}) = -\beta J \frac{q_\parallel}{2} \sum_{n=-N}^{+N} x_n^2 - \beta J \frac{q_\perp}{2} \sum_{n=-N}^{+N} x_n (x_{n-1} + x_{n+1})$$

$$+ \sum_{n=-N}^{+N} \tfrac{1}{2} \left[(1+x_n)\ln(1+x_n) + (1-x_n)\ln(1-x_n)\right], \qquad (4)$$

with

$$\lim_{N \to \pm\infty} x_N = \pm x_B.$$

In Eq. (4), $q_\parallel$ is the planar coordination number and $q_\perp$ is half the interplanar coordination number, so that we have the following identity: $q = q_\parallel + 2q_\perp$.

The local magnetization x_n's are then determined by finding the minimum of the free energy Eq. (4) as the solutions of the set of non-linear equations

$$-\beta J q_\| x_n - \beta J q_\perp (x_{n-1} + x_{n+1}) + \text{arctanh}(x_n) = 0. \tag{5}$$

The set of equations Eq. (5) can then be solved iteratively using the NR algorithm. (In this particular case, the Hessian matrix is tridiagonal).

For temperatures close to the transition temperature, it is useful to rewrite Eq. (5) by making the second order finite differences explicit as

$$-\beta J (q_\| + 2q_\perp) x_n - \beta J q_\perp (x_{n-1} + x_{n+1} - 2x_n) + \text{arctanh}(x_n) = 0. \tag{6}$$

In this range of temperature, the boundary region becomes much larger than the lattice spacing, so that it is more interesting to consider the envelope solution than the local magnetization. By expanding the inverse hyperbolic tangent in Eq. (6) to third order, it is easy to see that the envelope of the local magnetization is given by

$$x_n(y) = x_B \tanh\left(\frac{y}{2\lambda}\right), \tag{7}$$

where y is a continuous "distance" parameter. The boundary width λ, which is also the bulk correlation length, behaves near $t \approx 0$ as

$$\lambda \approx t^{-\frac{1}{2}}.$$

The excess free energy, or the surface tension σ, is computed as the difference between the system minimum free energy and the bulk free energy. It can be shown that σ vanishes near the critical temperature as

$$\sigma \approx t^{\frac{3}{2}}.$$

Equations (6) and (7), are representative of the well known "gradient square approximation" and are characteristic of any mean field approximation in the sum method based on a finite size cluster.

III. The Scalar Product (SP) Method

In the previous section we saw that, in setting up the CVM sum method, we first need to solve the bulk problem. We now review a method derived by R. Kikuchi in the 70's,[3-5] which also allows to evaluate accurately the surface tension from bulk information at almost no extra computing labor.

The method was called the scalar product method (SP) and was based upon the pioneering work of Clayton and Woodbury.[7] The SP and the corresponding SP-CVM approximation differ in spirit from the S-CVM approximation. In the SP method we treat analytically the surface tension problem through a transfer-matrix formalism, then ultimately the CVM approximation is introduced. The method is different from the S-CVM in which the approximation is introduced from the beginning. As we will see, the SP results in a totally different mean field critical behavior for σ. Reasoning based upon the Coherent Anomaly Method (CAM) of Suzuki[6] leads us to define a scheme in which the S and the SP-CVM bracket the exact solution.

We now present a simplified review of the basic SP formalism. A more sophisticated proof of the SP master equation (23) can be found in Ref. 8.

III.1 Basic Relations

It is well known that the use of the pair CVM approximation or the Bethe approximation, rather than the simple MFA (the point approximation), solves the one dimensional Ising model rigorously. This means an exact factorization of the total density
matrix in terms of reduced pair and point density matrices. In the same way, it is
easy to show that the density matrix for any $D \geq 2$ dimensional Ising model can be
factorized in terms of density matrices for two $(D-1)$ dimensional infinite clusters,
one being the slab cluster made from two adjacent $(D-1)$ planes and the other the
$(D-1)$ plane cluster itself.

In the following, we use ν_n to denote a possible configuration of plane n, and
we define $P_n(\nu_n)$ to be the probability of occurrence of the configuration ν_n. In the
same way, we define $P_{n+\frac{1}{2}}(\nu_n, \nu_{n+1})$ as the probability of simultaneously observing
configurations ν_n at plane n and ν_{n+1} at plane $n+1$. The interaction energy associated
with the planar configuration ν_n is denoted by $E_n(\nu_n)$ and the interplane interaction
energy is denoted by $E_{n+\frac{1}{2}}(\nu_n, \nu_{n+1})$.

Using the two consecutive plane clusters as the basic cluster, the system free
energy can be exactly written as

$$\beta \mathcal{F}_0 = - \sum_{n=-\infty}^{+\infty} \sum_{\nu_n} \left(\beta E_n(\nu_n) P_n(\nu_n) + P_n(\nu_n) \ln P_n(\nu_n) \right)$$

$$- \sum_{n=-\infty}^{+\infty} \sum_{\nu_n} \sum_{\nu_{n+1}} \beta E_{n+\frac{1}{2}}(\nu_n, \nu_{n+1}) P_{n+\frac{1}{2}}(\nu_n, \nu_{n+1})$$

$$+ \sum_{n=-\infty}^{+\infty} \sum_{\nu_n} \sum_{\nu_{n+1}} P_{n+\frac{1}{2}}(\nu_n, \nu_{n+1}) \ln P_{n+\frac{1}{2}}(\nu_n, \nu_{n+1}). \tag{8}$$

The probability variables $P_{n+\frac{1}{2}}(\nu_n, \nu_{n+1})$ and $P_n(\nu_n)$ are not independent of each
other, and obey coherence and constraint equations. These constraints are of two
kinds:

- Overlapping constraints.
  A planar cluster $P_n(\nu_n)$ represents the overlap cluster of two basic two plane
  clusters at positions $\nu = n - \frac{1}{2}$ and $\nu = n + \frac{1}{2}$, hence $P_n(\nu_n)$ should coherently be
  determined as the partial sum of $P_{n-\frac{1}{2}}(\nu_{n-1}, \nu_n)$ or of $P_{n+\frac{1}{2}}(\nu_n, \nu_{n+1})$. Thus we
  have

$$P_n(\nu_n) = \frac{1}{2} \left(\sum_{\nu_{n-1}} P_{n-\frac{1}{2}}(\nu_{n-1}, \nu_n) + \sum_{\nu_{n+1}} P_{n+\frac{1}{2}}(\nu_n, \nu_{n+1}) \right), \tag{9}$$

$$0 = \sum_{\nu_{n+1}} P_{n+\frac{1}{2}}(\nu_n, \nu_{n+1}) - \sum_{\nu_{n-1}} P_{n-\frac{1}{2}}(\nu_{n-1}, \nu_n). \tag{10}$$

- Normalization constraint.

  The variables $P_{n+\frac{1}{2}}(\nu_n, \nu_{n+1})$ which are defined as probabilities should sum to unity, so that we also have

$$\sum_{\nu_n} \sum_{\nu_{n+1}} P_{n+\frac{1}{2}}(\nu_n, \nu_{n+1}) = 1. \tag{11}$$

Associating the Lagrange multipliers $\alpha_n(\nu_n)$ to Eq. (10) and $\lambda_{n+\frac{1}{2}}$ to the normalization constraints, Eq. (11), we form the "augmented" free energy $\beta\mathcal{F}$ by adding the constraint contributions to the bare free energy $\beta\mathcal{F}_0$,

$$\beta\mathcal{F} = \beta\mathcal{F}_0 + \sum_{n=-\infty}^{+\infty} \sum_{\nu_n} \alpha_n(\nu_n) \left(\sum_{\nu_{n+1}} P_{n+\frac{1}{2}}(\nu_n, \nu_{n+1}) - \sum_{\nu_{n-1}} P_{n-\frac{1}{2}}(\nu_{n-1}, \nu_n) \right)$$

$$+ \sum_{n=-\infty}^{+\infty} \lambda_{n+\frac{1}{2}} \left(1 - \sum_{\nu_n} \sum_{\nu_{n+1}} P_{n+\frac{1}{2}}(\nu_n, \nu_{n+1}) \right). \tag{12}$$

From the explicit minimization of $\beta\mathcal{F}$, we then obtain

$$P_{n+\frac{1}{2}}(\nu_n, \nu_{n+1}) = \exp\left(\beta\lambda_{n+\frac{1}{2}}\right) h_n(\nu_n)\, \Gamma(\nu_n, \nu_{n+1})\, g_{n+1}(\nu_{n+1}), \tag{13}$$

$$g_n(\nu_n) = (P_n(\nu_n))^{\frac{1}{2}} \exp\left(+\beta\alpha_n(\nu_n)\right), \tag{14}$$

$$h_n(\nu_n) = (P_n(\nu_n))^{\frac{1}{2}} \exp\left(-\beta\alpha_n(\nu_n)\right), \tag{15}$$

$$\Gamma(\nu_n, \nu_{n+1}) = \exp\left(\tfrac{1}{2}\beta E_n(\nu_n) + \beta E_{n+\frac{1}{2}}(\nu_n, \nu_{n+1}) + \tfrac{1}{2}\beta E_{n+1}(\nu_{n+1})\right).$$

Treating the components $g_n(\nu_n)$ as a column vector $|g_n\rangle$ and the components $h_n(\nu_n)$ as a row vector $\langle h_n|$, we see that when the normalization condition is fulfilled the scalar product of these two vectors is unity,

$$\langle h_n | g_n \rangle = 1. \tag{16}$$

When the free energy is a minimum, $\beta\mathcal{F}$ can be further simplified by subtracting its derivative times the basis probability vector to obtain

$$\beta\mathcal{F} = \beta\mathcal{F} - \sum_{n=-\infty}^{+\infty} \sum_{\nu_n} \sum_{\nu_{n+1}} \frac{\partial \beta\mathcal{F}}{\partial P_{n+\frac{1}{2}}(\nu_n, \nu_{n+1})} P_{n+\frac{1}{2}}(\nu_n, \nu_{n+1})$$

$$= \sum_{n=-\infty}^{+\infty} \lambda_{n+\frac{1}{2}}. \tag{17}$$

Therefore, the Lagrange multiplier $\lambda_{n+\frac{1}{2}}$ has the meaning of a local free energy. Equations (13) and (16) are the two corner stones of the SP method. Indeed, substituting Eq. (13) into the coherence definition of the planar cluster leads to the induction relations,

$$|g_n\rangle = \exp\left(\beta\lambda_{n+\frac{1}{2}}\right) \Gamma |g_{n+1}\rangle, \qquad \langle h_n| = \exp\left(\beta\lambda_{n-\frac{1}{2}}\right) \langle h_{n-1}| \Gamma. \tag{18}$$

In Eq. (18) we have assembled the components of the energy part $\Gamma(\nu_n, \nu_{n+1})$ into a spatially invariant matrix Γ. These equations allow, through successive iterations, to propagate the far field information to the boundary centre, and for example we have

$$|g_0\rangle = \exp\left(\beta \sum_{j=0}^{m-1} \lambda_{j+\frac{1}{2}}\right) \Gamma^m |g_m\rangle. \tag{19}$$

We now introduce a symmetry breaking operator in the Hamiltonian in such a way that the Ising IPB is generated. Let $H^{\pm}$ be a vanishingly small external magnetic field operating on each side of the boundary, so that the matrix Γ splits into $\Gamma_{\pm}$. In the far field region, let $|g_+\rangle$ and $\langle h_-|$ be the bulk eigenvectors of matrices Γ_+ and Γ_-, respectivly, associated with the eigenvalue $\exp(\beta\lambda_0)$. By definition these eigenvectors are invariant under lattice translations, so that Eq. (19) leads to

$$|g_+\rangle = \exp(m\beta\lambda_0) \Gamma_+^m |g_+\rangle, \qquad \langle h_-| = \exp(m\beta\lambda_0)) \langle h_-| \Gamma_-^m. \tag{20}$$

Finally, since in the far field region we have $\lim_{m\to\infty} |g_m\rangle = |g_+\rangle$, by back substituting Eq. (20) into Eq. (19), and by forming the normalized scalar product $\langle h_0|g_0\rangle$, we end up with the following SP master equation

$$\exp(-a\beta\sigma) = \langle h_-|g_+\rangle, \tag{21}$$

where a is the sectional area, and the surface tension σ is defined as

$$a\sigma = \lim_{m\to\infty} \sum_{j=-m+\frac{1}{2}}^{m-\frac{1}{2}} (\lambda_j - \lambda_0). \tag{22}$$

This equation allows a simple interpretation since λ_j is the local free energy as shown in (17). In the case of the nearest neighbor Ising model, a final touch in the SP formalism is brought by noticing that in the far field domain, the Lagrange multipliers $\{\alpha\}$ in Eqs. (14) and (15) vanish because in the bulk any plane is a mirror and the cluster made of two infinitely large adjacent planes is symmetric with respect to the exchange of the plane configurations: $\nu_n \rightleftharpoons \nu_{n+1}$, so that we have

$$\exp(-a\beta\sigma) = \langle P_-|P_+\rangle^{\frac{1}{2}}. \tag{23}$$

The surface tension given by Eq. (23) is fully determined from the bulk phases as the normalized scalar product of two infinite vectors accounting for the probability of occurrence of planar configurations. The scalar product should now be approximated using the CVM.

III.2 CVM Approximations to the SP

We are now left with the problem of calculating the scalar product for σ in an infinite $D-1$ dimensional space using CVM approximations. First we notice that Eq. (23) is the sum of positive terms and have the same structure as an ordinary partition function for a given Hamiltonian $\mathcal{H}$. The usual definition for the free energy is

$$\exp(-\beta\mathcal{F}) = \sum_\nu \exp(-\beta\mathcal{H}(\nu)). \tag{24}$$

It is well known[9] that, provided the interactions are of finite range and contained in a maximal cluster α_d, the CVM approximation $\beta \mathcal{F}_{\alpha_d}$ for the free energy writes

$$\beta \mathcal{F}_{\alpha_d} = \sum_{\gamma \supseteq \alpha_d} a_\gamma \, \beta \mathcal{F}_\gamma, \qquad \beta \mathcal{F}_\gamma = \mathrm{Tr}_\gamma \, \rho_\gamma \beta \mathcal{H}_\gamma + \mathrm{Tr}_\gamma \, \rho_\gamma \ln \rho_\gamma. \tag{25}$$

As usual the $\{a\}$ are the CVM coefficients, ρ_γ is the reduced density matrix associated with the cluster γ whose bare Hamiltonian is $\beta \mathcal{H}_\gamma$ and $-\mathrm{Tr}_\gamma \, \rho_\gamma \ln \rho_\gamma$ is the entropy of the cluster γ.

It is interesting to rewrite the bare Hamiltonian $\beta \mathcal{H}_\gamma$ as $\ln(\tilde{\rho}_\gamma)$ where $\tilde{\rho}_\gamma = \exp(\beta \mathcal{H}_\gamma)$. This transformation is obviously insignificant from a computational point of view, but it shows that as soon as the Hamiltonian model is defined, $\tilde{\rho}_\gamma$ is nothing but the *a priori probability* (that is the Boltzmann factor) of finding the cluster γ in given configurations. The CVM approximation can thus be understood as an approximate method to construct lattice configurations matching an assigned subset of *a priori probabilities*, picking up the configuration that is most likely to occur at thermal equilibrium.

Let us now examine how the CVM approximation is brought into the scalar product formalism. Our starting point is the D dimensional Ising problem which is to have been studied with the bulk CVM approximation α_d. From the bulk CVM free energy minimization, we obtain a subset of planar probabilities. When we let α_{d-1} denote the maximal cluster corresponding to the cut of cluster α_d by a plane parallel to the boundary, all the reduced density matrices associated with cluster α_{d-1} and its subclusters are known. In the scalar product definition of σ, the sum over the configurations is macroscopic, and we are faced with the problem of approximating the number of ways to construct the infinite planar lattice subject to the assigned set of probabilities for local cluster. Therefore, $\rho_{\alpha_{d-1}}^I$ will play the role of the Boltzmann factor when building the planar lattice in phase I, and by analogy to the bulk case, the SP-CVM surface tension will be given by

$$a\beta\sigma = -\sum_{\beta \subseteq \alpha_{d-1}} a_\beta \, \mathrm{Tr}_\beta \, \rho_\beta \left(\tfrac{1}{2} \ln \left(\tilde{\rho}_\beta^I \, \tilde{\rho}_\beta^{II} \right) - \ln \rho_\beta \right), \tag{26}$$

with

$$\tilde{\rho}_\beta^I = \mathrm{Tr}_{\alpha_d/\alpha_{d-1}}^I \, \rho_{\alpha_d}^*, \tag{27}$$

where $\rho_{\alpha_d}^*$ is the density matrix for cluster α_d when the bulk free energy is a minimum, and the trace operates on the exterior of α_{d-1} in α_d. The surface tension for the thermal equilibrium of the IPB is obtained by minimizing Eq. (26) with respect to the planar correlation functions. In summary, the CVM-SP operates as follows:

- Given a cluster α_d for the bulk phase, we minimize the corresponding CVM free energy.
- At the minimum, we form the IPB "energies" for the planar cluster α_{d-1} associated with α_d.
- We minimize the CVM surface tension

To make the representation concrete, let us treat in some details the CVM pair approximation from which some general features of the SP-CVM will be drawn. The pair CVM approximation for the free energy of the Ising lattice of coordination number q is given by

$$\beta \mathcal{F} = \frac{q}{2}\beta\epsilon 4 y_{12} + \frac{q}{2}\sum_{i,j=1}^{2} y_{ij}\ln y_{ij} + (1-q)\sum_{i=1}^{2} x_i \ln x_i \,,$$

where

$$\sum_{i=1}^{2} x_i = 1, \quad y_{11} = x_1 - y_{12}, \quad y_{21} = y_{12}, \quad y_{11} = x_1 - y_{12}\,,$$

and the indices $i = 1,2$ correspond to spin up and spin down respectively. The equilibrium values of the two basis variables x_1 and y_{12} are derived by equating the derivatives of the free energy to zero. After some algebraic transformations we obtain

$$x_1 = \tfrac{1}{2}y_{12}^{-1} = 2\left(e^{2\beta\epsilon}+1\right), \tag{28}$$

$$x_1 = \frac{\exp\left(\frac{q}{2}\theta\right)}{\exp\left(\frac{q}{2}\theta\right) + \exp\left(-\frac{q}{2}\theta\right)} \; y_{12}^{-1} = 2\left[1 + e^{2\beta\epsilon}\cosh\left((q-1)\theta\right)\right], \tag{29}$$

where the auxiliary variable θ and the critical temperature are defined by

$$e^{2\beta\epsilon} = \frac{\sinh\left(\frac{q}{2}\theta\right)}{\sinh\left(\left(\frac{q}{2}-1\right)\theta\right)}\,,$$

$$\frac{kT_c}{\epsilon} = \frac{2}{\ln\left(\frac{q}{q-2}\right)}\,.$$

Equations (28) and (29) correspond to the disordered and ordered solutions, respectively.

The bulk solutions are now denoted by a hat. They are to be distinguished from the planar variables, written without a hat, still to be determined from the minimization of the surface tension σ given by

$$\beta\sigma = -\frac{q_{\|}}{2}\sum_{i,j=1}^{2} E_{ij}y_{ij} + (q_{\|}-1)\sum_{i=1}^{2} E_i x_i + \frac{q_{\|}}{2}\sum_{i,j=1}^{2} y_{ij}\ln y_{ij} + (1-q_{\|})\sum_{i=1}^{2} x_i \ln x_i \,,$$

with

$$E_{11} = E_{22} = \tfrac{1}{2}\ln\left(\hat{y}_{11}\hat{y}_{22}\right), \quad E_{21} = E_{12} = \ln\hat{y}_{12}, \quad E_1 = E_2 = \tfrac{1}{2}\ln\left(\hat{x}_1\hat{x}_2\right)\,.$$

Back substituting the "ordered" bulk solutions given by Eq. (29) in the previous definitions for the IPB energies, we obtain finally

$$\beta\sigma = E_\theta + \frac{q_{\|}}{2}\beta\epsilon 4 y_{12} + \frac{q_{\|}}{2}\sum_{i,j=1}^{2} y_{ij}\ln y_{ij} + \left(1-q_{\|}\right)\sum_{i=1}^{2} x_i \ln x_i \,,$$

with

$$E_\theta = \tfrac{1}{2}\left(q_{\|}-1\right)\ln(\hat{x}_1\hat{x}_2) - q_{\|}\ln\left(\hat{y}_{12}e^{2\beta\epsilon}\right),$$

where we collected in E_θ terms which are independent of the variables y's and x's.

Beyond the result of the minimization of σ, it is important to make the following remarks:

- Comparison of the $\beta \mathcal{F}$ and $\beta \sigma$ expressions makes us to see that the bulk factors $\hat{x}_1$ and $\hat{y}_{12}$ given, the surface tension σ has strictly the same form as the bulk free energy, apart from the factor $E_\emptyset$ which was not present in the bulk problem. The presence of $E_\emptyset$, which implicitly depends on the temperature through the bulk variables, is quite general in the SP-CVM, and is needed to ensure that σ goes to zero at the bulk critical temperature. More generally, when the cluster size is increased, longer range interactions appear in the form of *a priori probabilities* which were not in the bulk Hamiltonian.

- The factor $E_\emptyset$ being irrelevant in the minimization of σ, the solutions Eq. (28) and Eq. (29) for the bulk phases apply. We want to call attention to the possibility for the IPB boundary itself to undergo a phase transition. In our example, the boundary will be unstable at the temperature $kT_{\mathrm{IPB}}/\epsilon = 2/\ln\left(\frac{q_\parallel}{q_\parallel - 2}\right)$, which depends on the value of $q_\parallel$. From this example we expect boundary phase transition whenever the space dimensionality D of the bulk is greater then 3. This point will be discussed in the next section.

- The ground state boundary energy is given by expanding σ for temperatures close to zero and we find

$$\lim_{T \to 0} \frac{a\sigma}{\epsilon} = q - q_\parallel \, . \tag{30}$$

- For temperatures close to the bulk transition temperature, the mean field critical behavior of SP-CVM surface tension σ is given by

$$\lim_{T \to T_c^-} \frac{a\sigma}{\epsilon} \approx t \, , \tag{31}$$

where t have been defined in section II.

IV. Applications

We now apply the previous formalism of the SP-CVM and S-CVM in the cases of Ising IPB of the $\langle 10 \rangle$ and $\langle 11 \rangle$ orientations in the $2D$ square lattice and of the $\langle 100 \rangle$ orientation in the $3D$ cubic lattice. This choice of applications has been driven by the following considerations:

- In the $2D$ square lattice, exact results are available for both orientations,[10,11] allowing us to evaluate and scale our approximations.

- In the $3D$ cubic lattice, approximate but very accurate results are known from Monte Carlo simulations[12] and from low temperature series expansion.[13] However, there is no rigorous results corresponding to Onsager's. The work on this lattice also suggests the future direction we can take to study further $3D$ lattices, where σ's have not been investigated.

- The chosen cases correspond to bulk lattices of low coordination numbers, hence constituting a severe test for the CVM

In this section, we will show how a series of CVM approximations improve upon individual results. Several strategies can be followed and for example one might extrapolate either a series of SP-CVM approximations or a series of S-CVM approximations.

Another way to look at the problem is to recognize that the SP-CVM, which involves only bulk variables, is the companion method of the S-CVM, and one may consider a series of joint CVM approximations. More precisely, assume we may argue from the Coherent Anomaly Method (CAM) analysis that the exact critical exponent for the surface tension is intermediate between 1 (SP) and 1.5 (S). Thus, close to the transition temperature, the SP-CVM would overestimate the exact surface tension while the S-CVM method would underestimate it. We may expect that this result still holds at lower temperatures and then it becomes valuable to consider simultaneously a series of SP and S-CVM that accurately brackets the exact solution.

Since the above argument is based upon the difference of mean field critical behavior for σ between the two CVM methods, we start this section by summarizing some of the critical properties of the CVM sequence we use in this work.

IV.1 CAM Analysis of the Cluster Sequence

The CAM theory of Suzuki[6] is a genuine analytical continuation method of a series of mean field results. The method is based on the property that close to the transition point MFA behaves "classically." However when we consider the singular behavior of a sequence of MFA parametrized in terms of the critical temperature $T_c(MF)$, we see that an envelope function exists which does not take the "normal" form, for example it changes from simple pole to a branch point.

Let $\Phi(T)$ be any physical quantity of interest, for example the magnetic susceptibility or the surface tension. The "classical" MFA behavior for $\Phi(T)$ leads to a unique critical exponent $\gamma(\Phi, MF)$, say independent of the system dimensionality. Close to the transition point, $\Phi(T)$ behaves as

$$\Phi(T) \approx \bar{\Phi}(T_c(MF)) \left(\frac{T - T_c(MF)}{T_c(MF)} \right)^{\gamma(\Phi, MF)} . \tag{32}$$

The amplitude factor $\bar{\Phi}(T_c(MF))$ was called the Coherent Anomaly Factor (CAF). Suzuki has shown that this factor is sensitive to the difference between the mean field transition temperature $T_c(MF)$ and the exact transition temperature $T_c^\star$. When we consider a canonical series of MFA in which $T_c(MF) - T_c^\star$ monotonically decreases to zero, the CAF can be written as

$$\bar{\Phi}(T_c(MF)) \approx A(T_c^\star) \left(\frac{T_c(MF) - T_c^\star}{T_c^\star} \right)^{\Delta\gamma(\Phi)} , \tag{33}$$

$$\Delta\gamma(\Phi) = \gamma^\star(\Phi) - \gamma(\Phi, MF) . \tag{34}$$

The exponent $\Delta\gamma(\Phi)$ represents the difference between the exact critical exponent $\gamma^\star$ for the physical quantity Φ and the corresponding MFA critical exponent $\gamma(\Phi, MF)$. The knowledge of $T_c(MF)$ and $\bar{\Phi}(T_c(MF))$ from a series of three MFA is enough to determine both $T_c^\star$ and $\gamma^\star$.

There are many ways in which we may construct a series of MFA. Among the MFA, the CVM is attractive because small sized cluster approximations determine the Curie Temperature fairly close to the exact value, so that we may expect significant variations of the CAF from different approximations. What we have to pay special attention to is that the series of approximations are to satisfy the "canonical series" requirement of Suzuki, *i.e.*, the condition that the Curie temperatures of the series systematically converge to $T_c^\star$. In 2-dimensional lattice treatments of the CVM, it is

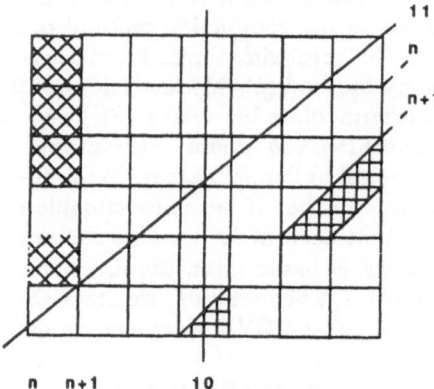

Figure 1. For the $\langle 10 \rangle$ IPB orientation, clusters made of n consecutive square parallel to the boundary are used. Similarly for the $\langle 11 \rangle$ we used n step clusters made of $2n - 1$ consecutive right angle triangles. Represented are the cases $n = 1$ and $n = 3$.

Table I

$T_c(n)$ and coherent anomaly factor $\bar{\chi}\left(T_c\left(n\right)\right)$ for the bulk magnetic susceptibility for the n square approximations (columns 1 and 2) and for the n step approximations (columns 3 and 4).

n	$T_c(n)$	$\bar{\chi}(T_c(n))$	$T_c(n)$	$\bar{\chi}(T_c(n))$
1	2.425666	1.416973	2.425666	1.416973
2	2.376130	1.849774	2.364830	1.981231
3	2.350312	2.250939	2.337386	2.538502
4	2.334538	2.629717	2.322445	3.039093

possible to design hierarchical sequence considering 1-dimensionally extended clusters of the ladder shape. The clusters we used in the $2D$ square lattice are shown in Fig. 1. It was proved by Schijper[14] that they form the canonical series. Associated to the $\langle 10 \rangle$ IPB orientation, we selected a sequence of n consecutive square clusters with $n = 1, 2, 3, 4$. Note that this n is different from n used to indicate lattice plane position in previous sections. For the $\langle 11 \rangle$ IPB orientation we have chosen a sequence of n step clusters made with $2n - 1$ right angle triangles sharing an edge ($n = 1, 2, 3, 4$). (Note the same sequence can be used in the triangular lattice replacing the right-angled triangles by equilateral triangles).

In Table I, we list the values of the Curie temperature and of the magnetic susceptibility CAF for the two sequences.

The convergence of $T_c(n)$ can be observed when we consider $T_c(n)$ as a function of the inverse of the sequence index n (Neville's extrapolation[15]). From the CAM analysis of the n square values in Table I, we estimate $T_c^\star$ as 2.2692 ± 0.0006 to be compared with Onsager's[10] exact value $T^\star = 2/\ln\left(1 + \sqrt{2}\right) = 2.269185$, and we find the critical exponent γ for the magnetic susceptibilty equals 1.748 ± 0.008 instead of

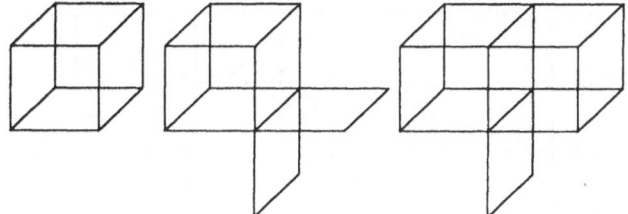

Figure 2. Clusters used for the present analysis of the $3D$ cubic lattice. Approximation $n = 1$ corresponds to the cube, $n = 2$ to the cube plus a cross shaped 10 points cluster and $n = 3$ to a double cube plus the 10 points cluster.

Table II

$T_c(n)$, coherent anomaly factor for the bulk magnetic susceptibility $\bar{\chi}(T_c(n))$ and for the SP surface tension $\bar{\sigma}(T_c(n))$ for cubic lattice.

n	$T_c(n)$	$\bar{\chi}(T_c(n))$	$\bar{\sigma}(T_c(n))$
1	4.580988	0.416973	3.65364
2	4.560239	0.453057	3.33375
3	4.544872	0.496270	3.08290

the exact 1.75. The CAM analysis of the data for the n step sequence in Table I fully leads to almost the same results. T_c^* being known with enough accuracy, we may now estimate the critical exponent related to the IPB.

As was seen in section I, the local magnetization in the S-CVM approximation varies practically as a hyperbolic tangent. By recording the boundary width in a temperature domain $T/T_c(n)$ between 0.99 and 0.999, we can determine the coherent anomaly factor $\bar{\lambda}(T_c(n))$ for the correlation length by fitting the data to the usual mean field behavior: $\lambda(T) \approx \bar{\lambda}(T_c(n))t^{-0.5}$. The results we obtain from the $n = 1, 2, 3$ square are respectively $\ln(\bar{\lambda}(n)) = 0.28528, 0.36060, 0.41786$, and we estimated the critical exponent ν for the correlation length to be 0.9641 instead of 1.

Along the same line, we estimated the surface tension critical exponent (from the S-CVM) μ to be 1.10 instead of 1. The discrepancy between the two values is not that large owing to the numerical inaccuracy of the S-CVM method close to $T_c(n)$ where more than 300 planes are required (corresponding to around 30,000 variables). A much better value, $\mu = 1.0052$ is achieved by analyzing the surface tension from the SP-CVM for the $n = 1, 2, 3, 4$ squares sequence ($\bar{\sigma}(n) = 6.4322, 5.7365, 5.4596, 5.3219$).

In the case of the 3 dimensional system, the existence of the canonical hierarchy sequences still has not been proved. A probably good sequence[16,17] would consider the tiling of cubes along a $2D$-space of slab shape. The first cluster of the sequence would be the cube, or $2 \times 2 \times 2$, the next cluster would be made with four cubes, or $3 \times 3 \times 2$ cluster, and so on. Since the number of variables grows too rapidly in such a sequence, we have used instead the clusters shown in Fig. 2. The relevant critical data for these clusters are listed in Table II, and from CAM we determined $T_c^* = 4.51 \pm 0.01$ and $\gamma = 1.2586 \pm 0.0054$ to be compared to the commonly accepted values of 4.5115 (Ref. 18) and 1.25 (Ref. 19). Finally, from the SP-CVM coherent anomaly factor

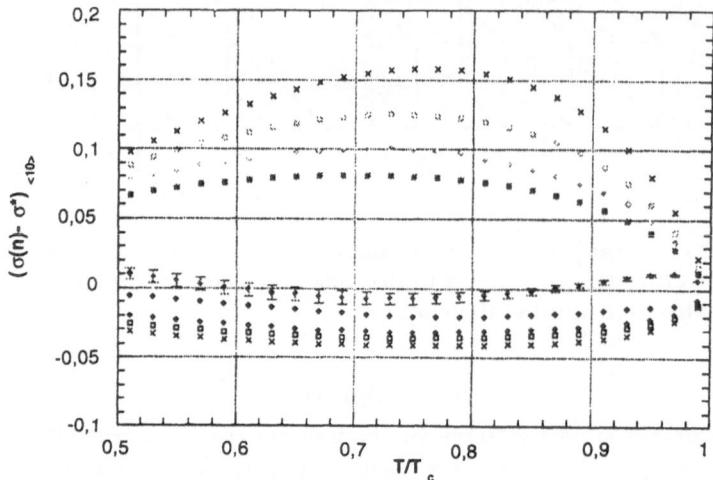

Figure 3. Difference between the CVM approximation for σ and Onsager's exact result $\sigma^\star$ of the $\langle 10 \rangle$ orientation IPB. The positive curves from the top to the bottom are for the $n = 1, 2, 3, 4$-square approximations. The negative curves from the bottom to the top are for the $n = 1, 2, 3$-square approximations. The two inner most curves correspond to the extrapolated SP-CVM (empty diamonds with error bar) and to the extrapolated S-CVM (filled diamonds).

for the surface tension, we estimated $\mu = 1.239 \pm 0.015$ in close agreement with the other estimate. Although it is unlikely that the three approximations in Fig. 2 form a canonical series, close agreements of the estimated γ and μ suggest some validity of the analysis.

IV.2 CVM and Extrapolation

Figure 3 plots $a \left(\sigma(n) - \sigma^\star \right) / J$ against the reduced temperature $T/T_c(n)$ for the n-square approximation, where $\sigma^\star$ is the Onsager[10] rigorous result for the $\langle 10 \rangle$ surface tension given by

$$\frac{a\sigma^\star}{J} = \frac{kT}{J} \ln \left[\exp \left(\frac{2J}{kT} \right) \tanh \left(\frac{J}{kT} \right) \right]. \tag{35}$$

Figure 4 plots the same information for the n-step approximation, but now $\sigma^\star$ is the Fisher and Ferdinand[11] rigorous result for the $\langle 11 \rangle$ surface tension given by

$$\frac{a\sigma^\star}{J} = \sqrt{2} \frac{kT}{J} \ln \left[\sinh \left(\frac{2J}{kT} \right) \right]. \tag{36}$$

In figures 3 and 4, the positive curves refer to the SP-CVM and the negative curves to the S-CVM. The SP-CVM curves from the top to the bottom correspond to the $n = 1, 2, 3, 4$ square or step approximations, respectively, while the S-CVM curves from the bottom to the top are for the $n = 1, 2, 3$ approximations. In the $2D$ square, a plane parallel to the boundary actually is a line. The CVM clusters involved in the SP surface tension are then linear chains of $n + 1$ points. The boundary never

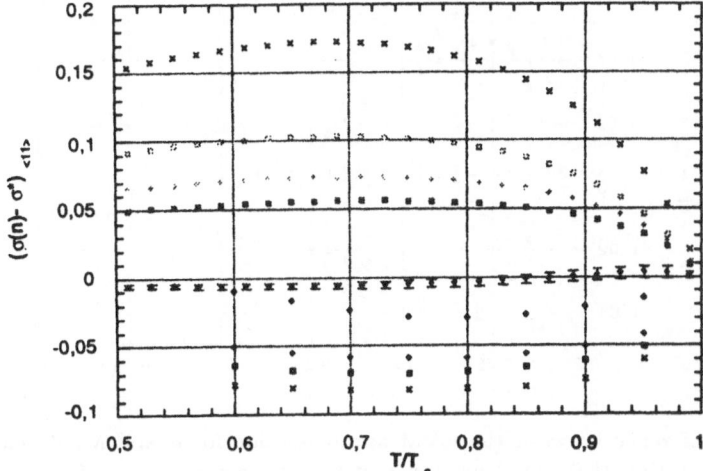

Figure 4. Same as figure 3 for the $\langle 11 \rangle$ orientation IPB.

undergoes a phase transition in itself and remains in a "disordered" state from the bulk Curie temperature T_c to $T = 0$. The following general properties are observed:

- For a given cluster approximation, the S-CVM is more accurate than the SP-CVM. In order to understand this difference, the best way is to consider the low temperature expansion of σ. It will be reported in a separate paper.
- As the approximation improves, the SP and S-CVM bracket the exact curve more and more efficiently, since in the limit of infinitely extended clusters both methods converge to the same result. The SP-CVM being an explicit function of the bulk phases meeting at the boundary, this result suggests that the surface tension is irrelevant of the precise boundary width. This is in line with Onsager's exact calculation of σ, in which the boundary profile and the boundary width are not in the theory.

In figures 3 and 4, the innermost two curves are the extrapolated SP-CVM and S-CVM, respectively. For the extrapolations of SP, we indicated the error bars also.

Despite the fact that the n-square approximation has been proved to converge to the exact solution,[14] how convergence proceeds with n is still unknown. Apart from the range close to the Curie temperature where the CAM applies, there is no unique method of extrapolating the sequence of CVM results to the rigorous limit. For reduced temperatures not too close to 1, one might expect a switch from critical to mean-field behavior, which may support polynomial or rational extrapolation. Making use of T_c^* known from CAM, we chose a reasonable extrapolation parameter defined as $T_c(n) - T_c^*$ which measures in some sense the distance between the approximation and the exact result. As it can be seen in the Figs. 3 and 4, a significant improvement of the results is made by extrapolating the series of CVM approximations. Extrapolation of the SP series is closer to the rigorous limit than the S series.

Figure 5 plots the similar information as Figs. 3 and 4 in the case of the cubic $\langle 100 \rangle$ orientation IPB. For comparison purposes, the approximate but fairly accurate values of σ^* were taken from tabulated Monte Carlo data[12] (in fact twice the listed value). The plotted curves are for the cube and the cube-plus-10-points clusters.

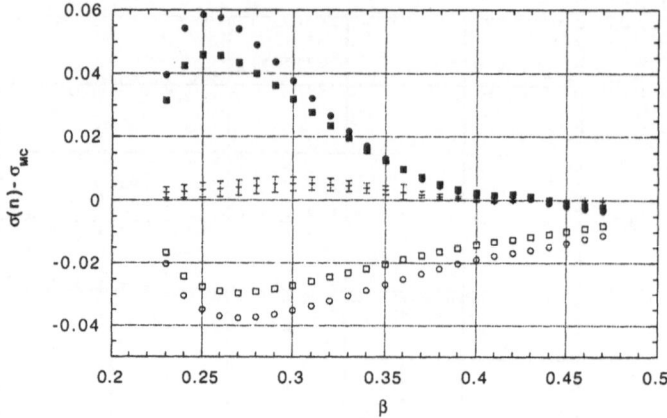

Figure 5. Difference between the CVM approximation for σ and Monte carlo results of the $\langle 100 \rangle$ orientation IPB. The two top and bottom curves are for the $n = 1, 2$ cluster approximation (see Fig. 2). The middle curve corresponds to the extrapolated arithmetic average of the SP-CVM and of the S-CVM.

In the $3D$ cubic lattice, a plane parallel to the boundary is the $2D$ square lattice. Therefore, the boundary may itself undergo a phase transition at a temperature T lower than the bulk Curie temperature, the boundary switching from a "disordered" to an "ordered" state as reported before.[20] Such phase transition is evidently felt in the SP-CVM, and an instability temperature for the surface tension can be defined, below which we should follow the stable branch for σ. The curves for the $3D$ case in Fig. 5 display the general features we discussed in the $2D$ case, but it can be noticed that individual results are almost one order of magnitude better than those in the $2D$ cases. Figure 5 shows that the extrapolated curve almost matches the MC data in the full range of temperature.

Rather than individually considering the SP and S surface tensions, in extrapolating the results of the two first clusters in Fig. 2, we construct their arithmetic mean and consider

$$\tilde{\sigma}(n) = \frac{\sigma^{SP}(n) + \sigma^{S}(n)}{2}, \qquad \tilde{\delta}(n) = \frac{\sigma^{SP}(n) - \sigma^{S}(n)}{2}, \qquad (37)$$

where n correspond a given CVM approximation. At given temperature and approximation, $\tilde{\sigma}(n)$ gives an estimate of σ and $\tilde{\delta}(n)$ a confidence interval, which can be extrapolated. The main advantage of averaging is that we can take advantage of the bracketing feature of the SP and S approaches.

We did the similar averaging in the $2D$ case too. The results are excellent, but have not been represented in Figs. 3 and 4 not to confuse the plots.

V. Conclusion

Extended defects such as IPB or APB are customarily studied within the S-CVM framework which can compute the boundary profile in addition to the excess free energy. When only the surface tension σ is of interest, the scalar product SP-CVM method can be used instead. The SP-CVM evaluates σ as the scalar product of two bulk vectors and almost does not involve extra work in excess of calculating the bulk phases. For this reason, a series of cluster approximations can be run and their results can be extrapolated to the rigorous limit. Since computation of bulk phase thermodynamics is a part of any S-CVM method, it is worthwhile to simultaneously consider the SP-CVM. The values of the surface tension obtained in the two methods bracket the exact solution. It is then safer to define the CVM approximation to the surface tension as the arithmetic average of σ^{SP} and σ^S and to define the corresponding confidence interval. This new scheme is expected to be very efficient in $3D$ cases even for small sized clusters, and may well be useful when studying various cases of interest including anisotropy effects.

References

1. R. Kikuchi, *Phys. Rev.* **79**, 718 (1950); **81**, 988 (1951).
2. R. Kikuchi and J. W. Cahn, *Acta Metall.* **27**, 1337 (1979).
3. R. Kikuchi, *J. Chem. Phys.* **57**, 777 (1972).
4. R. Kikuchi, *J. Chem. Phys.* **57**, 783 (1972).
5. R. Kikuchi, *J. Chem. Phys.* **65**, 4545 (1976).
6. M. Suzuki, *J. Phys. Soc. Japan* **55**, 4205 (1986).
7. D. B. Clayton and J. W. Woodbury Jr., *J. Chem. Phys.* **55**, 3895 (1971).
8. P. Cenedese and R. Kikuchi, *Physica A* (in preparation)
9. F. Ducastelle, *Order and Phase Stability in Alloys*, edited by F. R. De Boer and D. G. Pettifor, (North-Holland, 1991).
10. L. Onsager, *Phys. Rev.* **65** , 117 (1944).
11. M. E. Fisher and A. E. Ferdinand, *Phys. Rev. Lett.* **19**, 169 (1967).
12. M. Hasenbusch and K. Pinn, *Physica A* **203**, 189 (1994).
13. H. Arisue, *Physics Letters B* **313**, 187 (1993).
14. A. G. Schlijper, *J. Stat. Phys.* **35**, 285 (1984); **40**, 1 (1985).
15. D. S. Gaunt and A. J. Guttmann, *Phase transitions and critical phenomena*, Vol. 3, edited by C. Domb and M. S. Green, (Academic Press)
16. A. Finel, *Thèse d'état Université Paris VI*, (1987).
17. R. Kikuchi, *Prog. Theor. Phys.* Suppl. **115**, 1 (1994).
18. G. S. Pawley, R. H. Swendsen, D. J. Wallace, and K. G. Wilson, *Phys. Rev. B* **29**, 4030 (1984).
19. J. C. Le Guillou and J. Zinn-Justin, *Phys. Rev. B* **21**, 3976 (1980).
20. R. Kikuchi, *J. Chem. Phys.* **66**, 3352 (1977).

Ordering of Oxygen Atoms in YBa$_2$Cu$_3$O$_{6+x}$: Long-Range Coulomb Repulsions

A. A. Aligia and J. Garcés

Centro Atómico Bariloche
Comisión Nacional de Energía Atómica
8400 Bariloche
ARGENTINA

Abstract

On the basis of strong-coupling calculations of the electronic structure, which are able to explain the dependence of the superconducting critical temperature T_c and other properties on the O content x, we have proposed a simple structural model for the O ordering in the basal CuO$_x$ planes of YBa$_2$Cu$_3$O$_{6+x}$: any two O atoms of these planes repel each other with a screened Coulomb repulsion, but the repulsion between second-neighbor O atoms with a Cu in between is reduced by a quantity ΔE due to charge transfer effects. For calculated values of ΔE and the screening length λ, the model explains practically all the observed diffraction patterns. For $1/2 < x < 3/4$, the ground state x as a function of O chemical potential displays a behavior which resembles a complete devil's staircase. The composition-temperature phase diagram of the model, and in particular the effect of repulsions beyond second nearest-neighbors, is studied using the 3×3 point approximation of the CVM. Since this approximation fails at low enough temperatures for $1/3 < x < 1/2$, the particular case $x = 3/8$ is studied with another CVM approximation. The results are in semiquantitative agreement with experiment. At low enough temperatures for $1/2 < x < 3/4$, the model is mapped into an effective one-dimensional Ising model which explains the observed split diffuse diffraction peaks.

I. Introduction

The layered atomic structure of YBa$_2$Cu$_3$O$_{6+x}$ can be divided into different (eventually corrugated) planes.[1] The most interesting are the CuO$_2$ ones (two per unit cell),

Theory and Applications of the Cluster Variation and Path Probability Methods
Edited by J.L. Morán-López and J.M. Sanchez, Plenum Press, New York, 1996

which are responsible for the superconductivity, and the CuO_x planes, which act as reservoirs of charge. Both types of planes consist of a nearly square lattice of Cu atoms but, while in the CuO_2 planes each position between two Cu atoms is occupied by an O atom, in the CuO_x planes only a fraction $x/2$ of the possible O positions is occupied. The subject of the present study is the ordering of these O atoms.

The electronic and atomic structure of $YBa_2Cu_3O_{6+x}$ are self-consistently related:[2-5] the ground-state atomic structure corresponds to the minimum electronic energy and, on the other hand, it determines the distribution of electrons among the different planes. Experimentally, it has been confirmed that the superconducting critical temperature T_c increases with a thermal treatment which favors ordering of full and empty parallel CuO chains.[6,7] A simple explanation of this relation is given at the beginning of the next section. It is based on an extreme "strong-coupling" approach in which covalency is neglected, but the effect of correlations, particularly on-site and nearest-neighbor repulsions, are treated exactly. The effect of covalency can be taken into account by different methods developed for high-T_c systems, as explained in section II. Using these methods, the main electronic properties of $YBa_2Cu_3O_{6+x}$ can be explained and our structural model can be justified.

An alternative structural model for $YBa_2Cu_3O_{6+x}$, the asymmetric next-nearest-neighbor Ising (ASYNNNI) model,[8] assumes that only three short-range O–O interactions are different from zero and takes these parameters from fitting of total energies calculated with LMTO-ASA first-principle calculations.[9] Since the *ab initio* calculations treat the correlations in the Hartree-Fock approximation, this is a "weak-coupling" approach. Similar approaches were successfully applied to metallic alloys.[10] However, in the case of high-T_c compounds, the magnitude of several interactions is much larger than the hopping terms. These energies were calculated by constrained-density-functional calculations.[11] The most important interactions are the Cu and O on-site Coulomb repulsions, $U_d \sim 10.5$ eV and $U_p \sim 4$ eV, respectively. The Cu–O nearest-neighbor (NN) interatomic repulsion $U_{pd} \sim 1.2$ eV is also crucial, as we show in Section II. In contrast, the most important hoppings, the Cu–O and O–O NN ones, $t_{pd} \sim 1.3$ eV and $t_{pp} \sim 0.65$ eV, respectively, are nearly an order of magnitude smaller than U_d. As a consequence, the *ab initio* calculations are not even able to obtain the insulator character of the ground state of the parent compounds of high-T_c systems. In addition, even in the metallic case, the correlation energy (the difference between the energy calculated in Hartree Fock and the exact energy) is of the order of 0.6 eV per unit cell in $YBa_2Cu_3O_{6+x}$ and depends on the O ordering.[4,12] This energy is much larger than all the parameters of the ASYNNNI and thus, the LMTO-ASA results for them[9] are not reliable.

A crucial issue for the structural model is the nature and number of interactions which should be included. One usually likes simple models with few interactions and/or parameters. The ASYNNNI models includes interactions up to second NN O atoms which are at a distance of one lattice parameter ($a \sim b \sim 3.8$ Å). However, the high-T_c systems have a very low number of carriers and then, long range Coulomb interactions are poorly screened. For the maximum possible hole doping ($x = 1$), $YBa_2Cu_3O_{6+x}$ has 0.0029 holes/Å^3.[13-15] Thus, in average, the carriers are separated by ~ 7 Å. This suggest that O–O repulsions at distances $2a$ cannot be neglected. In fact these repulsions are crucial[16] in the explanation of Khachaturyan and Morris[17] of the split diffuse diffraction peaks observed for $x \sim 0.6$.[18,19] This theory has been extended to all compositions and temperatures by one of us,[20] as is explained in section V.3.

The considerations presented above support our strong-coupling approach to the electronic structure, discussed briefly in section II, and provide some justification to our simple structural model, presented in section III. The results for the ground state of the atomic structure are presented in section IV. Section V contains the thermodynamic calculations using the CVM and is subdivided in three parts: subsection V.1 is devoted to the high temperature region, subsection V.2 to the special case $x = 3/8$, and the subsection V.3 to the quasi one-dimensional ordering. Section VI contains a discussion.

II. The Electronic Structure

In this section, we explain briefly the assumptions, methods and main results of our "strong-coupling" approach to the electronic structure of YBa$_2$Cu$_3$O$_{6+x}$. We also provide a simple argument, which allows a qualitative understanding of the mutual interdependence of O ordering in the CuO$_x$ planes and charge transfer to the superconducting CuO$_2$ planes. Alternative "strong-coupling" points of view were given by Latgé et al.[21] and Uimin et al.[22]

Each Cu atom of the CuO$_x$ planes has one O NN atom above the plane (on a line perpendicular to the latter passing through the Cu atom) and another one below it (see Fig. 1 or Ref. 1). The electronic structure calculations,[23] suggest that, at least to a first approximation, these CuO$_{2+x}$ structures do not have hopping with the CuO$_2$ planes. This means that we can calculate the electronic structure of both entities separately, with the condition that the Fermi level ϵ_F is the same.

The relation between electronic and atomic structure is easily visualized taking the limit of zero hopping in the CuO$_{2+x}$ subsystem.[3,4] Although this limit is not quantitatively valid, it displays the crucial effects of the Cu intratomic (U_d) and Cu–O interatomic (U_{pd}) repulsions. Repulsions beyond Cu–O NN are neglected for the moment. The right side of Fig. 1 shows the structure which, for $x = 1/2$, minimizes the total Coulomb repulsion energy under the assumption that all atoms related by symmetry operations in the tetragonal phase (for example all Cu atoms of the CuO$_x$ planes) have the same charge.[2,3,24] It is a deformed hexagonal structure and we call it HS. However, the experimentally observed structure at low temperature[25–28] is the "chain structure" (CS) represented on the left of Fig. 1. Let us assume that for both structures, CS and HS, all O ions in the CuO$_{2+x}$ subsystem are present as O^{-2} (the argument works equally well if more realistic smaller charges[4,24] are used). For CS there are two types of Cu atoms in the CuO$_x$ planes: the four-fold coordinated and the two-fold coordinates ones. The energy necessary to put the first hole in the former (latter) ones, converting the Cu$^+$ in Cu^{+2} ions is $\epsilon_{Cu} - 8U_{pd}$ ($\epsilon_{Cu} - 4U_{pd}$), where the second term represents the effect of the attraction of the added hole with its NN O^{-2} ions. Instead, for the HS, all Cu ions are three-fold coordinated and the hole energy level is at $\epsilon_{Cu} - 6U_{pd}$. The energy levels for both structures are represented in Fig. 2. The energy necessary to put the second hole in any Cu atom (Cu$^{+2} \rightarrow$ Cu^{+3} transition) lies an energy $U_d \sim 10.5$ eV above the corresponding one for adding the first hole, and can be neglected (this fact is not properly taken into account in usual first-principles calculations).

Depending on the position of the Fermi energy ϵ_F, four different situations can occur. If $\epsilon_F > \epsilon_{Cu} - 4U_{pd}$ all levels of Fig. 2 are occupied, all Cu ions are thus Cu^{+2}, the energies for both the CS and the HS are the same, and the CuO$_2$ planes should be in the insulating (undoped) state to satisfy the balance of charges (in YBa$_2$Cu$_3$O$_{6+x}$,

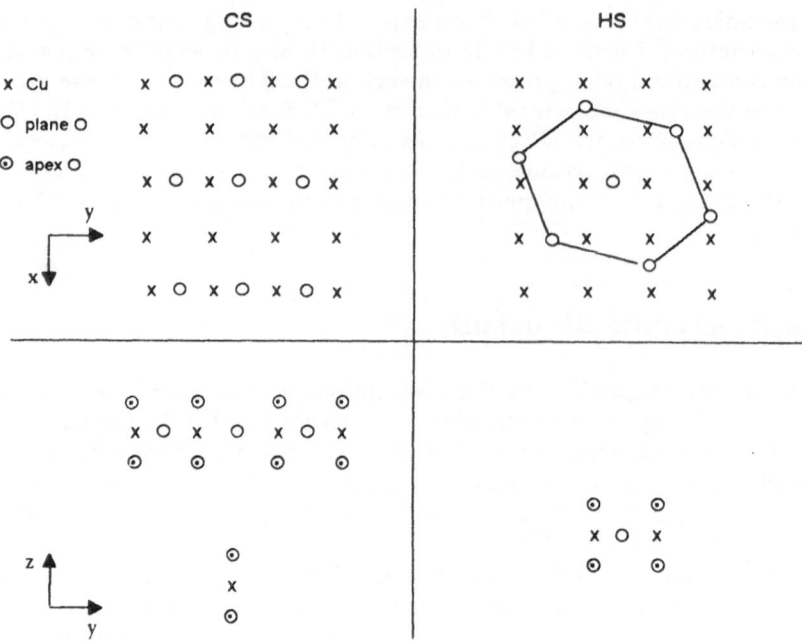

Figure 1. Top: top view of the CuO₂ planes for ordering in infinite CuO chains (left) and for the superstructure which minimizes the Coulomb repulsion energy between equally charged O atoms (right) for $x = 1/2$. Bottom: side view of the CuO₂ planes showing the apex O atoms and the atomic structures which are electronically "disconnected" from the rest of the system.

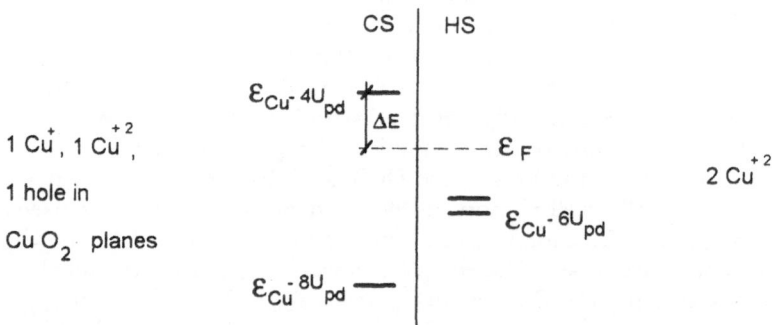

Figure 2. Energy labels of the structure shoen in Fig. 1 in the limit of vanishing hopping energies $t_{pd} = t_{pp} = 0$. The dashed line is a possible position of the Fermi level ϵ_F of the CuO₂ planes. The corresponding charge distribution is also indicated.

Y is Y^{+3}, Ba is Ba^{+2} and then, if all Cu ions are Cu^{+2}, the O ions should be O^{-2} for $x = 0.5$). Instead, in the more realistic situation in which $\epsilon_{Cu} - 6U_{pd} < \epsilon_F < \epsilon_{Cu} - 4U_{pd}$, a transfer of holes from the two-fold coordinated Cu ions (left as Cu^{+}) to the superconducting CuO₂ planes takes place only for the CS. At the same time, and as consequence of the charge transfer, only the CS gain an energy ΔE per Cu. Thus, although the HS minimize the long-range O–O repulsion, the CS can be the ground state as experimentally observed.[25-28] This picture can also explain the increase of superconductivity with annealing, since the CS is favored by this process,[6,7] and the superconducting critical temperature T_c increases with the number of holes h in each CuO_2 plane for $h < 0.2$.[29] In addition, the observed composition dependence of the amount of Cu^+, which varies as $1 - 2x$ for $x < 0.2$ and as $1 - x$ for $x > 0.4$,[30-32] is consistent with the above picture if a transition from the HS to the CS takes place for $x \sim 0.3 + 0.1$.

However, the above picture (presented first in Ref. 3) is only qualitatively valid. It cannot explain the plateau in h[13,14] and T_c.[14,33,34] For example, for $x = 0.5$ it predicts $h = 0.25$, while experimentally $h \sim 0.11$.[13,14] Thus, the analysis of the resistivity based on similar pictures[35] cannot be quantitatively valid.[36]

For a quantitative calculation of the electronic energy, we have started from a three-band Hubbard model[37,38] for the CuO_2 planes. This model contains $3d_{x^2-y^2}$ orbitals at the Cu sites and $2p_\sigma$ orbitals at the O sites (the subscript σ means that the orbitals point towards their NN Cu sites). For the CuO_{2+x} subsystem, we have to extend the model to take into account adequately the variable O content (redefinition of the vacuum state[2]). The appropriate extension is written in Eqs. (1) and (3) to (6) of Ref. 4 and will not be reproduced here.

As it is clear from Fig. 1, the CuO_{2+x} subsystem consists of isolated CuO_2 3-atom clusters (without NN hopping to other atoms), isolated Cu_2O_5 7-atom clusters, and infinite CuO_3 chains. The finite size clusters are diagonalized exactly. The Hamiltonian of the infinite chains and that of the CuO_2 planes is mapped at low energies into a generalized one-band Hubbard model[39] using a cell-perturbation method. This method allows us to treat exactly the highest energy in the problem, namely U_d. As a consequence, the antiferromagnetic Hartree-Fock approximation (AFHFA) near half filling[40] or the slave-boson mean-field approximation (SBMFA)[39] are good approximations for the total energy of the system. As an example, a recently exactly solved extended Hubbard chain displays a Mott transition at $U/t = 4$,[41] while the SBMFA gives $U/t = 16/\pi = 5.08$[39] and this parameter is more sensitive than the energy. For convenience, we have used the AFHFA for the CuO_3 chains[5] and the SBMFA for the CuO_2 planes. From the energy of the latter, using Thomas-Fermi theory of screening for an isotropic medium,[42] the value $h = 0.12$ for $0.5 < x < 0.89$[13] and the size of the unit cell,[1] we obtain the following value of the screening length (details are given in Refs. 4 and 12)

$$\lambda = 1.68 \text{ Å} = 0.62a_0 \quad \text{for} \quad 0.5 < x < 0.89, \tag{1}$$

where $a_0 = a/2$ is the minimum possible O–O distance. Since the isotropic Thomas-Fermi theory is a crude approximation, the resulting value of λ is only indicative and we believe that it is a lower bound.

As far as the electronic structure is concerned, the structures composed of perfect Cu–O chains in the CuO_x planes are composed of a fraction x of CuO_3 chains per Y atom, and $1 - x$ of CuO_2 3-atom clusters. Instead, it is easy to see that for structures composed of isolated O atoms in the CuO_x planes, like the deformed hexagonal structures HS, for $x < 0.5$, a fraction $2x$ of the Cu atoms of the CuO_x planes lie in 7-atom Cu_2O_5 clusters and the rest are in 3-atom CuO_2 clusters. As we shall see in section IV, for the HS and compositions $1/2 < x < 2/3$ (and for simplicity we extrapolate this behavior up to $x = 1$), the respective fractions of infinite chains, 7-atom and 3-atom clusters are $2x - 1$, $2 - 2x$ and zero, respectively. Taking into account this information the problem of calculating the electronic structure of $YBa_2Cu_3O_{6+x}$ for the CS or the

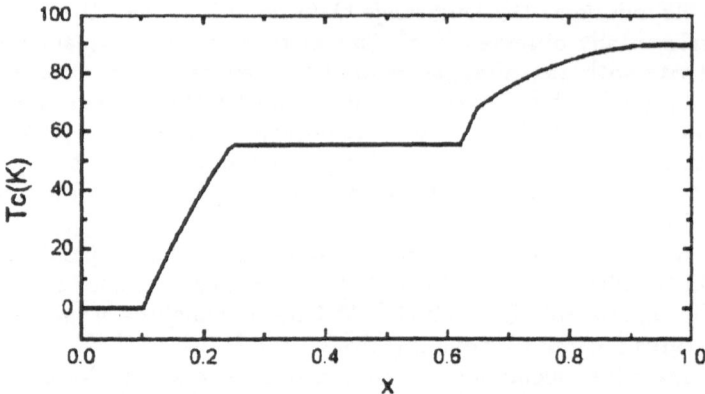

Figure 3. T_c as a function of O content for the CS.

HS reduces to distribute $2x$ holes (each neutral O atom added to the system brings two holes) into the CuO_2 planes, and the corresponding fractions of CuO_3 chains, 3-atom and 7-atom clusters, in such a way as to minimize the total energy. To simplify the minimization, we took the form used in Ref. 43 for the dependence of the energy of the CuO_2 planes with the number of holes h. This form has been derived from a high-temperature expansion of a realistic t–J model. The extended t–J model also describes accurately the low-energy physics of the three-band Hubbard model.[44] The on-site energy for the CuO_2 planes was shifted in such a way that $h = 0.25$ for $x = 1$ as experimentally reported.[15] The parameters for the CuO_{2+x} subsystem were assumed similar to those derived for La_2CuO_4:[11] $U_d = 10$ eV, $U_p = 5$ eV, $\Delta = 3$ eV (Cu–O charge-transfer energy for $x = 1$), $t_{ch} = 1.5$ eV (Cu–O hopping along the chains), Cu apex O hopping $t_{ap} = 1.3t_{ch}$,[4] $U_{pd} \sim 1.5$ eV, and the apex on-site energy was chosen so that when all Cu ions are +2, it costs $\Delta_{ap} \sim 1$ eV more energy to add a hole on an apex O^{-2} ion than in an O^{-2} ion of the CuO_x plane. Since the values of U_{pd} and Δ_{ap} are not well established, we have chosen three different pairs of values which can explain the observed T_c *vs.* x: (a) $\Delta_{ap} = 0$, $U_{pd} = 1.7$ eV, (b) $\Delta_{ap} = 0.8$ eV, $U_{pd} = 1.55$ eV and (c) $\Delta_{ap} = 1.5$ eV, $U_{pd} = 1.45$ eV. For the three pairs, the resulting functions $h(x)$ are very similar. Assuming for the function $T_c(h)$ the form of an inverted parabola with maximum 90 K at $h = 0.23$ and passing through $T_c(0.1) = 0$,[29] we obtain the $T_c(x)$ for the CS shown in Fig. 3 for parameters (b) (for (a) and (c) the curve is practically the same).

The shape of the curve can be explained at follows: for small x, holes enter only in the CuO_2 planes. Thus, h and ϵ_F grow until, for $x \sim 0.2$, the latter reaches the energy necessary to put the first hole in a 3-atom CuO_2 clusters. For larger x, ϵ_F remains pinned at this value until, for $x \sim 0.6$, all CuO_2 clusters $(1 - x$ per Y atom) have already one hole. Then ϵ_F, h and T_c start to increase again. For $x \sim 0.7$ the infinite CuO_3 chains begin to participate in the sharing of holes, and the increase in ϵ_F, h and T_c is slower. For the HS the behavior of T_c is qualitatively similar, but $h = T_c = 0$ for $x < 0.5$ since all holes go into 7-atom Cu_2O_5 clusters. For $x > 0.4$, Fig. 3 is in qualitative agreement with experiment.[14,33,34] For smaller values of x, experimentally $T_c = 0$, which is probably due to a transition to the HS, as will be discussed later.

Note that this explanation of $T_c(x)$, in contrast to alternative approaches,[21,22] works for a perfect system in which the CuO_3 chains are infinite, and does not require the presence of defects in them. Our result for T_c is not very sensitive to the presence

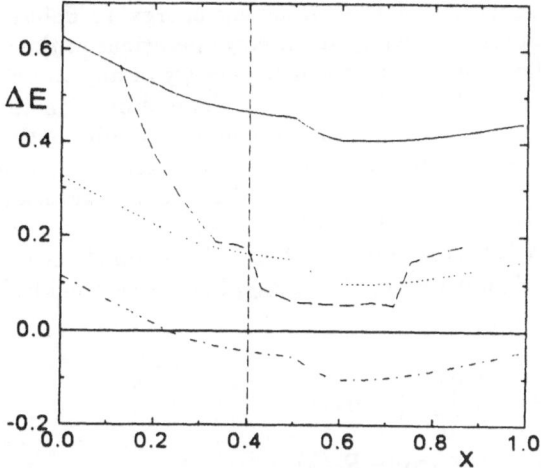

Figure 4. Full lines: ΔE as a function of O content for (from top to bottom) $\Delta_{ap} = 0$, 0.8 and 1.5 eV. Other parameters are indicated in the text. The dashed line is the boundary between HS and CS calculated with the structural model.

or abscence of a few percent of O vacancies in the CuO$_3$ chains. In contrast, the low frecuency optical conductivity, $\sigma(\omega)$,[15] of the chains is strongly affected by these vacancies. For $\omega < 14$ eV, $\sigma(\omega)$ is very well approximated by 0.28 (eV/ω)^2 exp(-0.35 eV/ω), which results from our model assuming the presence of 6.3% O vacancies.[45]

For the structural model explained in the next section, it is important to calculate the quantity ΔE defined as the difference of energy between the HS and the CS divided by the difference in second NN O atoms with a Cu in between. ΔE is shown in Fig. 4 as a function of x, for the set of parameters (a), (b), and (c) given above. While ΔE vs. x is rather flat, ΔE decreases markedly with increasing on-site energy of the apex O ions. Note that in spite of the argument given at the beginning of this section, ΔE is predominantly negative for large values of Δ_{ap}. This is because (as it has been shown[12] in an example comparing the energy of a system composed of a CuO$_2$ cluster and a CuO$_4$ cluster with another of two CuO$_3$ clusters) covalency, for large U_d, favors three-fold coordination. It has also been rigorously shown that if the apex O are all O^{-2} (large Δ_{ap}) and in absence of charge transfer (in the insulating system) $\Delta E < 0$.[46] It is the possibility of transfering a hole from a two-fold coordinated Cu^{+2} (leaving it as Cu$^+$) to the superconducting CuO$_2$ planes or the apex O atoms, driven by U_{pd}, which can give rise to a positive ΔE and stabilize the CS. Even for positive ΔE, the CS are not always stable, since the electronic energy so far calculated has not included the O–O repulsions. The effect of the latter and its competition with a positive ΔE is the subject of the rest of this work.

III. The Structural Model for CuO$_x$ Planes

As already mentioned in section I, the average distance between two NN carriers in YBa$_2$Cu$_3$O$_{6+x}$ is always larger than 7 Å. Even for the O content x optimum for superconductivity, the system is a bad metal in the normal phase. This suggests starting from a calculation of the Madelung energy to determine the O ordering. Assuming that all atoms related by symmetry operations in the tetragonal phase have the same charge, the difference between the energy of any two structures which differ only in the O ordering of the CuO_z planes, depend only on the repulsions between O atoms of these planes. Taking into account in addition the dielectric screening (atomic displacements and polarizations around O ions), electronic screening (in the isotropic Thomas-Fermi approximation), and the charge-transfer effects due to Cu–O NN repulsion U_{pd} explained in the previous section, leads to the model proposed in Ref. 2: any two i-th NN O ions of the CuO_z planes repel each other with a screened Coulomb repulsion V_i, but the repulsion V_{2Cu} between second NN O atoms in between is reduced by ΔE

$$H = \frac{1}{2} \sum_{ij} V_i n_j n_{j+i}, \tag{2}$$

$$V_i = V \frac{a_0 \exp(-R_i/\lambda)}{R_i}, \quad \text{but} \quad V_{2Cu} = V_2 - \Delta E, \tag{3}$$

$$V = \frac{q^2}{\epsilon a_0}, \tag{4}$$

Here $n_j = 0, 1$ is the O occupation of site j. The i-th NN's of site j are labeled by $j + i$, R_i is the distance between any two i-th NN. $a_0 = 2.76$ Å $= a/\sqrt{2}$ is the lattice parameter of the sublattice, assumed square, of all O positions. λ is the screening length, $q = 1.7$ is the charge of the O ions[24] and we use $\epsilon = 14.7$.[24] This leads to $V = 1.025$ eV.

The model is justified in more detail in Refs. 3, 24 and 48. However, it is an oversimplification of the actual physics in the system and the conclusions derived from it can only have qualitative validity. A critique is given in Ref. 4. In spite of this, we will show that the model not only shows interesting physics, but is able to explain most of the structural data for reasonable values of ΔE (in the interval $(0.03, 0.18)$ eV) and λ (larger than $4a_0$ in the semiconducting phase $x < 0.4$ and ~ 1 in the metallic phase $x > 0.5$). These values imply that at least for $x < 0.4$, V_{2Cu} is positive (repulsive). The sign of V_{2Cu} has been a matter of controversy (Refs. 12, 46 and references therein). However it has been rigorously proven that in the semiconducting phase $V_{2Cu} > 0$ if the apex O ions are (as the experience suggests) nearly O^{-2} (Ref. 46).

IV. Ground State of the Atomic Structure

For models with a small number of short-range interactions, the method of geometrical inequalities of Kaburagi and Kanamori[49] or the cluster method of Allen and Cahn[50] are in general useful for finding the ground state of Ising models. However, they are not suitable for our model (Eqs. (2) to (4)) due to the nature and large number of the relevant interactions. An example of a failure of the second method will be given in section V.1. Monte Carlo calculations were useful for example, to explain the kinetics of structural transformations in $YBa_2Cu_3O_{6+x}$ using our model supplemented with long-range elastic energies.[48,51] However, Monte Carlo calculations present difficulties at low temperatures: the "simulated annealing" procedure is characterized by extremely slow kinetics and complete ordering is almost never obtained.[52–54]

In view of the above mentioned difficulties, we have directly compared the energy of a large number of superstructures, which include all those proposed on the basis of

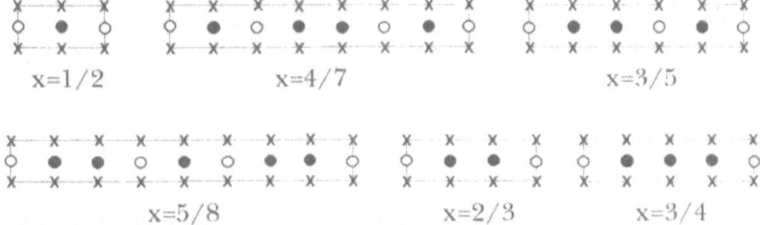

$x=1/2$ $x=4/7$ $x=3/5$

$x=5/8$ $x=2/3$ $x=3/4$

Figure 5. Examples of the ground state of the model for $f < 0$ ($V_{2Cu} < 0$) and different compositions. Crosses denote Cu atoms, solid circles represent O atoms, and open circles denote vacant sites, which if filled, complete the structure for $x = 1$. The corresponding CS for compositions $1 - x$ are obtained interchanging solid and open circles.

experimental information. The set of calculated structures include: (a) all structures whose unit cell or a multiple of it is a rectangle of largest side $< 3a$ and area $< 8a^2$, (b) all structures with unit cell $1 \times n$ (which we call CS) for $n \leq 16$, (c) all simple lattices of O ions with $x > 1/8$, and (d) all structures with $1/2 < x < 7/8$ formed adding simple lattices of O vacancies to the structure of $x = 1$ (called OI^1). To obtain one and only one representative of all structures equivalent by symmetry, we used an extension of a group-theoretical method developed by H. Bonadeo,[55] described briefly in Ref. 3. We cannot guarantee that the ground state does not correspond to a structure of larger unit cell than those studied. This is in fact the case for $x = 3/8$, as will be discussed in detail in section V.2. However, as it becomes clear below, this study is enough to determine the general features of the ground state, and only minor deviations from it are expected.

For simplicity, we begin our discussion assuming first that λ and ΔE (see Eqs. (2) and (3)) are independent of O content x. We also define for convenience the ratio

$$f = \frac{V_{2Cu}}{V_2} = 1 - \frac{\Delta E}{V_2}. \tag{5}$$

The general features of the ground state (GS) can be summarized as follows.[2,3,24] For $f < 0$ (attractive V_{2Cu}), and any x, the GS consists of perfect infinite CuO chains (CS). The problem of the order in the direction perpendicular to the chains takes the form of a one-dimensional Ising model with interactions W_n satisfying $2W_n < W_{n-1} + W_{n+1}$ for $n > 2$. This problem has been solved exactly.[56,57] In the context of YBa₂Cu₃O₆₊ₓ, the relevant CS structures with $x \neq 1/2$ were proposed first in Ref. 2. Some examples are given in Fig. 5.

In the symmetric case ($f = 1$, $V_{2Cu} = V_2$) of uncorrected screened Coulomb repulsions, for all compositions of the form $x = 2/n$ with $n > 2$ integer, the GS consists of a simple lattice of O ions. The unit vectors of these lattices are given in Ref. 3. They correspond to the smallest possible deformation (imposed by the O sublattice) of a perfect simple hexagonal lattice with the same O concentration. We call these superstructures HS. For $x = 1 - 1/n$, for $n > 3$ integer, the GS consist of as simple lattice of O vacancies added to the $x = 1$ OI structure. The simple lattices of additional vacancies for $x = 1 - 1/n$, and O atoms for $x = 2/n$, have the same form (the former rotated 45° and expanded by a factor $\sqrt{2}$) as demonstrated in previous ground-state studies.[3,24] Examples of these structures are shown in Fig. 6. They can be constructed minimizing the number of first NN O atoms, then the number of second NN without increasing the number of first NN, and so on.

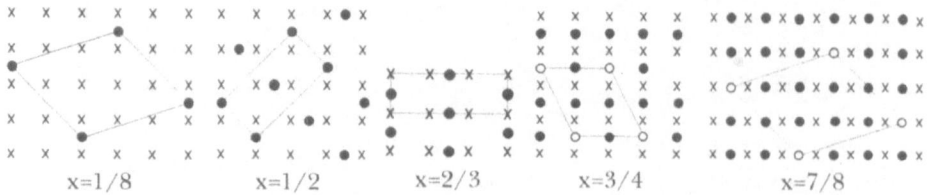

Figure 6. Examples of the GS of the model for $f = 1$ ($\Delta E = 0$) and different compositions. The meaning of the different symbols is the same as in Fig. 5. We call these structures HS.

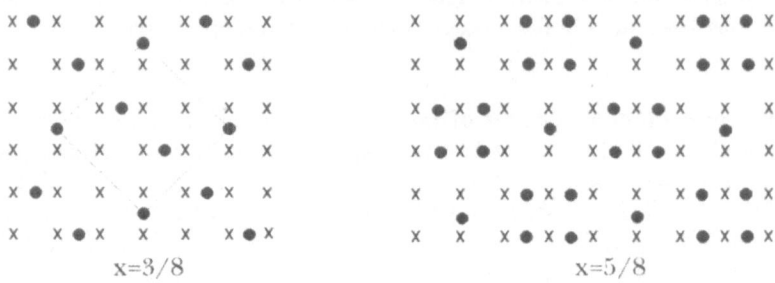

Figure 7. Structures of minimum energy among all those studied (see text) for a bare Coulomb repulsion between ant two ions (model Eqs. (2)–(4) for $\Delta E = 0$ and large λ), for $|x - 1/2| = 1/8$.

In general, for small screening length ($\lambda < a_0$), and $f = 1$, only the above mentioned HS with compositions $x = 2/n$ or $x = 1 - 1/n$ are stable. For other compositions there is phase separation. However, for large values of λ (as expected in the semiconducting phase), other GS structures appear. For simplicity we extend the denomination "HS" also to these GS structures or slightly excited structures for $f = 1$. In particular among the set of studied structures we find that those of minimum energy for $x = 3/8$ and $x = 5/8$ have the form shown in Fig. 7. The latter structure describes the hole-doped O atoms of CuO_2 planes in the GS of the electronic structure of the three-band Hubbard model, in the "atomic limit" (zero hopping), for particular parameters,[48] but is not relevant for O ordering and will not be discussed here any further. The HS obtained for $x = 3/8$, although fully compatible with x-ray experiments at a similar compositions,[59] is actually a slightly excited structure (it has too many third NN), as discussed in more detail in section V.2.

While all the CS are characterized by the fact that each O atom has two second NN O atoms with a Cu in between, the HS with $x < 1/2$ have no second NN O atoms, and the HS with $x > 3/4$ have no second NN additional vacancies added to the $x = 1$ structure. While for $f < 0$ the GS is a CS and for $f = 1$ the GS is an HS, for a small range of intermediate values of f (which depends on x), the GS structure displays short O–Cu–O– $\cdots$ –O chains, in most cases involving only two second NN O atoms, or similar structures of vacancies added to the $x = 1$ structure (see Fig. 8). We call these structures PS.

At constant screening length λ, the (x, f) phase diagram has the form shown in Fig. 9 for two values of λ. The regions $0 < x < 1/8$ and $7/8 < x < 1$ are empty because we have not considered PS and HS in these regions. More details are given in Ref. 3.

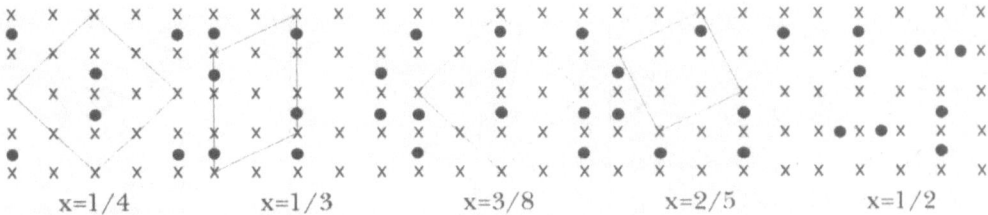

x=1/4   x=1/3   x=3/8   x=2/5   x=1/2

Figure 8. Structures denoted by PS in the text, which are the GS for particular parameters. Symbols as in Fig. 5. By interchanging solid and open circles the corresponding structures for compositions $1 - x$ are obtained.

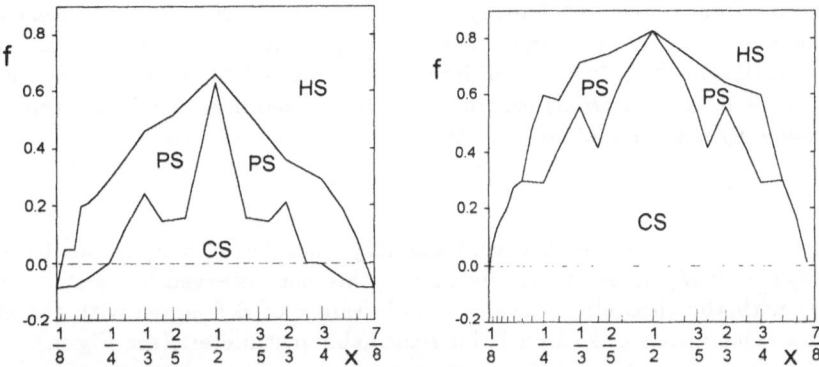

Figure 9. Ground state of the system as a function of composition x and reduction factor $f = V_{2Cu}/V$ for $\lambda = a_0$ (left) and $\lambda = 4a_0$ (right).

We can see from Fig. 9 that if $f \sim -1$, for all x, as in the ASYNNNI model,[8] only CS are present in the ground state. The richest situation is for small positive f. This case is particularly interesting, because, as explained before,[2,3] it predicts HS for small or large values of x, and CS for x near 1/2, as experimentally observed[18,25−28,34,60−62] (there is a criticism to some of this experimental results, which is addressed below). However, while Fig. 9 gives a picture of the expected ground states, it is not realistic to take constant values of λ and f. The screening length λ should be much larger in the semiconducting phase $x < 0.4$.[14,65] This and Eqs. (3) and (5) imply that f is larger in this phase, since ΔE, according to the results of section II, does not depend very much on x. On the other hand, while Fig. 9 is rather symmetric around $x = 1/2$, the experimental evidence[18,26] favors the CS predicted in Ref. 2 for $x = 5/8$ (see Fig. 5), x-ray[59] and neutron[63,64] experiments for $x = 3/8$ favor HS. In addition, the PS, which were not observed experimentally, probably appear in the GS for certain parameters due to the arbitrary assignment of the whole correction energy ΔE, to the interaction V_{2Cu}, and are not important.

Motivated by the above discussion, we have calculated[4,12] the phase diagram including only CS and HS, taking ΔE instead of f as a parameter, and using the value of the screening length $\lambda = 1.68$ Å calculated in section II for $1/2 < x < 7/8$ and $\lambda = 10a_0$ in the semiconducting region $x < 3/7$ (the results are insensitive to λ for $\lambda > 4a_0$.[24]). The result, is shown in Fig. 10, together with an indication of the experimental results. We see that for $\Delta E \sim 0.05V \sim 0.05$ eV, all the diffraction

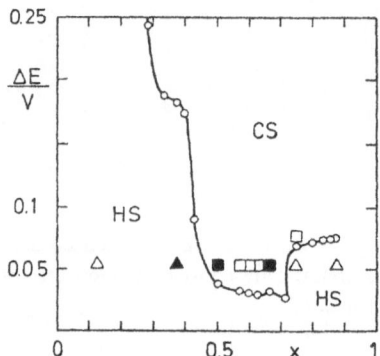

Figure 10. Ground state of $YBa_2Cu_3O_{6+x}$ as function of ΔE and x. For each composition corresponding to a circle and larger values of the ground state is a CS. Squares (triangles) denote experiments[18,25–28,34,59–63] which support the CS (HS) for given composition. Open symbols refer to (controversial) electron-diffraction experiments, and solid symbols denote confirmation by neutron and x-ray measurements.

experiments, except one of the contradictory ones for $x = 3/4$, can be explained. For $\Delta E/V < 0.18$, at most two diffraction patterns observed by electron diffraction disagree with the theoretical results. These values of ΔE agree with the electronic-structure calculations of section II for reasonable parameters (see Fig. 4).

Also, a transition from an insulator HS to a superconducting CS from $x \sim 0.45$ as x increases, is consistent with the electronic structure calculations and experimental results on resistivity,[14,66] composition dependence of the amount of Cu^+,[30,31] Raman spectra,[67] and nuclear quadrupole resonance.[68] In addition, the photodoping experiments for $x \sim 0.4$[69,70] can be interpreted as a light induced transition from HS to CS, and return to HS after the illumination ceases.[4,12,36]

As we have explained above, the agreement between theoretical (for some values of ΔE) and experimental results is encouraging. However, there is some controversy about both results that we should address. The theoretical discussion was concentrated on the sign of V_{2Cu} (Refs. 12, 46 and references therein). A positive V_{2Cu} is required to obtain HS in the GS, while for negative V_{2Cu} only CS exist in the GS for any x. It has been shown recently that in the semiconducting phase, the CS have larger energy than the HS, if the hole occupation in apex O atoms can be neglected.[46] On the other hand, experimental results obtained by electron diffraction might correspond to metastable states due to the local heating caused by the electron beam.[26] Also, very recent electron diffraction measurements are interpreted as showing that the CuO_2 planes have a distortion which gives rise to a $2\sqrt{2} \times 2\sqrt{2}$ superstructure.[71] The authors suggest that previous observations of unit cell multiple of $2\sqrt{2} \times 2\sqrt{2}$, which are compatible with order in HS in the CuO_x planes, are in fact due to the distortion in the CuO_2 planes and not to O ordering in the CuO_x planes. However, it is very unlikely that a detailed explanation of all the intensities of the numerous reflections investigated by neutron diffraction for $x \sim 3/8$[63] can be given in terms of distortions in the CuO_2 planes. Instead, as we show in section V.2, all these intensities are in excellent agreement with O ordering in an HS.[64]

To end this section, we would like to discuss an issue, which although probably academic for $YBa_2Cu_3O_{6+x}$,[72] is of general interest. At zero temperature, the third

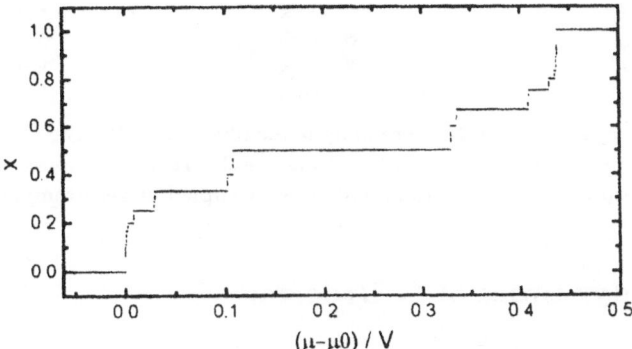

Figure 11. O composition as a function of the O atomic chemical potential for $\lambda = a_0$ and ΔE sufficiently large to stabilize the CS.

principle of thermodynamics imposes that only definite compounds (occupation zero or one at each site) exist. Let us assume that the compositions or these compounds x_i are ordered in such a way that $x_1 < x_2 < x_3 < \cdots < x_N$. Between any two consecutive compositions there should be phase separation. However phase separation in a model like ours (without movement of cations) implies violation of charge neutrality and is inhibited by Coulomb interactions, at least for infinite screening length λ. This suggests that the number of different ground state structures N should be infinite and that there is a stable ground state for each composition. This is the case of the one-dimensional (1D) Ising model with interactions W_n at distance na satisfying $2W_n < W_{n-1} + W_{n+1}$, for $n > 1$.[56,57] Our model for $f \sim -1$ (or values realistic for $YBa_2Cu_3O_{6+x}$ in the range $0.5 < x < 3/4$ or larger) reduces to this 1D model. Our model for $f \sim -1$ has also been studied by Adelman *et al.*[73] at finite temperatures. These authors have shown that the O composition x as a function of O chemical potential μ has a form which suggests a complete devil's staircase. Other studies at zero temperature[5,72] also support this statement. In Fig. 11 we show the function $x(\mu)$ for $\lambda = a_0$, assuming the CS are stable. This assumption is probably valid only in a limited range of compositions. Although a finite number of different compositions are represented in the figure (81 due to our restriction that the unit cell is shorter than $17a$), the behavior of the curve when all compositions are allowed is easy to imagine. Since we find that for all superstructures the most intense reflections correspond to wave vectors $(2\pi/a)x$ and $(2\pi/a)(1-x)$, Fig. 11 also represents the main wave vector of the structure as a function of μ.

For $\lambda < a_0$, the CS with $x \neq 0$ and $x \neq 0$, which are stable over a not negligible interval of μ, are those with $x = 1/n$ with $n < 5$, $x = 2/3$, and those obtained from the previous one by replacing x by $1 - x$. The CS with $x = 5/8$, which agrees with an observed diffraction pattern,[18] has a tiny stability range and disappears together with the devil's staircase behavior when a phenomenological term is added to the free energy to take into account volume relaxation.[72] The relaxation of a free variable has the effect of an effective attractive interaction and only a few CS are stable at zero temperature for any finite λ. The above mentioned diffraction patterns are probably due to disorder perpendicular to the chains, of thermal or other origin,[17,20] as explained in section V.3. This also explains naturally the width of the peaks.

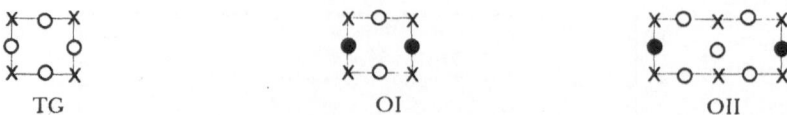

TG   OI   OII

Figure 12. Unit cells of the stable thermodynamic phases of $YBa_2Cu_3O_{6+x}$ at intermediate or high temperatures. Crosses denote Cu atoms, circles represent possible O positions and solid circles correspond to the positions which are occupied at zero temperature.

V. Thermodynamics of O Ordering

In the calculations of the previous section we have taken into account the interactions between any two O ions at all distances (in practice the interactions at distances larger than several times the screening length λ have a negligible contribution to the total energy). This is not possible in the CVM calculations of the energy. The number of interactions that can be taken into account is limited by the size of the basic cluster. We have chosen a square cluster containing nine O positions.[2,74,75] This 3×3 point CVM approximation, introduced by us several years ago,[2] was the most accurate used in the problem until very recently, when a 12-point approximation was used.[76] The 3×3 point approximation was also used by Morán-López and Sanchez,[77] but these authors have not studied the effect of interactions beyond second NN. The number of non-equivalent interactions that are included in the cluster is 8 (6 in our model: V_i, $i = 1, 2, \ldots, 5$ and V_{2Cu}). The diagonal of the cube is $2a$. As has been mentioned in the introduction, and it will become clear in part V.3, interactions at distance $2a$ are essential to explain observed split diffuse diffractions peaks.[16,20] The 4-and 5-point approximation of the CVM[8] can not take these interactions into account.

In spite of the relatively large size of the 3×3 point cluster, the fact that the interactions beyond $2a$ (fifth NN) are cut off reduces considerably the number of different structures present in the ground state (GS). Moreover, the cut-off procedure can lead to spurious GS with too large a 'number of O atoms at distances just above the cut-off distance. This effect is discussed in more detail elsewhere.[58,75] Taking into account these shortcomings, we have repeated the GS analysis of the previous section and this allowed us to identify a *region of failure* of the 3×3 point approximation for $1/3 < x < 1/2$ and low temperatures. This is due to the fact that at very low temperatures, the CVM works as a method of finding the ground state similarly to the cluster method of Allen and Cahn,[50] and this method can lead to impossible structures, with less energy than the true GS.[50,74] This will be explained in more detail in subsection V.1.

Due to the above mentioned limitations, the 3×3 point calculations presented in subsection V.1 are not realistic for low temperatures (below ~ 300 K). Since to simplify the calculations, we assumed infinite NN repulsion V_1, the calculations are also quantitatively incorrect at high temperatures (above ~ 700 K). In the intermediate temperature region, there are predominantly three phases observed in $YBa_2Cu_3O_{6+x}$: the tetragonal TG, orthorhombic OI and double cell orthorhombic OII, which correspond to ideal compositions 0, 1 and 1/2 respectively. The unit cell of these phases are represented in Fig. 12.

In subsection V.2 we consider the special case $x = 3/8$, for which the 3×3 point approximation fails, and the results are used to explain neutron-diffraction experiments.

In subsection V.3 we consider the region of compositions (or parameters) for which the CS are stable. This allows to map the problem into a one-dimensional Ising model, and we retain only interactions up to a distance $2a$. The neutron scattering of this model is calculated using generating functions, and agrees very well with observed split diffuse electron diffraction peaks.

V.1 The Phase Diagram at Intermediate Temperatures

Here we calculate the phase diagram using the 3×3 point approximation of the CVM, taking into account only the phases TG, OI and OII. The unit cell of each of these phases is represented in Fig. 12.

The 3×3 point approximation has been described in detail in Ref. 75. In the pictorial representation,[78] the entropy $S = k_B \ln \Omega$ of the OII phase can be expressed as

$$\Omega = \frac{\left\{ \begin{smallmatrix} \bullet & \circ & \circ \\ & \times & \\ \circ & \bullet & \circ \end{smallmatrix} \right\}^4 \left\{ \begin{smallmatrix} \circ & \circ & \bullet \\ & \times & \\ \circ & \circ & \circ \end{smallmatrix} \right\}^4}{\left\{ \begin{smallmatrix} \bullet & \circ & \circ \\ \times & & \\ \circ & \bullet & \circ \\ & \times & \\ \circ & \circ & \bullet \end{smallmatrix} \right\} \left\{ \begin{smallmatrix} \circ & \circ & \bullet \\ \times & & \\ \circ & \circ & \circ \\ & \times & \\ \bullet & \circ & \circ \end{smallmatrix} \right\} \left\{ \begin{smallmatrix} \circ & \bullet & \circ \\ & \times & \\ \circ & \circ & \circ \end{smallmatrix} \right\}^2 \left\{ \begin{smallmatrix} \bullet & \circ \\ \times & \\ \circ & \bullet \end{smallmatrix} \right\} \left\{ \begin{smallmatrix} \circ & \circ \\ \times & \\ \circ & \circ \end{smallmatrix} \right\} \left\{ \begin{smallmatrix} \circ & \circ \\ \bullet & \circ \end{smallmatrix} \right\}^2} \tag{6}$$

The entropy of the other phases can be easily derived from this expression.

The mathematical algorithm used is a modification of Kikuchi's natural Iteration Method (NIM).[79] Our basic variables are the probabilities p_i of configurations i of the basic cluster and subclusters. The configurations with NN O atoms, which only enter at high temperatures, were eliminated from the beginning. We had to replace the "minor iterations" of the NIM[79] (which did not converge in our case) by numerical subroutines to solve the equations for the p_i imposed by symmetry.[74] This method allows us to reach very low temperatures, including $T = 0$ (except for particular values of the parameters). Comparing the energy and the resulting cluster configurations at $T = 0$ with our ground state analysis, for $x \sim 0.4$, the failure of the approximation is put in evidence. To illustrate the origin of this failure, let us assume that $f < 0$ (attractive V_{2Cu}). Then, the ground state for $x = 1/3$ is a CS with every third CuO chain full of O and no O ions elsewhere, as shown in Fig. 13. This phase is called $\overline{OIII}$.

The CVM program in the tetragonal (TG) phase correctly gives only six basic cluster configurations, with nonvanishing probability $p_i = 1/6$ at $T = 0$ (see Fig. 6). The resulting CVM energy coincides with the ground state energy. The entropy is of course wrong, but this is due to the fact that the TG phase is not stable at $T = 0$. Now let us increase the chemical potential μ from its value at the GS for $x = 1/3$. The O occupation should increase. In the real system, for $x > 1/3$, at lest fifth NN O atoms (at a distance $2a$) are present in any low-energy structure. However in the CVM method for $1/3 < x < 0.4$, what happens is that the probability p_0 for the configuration without O atoms decreases and no new configurations appear. The situation is the worst for $x = 0.4$, where $p_0 = 0$. As a consequence the energy at $T = 0$

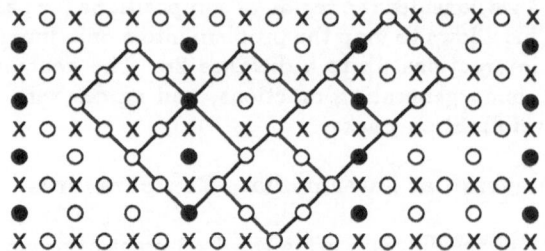

Figure 13. Ground state for $x = 1/3$, $f < 0$ and different configurations of the 3×3 basic cluster which persist at $T = 0$ in the CVM approximation for the tetragonal phase.

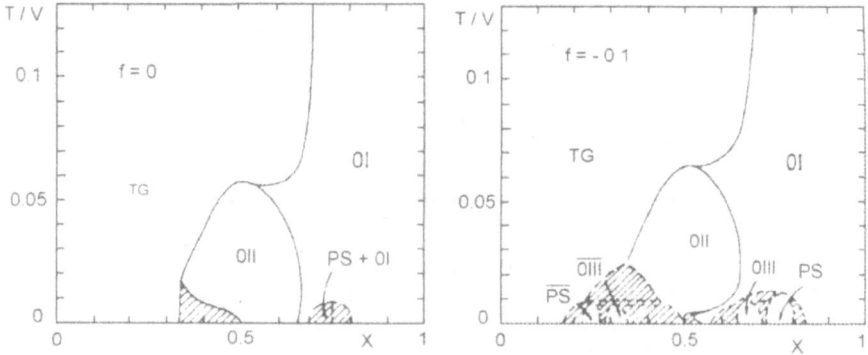

Figure 14. Phase diagram of the model Eqs. (2)–(5) in the 3×3 point approximation of the CVM, for $\lambda = a_0 = a/\sqrt{2}$ and two different values of f. The hatched areas represents coexistence of two phases. The PS are $2\sqrt{2} \times 2\sqrt{2}$ structures (see Fig. 8). The entropy of the PS was neglected at the left. The dashed line at the right is a possible modification of the phase diagram, based on our ground state analysis, when PS and OIII (1×3) phases are included.

of the TG phase is less than the true GS energy for $1/3 < x < 1/2$. Since the method chooses the phase of lowest energy and there is at least one phase with energy lower than the GS energy, the failure of the method is not particularly related with the TG phase.

The resulting phase diagram for our model with constant f and λ is shown in Fig. 14. Due to the cut off of interactions beyond fifth NN, the values of f should be reduced with respect to the corresponding ones when repulsions at arbitrary distances are included. In particular $f = -0.1$ in Fig. 14 corresponds to slightly positive f in the previous section.

Quantitatively, the phase diagram is similar to the corresponding one of the ASYNNNI model[8,80,81] (shown in Fig. 15), except for the absence of an orthorhombic phase $\overline{OI}$ at low x and T (not observed experimentally), and the character of the OII-TG transition. In our model this transition seems to be of second order (although we can not distinguish it from a first order transition with a very narrow two-phase field with $\Delta x \sim 0.01$), except in the region of failure of the approximation (low T and $1/3 < x < 1/2$). For a comparison of experimental data with theoretical results with the ASYNNNI, see Ref. 82. For $V \sim 1$ eV estimated in the previous section, the

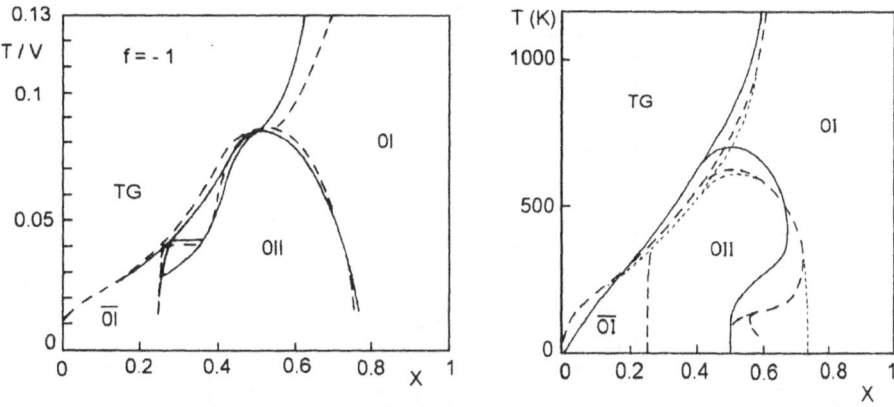

Figure 15. The x–T phase diagram. Left: for $V_1 = +\infty$, $V_2 = -V_{2Cu} = \exp\left(-\sqrt{2}\right)/\sqrt{2}$, and other $V_i = 0$. Dashed line corresponds to Ref. 80. Right: Dotted line: ASYNNNI model with first-principle parameters $V_1 = 0.375$ eV, $V_2 = 0.060$ eV and $V_{2Cu} = -0.131$ eV. Dashed line: effect of repulsions V_3, V_4 and V_5 for $\lambda = 0.2787a_0$. Full line: effect of V_3, V_4 and V_5 for $\lambda = 0.62a_0$. At the right, the TG–OII two phase field was replaced by the line of local instability of the OII phase.

transition temperatures shown in Fig. 14 are in good agreement with experiment. In view of the simplified character of the model, we do not pretend a quantitative explanation of the phase diagram. If a (more realistic) composition dependent screening length λ is considered (very large in the semiconducting phase $x < 0.4$), the TG-OII boundary is shifted towards $x = 0.5$, as shown in Ref. 83. In Fig. 15 we show results of the ASYNNNI model, which includes only V_1, V_2 and $V_{2Cu} < 0$, to display, for comparison, the effect of the repulsions beyond second NN. For simplicity we took infinite V_1, but this does not alter the OII-TG and OII-OI boundaries.[83]

At the left of Fig. 15, the first low-temperature results with the 4-and 5-point CVM approximation[80] are also shown for comparison. More accurate methods[82] suggest that there should be no first-order TG-OII transition and one might expect a significant reduction or elimination of the corresponding two-phase field when more accurate CVM approximations are used. We see that this is not the case of the 3×3 point approximations. However, there is some improvement and reduction of the transition temperatures, when the 12-point CVM approximation is used.[76] The transition temperature OII-TG for $x = 1/2$, T_0, for fixed λ, increases with decreasing $V_{2Cu} = fV_2$ (for example $T_0 = 0.032$ for $f = 0.4$, $\lambda = a_0$ and $T_0 = 0.059$ for $f = 0$, $\lambda = a_0$) and decreases when repulsions beyond second NN are cut off. From these two competing effects, the first one dominates in the change of parameters from Fig. 14 to Fig. 15 left. Fig. 15 right displays the second effect. In this figure, for simplicity and since the TG-OII transition is second order when more accurate methods are used,[82] we have calculated the transition using the Hessian matrix formed by the second derivatives of the thermodynamic potential with respect to independent cluster probabilities.[75]. Negative values of this matrix signal the local instability of the OII phase. The value $\lambda = 0.2787a_0$ was obtained assuming that the *ab initio* calculated V_1 and $V_2{}^9$ satisfy Eq. (3). The more realistic (but perhaps still small) value $\lambda = 0.62a_0$ was calculated in section II. We see that even small interactions beyond second NN affect significantly the phase diagram, particularly at low temperatures.

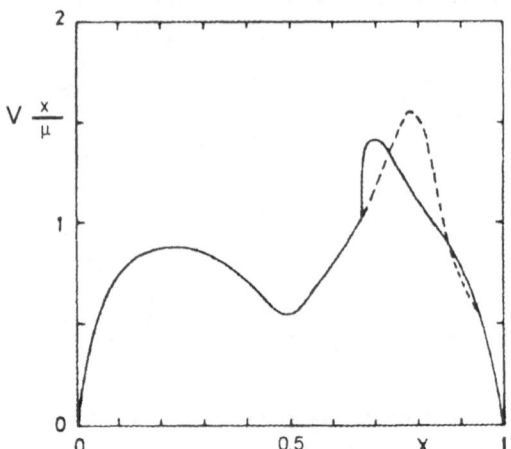

Figure 16. Derivative of the O content x with respect of the O chemical potential μ as a function of x for $\lambda = 1$, $f = 0$ and $T = 0.07V$. Dashed line corresponds to the metastable extension of the TG phase inside of IO phase.

In Refs. 74 and 75 different correlation functions were calculated. Here, we would like to present also our result for the experimentally measured[84,85] $\partial x/\partial \mu$ where μ is the O chemical potential. The result is shown in Fig. 16.

The structure around $x = 0.7$ is in qualitative agreement with experiment. However, according to Schleger, although our model agrees with experiment better than the ASYNNNI,[86] the results are not quantitatively valid.[86,87] Schleger, Hardy and Casalta have obtained a good quantitative agreement with the measured $\partial x/\partial \mu$, introducing in the model spin and charge degrees of freedom.[87] An alternative explanation of these degrees of freedom, which we believe is more realistic, can probably be given in terms of the holes in the isolated CuO_2 clusters mentioned in Section II.

V.2 Structures for x ~ 3/8 at Low Temperatures

The present study was motivated by detailed neutron scattering studies of Sonntag *et al.*[73] These authors measured all superstructure reflections with wave vectors $(2\pi/a)(h, k, 0)$ with $\theta < h$, $k < 7/4$ and scattering angle $15° < 2\theta < 90°$. The observed intensities I_{obs} are symmetric under interchange of h and k. The I_{obs}, together with the corresponding intensities for three different proposed structures A, B, E (with scale factor adjusted to reproduce the two most intense I_{obs}), are shown in Table I.

The symbol $-$ for I_F (I_{obs}) means that the corresponding intensity is zero (less than the statistical error 10). The asterisk in the fourth row means that I_F is reduced by a factor $1 - 2p$ defined below. The underlined intensities correspond to maximum possible intensity of the corresponding structure (the reflection from all atoms interfere constructively). The structures are represented in Fig. 17. This experimental information imposes severe constraints to the structures which try to explain it. For example, second NN interfere destructively for the two most intense observed peaks (2 and 14) and this rules out any stable or metastable structure predicted by the ASYNNNI model.[12,53,54] However, this fact suggest that structures which minimize the Coulomb repulsion between O ions, particularly taken into account that the sys-

Table I

Neutron scattering measurements[63] and calculated intensities for different superstructures.

number	1	2	3	4	5	6	7	8	9	10	11	12	13	14	15
$4h, 4k$	0,2	2,2	1,3	3,3	2,4	1,5	3,5	0,6	2,6	5,5	1,7	4,6	3,7	6,6	5,7
I_{obs}	40	420	50	–	–	–	–	–	30	–	–	–	–	150	–
I_F	–	420	41*	–	–	–	22*	–		21*	–	19*	–	150	16
I_B	65	420	41	31	29	26	22	22	191	19	19	18	18	150	16
I_C	65	420	41	31	29	235	22	22	21	19	19	18	163	150	16

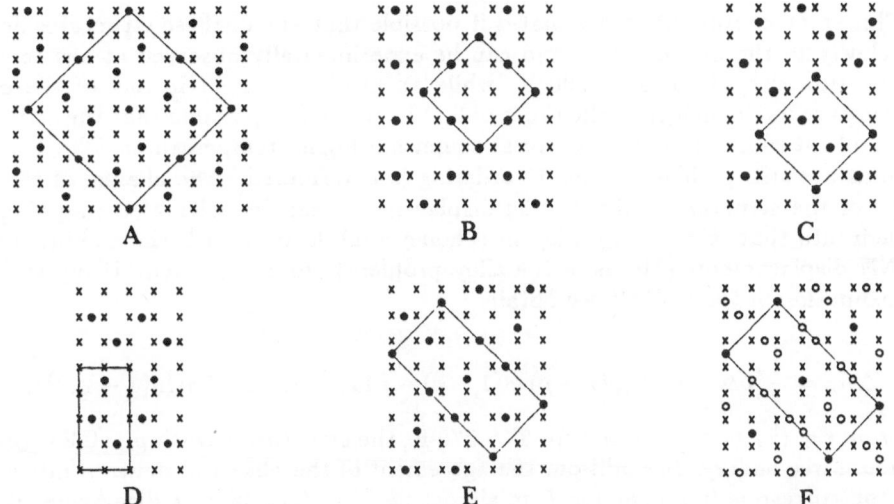

Figure 17. Six superstructures of the CuO_x planes which have nearly the ground-state energy of the model for symmetric screened Coulomb repulsions ($V_{2Cu} = V_2$). Crosses denote Cu atoms and open circles in F represent sites that are occupied by O atoms with probability p $(1 - p)$ if the corresponding site is empty (occupied) in E.

tem is semiconducting for $x \sim 3/8$, have a chance to explain the experiments. Several structures with very low Coulomb energy are represented in Fig. 17.

B is the structure proposed by the experimentalists,[63] C is the structure of minimum energy among all those of unit cell $2\sqrt{2} \times 2\sqrt{2}$ obtained in the previous section, and which best explains x-ray diffraction experiments.[59] A, and C to E were obtained "by hand" looking for structures which minimize successively the number of first, second and further NN. Finally F was constructed from E noting that certain atoms of E can be displaced with a small cost in energy Δ ($\Delta = 0.0299V$ for $\lambda = 10$), due to the fact that the total number of NN at distances less than $\sqrt{13} = 3.6$ is not altered. A is the structure of minimum energy. However, as shown in Fig. 18, the difference of energy between any two of these structures is very small, and saturates for large λ. Remember that $V \sim 1.025$ eV $= 12\,100$ K.

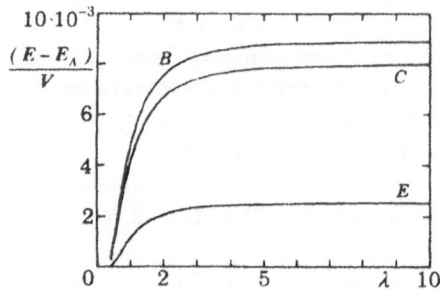

Figure 18. Difference between the energy per Cu of the CuO$_x$ planes of different super-structures shown in Fig. 17 and that of A as a function of screening length. E_D is not shown in the figure because $|E_E - E_D| < 0.0003V$.

This small energy difference makes it possible that any of these structures decays very slowly to the ground state and can be experimentally observed at low temper-atures. However, since $\Delta \sim 300$ K, while an O displacement in any of the other structures costs an energy of the order of 900 K, it is quite possible that the structure F is stable at room temperatures, or at somewhat higher temperatures, for which the O atoms are still mobile and the O ordering is determined. The change of the free energy of the structure F due to the displacements can be calculated mapping the problem into that of a binary alloy in a honeycomb lattice, with the condition that two NN displacements (atoms in the alloy problem) are not allowed. Using the pair approximation of the CVM[78] we obtain

$$\Delta F_F = \frac{1}{4}\left[p\Delta - k_B T\left(2(1-p)\ln(1-p) - p\ln p - (3/2 - 3p)\ln(1-2p)\right)\right]. \quad (7)$$

From Eq. (7) we obtain that for $T > 370$ K, the structure F with $p \sim 0.2$ is that of minimum free energy. In addition, the agreement of the observed neutron intensities with the corresponding ones for F is almost perfect. The largest disagreement (for the first peak) can be explained by the presence of 0.96% of the OII phase, which is frequently found in diffraction electron experiments for $0.3 < x < 0.7$.[18,26,60] For all other wave vectors $|I_{obs} - I_F|$ is smaller or of the order of the statistical error 10. Instead, I_B has a strong discrepancy with I_{obs} for peak 9, I_C for peaks 6 and 13, and the other structures shown in Fig. 17 cannot explain peak 2.

V.3 One-Dimensional O Ordering

In our model at least for $1/2 \leq x < 3/4$, and in the ASYNNNI model for all x,[8,88] the correlation length along the chains becomes large at low enough temperatures and the systems enters into a quasi one-dimensional (1D) regime in which the physics is described approximately by a 1D Ising model,[8,72] which describes the order in the direction perpendicular to the planes. In the case of the ASYNNNI, only NN interactions in this Ising model are important. Here we show that the split diffuse diffraction peaks observed in YBa$_2$Cu$_3$O$_{6+x}$ for $1/2 < x < 3/4$[18,19] can be explained using a 1D Ising model with first and second NN repulsions W_1 and W_2.[20] Previously Khachaturyan and Morris explained some of the observed peaks assuming a random faulting of the OII phase.[17] This theory has been criticized because of the absence of a statistical model for O ordering and the restricted range of compositions solved

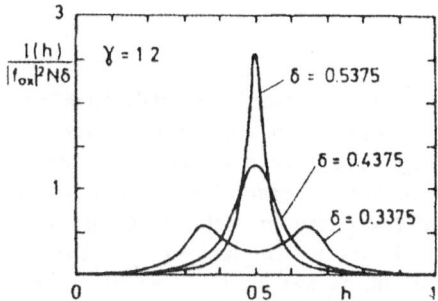

Figure 19. Intensity as a function of wave vector for $\gamma = 1.2$, and several values of $\delta = 1 - x$.

$(1/2 < x < 2/3)$.[89] In fact, the theory of Khachaturyan and Morris is the zero-temperature limit of the one presented here, which is valid for all compositions.

To obtain analytical expressions, we assume that $W_1 \gg T$. For $x > 1/2$ this implies that two empty CuO chains (Cu-vacancy chains) cannot be NN. This allows to represent any possible structure by a sequence of two strips: $s1$ composed of a CuO chain, and $s2$ composed of a Cu-vacancy chain followed by a CuO chain. The problem can then be mapped into an effective 1D Ising model with first NN repulsion W_2 between two $s2$. The energy per strip is given by yW_2 where y is the probability of finding two NN $s2$ strips. Calling $z = (1 - x)/x$ the probability of finding one $s2$ strip, the solution of the effective 1D model with the pair approximation of the CVM gives y

$$y = z - \frac{\gamma}{2} + \left[\left(z - \frac{\gamma}{2}\right)^2 + z^2(\gamma - 1)\right]^{1/2}, \qquad \gamma = \frac{1}{1 - \exp(-W_2/T)}, \qquad (8)$$

With this value of y and the elegant method of the generating functions, the scattering intensity $I(h)$ for wave vectors $(2\pi/a)(h, 0, 0)$ can be calculated[20]

$$I(h) = \frac{N |f_{ox}|^2 (1 - x)(1 - \alpha)(1 + \alpha - \beta)}{4\alpha \cos^2(2\pi h) + 2\beta(1 + \alpha)\cos(2\pi h) + (1 - \alpha)^2 + \beta^2}, \qquad (9)$$

where N is the total number of unit cells, f_{ox} is the oxygen scattering factor, $\alpha = \beta - y/z$, and $\beta = (z - y)/(1 - z)$. In the case that the CS were stable for $x < 1/2$, $I(h)$ would be symmetric under interchange of x by $1 - x$. For $x > 4/7$, and low enough temperatures, there are two diffuse diffraction peaks with intensity maxima at

$$h_{max} = \frac{1}{2\pi} \arccos\left[\frac{x\gamma}{4(2x - 1)}\right]. \qquad (10)$$

In Fig. 19 we show the result for $I(h)$ corresponding to the experimental results for O deficiencies $\delta = 1 - x$, of 0.27, 0.35 and 0.43.[19] According to Ref. 90, we replaced δ by $5\delta/4$ in the theoretical curves. The agreement with experiment is very satisfactory.

VI. Discussion

A complete understanding of the electronic and structural properties of $YBa_2Cu_3O_{6+x}$ requires a self-consistent theory of the electronic structure and the oxygen ordering in the CuO_x planes. Simple arguments showing the relationship between O ordering and charge transfer to the superconducting CuO_2 planes, and the important role of correlations, were presented in Refs. 3, 4 and at the beginning of section II. Such a self-consistent theory is still lacking and the origin of several electronic properties of high-T_c systems, as well as the stability of several structures observed in $YBa_2Cu_3O_{6+x}$ (see Section IV) is still debated. The main difficulty in the development of such a theory is the presence of very strong correlations in the problem. Very few exact results exist for strongly correlated systems (see for example Ref. 41) and most of our present knowledge of these systems is based on exact calculations in finite clusters and several analytical approximations, like those mentioned in section II.[2,5,11,12,21,22,37-40,43-46] However, these treatments have made important progress since the discovery of high-T_c superconductivity and some basic features of the high-T_c systems are well understood at present.

On the basis of this knowledge, we have calculated the hole count in $YBa_2Cu_3O_{6+x}$ and from it the superconducting critical temperature T_c. These quantities, as well as the dependence of the amount of Cu^+ on x, and the qualitative effects of uniaxial pressure and substitution of Y by rare earths discussed elsewhere,[4] are in good agreement with experiment. This information is used to construct an oversimplified lattice-gas model for the O ordering in the CuO_x planes. In spite of its simplicity, the model is able to explain practically all observed diffraction patterns, as discussed in sections IV and V.2 and V.3, and the main features of the thermodynamics (section V). Although our approach is not fully self-consistent, the electronic, structural and thermodynamic calculations harmoniously complement each other.

There are several aspects of our results which need to be improved. We have obtained that the "chain structures" CS are metallic, while "quasi hexagonal structures" HS are semiconducting for $x < 0.5$ (section II). It is also clear that the relative stability of the HS increases with decreasing x for $x < 0.5$ (Figs. 9 and 10). Thus, a transition from semiconducting HS to superconducting CS with increasing x is expected, and as discussed in detail in Ref. 4, several experimental evidence from resistivity,[14,66] photoconductivity,[69,70] composition dependence of the amount of Cu^+,[30,31] Raman spectra,[67] and nuclear quadrupole resonance[67] indicate that such a transition exists for $x \sim 0.4$. However, we are not able to calculate the exact point of this transition. For this, a more realistic model for the energy and the screening mechanisms is required. This problem is also reflected in the thermodynamics: our phase diagrams presented in section V have only two parameters λ and f independent of concentration. However, the screening length λ should vary with the number of carriers (which in turn depends on x), and the type of order. Eq. (5) implies that f is not constant either. Finally, as noted by Schleger *et al.*,[87] it seems necessary to add other degrees of freedom, absent in lattice-gas models, for a quantitative explanation of the thermodynamics. Degrees of freedom of this kind are expected from the isolated CuO_2 clusters mentioned in section II.

In summary, in spite of its limitations, an approach based on intra- and interatomic Coulomb repulsions is able to explain at least qualitatively the main electronic, structural and thermodynamic properties of $YBa_2Cu_3O_{6+x}$.

Acknowledgements

We would like to thank P. Vairus and H. Bonadeo for their help in sections II and IV respectively.

References

1. M. Guillaume, P. Allenspach, J. Mesot, B. Roessli, U. Staub, P. Fischer, and A. Furrer, *Z. Phys. B* **90**, 13 (1993).
2. A. A. Aligia, J. Garcés, and H. Bonadeo, *Phys. Rev. B* **42**, 10226 (1990).
3. A. A. Aligia, H. Bonadeo, and J. Garcés, *Phys. Rev. B* **43**, 542 (1991).
4. A. A. Aligia and J. Garcés, *Phys. Rev. B* **49**, 524 (1994).
5. P. Vairus, A. A. Aligia, and J. Garcés, *Physica C* **235-240**, 301 (1994).
6. H. Claus, S. Yang, A. P. Paulikas, J. W. Downey, and B.W. Veal, *Physica C* **171**, 205 (1990).
7. J. Kircher, E. Brücher, E. Schönherr, R. K. Kremer, and M. Cardona, *Phys. Rev. B* **46**, 588 (1992).
8. G. Ceder, M. Asta, W. C. Carter, M. Kraitchman, D. de Fontaine, M. E. Mann, and M. Sluiter, *Phys. Rev. B* **41**, 8698 (1990), references therein.
9. P. A. Sterne and L. T. Wille, *Physica C* **162-164**, 223 (1989).
10. J. M. Sanchez, *Phys. Rev. B* **48**, 14013 (1993); references therein.
11. M. S. Hybertsen, E. B. Stechel, M. Schluter, and D. R. Jennison, *Phys. Rev. B* **41**, 11068 (1990).
12. A. A. Aligia and J. Garcés, *Solid State Commun.* **87**, 363 (1993).
13. Z. Z. Wang, J. Clayhold, N. P. Ong, J. M. Tarascon, L. H. Greene, W. R. McKinnon, and G. W. Hull, *Phys. Rev. B* **36**, 7222 (1987).
14. O. E. Parfionov and A. A. Konovalov, *Physica C* **202**, 385 (1992).
15. L. D. Rotter, Z. Schlesinger, R. T. Collins, F. Holtzberg, C. Field, V. W. Welp, G. W. Crabtree, J. Z. Lin, Y. Fang, K. G. Vandervoort, and S. Fleshler, *Phys. Rev. Lett.* **67**, 2741 (1991).
16. A. A. Aligia, *Phys. Rev. Lett.* **65**, 2475 (1990).
17. A. G. Khachaturyan and J. W. Morris, Jr., *Phys. Rev. Lett.* **64**, 77 (1990).
18. R. Beyers, B. T. Ahn, G. Gorman, V. Y. Lee, S. S. P. Parkin, M. L. Ramírez, K. P. Roche, J. E. Vázquez, T. M. Gür, and R.A. Huggins, *Nature* **340**, 619 (1989).
19. L. E. Levine and M. Däumling, *Phys. Rev. B* **45**, 8146 (1992).
20. A. A. Aligia, *Phys. Rev. B* **47**, 15308 (1993).
21. A. Latgé, E. V. Anda, and J.L. Morán López, *Phys. Rev. B* **42**, 4288 (1990).
22. G. Uimin and J. Rossat-Mignod, *Physica C* **199**, 251 (1992).
23. L. F. Mattheis and D. R. Hamann, *Solid State Commun.* **63**, 395 (1987).
24. A. A. Aligia, J. Garcés, and H. Bonadeo, *Physica C* **190**, 234 (1992).
25. M. A. Alario-Franco, C. Chaillout, J. J. Capponi, J. Chenavas, and M. Marezio, *Physica C* **156**, 455 (1988).
26. J. Reyes-Gasga, T. Krekels, G. Van Tendeloo, J. Van Landuyt, S. Amelinckx, W. H. M. Bruggink and H. Verweij, *Physica C* **159**, 831 (1989).
27. R. M. Fleming, L. F. Schreemeyer, P. K. Gallagher, B. Batlogg, L. W. Rupp, and J.V. Waszczak, *Phys. Rev. B* **37**, 7920 (1988).
28. Y. P. Lin, J. E. Greedan, A. H. O'Reilly, J. N. Reimers, and C. V. Stager, *J. Solid State Chem.* **84**, 226 (1990).
29. J. J. Neumeier and H. A. Zimmermann, *Phys. Rev. B* **47**, 8385 (1993).

30. J. M. Tranquada, S. M. Heald, A. R. Moodenbaugh, and Y. Xu, *Phys. Rev. B* **38**, 8893 (1988).
31. H. Tolentino, A. Fontaine, F. Baudelet, T. Gourieux, G. Krill, J. Y. Henry, and J. Rossat-Mignod, *Physica C* **192**, 115 (1992).
32. W. W. Warren, Jr., R. E. Walstedt, G. F. Brennert, R. J. Cava, B. Batlogg, and L.W. Rupp, *Phys. Rev. B* **39**, 831 (1989).
33. R. J. Cava, B. Batlogg, C. H. Chen, E. A. Rietman, S. M. Zahurak, and D. Werder, *Phys. Rev. B* **36**, 5719 (1987).
34. C. N. R. Rao, R. Nagarajan, A. K. Ganguli, G. N. Subbana, and S. V. Bhat, *Phys. Rev. B* **42**, 6765 (1990).
35. N. Chandrasekhar, O. T. Valls, and A. M. Goldman, *Phys. Rev. Lett.* **71**, 1079 (1993); references therein. The explanation proposed for the observed electric-field effects in unlikely.[36]
36. A. A. Aligia, *Phys. Rev. Lett.* **73**, 1561 (1994).
37. V. J. Emery, *Phys. Rev. Lett.* **58**, 2794 (1987).
38. P. B. Littlewood, C. M. Varma, and E. Abrahams, *Phys. Rev. Lett.* **63**, 2602 (1989); references therein.
39. M. E. Simón and A. A. Aligia, *Phys. Rev. B* **48**, 7471 (1993); references therein.
40. H. Hasegawa, *Phys. Rev. B* **41**, 9168 (1990).
41. L. Arrachea and A. A. Aligia, *Phys. Rev. Lett.* **73**, 2240 (1994).
42. N. W. Aschcroft and N. D. Mermin, *Solid State Physics* (Holt, Rinehart and Winston, New York, 1976).
43. R. Fehrenbacher and T. M. Rice, *Phys. Rev. Lett.* **70**, 3471 (1993).
44. C. D. Batista and A. A. Aligia, *Phys. Rev. B* **48**, 4212 (1993); A. A. Aligia, M. E. Simón, and C.D. Batista, *Phys. Rev. B* **49**, 13061 (1994).
45. A. A. Aligia, E. Gagliano, and P. Vairus, *Phys. Rev. B* **52**, 13061 (1995).
46. A. A. Aligia, *Europhys. Lett.* 26, 153 (1994).
47. G. A. Samara, W. F. Hammetter, and E. L. Venturini, *Phys. Rev. B* **41**, 8974 (1990).
48. S. Semenovskaya and A. G. Khachaturyan, *Phys. Rev. B* **46**, 6511 (1992).
49. M. Kaburagi and J. Kanamori, *Prog. Theor. Phys.* **54**, 30 (1979).
50. S. M. Allen and J. W. Cahn, *Acta Metall.* **20**, 423 (1972).
51. S. Semenovskaya and A. G. Khachaturyan, *Phys. Rev. Lett.* **67**, 2223 (1991).
52. A. A. Aligia, A. G. Rojo, and B. Alascio, in *Progress in High Temperature superconductivity*, edited by R. Nicolsky (World Scientific, Singapore, 1988), Vol.9, p. 406.
53. Zhi-Xiong Cai and S. D. Mahanti, *Phys. Rev. B* **40**, 6558 (1989).
54. C. P. Burmester and L. T. Wille, *Phys. Rev. B* **40**, 8795 (1989).
55. H. Bonadeo, *Can. J. Phys.* **62**, 904 (1984).
56. V. L. Pokrovsky and G. V. Uimin, *J.Phys. C* **11**, 3535 (1978).
57. J. Hubbard, *Phys. Rev. B* **17**, 494 (1978).
58. A. A. Aligia and J. Garcés, *Phys. Rev. B* **52**, 6227 (1995).
59. R. Sonntag, Th. Zeiske, and D. Hohlwein, *Physica B* **180-181**, 374 (1992).
60. D. J. Werder, C. H. Chen, R. J. Cava, and B. Batlogg, *Phys. Rev. B* **38**, 5130 (1988); references therein.
61. V. Plakhty, A. Stratilatov, Yu. Chernekov, V. Federov, S. K. Sinha, C. K. Loong, B. Gaulin, M. Vlasov, and S. Moshkin, *Solid State Commun.* **84**, 639 (1992).
62. P. Schleger, H. Casalta, R. Hadfield, H. F. Poulsen, M. von Zimmermann, N. H. Andersen, J. R. Schneider, R. Liang, P. Dosanjh, and W. N. Hardy, *Physica C* **241**, 103 (1995).

63. R. Sonntag, D. Hohlwein, T. Brückel, and G. Collin, *Phys. Rev. Lett.* **66**, 1497 (1991).
64. A. A. Aligia, *Europhys. Lett.* **18**, 181 (1992).
65. Y. Tokura, J. B. Torrance, T. C. Huang, and A. I. Nazzal, *Phys. Rev. B* **38**, 7156 (1988).
66. H. Shaked, B. W. Veal, J. Faber Jr., R. L. Hitterman, U. Balachandran, G. Tomlis, H. Shi,, L. Morss, and A. P. Paulikas, *Phys. Rev. B* **41**, 4173 (1990).
67. G. Burns, F. H. Dacol, C. Field, and F. Holtzberg, *Solid State Commun.* **77**, 367 (1991).
68. V. S. Kasperovich and E. V. Charnaya, *Fiz. Tverd. Tela* **34**, 2040 (1992) [*Sov. Phys. Solid State* **34**, 1089 (1992)].
69. V. I. Kudinov, I. L. Chaplygin, A. I. Kirilyuk, N. M. Kreines, R. Laiho, and E. Lädheranta, *Phys. Lett. A* **157**, 290 (1991).
70. G. Nieva, E. Osquiguil, J. Guimpel, M. Maenhoudtt, B. Wuyts, Y. Bruynseraede, M. B. Maple, and I. K. Schuller, *Phys. Rev. B* **46**, 14249 (1992).
71. T. Krekels, S. Kaesche, and G. Van Tendeloo, *Physica C* **248**, 317 (1995).
72. A. A. Aligia, J. Garcés, and J. P. Abriata, *Physica C* **221**, 109 (1994).
73. D. Adelman, C. P. Burmester, L. T. Wille, P. A. Sterne, and R. Gronsky, *J. Phys.* 4, L585 (1992).
74. A. A. Aligia and J. Garcés, *Phys. Rev. B* **44**, 7102 (1991).
75. A. A. Aligia and J. Garcés, *Physica C* **194**, 223 (1992).
76. G. Grigelionis, S. Lapinskas, A. Rosengren, and E. E. Tornau, *Physica C* **242**, 183 (1995).
77. J. L. Morán López and J. M. Sanchez, *Physica C* **210**, 401 (1993).
78. R. Kikuchi, *Phys. Rev.* **81**, 988 (1951).
79. R. Kikuchi, *J. Chem. Phys.* **65**, 4545 (1976).
80. V. E. Zubkus, S. Lapinskas, and E. E. Tornau, *Physica C* **159**, 501 (1989).
81. R. Kikuchi and J. S. Choi, *Physica C* **160**, 347 (1989).
82. D. K. Hilton, B. M. Gorman, P. A. Rikvold, and M. A. Novotny, *Phys. Rev. B* **46**, 381 (1992).
83. V. E. Zubkus, E. E. Tornau, S.Lapinskas, and P. J. Kundrotas, *Phys. Rev. B* **43**, 13112 (1991).
84. W. R. McKinnon, M. L. Post, L. S. Selwyn, G. Pleizier, J. M. Tarascon, P. Barboux, L. H. Greene, and G. W. Hull, *Phys. Rev. B* **38**, 6543 (1988).
85. P. Schleger, W. N. Hardy, and B. X. Yang, *Physica C* **176**, 261 (1991).
86. P. Schleger, Ph.D. Thesis, The University of British Columbia, 1992.
87. P. Schleger, W. N. Hardy, and H. Casalta, *Phys. Rev. B* **49**, 514 (1994).
88. V. M. Matic, *Physica C* **230**, 61 (1994).
89. D. de Fontaine and S. C. Moss, *Phys. Rev. Lett.* **67**, 527 (1991).
90. T. Krekels, H. Zou, G. Van Tendeloo, D. Wagener, M. Buchgeister, S. M. Hosseini, and P. Herzog, *Physica C* **196**, 363 (1992).

Application of the CVM Continuous Atomic Displacement Formulation to the Fracture Problem

Kin-ichi Masuda-Jindo,[1] Ryoichi Kikuchi,[2] and Robb Thomson,[3]

[1] Department of Materials Science and Engineering
Tokyo Institute of Technology
Nagatsuta, Midori-ku, Yokohama 227
JAPAN

[2] Department of Materials Science and Engineering
University of California
Los Angeles, CA 90095-1595
U.S.A.

[3] Institute for Materials Science and Engineering
National Institute of Standards and Technology
Gaithersburg, MD, 20899
U.S.A.

Abstract

A new formulation of the CVM which allows continuous atomic displacement from lattice points is applied to the fracture problems of materials. We formulate atomic displacements in a cracked lattice using the pair approximation of the CVM, and study the fundamental aspects of the crack growth and crack stability relation. For simplicity, we focus our attention on a double ended smallest crack in the $2D$ square lattice. The crack is opened by an external load F which is a pair of oppositely directed forces applied to the two neighboring atoms (top and bottom) perpendicular to the cleavage plane at the center of the crack region. The critical stress intensity factor K_I and the lattice trapping of the cracks in the square lattice are calculated for finite temperatures, and compared with those obtained by the lattice Green's function method at zero temperature.

Theory and Applications of the Cluster Variation and Path Probability Methods
Edited by J.L. Morán-López and J.M. Sanchez, Plenum Press, New York, 1996

I. Introduction

It is well known that the fracture behavior of materials depends strongly on the material parameters, such as temperature, strain rate, and external chemical environment.[1] The first two are usually linked effects. The third one has been discussed extensively by using the lattice Green's function approach.[1-6] In the present study, we focus our attention on the thermal (temperature) effects on the fracture behavior by using the Cluster Variation Method (CVM). In the existing CVM formulations, atoms are placed on lattice points. We introduce a new formulation in which atoms can be displaced from a lattice point.[7,8] The probability of finding an atom displaced at r in dr is written as $f(r)\,dr$, and the corresponding pair probability is written as $g(r_1, r_2)\,dr_1\,dr_2$. Here, r_1 and r_2 denote atomic displacements from the reference atom positions R_1 and R_2. Using the pair approximation of the CVM (regarding an atom at r as a species r), we formulate the atomic displacement around a crack at finite temperatures, and investigate the fundamental aspects of the crack-tip processes, *e.g.*, crack stability relation and crack growth.

Hsieh and Thomson[2] employed a lattice-statics approach for a discrete two-dimensional ($2D$) crack at absolute zero temperature. In that work, the force was taken to be linear up to an arbitrary displacement and set equal to zero for larger displacements. The lattice trapping of cracks has been found to be of great significance and the lattice trapping parameter $A_N = (F_+ - F_-)/F_-$ becomes as large as 1.3 (130%) by using the bond snapping force laws. The lattice theory of fracture can be applicable to important problems such as the relationships between the limits of lattice trapping, Griffith's criterion for fracture, and the fundamental role of surface energy in brittle fracture. We compare the present crack calculations at finite temperatures with those at absolute zero temperature obtained by the lattice Green's function approach. In Fig. 1, we show the atom positions, bond indices (details are given in section II.1) and external force F applied to the prescribed force F using the continuous CVM formalism.

II. Principle of Calculations

The easiest approach to atomic phenomena at crack tips is to consider an infinite lattice of atoms in which a crack is formed by splitting the bonds between rows in two dimensions. A force is then applied to the center atom pair which is required to hold the cracked atoms apart. For zero temperature calculations, we use a lattice Green's function method for the atomistic displacements of the crack. The atomic geometry of a crack in two-dimensional ($2D$) square lattice is shown in Fig. 2. For simplicity, we consider the smallest crack with one broken and two non-linear bonds.

When the external force $F = 0$, the system is homogeneous and reduces to the previous treatments[1,2] for the perfect lattice, for which the basic solution is

$$g(r_1, r_2) = \exp\left(\frac{\beta\lambda}{\omega}\right) \exp\left(-\beta\varepsilon(r_1, r_2)\right) [f(r_1)fr_2]^{\frac{2\omega-1}{2\omega}} \Gamma_{\text{corr}}, \qquad (1)$$

where Γ_{corr} is a correction factor, ω is a half of the coordination number Z, *e.g.*, $\omega = 2$ for $2D$ square lattice.

When the external force F is applied, first we have to identify a lattice point location. But the basic equations remains practically the same as in Eq. (1) as long as the external force F is not large, except the atomic pair PQ in Fig. 1. We need

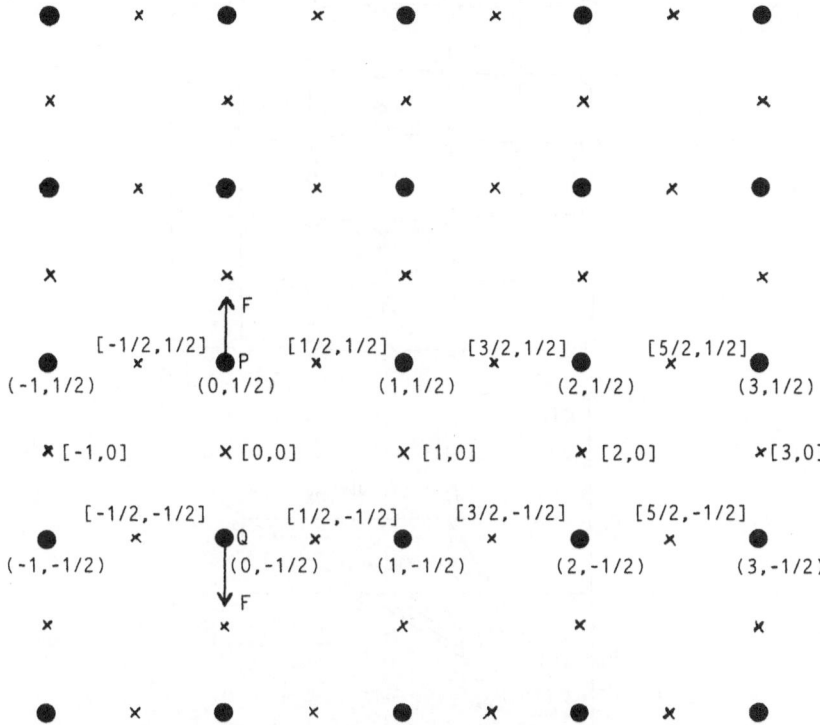

Figure 1. Atomic positions and bond indices in $2D$ square lattice.

the work done by the external force F on the pair so that the equation corresponding to (1) is

$$g(\mathbf{r}_P, \mathbf{r}_Q) = \exp\left(\frac{\beta\lambda_{PQ}}{2}\right)\exp\left(-\beta\varepsilon(\mathbf{r}_P, \mathbf{r}_Q) + \beta F(y_P - y_Q)\right)$$
$$\times \left[f(\mathbf{r}_P)f(\mathbf{r}_Q)\right]^{\frac{3}{4}}\Gamma_{\text{corr}}, \qquad (2)$$

where y_P is the y axis of P. Since $(y_P - y_Q)F$ is the work done by F on the pair, it is the increase in the energy of the pair. When the external force F is positive and is pulling P and Q away, $(y_P - y_Q) > 0$ is more plausible than $(y_P - y_Q) < 0$, indicating the sign $+$ in front of the force term. Eqs. (1) and (2) are the basic heuristic description of the equations we work with. In the following sections we describe the details of the formulation.

II.1 Notation

The position of the reference lattice site is expressed with a vector $\mathbf{m} \equiv (m, n)$. As shown in Fig. 1, the horizontal coordinate m is an integer not smaller than -1 (in units of the lattice constant of the perfect lattice):

$$m = -1, 0, 1, \ldots . \qquad (3)$$

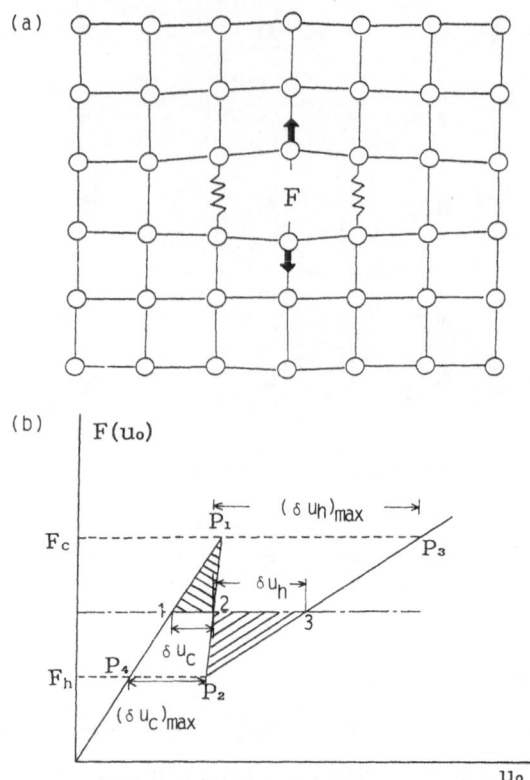

Figure 2. Atomic geometry of the smallest crack in $2D$ square lattice (a), and graphical solution of the bilinear force law (b). The external force is plotted as a function of atomic displacement at the point of application of the force, u_0.

The vertical coordinate n is a half integer,

$$n = \ldots, -h, h, 3h, 5h, \ldots, \tag{4}$$

where we define

$$h \equiv \tfrac{1}{2}. \tag{5}$$

By symmetry, the minimum of n is $-h$ in actual computations. On the other hand, the position of the bond is indicated by the midpoint of the bond, as shown in Fig. 1 with a cross. The horizontal bonds are expressed as

$$(-h, n),\ (h, n),\ (3h, n),\ (5h, n), \ldots, \tag{6}$$

where n is one of the vertical coordinates in (4), and a half integer in the range $-h \leq n$. The position of the vertical bond is written as

$$(m, 0),\ (m, 1),\ (m, 2), \ldots, \tag{7}$$

where m is one of the integers in the range $-1 \leq m$ as in (3).

The point probability function at $\mathbf{m}$ is denoted by $f(\mathbf{m}; \mathbf{r})$. On the other hand, there are two kinds of pair probability functions for the system including lattice defects such as cracks, *i.e.*, horizontal (x-direction) and vertical (y-direction) probability functions. A pair function in the x-direction is written as $g_x(m_x; r_l, r_r)$, where r_l is for the left and r_r is for the right point. A pair function in the y-direction is written as $g_y(m_y; r_b, r_t)$, where r_b is for the bottom and r_t is for the top position.

A unit vector in the x direction is written as $\hat{i}$, and that in the y direction is $\hat{j}$. The bond position and the point position are connected as follows. For instance, the four bonds next to a point m are

$$
\begin{aligned}
\text{left:} \qquad & m_x = m - h\hat{i}, \quad m_y = m, \\
\text{right:} \qquad & m_x = m + h\hat{i}, \quad m_y = m, \\
\text{bottom:} \qquad & m_x = m, \quad m_y = m - h\hat{j}, \\
\text{top:} \qquad & m_x = m, \quad m_y = m + h\hat{j}.
\end{aligned}
\tag{8}
$$

II.2 Reduction Relation and Constraints

A point probability function $f(m; r)$ can be written by reducing one of its four neighbor pairs in Eq. (8),

$$
f(m; r) = \int dr_r \, g_x(m + h\hat{i}; r, r_r)
\tag{9}
$$

$$
= \int dr_l \, g_x(m - h\hat{i}; r_l, r)
\tag{10}
$$

$$
= \int dr_t \, g_y(m + h\hat{j}; r, r_t)
\tag{11}
$$

$$
= \int dr_b \, g_y(m - h\hat{j}; r_b, r).
\tag{12}
$$

These four equations induces three constraint relations among the pair probability functions g's. In the free energy expression, we write these constraints using Lagrange multipliers α, γ and κ.

From Eqs. (9) and (10) for a horizontal pair distribution function g_x, one can obtain the following relationship

$$
\sum_m C_\alpha(m) \equiv \sum_m \int dr \, \alpha(m; r) \int dr_2 \, (g_o(m + h\hat{i}; r, r_2) - g_x(m - h\hat{i}; r_2, r))
$$

$$
= \sum_m \int dr_1 \int dr_2 \, (\alpha(m; r_1) - \alpha(m + \hat{i}; r_2)) \, g_x(m + h\hat{i}; r_1, r_2).
\tag{13}
$$

From Eqs. (11) and (12) for a vertical pair distribution function g_y, one can get

$$
\sum_m C_\gamma(m) = \sum_m \int dr_1 \int dr_2 \, (\gamma(m; r_1) - \gamma(m + \hat{j}; r_2)) \, g_y(m + h\hat{j}; r_1, r_2).
\tag{14}
$$

In order to make the Lagrange terms symmetric, we equate $(9) + (10) = (11) + (12)$, use the Lagrange multiplier κ, and then make a transformation of the type similar to that in (13) to obtain

$$\sum_m C_\kappa(m) \equiv \sum_m \int dr\ \kappa(m;r) \int dr_2\ (g_x(m+h\hat{i};r,r_2) + g_x(m-h\hat{i};r_2,r))$$

$$- \sum_m \int dr\ \kappa(m;r) \int dr_2\ (g_y(m+h\hat{j};r,r_2) + g_y(m-h\hat{j};r_2,r))$$

$$= \sum_m \int dr_1 \int dr_2\ (\kappa(m;r_1) + \kappa(m+\hat{i};r_2))\ g_x(m+h\hat{i};r_1,r_2)$$

$$- \sum_m \int dr_1 \int dr_2\ (\kappa(m;r_1) + \kappa(m+\hat{j};r_2))\ g_y(m+h\hat{j};r_1,r_2). \quad (15)$$

II.3 The Internal Energy

We consider an ensemble made of M atoms. The nearest-neighbor pair potential is written as $\varepsilon((r(m_1), r(m_2))$ where m_1 and m_2 are nearest neighbors. The energy E, per single atom, due to the pair interaction potential ε is then given by

$$\frac{E_\varepsilon}{M} = \sum_{m_x} \int dr_l \int dr_r\ (\varepsilon(r_l, r_r)\, g_x(m_x; r_l, r_r))$$

$$+ \sum_{m_y} \int dr_b \int dr_t\ (\varepsilon(r_b, r_t)\, g_y(m_y; r_b, r_t)), \quad (16)$$

where m_x is the center point of a bond in the x-direction, r_l and r_r are the left- and right-sides of the bond. For a bond in the y-direction, the center point of a bond is m_y, and the bottom and top points are r_b and r_t, respectively.

When there is an external force F applied to the center pair of atoms at $(0, h)$ and $(0, -h)$, as shown in Fig. 1, the work done by F on the system is

$$\frac{E_w}{M} = F \int dr\ yf(m = (0, h); r = (x, y))$$

$$- F \int dr\ yf(m = (0, -h); r = (x, y)). \quad (17)$$

The first integral is for the work done on the upper atom, while the second integral is for the bottom atom. When one takes into account the statistical symmetry of the upper and the bottom halves of the system, one can rewrite the above Eq. (17) as

$$\frac{E_w}{M} = 2F \int dr_{(0,h)}\ y_{(0,h)} f(m = (0, h); r_{(0,h)}). \quad (18)$$

II.4 Evaluation of Entropy

The entropy of the system containing lattice defects (cracks) is derived from the general expression of the CVM, which is written for the pair approximation as

$$\exp\left(\frac{S}{k}\right) = \prod_{\text{points}} W_M \prod_{\text{pairs}} G_M, \quad (19)$$

where W_M is the number of ways of distributing point species to M points in the ensemble, and G_M is the correlation correction factor to take into account correlation between two points of a pair, formulated for the M systems in an ensemble. The second product goes over all pairs in the system. The details of the correlation correction factor is given in a recent publication by one of the present authors.[10] The advantage of using the correlation correction concept and the ensemble point of view is that the expression (19) can be used when each bond is statistically non-equivalent. W_M and G_M are written as

$$W_M = \frac{M!}{\{\text{Point}\}_M},$$

$$G_M = \frac{\{\text{Left Point}\}_M \, \{\text{Right Point}\}_M}{\{\text{Pair}\}_M \, M!}. \tag{20}$$

It is convenient to distribute the "point" product in Eq. (19) over pairs. Since each point is shared by four pairs, we can rewrite (19) as

$$\exp\left(\frac{S}{k}\right) = \prod_{x\text{-pairs}} \left((W_M(\text{L-Point}))^{1/4} (W_M(\text{R-Point}))^{1/4} G_M(x\text{-Pair}) \right)$$

$$\times \prod_{y\text{-pairs}} \left((W_M(\text{B-Point}))^{1/4} (W_M(\text{T-Point}))^{1/4} G_M(y\text{-Pair}) \right), \tag{21}$$

where L-, R-, B- and T- stand for Left-, Right-, Bottom- and Top-, respectively. Equation (21) can be rewritten as

$$\exp\left(\frac{S}{k}\right) = \prod_{x\text{-pairs}} \frac{(\{\text{L-Point}\}_M)^{3/4} (\{\text{R-Point}\}_M)^{3/4}}{\{x\text{-Pair}\}_M \, M!}$$

$$\times \prod_{y\text{-pairs}} \frac{(\{\text{B-Point}\}_M)^{3/4} (\{\text{T-Point}\}_M)^{3/4}}{\{y\text{-Pair}\}_M \, M!}. \tag{22}$$

Taking the logarithm of Eq. (22), we derive the entropy expression

$$\frac{S_M}{kM} = \sum_{m_x} \frac{3}{4} \left(\int dr_l \, L(f(m_x - h\hat{\mathbf{i}}; r_l)) + \int dr_r \, L(f(m_x + h\hat{\mathbf{i}}; r_r)) \right)$$

$$+ \sum_{m_x} \left(-\int dr_l \int dr_r \, L(g_x(m_x; r_l, r_r)) + \frac{1}{2} \right)$$

$$+ \sum_{m_y} \frac{3}{4} \left(\int dr_b \, L(f(m_y - h\hat{\mathbf{j}}; r_b)) + \int dr_t \, L(f(m_y + h\hat{\mathbf{j}}; r_t)) \right)$$

$$+ \sum_{m_y} \left(-\int dr_b \int dr_t \, L(g_x(m_y; r_b, r_t)) + \frac{1}{2} \right). \tag{23}$$

where $L(f)$ is a functional defined as

$$L(f) \equiv f \ln f - f. \tag{24}$$

In Eq. (23), the $\frac{1}{2}$ terms come from $L(M!)$ in Eq. (22).

II.5 The Free Energy

The free energy is written using (16) and (17) for the energy, (23) for the entropy, the constraint terms C_α in (13), C_γ in (14), C_κ in (15), and the normalization terms.

$$
\begin{aligned}
\Psi \equiv \frac{\beta F}{M} = \beta \Bigg(& \sum_{m_x} \int dr_l \int dr_r \, (\varepsilon(r_l, r_r) g_x(m_x; r_l, r_r)) \\
& + \sum_{m_y} \int dr_b \int dr_t \, (\varepsilon(r_b, r_t) g_y(m_y; r_b, r_t)) \Bigg) \\
& + \beta F \Bigg(\int dr \, yf(m = (0, h); r = (x, y)) \\
& \qquad - \int dr \, yf(m = (0, -h); r = (x, y)) \Bigg) \\
& - \sum_{m_x} \frac{3}{4} \Bigg(\int dr_l \, L(f(m_x - h\hat{i}; r_l)) + \int dr_r \, L(f(m_x + h\hat{i}; r_r)) \Bigg) \\
& - \sum_{m_x} \Bigg(- \int dr_l \int dr_r \, L(g_x(m_x; r_l, r_r)) + \frac{1}{2} \Bigg) \\
& - \sum_{m_y} \frac{3}{4} \Bigg(\int dr_b \, L(f(m_y - h\hat{j}; r_b)) + \int dr_t \, L(f(m_y + h\hat{j}; r_t)) \Bigg) \\
& - \sum_{m_y} \Bigg(- \int dr_b \int dr_t \, L(g_x(m_y; r_b, r_t)) + \frac{1}{2} \Bigg) \\
& + \sum_m \int dr_1 \int dr_2 \, (\alpha(m; r_1) - \alpha(m + \hat{i}; r_2)) g_x(m + h\hat{i}; r_1, r_2) \\
& + \sum_m \int dr_1 \int dr_2 \, (\gamma(m; r_1) - \gamma(m + \hat{j}; r_2)) g_y(m + h\hat{j}; r_1, r_2) \\
& + \sum_m \int dr_1 \int dr_2 \, (\kappa(m; r_1) + \kappa(m + \hat{i}; r_2)) g_x(m + h\hat{i}; r_1, r_2) \\
& - \sum_m \int dr_1 \int dr_2 \, (\kappa(m; r_1) + \kappa(m + \hat{j}; r_2)) g_y(m + h\hat{j}; r_1, r_2) \\
& + \beta \sum_{m_x} \lambda(m_x) \Bigg(1 - \int dr_l \int dr_r \, g(m_x; r_l, r_r) \Bigg) \\
& + \beta \sum_{m_y} \lambda(m_y) \Bigg(1 - \int dr_b \int dr_t \, g(m_y; r_b, r_t) \Bigg).
\end{aligned}
\tag{25}
$$

II.6 Minimization of Free Energy Ψ

We now minimize the free energy Ψ given in the previous subsection II.5 with respect to the pair distribution functions g_x and g_y, in order to obtain the thermodynamically equilibrium solutions. For the atom pairs away from $(0, h)$ and $(0, -h)$, we have

$$\frac{\partial \Psi}{\partial g_x(\mathbf{m}_x; \mathbf{r}_l, \mathbf{r}_r)} \equiv \beta\varepsilon(\mathbf{r}_l, \mathbf{r}_r) - \tfrac{3}{4}\ln\left(f(\mathbf{m}_x - h\hat{\mathbf{i}}; \mathbf{r}_l)\,f(\mathbf{m}_x + h\hat{\mathbf{i}}; \mathbf{r}_r)\right)$$

$$+ \ln g_x(\mathbf{m}_x; \mathbf{r}_l, \mathbf{r}_r) - \beta\lambda(\mathbf{m}_x)$$
$$+ \left(\alpha(\mathbf{m}_x - h\hat{\mathbf{i}}; \mathbf{r}_l) - \alpha(\mathbf{m}_x + h\hat{\mathbf{i}}; \mathbf{r}_r)\right)$$
$$+ \left(\kappa(\mathbf{m}_x - h\hat{\mathbf{i}}; \mathbf{r}_l + \kappa(\mathbf{m}_x + h\hat{\mathbf{i}}; \mathbf{r}_r)\right) = 0, \tag{26}$$

$$\frac{\partial \Psi}{\partial g_y(\mathbf{m}_y; \mathbf{r}_b, \mathbf{r}_t)} \equiv \beta\varepsilon(\mathbf{r}_b, \mathbf{r}_t) - \tfrac{3}{4}\ln\left(f(\mathbf{m}_y - h\hat{\mathbf{j}}; \mathbf{r}_b)\,f(\mathbf{m}_y + h\hat{\mathbf{j}}; \mathbf{r}_t)\right)$$

$$+ \ln g_y(\mathbf{m}_y; \mathbf{r}_b, \mathbf{r}_t) - \beta\lambda(\mathbf{m}_y)$$
$$+ \left(\gamma(\mathbf{m}_y - h\hat{\mathbf{j}}; \mathbf{r}_b) - \gamma(\mathbf{m}_y + h\hat{\mathbf{j}}; \mathbf{r}_t)\right)$$
$$- \left(\kappa(\mathbf{m}_x - h\hat{\mathbf{i}}; \mathbf{r}_l) + \kappa(\mathbf{m}_x + h\hat{\mathbf{i}}; \mathbf{r}_r)\right) = 0. \tag{27}$$

In minimizing the free energy Ψ with respect to the atom pairs involving the point $(0, h)$ or $(0, -h)$, there are several equivalent but different ways of differentiation. The question is how to write the $f(\mathbf{m} = (0, h); \mathbf{r})$ and $f(\mathbf{m} = (0, -h); \mathbf{r})$ terms in (25).

One of the possible ways, which leads to the simplest looking set of equations, is to treat

$$f(\mathbf{m} = (0, h); \mathbf{r}) = \int d\mathbf{r}_b \; g_y(\mathbf{m} = (0, 0); \mathbf{r}_b, \mathbf{r}), \tag{28}$$

$$f(\mathbf{m} = (0, -h); \mathbf{r}) = \int d\mathbf{r}_t \; g_y(\mathbf{m} = (0, 0); \mathbf{r}, \mathbf{r}_t). \tag{29}$$

When we use them, (26) and (27) hold for all pairs except the central pair located at $\mathbf{m} = (0, 0)$. The equation for this pair is the same as (29) except that the energy term $\beta\varepsilon(\mathbf{r}_b, \mathbf{r}_t)$ is replaced by

$$\beta\varepsilon(\mathbf{r}_b, \mathbf{r}_t) \to \beta\varepsilon(\mathbf{r}_b, \mathbf{r}_t) + 2\beta F\, y_t. \tag{30}$$

We take the exponentials of (26) and (27) and write the following for all pairs except the central pair,

$$g_x(\mathbf{m}_x; \mathbf{r}_l, \mathbf{r}_r) = \exp\left(\beta\lambda(\mathbf{m}_x)\right) \exp\left(-\beta\varepsilon(\mathbf{r}_l, \mathbf{r}_r)\right)$$
$$\times \left(f(\mathbf{m}_x - h\hat{\mathbf{i}}; \mathbf{r}_l)\,f(\mathbf{m}_x + h\hat{\mathbf{i}}; \mathbf{r}_r)\right)^{3/4}$$
$$\times \exp\left(-\alpha(\mathbf{m}_x - h\hat{\mathbf{i}}; \mathbf{r}_l) + \alpha(\mathbf{m}_x + h\hat{\mathbf{i}}; \mathbf{r}_r)\right)$$
$$\times \exp\left(-\kappa(\mathbf{m}_x - h\hat{\mathbf{i}}; \mathbf{r}_l) - \kappa(\mathbf{m}_x + h\hat{\mathbf{i}}; \mathbf{r}_r)\right), \tag{31}$$

$$g_y(\mathbf{m}_y; \mathbf{r}_b, \mathbf{r}_t) = \exp\left(\beta\lambda(\mathbf{m}_y)\right) \exp\left(-\beta\varepsilon(\mathbf{r}_b, \mathbf{r}_t)\right)$$
$$\times \left(f(\mathbf{m}_y - h\hat{\mathbf{j}}; \mathbf{r}_b)\,f(\mathbf{m}_y + h\hat{\mathbf{j}}; \mathbf{r}_t)\right)^{3/4}$$
$$\times \exp\left(-\gamma(\mathbf{m}_y - h\hat{\mathbf{j}}; \mathbf{r}_b) + \gamma(\mathbf{m}_y + h\hat{\mathbf{j}}; \mathbf{r}_t)\right)$$
$$\times \exp\left(\kappa(\mathbf{m}_y - h\hat{\mathbf{j}}; \mathbf{r}_b) + \kappa(\mathbf{m}_y + h\hat{\mathbf{j}}; \mathbf{r}_t)\right). \tag{32}$$

For the central pair at $m = (0,0)$, the energy term $\beta\varepsilon(r_b, r_t)$ is replaced as in Eq. (30);

$$\beta\varepsilon(r_b, r_t) \to \beta\varepsilon(r_b, r_t) + 2\beta F\, y_t \,. \tag{33}$$

The horizontal pair equation (31) holds for all horizontal pairs without exception.

When the free energy Ψ derived in the subsection II.5 is a minimum and Eqs. (28) and (29) hold, we subtract from Ψ two zero terms, to transform as

$$\Psi = \Psi - \sum_{m_x} g(m_x; r_l, r_r)\frac{\partial\Psi}{\partial g_x(m_x; r_l, r_r)} - \sum_{m_y} g_y(m_y; r_b, r_t)\frac{\partial\Psi}{\partial g_y(m_y; r_b, r_t)} \,, \tag{34}$$

which becomes

$$\Psi \equiv \frac{\beta F}{M} = \beta\sum_{m_x}\lambda(m_x) + \beta\sum_{m_y}\lambda(m_y) \,, \tag{35}$$

or,

$$\frac{F}{M} = \sum_{m_x}\lambda(m_x) + \sum_{m_y}\lambda(m_y) \,. \tag{36}$$

The above λ corresponds to $\omega\lambda$ in (4.1) of Kikuchi and Chen.[9]

Finally, we check the consistency of the formulations for homogeneous and inhomogeneous systems. When the external forces F goes to zero, the system becomes homogeneous, and hence the equations are expected to reduce to those of the perfect $2D$ lattice. The main difference in the formulations of homogeneous and inhomogeneous systems is the constraint terms:

1) In the present model of the $2D$ square lattice, the equalities of (9)=(10) and (11)=(12) are both equivalent to a 90° rotation in the homogeneous system operated twice.

2) The equality (9)=(10), or (9)+(10) = (11)+(12), which is to be satisfied in the $F \to 0$ limit, is in a 180° rotation condition taken into account in the formulation of the homogeneous system.

Because of these two equivalences, the formulation of inhomogeneous system reduces to that of the homogeneous system in the limit of $F \to 0$.

III. Results and Discussions

Firstly, we briefly discuss the crack properties at absolute zero temperature obtained by the lattice Green's function method.[1-7] We assume that an external force F is applied at a crystal site in a direction perpendicular to the cleavage plane and that any other boundary conditions are fixed displacement conditions. Further we consider the case where only one bond at the crack tip is stretched into its nonlinear cohesive range.

Using the Green's function of the cracked lattice D, the displacement u_0 of the atom at the center of a crack is given by

$$u_0 = \bar{D}_{00}F - 2D_{0N}f \,. \tag{37}$$

On the other hand, atomic displacement u_N (N indicates the site of the nonlinear bond, and $N = 1$ for the smallest crack) for the atom at the tip of the crack can be given by

$$u_N = D_{0N}F + D_{NN}f + D_{-NN}f .\tag{38}$$

Here D_{ij} is the Green's function matrix of the cracked lattice, and it is symmetric in i and j. D_{ij} can be obtained by solving the Dyson equation for the lattice

$$D_{ij} = G_{ij} + \sum_{m,n} G_{im}\,\delta\Phi_{mn}\,D_{nj} ,\tag{39}$$

where $\delta\Phi_{mn}$ denotes change in the force constant matrix due to the cracked bonds. F is the externally applied force on the central atom pair, and f is the force holding the atom pair together at the crack tip. These crack tip forces are the nonlinear cohesive bond forces, $f(u_N)$. By symmetry, all the displacements below the crack line are the negative of their counterparts above the crack line. The Greens function matrix is a set of constants which gives the compliance of the linear lattice system, and the method for computing these constants has been developed by Hsieh and Thomson.[3] For simplicity, we will use the bilinear force law for simulating the nonlinear cohesive bond force with linear region for $0 \le u \le \xi$ and the attractive cohesive bond region for $\xi \le u \le R$. The bond snapping is obtained by allowing $R \to \xi$. The equation for this force law is

$$f = 2B \begin{cases} u & \text{if } 0 \le u \le \xi; \\ (R - u)/(R - \xi) & \text{if } \xi \le u \le R; \\ 0 & \text{if } u \ge R. \end{cases}\tag{40}$$

The crack growth from size N (total lenght $2N + 1$) to $N + 1$ ($2N + 3$) generally occur by means of thermal fluctuations, and the activation energy for making such a jump at a specified value of the external force, F, is given by the expression

$$2\Delta E = \int_1^2 F(u_0)\,du_0 - F\,\Delta u_0 .\tag{41}$$

The integration is over the compliance curve from point 1 to point 2 in Fig. 2a. The initial and final points are given by the intersection of the horizontal line with the compliance curves. A backward fluctuation will occur for a jump from point 3 to point 2. The forward and backward activation energies are obtained most easily by a graphical method. These energies are shown as the cross-hatched areas in Fig. 2 and are given by

$$\Delta E_f = \tfrac{1}{2}(du_c)_{\max}\,[F_c^2/(F_c - F_h)](1 - F/F_h)^2 ,\tag{42}$$

$$\Delta E_b = \tfrac{1}{2}(du_h)_{\max}\,[F_h^2/(F_c - F_h)](F/F_h - 1)^2 ,\tag{43}$$

where $(du_c)_{\max}$ and $(du_h)_{\max}$ are given in Fig. 2a. By setting $\Delta E_f = \Delta E_b$, one can get the zero temperature Griffith value for the activation energy. We find that the zero temperature Griffith stress is given by the simple expression $F_G = \sqrt{F_c \cdot F_h}$.

As shown in Fig. 2b, as the force is increased, first a value of applied force, F_h, is reached, where two new solutions become possible (three in all) on the $N + 1$ branch of the crack corresponding to the tip atom jumping into the released portion of the bond function. Then, as the force increases beyond an applied force, F_c, no solutions remain

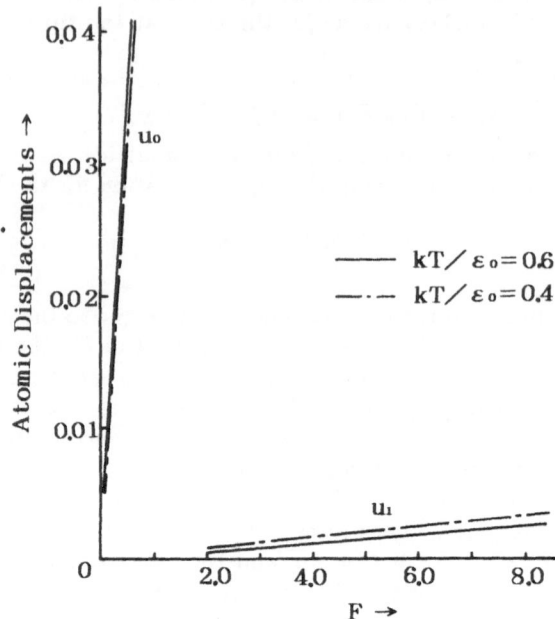

Figure 3. Atomic dispacements u_0 and u_1 at temperature $kT/\varepsilon_0 = 0.4$ (dot-dashed lines) and at $kT/\varepsilon_0 = 0.6$ (solid lines).

on the N branch where the force function is near its maximum value. For unstable loading configurations, F_c corresponds to spontaneous growth of the crack, and F_h corresponds to spontaneous healing of the crack. We shall now present a general discussion of this question based on our $2D$ results, which shows the narrow and restrictive conditions that must be met in order for the energy barriers to disappear for all possible external forces. This curve shows schematically the existence of lattice trapping of cracks, between the external loads F_c and F_h at zero temperature.

In order to obtain the thermodynamic driving force on the crack at finite temperatures, entropy terms must be added to J. This type of entropy effect can be taken into account efficiently within the present continuous CVM treatment. We have found that the lattice trapping decreases with increasing temperature. Figure 2b is a schematic drawing of u_0 as a function of the external force F.

In Fig. 3, we present the atomic displacements u_0 and u_1, by dot-dashed lines, as a function of the external load F at lower temperature $kT/\varepsilon_0 = 0.4$, where ε_0 denotes the pair interaction energy between nearest-neighbour atoms, with interatomic distance r_0, in the perfect lattice at absolute zero temperature. When one uses a Lennard-Jones potential, ε_0 is simply the energy parameter of the potential.[8] ASlso shown by solid lines in Fig. 3 are the atomic displacements u_0 and u_1 at higher temperatures $kT/\varepsilon_0 = 0.6$. One can see in Fig. 3 that the atomic displacement u_0 at higher temperature becomes larger compared to that at lower temperature. On the other hand, we have found that the atomic displacement u_1 at higher temperatures becomes smaller compared to that at lower temperature. This indicates that the stress intensity factor K_I becomes larger and the material becomes more ductile compared at those of zero temperature.

References

1. R. Thomson, *Solid State Phys.* **39**, 1 (1986).
2. R. Thomson, C. Hsieh, and V. Rana, *J. Appl. Phys.* **42**, 3154 (1971).
3. C. Hsieh and R. Thomson, *J. Appl. Phys.* **44**, 1051 (1973).
4. D. M. Esterling, *J. Appl. Phys.* **47**, 486 (1976).
5. E. R. Fuller, Jr. and R. M. Thomson, *Fracture Mechanics of Ceramics*, Vol.4, edited by R. C. Bradt, D. P. H. Hasselman, and F. F. Lange (Plenum Publishing Corporation, 1978); R. M. Thomson and E. R. Fuller, Jr., *Fracture Mechanics of Ceramics*, Vol.5, edited by R. C. Bradt, D. P. H. Hasselman, and F. F. Lange (Plenum Publishing Corporation, 1983) p. 253.
6. R. Thomson, V. K. Tewary, and K. Masuda-Jindo, *J. Mater. Res.* **2**, 619 (1987).
7. R. Thomson, S. J. Zhou, A. E. Carlsson, and V. K. Tewary, *Phys. Rev. B* **46**, 10613 (1992).
8. R. Kikuchi and A. Beldjenna, *Physica A* **182**, 617 (1992).
9. R. Kikuchi and L. -Q. Chen, *Computer Aided Innovation of New Materials II*, edited by M. Doyama, J. Kihara, M. Tanaka, and R. Yamamoto, (Elsevier Science Publishers B. V., 1993) p. 735.
10. R. Kikuchi, *Prog. Theor. Phys.* Suppl. **115**, 1 (1994).

Cluster Variation Method Calculations in Binary and Ternary *bcc* or *fcc* Phases

C. Colinet

Laboratoire de Thermodynamique et de Physico-Chimie Métallurgiques
UMR CNRS 5614, INPG/UJF
38402 Saint Martin d'Hēnes
FRANCE

Abstract

Usually, Cluster Variation Method calculations are performed at constant chemical potentials for the purpose of obtaining phase diagrams. However, it is sometimes useful to perform Cluster Variation Method calculations at constant composition in a single phase domain to look at the evolution of thermodynamic data and site occupations with temperature. This method allows also the comparison with results obtained with other models or using various Cluster Variation Method approximations. The resolution of the free energy minimization can be performed using the Natural Iteration Method initiated by Kikuchi or the Newton-Raphson procedure. When the Natural Iteration Method is used, it is necessary to introduce Lagrange parameters to take the constraints into account: normalization equations and fixed composition. These Lagrange parameters are related to the chemical potentials of the constituents. Some results obtained in binary and ternary systems possessing either *bcc* or *fcc* structure are shown, particularly the evolution of the free energy and the sites occupations with temperature and composition are displayed.

I. Introduction

The Cluster Variation Method (CVM) is now extensively used to perform phase diagram calculations specially in systems which display order-disorder transformations. Numerous calculations of prototype phase diagrams have been performed with the purpose of comparing either various CVM approximations, for example in *fcc* lattices to compare the tetrahedron and the tetrahedron-octahedron approximation, or to compare CVM and Monte-Carlo calculations (see Ref. 1). The input parameters are the cluster interactions: pair, triangle, tetrahedron, ..., interactions. These

Theory and Applications of the Cluster Variation and Path Probability Methods
Edited by J.L. Morán-López and J.M. Sanchez, Plenum Press, New York, 1996

313

interactions may be deduced from thermodynamic data (enthalpies of formation, temperatures of order-disorder transformations) or from diffuse scattering data. Since the last ten years *ab initio* calculations have been performed in a large number of alloys, the treatment of these results allow to obtain the cluster interactions and therefore to express the internal energy of the considered system.

In recent years, calculations of prototype phase diagrams in binary or ternary systems have shown that the subject was always open to research and discussion. For example Finel and Ducastelle[2] have calculated in binary *fcc* systems with antiferromagnetic first neighbor interactions the domain of the L' phase (P4/mmm) corresponding to three different occupancies of the simple cubic sublattices of the *fcc* lattice. More recently Marty[3] and Cenedese *et al.*[4] found that this L' phase was a ground state in ternary *fcc* systems for the $A_{0.5}B_{0.25}C_{0.25}$ composition and calculated the ternary coherent phase diagram within the CVM tetrahedron approximation. Our purpose in this work is to show that the domain of this L' phase extends from the limiting binary system to the ternary when antiferromagnetic first neighbor interactions occur in the three limiting binary systems.

In *bcc* binary alloys a four different occupancies of the *fcc* sublattices of the *bcc* lattice phase called F$\bar{4}$3m was observed by Inden[5] and Bell[6] when the ratio of the second neighbor pair interaction and the first neighbor pair interaction is greater than 2/3. Very recently Rubin[7] and Rubin and Finel[8] have calculated prototype *bcc* binary phase diagrams including this phase. The F$\bar{4}$3m phase can be a ground state in ternary systems for the composition $A_{0.5}B_{0.25}C_{0.25}$. Our purpose is to show the extention of the F$\bar{4}$3m phase from the binary to the ternary when the energetic conditions are fulfilled.

In this work we have performed calculations at constant composition for various temperatures, or at constant temperature for various compositions, to look at the site occupancies. In this case it is preferable to minimize the free energy of the system instead of minimizing the grand potential at fixed chemical potentials.

The outline of this paper is as follows: in section II we review some basic concepts of the CVM for multicomponent systems using the regular tetrahedron as the basic cluster in *fcc* phases. In section III the same is done using the irregular tetrahedron as the basic cluster in *bcc* phases. In section IV we explain how the L' phase based on *fcc* lattice extends from the binary to the ternary system when this phase is a ground state. In section V the stability of the F$\bar{4}$3m phase in binary *bcc* systems is presented. The development of this phase in a ternary system, where it is a ground state, is displayed. Finally section VI summarizes our conclusions.

II. Equations in the *fcc* Lattice

II.1 Tetrahedron Approximation

The simplest cluster including six first neighbor interactions is a regular tetrahedron as shown in Fig. 1. The sites called α, β, γ and δ define four interpenetrating simple cubic sublattices with a total number of $N/4$ lattice points each. The species A, B, C occupying the lattice points will be indicated by i, j, $\ldots$, indices. The mean occupation of a sublattice α by the species i is defined as $X_i^\alpha = N_i^\alpha/(N/4)$, where N_i^α is the number of species i on the sublattice α. The various superstructures can be defined by the equivalence of sublattice occupations

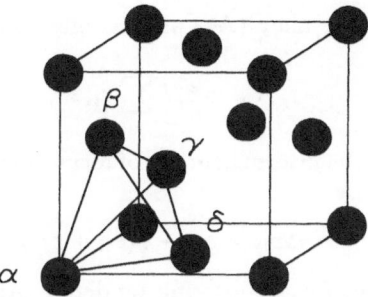

Figure 1. Configuration of the regular tetrahedron in the *fcc* lattice.

$$
\begin{aligned}
\text{A1:} &\quad X_i^\alpha = X_i^\beta = X_i^\gamma = X_i^\delta \,, \\
\text{L1}_0: &\quad X_i^\alpha = X_i^\beta \neq X_i^\gamma = X_i^\delta \,, \\
\text{L1}_2: &\quad X_i^\alpha = X_i^\beta = X_i^\gamma \neq X_i^\delta \,, \\
\text{L}': &\quad X_i^\alpha = X_i^\beta \neq X_i^\gamma \neq X_i^\delta \,.
\end{aligned}
\tag{1}
$$

Any tetrahedron in the *fcc* lattice always shares the vertices with each of the four sublattices, thus we have to consider only one type of tetrahedron. The basic variables of the treatment are the probabilities Z_{ijkl} that a tetrahedron takes a configuration $ijkl$ on the sublattice points α, β, γ and δ. The variables for subclusters, e.g. $U_{ijk}^{\alpha\beta\gamma}$ for the triangle, $Y_{ij}^{\alpha\beta}$ for the pairs of first neighbors and X_i^α for the point probabilities, are derived as linear combinations of the Z's by the reduction relations.

An additional quantity useful in characterizing the system is the mole fraction x_i of each species

$$
x_i = \tfrac{1}{4}(X_i^\alpha + X_i^\beta + X_i^\gamma + X_i^\delta).
\tag{2}
$$

II.2 Energy Expression

The internal energy can be written in terms of cluster interactions of the basic cluster, here tetrahedron interactions. Since in the *fcc* structure there is two tetrahedra per lattice point we may write

$$
U = 2N \sum_{ijkl} \varepsilon_{ijkl}\, Z_{ijkl} \,,
\tag{3}
$$

where ε_{ijkl} is the energy of the tetrahedron occupied by the species i, j, k and l. If we assume that only pairwise interactions ε_{ij} between next neighbors i and j are dominant, then the tetrahedron energy ε_{ijkl} can be expressed in terms of these pairwise interactions according to

$$
\varepsilon_{ijkl} = \tfrac{1}{2}(\varepsilon_{ij} + \varepsilon_{ik} + \varepsilon_{il} + \varepsilon_{jk} + \varepsilon_{jl} + \varepsilon_{kl}).
\tag{4}
$$

The fraction $\tfrac{1}{2}$ takes care of the fact that the first neighbor bonds are shared with two tetrahedra. It is common use to take the energy of the pure components in the same structure as the alloy (A1) as reference. This reference energy is for the constituent i,

$$U_0 = 6N \sum_i \varepsilon_{ii} x_i, \tag{5}$$

with this reference state, the internal energy of formation or of mixing can be rewritten in terms of $\Delta \varepsilon_{ij}$ defined by

$$\Delta \varepsilon_{ij} = \varepsilon_{ij} - \tfrac{1}{2}(\varepsilon_{ii} + \varepsilon_{jj}), \tag{6}$$

which take negative values for an ordering tendency. Alternatively some authors use either interchange energies defined by relation

$$W_{ij} = \varepsilon_{ii} + \varepsilon_{jj} - 2\varepsilon_{ij}, \tag{7}$$

or effective pair energies,

$$J_{ij} = \tfrac{1}{4}(\varepsilon_{ii} + \varepsilon_{jj} - 2\varepsilon_{ij}), \tag{8}$$

W_{ij} and J_{ij} take positive values for an ordering tendency or antiferromagnetic interactions.

II.3 Entropy Expression

In the case of the *fcc* lattice Golosov *et al.*[9] and van Baal[10] were the first authors to give the expression of the configurational entropy for a general ordered state using a regular tetrahedron as basic cluster,

$$\begin{aligned}
\frac{S}{k_B N} = &-\frac{5}{4} \sum_i \left[L(X_i^\alpha) + L(X_i^\beta) + L(X_i^\gamma) + L(X_i^\delta) \right] \\
&+ \sum_{ij} \left[L(Y_{ij}^{\alpha\beta}) + L(Y_{ij}^{\alpha\gamma}) + L(Y_{ij}^{\alpha\delta}) + L(Y_{ij}^{\beta\gamma}) + L(Y_{ij}^{\beta\delta}) + L(Y_{ij}^{\gamma\delta}) \right] \\
&- 2 \sum_{ijkl} L(Z_{ijkl}),
\end{aligned} \tag{9}$$

where k_B is the Boltzmann's constant and N the total number of lattice points. The function $L(x)$ is $x \ln x$.

II.4 Grand Potential

In a phase diagram calculation the chemical potentials are fixed. The thermodynamic function to be minimized in this instance is the grand potential GP defined as :

$$\mathrm{GP}(V, T, \mu_i) = U - TS - N \sum_i x_i \mu_i, \tag{10}$$

where μ_i are the chemical potentials of the species i. Note that in this equation and in the following we will consider molar thermodynamic data ($N = N_{Av}$). The size of the system is determined by the total number of atoms N. Since we are not treating vacancies, the chemical potentials are not all independent and we may introduce a new set of chemical potential μ_i^* (called effective chemical potentials) which fulfill the relation

$$\sum_i \mu_i^* = 0, \tag{11}$$

The equilibrium state at constant temperature is obtained by minimizing the effective grand potential defined by

$$\Omega = U - TS - N \sum_i x_i \mu_i^* .$$ (12)

In a system including n species it is easy to see that,

$$\mu_i^* = \mu_i - \frac{1}{n} \sum_i \mu_i ,$$ (13)

$$\Omega = \frac{N}{n} \sum_i \mu_i .$$ (14)

II.5 Minimization of Effective Grand Potential by Natural Iteration Method (NIM)

Combining Eqs. (3), (9) and (12) we obtain the CVM effective grand potential which must be minimized in order to yield the equilibrium state. Two different minimization algorithms have been used with success, the Newton-Raphson method and the Natural Iteration Method (NIM) developed by Kikuchi.[11] In the NIM the grand potential is minimized with respect to the tetrahedron probabilities using a Lagrange multiplier to take the normalization equation of the tetrahedron probabilities into account. The Newton-Raphson procedure works with a set of independent variables, Sanchez and de Fontaine,[12] and independently Aggarwal and Tanaka[13] proposed the use of correlation functions. Inden and Pitsch[1] have proposed a generalization to systems with any number of components. The grand potential is minimized with respect to these correlation functions using the Newton-Raphson method. This procedure was used with success by Marty and Cenedese[4] and Rubin and Finel[8,14] in binary and ternary systems. Compared with the NIM the Newton-Raphson method converges faster. In this work we have chosen the NIM to solve the minimization equations which can be written (see Ref. 15),

$$Z_{ijkl} = U_{ijkl}^{1/2} X_{ijkl}^{-5/8} \exp\left(\frac{-\varepsilon_{ijkl}}{k_B T}\right) \exp\left(\frac{\mu_i^* + \mu_j^* + \mu_k^* + \mu_l^*}{8 k_B T}\right) \exp\left(\frac{\lambda}{2 k_B T}\right),$$ (15)

here λ is a Lagrange parameter. U_{ijkl} and X_{ijkl} take account of the triangle and point probabilities

$$U_{ijkl} = U_{ijk}^{\alpha\beta\gamma} U_{ijl}^{\beta\gamma\delta} U_{jkl}^{\beta\gamma\delta} U_{ikl}^{\alpha\gamma\delta} ,$$ (16)

$$X_{ijkl} = X_i^\alpha X_j^\beta X_k^\gamma X_l^\delta .$$ (17)

The natural iteration algorithm will not be repeated here but may be found in the papers of Kikuchi and de Fontaine[15] or Kikuchi and Murray.[16] After the iteration has converged, the Lagrange multiplier λ is identified as

$$\lambda = \frac{\Omega}{N} .$$ (18)

II.6 Phase Diagram Calculation

The phase boundary between two phases is obtained when the grand potentials of the two phases are equal for given values of the effective chemical potentials. The procedure used to obtain the equilibrium between two phases has been described by Kikuchi and de Fontaine,[15] and Kikuchi and Murray.[16] In practice in binary systems we perform successive phase boundary calculations by increasing or decreasing temperature. In ternary systems isothermal sections are calculated by moving the effective chemical potential of one constituent.

II.7 Free Energy Minimization

When the minimization of the grand potential at fixed value of the effective chemical potential is performed the result of the calculation is the site occupancies and all the thermodynamic data, molar energy of formation, molar configurational entropy and molar free energy. The composition of the system for a given value of the effective chemical potential is obtained using the relations (2). However when it is desirable to obtain thermodynamic informations and site occupancies for a given composition, it is necessary to perform the minimization of the free energy. We will in the following explain the procedure in a ternary system. If the composition is constant the following relations must be fulfilled

$$\sum_{ijkl} f^A_{ijkl} Z_{ijkl} = 4x_A ,$$

$$\sum_{ijkl} f^B_{ijkl} Z_{ijkl} = 4x_B ,$$

$$\sum_{ijkl} f^C_{ijkl} Z_{ijkl} = 4x_C , \tag{19}$$

where f^A_{ijkl}, f^B_{ijkl}, f^C_{ijkl} are respectively the number of A, B and C atoms in the tetrahedron $ijkl$. It is important to quote that these three relations imply the normalization equation of the tetrahedron probabilities ($\sum_{ijkl} Z_{ijkl} = 1$). The molar free energy is minimized with respect to the tetrahedron probabilities introducing three Lagrange parameters λ_A, λ_B and λ_C to take into account the constraints due to the constant composition. It is equivalent to minimize the function:

$$\Phi = F + N\lambda_A \left(4x_A - \sum_{ijkl} f^A_{ijkl} Z_{ijkl} \right) + N\lambda_B \left(4x_B - \sum_{ijkl} f^B_{ijkl} Z_{ijkl} \right)$$

$$+ N\lambda_C \left(4x_C - \sum_{ijkl} f^C_{ijkl} Z_{ijkl} \right). \tag{20}$$

The minimization equations can be written in the same formalism as in the grand potential minimization,

$$Z_{ijkl} = U^{1/2}_{ijkl} X^{-5/8}_{ijkl} \exp\left(\frac{-\varepsilon_{ijkl}}{k_B T} \right) \exp\left(\frac{\lambda_A f^A_{ijkl}}{2k_B T} \right) \exp\left(\frac{\lambda_B f^B_{ijkl}}{2k_B T} \right) \exp\left(\frac{\lambda_C f^C_{ijkl}}{2k_B T} \right), \tag{21}$$

where U_{ijkl} and X_{ijkl} have been expressed previously [Eqs. (16) and (17)]. To solve the problem the NIM is used, but to obtain the Lagrange parameters from Eqs. (19) it is necessary to use a Newton-Raphson iteration procedure inside each major iteration. When the convergence of the procedure is obtained the chemical potentials are deduced from the Lagrange parameters;

$$\mu_A = 4\lambda_A, \quad \mu_B = 4\lambda_B, \quad \mu_C = 4\lambda_C. \tag{22}$$

When necessary, the effective chemical potentials and the grand potential Ω are calculated using the relations (13) and (14).

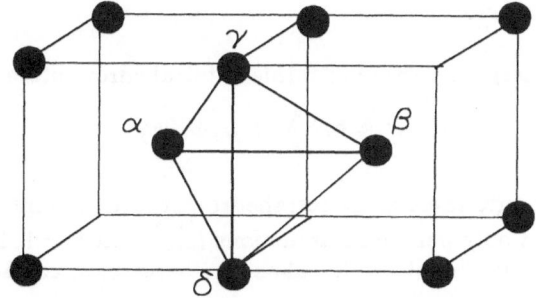

Figure 2. Configuration of the irregular tetrahedron in the bcc lattice.

The NIM is not very sensitive to the choice of the numerical values of the starting distribution. The free energy minimization does not conserve symmetries if many successive calculations are performed at a same composition for continuous variations of temperature or at a given temperature for continuous variations of composition, but an hysteresis is observed. In this way it is possible to observe transitions, but careful comparisons of the free energies of the possible phases must be done to determine the exact transition temperature. It is also possible to get the thermodynamic data of a metastable phase by imposing symmetry relations in the NIM solution.

III. Equations in the bcc Lattice

III.1 Tetrahedron Approximation

The simplest cluster including first and seconds neighbor interactions is an irregular tetrahedron as shown in Fig. 2. The sites called α, β, γ and δ define four interpenetrating fcc sublattices with a total number of $N/4$ lattice points each. The edges α–β and γ–δ are second neighbor bonds and the rest are first neighbor bonds. The mean occupation of a sublattice by the species i is defined as in the fcc lattice. The various superstructures can be defined by the equivalence of sublattice occupations

$$
\begin{aligned}
&\text{A2:} && X_i^\alpha = X_i^\beta = X_i^\gamma = X_i^\delta, \\
&\text{B2:} && X_i^\alpha = X_i^\beta \neq X_i^\gamma = X_i^\delta, \\
&\text{B32:} && X_i^\alpha = X_i^\gamma \neq X_i^\beta = X_i^\delta, \\
&\text{D0}_3\text{:} && X_i^\alpha = X_i^\beta \neq X_i^\gamma \neq X_i^\delta, \\
&\text{F}\bar{4}\text{3m:} && X_i^\alpha \neq X_i^\beta \neq X_i^\gamma \neq X_i^\delta.
\end{aligned}
\tag{23}
$$

Some ternary superstructures have got individual symbols: $L2_0$ for ternary B2 and $L2_1$ for ternary $D0_3$.

As in the fcc lattice, any tetrahedron always shares the vertices with each of the four sublattices. The basic variables of the treatment are the probabilities Z_{ijkl} that a tetrahedron takes a configuration $ijkl$ on the sublattice points α, β, γ and δ. The variables for subclusters e.g., $U_{ijl}^{\alpha\beta\delta}$ for the triangles, $V_{ij}^{\alpha\beta}$ and $V_{ij}^{\gamma\delta}$ for the pairs of second neighbors, $Y_{ik}^{\alpha\gamma}$ for the pairs of first neighbors and X_i^α for the point probabilities are derived as linear combinations of the Z's by the reduction relations.

III.2 Energy Expression

When the internal energy is written in terms of tetrahedron interactions we may write

$$U = 6N \sum_{ijkl} \varepsilon_{ijkl} Z_{ijkl}, \tag{24}$$

since in the *bcc* structure there is six tetrahedra per lattice point.

If we assume that only pairwise interactions ($\varepsilon_{ij}^{(k)}$ with $k = 1, 2$) between first and second neighbors are dominant the tetrahedron energy ε_{ijkl} can be written as

$$\varepsilon_{ijkl} = \tfrac{1}{6}\left(\varepsilon_{ik}^{(1)} + \varepsilon_{il}^{(1)} + \varepsilon_{jk}^{(1)} + \varepsilon_{jl}^{(1)}\right) + \tfrac{1}{4}\left(\varepsilon_{ij}^{(2)} + \varepsilon_{kl}^{(2)}\right). \tag{25}$$

The fractions $\tfrac{1}{6}$ and $\tfrac{1}{4}$ are introduced because first and second neighbor bonds are shared with 6 and 4 tetrahedra respectively.

III.3 Entropy Expression

In the case of the *bcc* lattice, Kikuchi and Sato[17] and Kikuchi and Van Baal[18] were the first authors to give the expression of the configurational entropy for a general ordered state using an irregular tetrahedron as basic cluster,

$$\frac{S}{k_{\mathrm{B}}N} = -6\sum_{ijkl} L(Z_{ijkl}) + 3\sum_{ijk}\left[L(U_{ijk}^{\alpha\beta\gamma}) + L(U_{ijk}^{\alpha\beta\delta}) + L(U_{ijkl}^{\alpha\gamma\delta}) + L(U_{ijk}^{\beta\gamma\delta})\right]$$

$$- \frac{3}{2}\sum_{ij}\left[L(V_{ij}^{\alpha\beta}) + L(V_{ij}^{\gamma\delta})\right]$$

$$- \sum_{ik}\left[L(Y_{ik}^{\alpha\gamma}) + L(Y_{ik}^{\alpha\delta}) + L(Y_{ik}^{\beta\gamma}) + L(Y_{ik}^{\beta\delta})\right]$$

$$+ \frac{1}{4}\sum_{i}\left[L(X_i^{\alpha}) + L(X_i^{\beta}) + L(X_i^{\gamma}) + L(X_i^{\delta})\right]. \tag{26}$$

III.4 Minimization of Grand Potential by NIM

The equilibrium state at constant temperature is obtained by minimization of the effective grand potential written for one mol of the constituents

$$\Omega = U - TS - N\sum_{i} x_i \mu_i^*. \tag{27}$$

The minimization equations can be written as

$$Z_{ijkl} = U_{ijkl}^{1/2} V_{ijkl}^{-1/4} Y_{ijkl}^{-1/6} X_{ijkl}^{1/24}$$

$$\times \exp\left(\frac{-\varepsilon_{ijkl}}{k_{\mathrm{B}}T}\right)\exp\left(\frac{\mu_i^* + \mu_j^* + \mu_k^* + \mu_l^*}{24k_{\mathrm{B}}T}\right)\exp\left(\frac{\lambda}{6k_{\mathrm{B}}T}\right), \tag{28}$$

where

$$U_{ijkl} = U_{ijk}^{\alpha\beta\gamma} U_{ijl}^{\alpha\beta\delta} U_{jkl}^{\beta\gamma\delta} U_{ikl}^{\alpha\gamma\delta}, \tag{29}$$

$$Y_{ijkl} = Y_{ik}^{\alpha\gamma} Y_{il}^{\alpha\delta} Y_{jk}^{\beta\gamma} Y_{jl}^{\beta\delta}, \tag{30}$$

$$V_{ijkl} = V_{ij}^{\alpha\beta} V_{kl}^{\gamma\delta}, \tag{31}$$

$$X_{ijkl} = X_i^\alpha X_j^\beta X_k^\gamma X_l^\delta. \tag{32}$$

After the iteration procedure has converged, the Lagrange multiplier λ is identified as

$$\lambda = \frac{\Omega}{N}. \tag{33}$$

III.5 Free Energy Minimization

The same procedure as proposed in *fcc* phases is used in *bcc* phases. In the case of a ternary phase the minimization equations are

$$Z_{ijkl} = U_{ijkl}^{1/2} V_{ijkl}^{-1/4} Y_{ijkl}^{-1/6} X_{ijkl}^{1/24}$$

$$\times \exp\left(\frac{-\varepsilon_{ijkl}}{k_B T}\right) \exp\left(\frac{\lambda_A f_{ijkl}^A}{6k_B T}\right) \exp\left(\frac{\lambda_B f_{ijkl}^B}{6k_B T}\right) \exp\left(\frac{\lambda_C f_{ijkl}^C}{6k_B T}\right). \tag{34}$$

The minimization equations are solved using the NIM with minor iterative procedure to get the three Lagrange parameters λ_A, λ_B and λ_C. When the convergence is obtained the chemical potentials of A, B and C are obtained by the relations

$$\mu_A = 4\lambda_A, \quad \mu_B = 4\lambda_B, \quad \mu_C = 4\lambda_C. \tag{35}$$

The procedure used in the calculations is the same as explained for *fcc* systems.

IV. Results in fcc Lattice

IV.1 Binary Systems

Finel and Ducastelle[2] presented the *fcc* prototype binary phase diagram with antiferromagnetic first neighbor interactions. The CVM in the tetrahedron and in the tetrahedron-octahedron approximation were presented. Finel and Ducastelle[2] introduced in the calculation of the phase diagram the L' phase where the site occupations are

$$X_i^\alpha = X_i^\beta \neq X_i^\gamma \neq X_i^\delta. \tag{36}$$

This phase has some unfamiliar characteristics, since, even at zero temperature, it does not correspond to a perfectly ordered state. The L'–L1$_0$ transition was found to be of second order, the L'–L1$_2$ transition was found of first order. Finel and Ducastelle[2] observed that the free-energy difference between L' and L1$_0$ is very small. To test these points we have performed free energy minimization at constant composition and temperature in a binary system with antiferromagnetic first neighbor interactions. The value of the pair interaction parameter is $\Delta\varepsilon_{AB} = -2000\, R\, J\, mol^{-1}$ where R is the perfect gas constant. In Table I the values of the molar free energy obtained for the composition $x_A = 0.58$ are reported as function of temperature for the L' and L1$_0$ surstructures. In Figs. 3a and 3b the sites occupancies are displayed. One observes clearly that in L' phase the occupancies of γ and δ sites are slightly different and become equal for the transition temperature which is about 600 K for the interaction parameter quoted above. The difference of the molar free energies of the L' and L1$_0$ surstructures is extremely small.

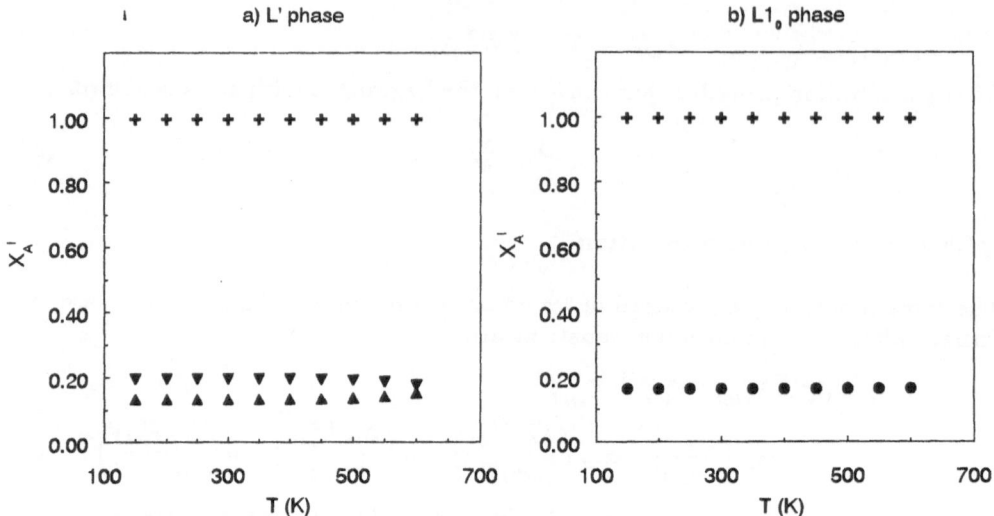

Figure 3. Sites occupations $X_A^\alpha = X_A^\beta$ (crosses), X_A^γ (up-triangles), X_A^δ (down-triangles) in the L′ binary phase (a), $X_A^\alpha = X_A^\beta$ (crosses) $X_A^\gamma = X_A^\delta$ (filled-circles) in the metastable L1$_0$ phase (b) between 150 K and 600 K. The interaction parameter in the binary system is $\Delta\varepsilon_{AB} = -2000\ R\,\mathrm{J\,mol^{-1}}$.

Table I

Free energy values of the L′ and L1$_0$ phases as a function of temperature for the composition $x = 0.58$. The interaction parameter is $\Delta\varepsilon = -2000\ R\,\mathrm{J\,mol^{-1}}$.

T(K)	L′ phase	L1$_0$ phase
	F(J/mol)	
150	−61433.00	−61432.63
200	−61512.43	−61511.94
250	−61581.86	−61691.26
300	−61671.29	−61670.60
350	−61750.73	−61750.01
400	−61830.20	−61829.53
450	−61909.75	−61909.22
500	−61989.47	−61989.14
550	−62069.49	−62069.37
600	−62149.98	−62149.98

Table II

Free energy values of the L' and L1$_0$ phases as a function of composition at $T = 150$ K. The interaction parameter is $\Delta\varepsilon = -2000\ R\,\mathrm{J\,mol}^{-1}$.

x	F(J/mol)	
	L' phase	L1$_0$ phase
0.6	−60119.36	−60116.84
0.595	−60448.60	−60447.12
0.59	−60777.41	−60776.50
0.585	−61105.58	−61105.01
0.58	−61443.00	−61432.63
0.575	−61759.62	−61759.38
0.57	−62085.41	−62085.25
0.565	−62410.33	−62410.23
0.56	−62734.35	−62734.29
0.55	−63057.46	−63379.58
0.545	−63700.74	−63700.72
0.54	−64020.81	−64020.80
0.535	−64339.73	−64339.72
0.53	−64657.40	−64657.40

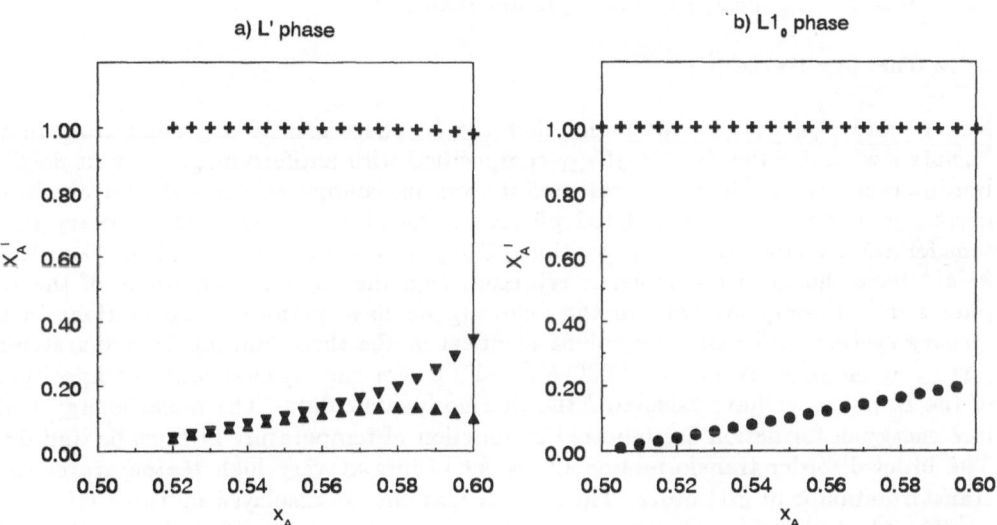

Figure 4. Sites occupations $X_A^\alpha = X_A^\beta$ (crosses), X_A^γ (up-triangles), X_A^δ (down-triangles) in the L' phase (a) and $X_A^\alpha = X_A^\beta$ (crosses), $X_A^\gamma = X_A^\delta$ (filled-circles) in the metastable L1$_0$ phase (b) at $T = 150$ K as a function of composition. The interaction parameter in the binary system is $\Delta\varepsilon = -2000\ R\,\mathrm{J\,mol}^{-1}$.

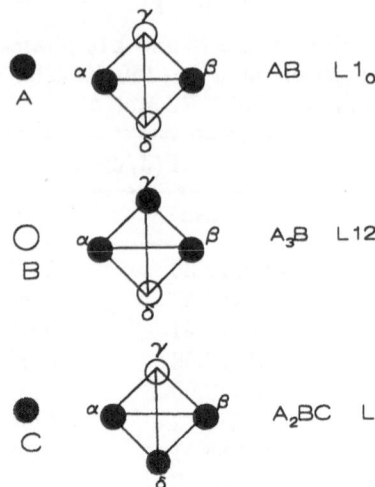

Figure 5. Tetrahedron configurations in binary and ternary *fcc* ordered phases.

The comparison of the L' and L1$_0$ phases has also been done at constant temperature ($T = 150$ K) as function of the composition. In Table II the values of the molar free energy of the two phases have been reported as function of composition. In Figs. 4a and 4b the site occupancies in phases L' and L1$_0$ are displayed. Looking at these results it is clear that the L' phase becomes more stable when the composition increases, however the two phase domain L'/L1$_2$ becomes more stable than L' phase alone until the x_A molar fraction is greater than 0.58.

IV.2 Ternary System

Marty[3] and Cenedese *et al.*[4] found that the L' phase can be a ground state in a ternary system for the $A_{0.5}B_{0.25}C_{0.25}$ composition with antiferromagnetic first neighbor interactions. In Fig. 5 are presented the various configurations of the tetrahedron corresponding to perfectly ordered phases in the binaries and in the ternary system for the $A_{0.5}B_{0.25}C_{0.25}$ composition. Using a very elegant formalism, Cenedese *et al.*[4] have shown in an isometric representation the stability conditions of the L' phases in a ternary system. In the following we have performed calculations in a ternary system with pair interactions identical in the three limiting binary systems ($\Delta\varepsilon_{AB} = \Delta\varepsilon_{AC} = \Delta\varepsilon_{BC} = -2000 \; R\,\mathrm{J\,mol^{-1}}$). For the stoichiometric composition of the L' phase we have calculated the thermodynamic data. The molar energy and free energy of formation are reported as function of temperature in Figs. 6a and 6b. The order-disorder transformation L' $\rightarrow$ A1 occurs at very high temperature; the transformation is of first order. The site occupations are displayed in Table III.

The phase diagram calculations performed by Cenedese *et al.*[4] lead to isothermal sections at relatively high temperatures, then the L' phase in the binaries does not appear. We have performed calculations at very low temperatures along the line which joins the point $x_A = 0.5$, $x_B = 0.25$, $x_C = 0.25$ in the ternary system to the point $x_A = 0.58$, $x_B = 0.42$ in the limiting binary system to verify the stability of the L' phase at low temperatures and to observe the evolution of the site occupancies. The results are presented in Table IV. We have also performed calculations for $x_B = x_C$ and increasing x_A. The results presented in Table V show the evolution of the site

Table III

Site occupations in the L$'$ phase as function of temperature for the composition $x_A = 0.5$, $x_B = x_C = 0.25$. The interaction parameters in the three limiting binary systems are the same ($\Delta\varepsilon = -2000\ R\,J\,mol^{-1}$).

		Site occupations			
		α	β	γ	δ
200 K	A	0.9959	0.9959	0.1253	0.1837
	B	0.0037	0.0037	0.8380	0.7837
	C	0.0004	0.0004	0.0267	0.0325
600 K	A	0.9946	0.9946	0.0141	0.1694
	B	0.0021	0.0021	0.9472	0.4158
	C	0.0033	0.0033	0.0387	0.4148
1000 K	A	0.9977	0.9977	0.0012	0.0482
	B	0.0016	0.0016	0.9971	0.0948
	C	0.0006	0.0006	0.0017	0.8570
1400 K	A	0.9924	0.9924	0.0076	0.0076
	B	0.0038	0.0038	0.9874	0.0050
	C	0.0038	0.0038	0.0050	0.9874
1700 K	A	0.9627	0.9627	0.0373	0.0373
	B	0.0186	0.0186	0.9396	0.0231
	C	0.0186	0.0186	0.0231	0.9396
1800 K	A	0.9250	0.9250	0.0750	0.0750
metastable	B	0.0375	0.0375	0.8793	0.0457
	C	0.0375	0.0375	0.0457	0.8793

occupancies. For x_A greater 0.683, the L2$_1$ phase is more stable than the L$'$ phase. As the transition is of first order, the two phase domain between L$'$ and L2$_1$ extends from $x_A = 0.63$ to $x_A = 0.71$.

Cenedese et al.[4] pointed out the presence of a L$''$ phase for the equiatomic composition of A, B and C when the interaction parameters are the same in the three limiting binary systems. This phase is characterized by different occupations of the sites α, β, γ, and δ. We have checked this possibility and performed calculations for this composition at various temperatures. The site occupancies are reported in Table VI. The molar energy and free energy of formation of this phase and of the disordered A1 phase are reported on Figs. 7a and 7b. It appears that this phase is more stable than A1 in a large temperature domain. However it seems necessary to look at the possibility of a three L$'$ phases equilibrium. It seems that this L$''$ phase appears only in the case of three limiting binaries possessing the same pair interactions.

In the course of the present study we have considered systems where the interaction parameters were different in the three limiting binaries provided that the L$'$ phase was a ground state for the composition $x_A = 0.5$, $x_B = x_C = 0.25$. In many cases the congruent transformation L$'$ $\rightarrow$ A1 does not occur, according to the interaction parameters the L1$_2$ or L1$_0$ phase appears more stable when increasing temperature.

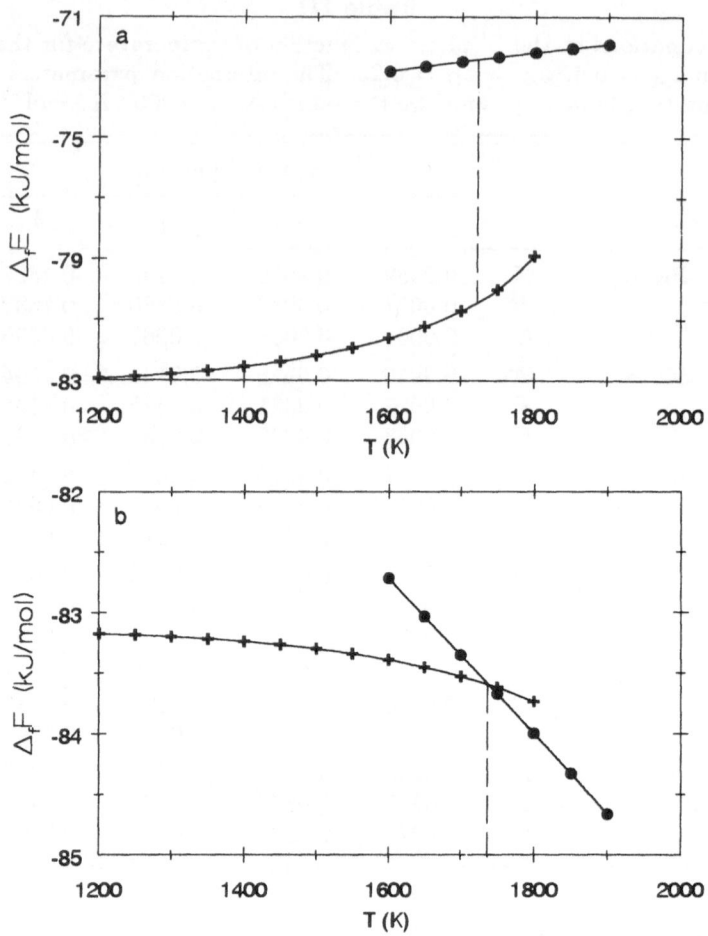

Figure 6. Molar energy (a) and free energy (b) of formation of the L' (crosses) and A1 (filled-circles) phases as function of temperature in a ternary A-B-C system. The composition is $x_A = 0.5$, $x_B = 0.25$, $x_C = 0.25$. The interaction parameters in the three limiting binary systems are the same $\Delta\varepsilon = -2000\ R\,\mathrm{J\,mol^{-1}}$.

However it is necessary to perform phase diagram calculations to investigate the possible occurrence of a two phase domain.

V. Results in bcc Lattice

V.1 Binary Systems

In binary *bcc* phases four different ordered phases B2, B32, D0$_3$ and F$\bar{4}$3m appear if first and second neighbor pair interactions are taken into account. When these pair interactions are negative, $\Delta\varepsilon_{AB}^{(1)} < 0$ and $\Delta\varepsilon_{AB}^{(2)} < 0$ (or $J^{(1)} > 0$ and $J^{(2)} > 0$) the ordered phases which appear depend of the ratio $\Delta\varepsilon_{AB}^{(2)}/\Delta\varepsilon_{AB}^{(1)}$. If $\Delta\varepsilon_{AB}^{(2)}/\Delta\varepsilon_{AB}^{(1)} < 2/3$

Table IV

Site occupations along the line which joins the point $x_A = 0.5$, $x_B = 0.25$, $x_C = 0.25$ in the ternary fcc system to the point $x_A = 0.58$, $x_B = 0.42$ in the limiting binary A-B system at $T = 150$ K. The interaction parameters are $\Delta\varepsilon_{AB} = \Delta\varepsilon_{AC} = \Delta\varepsilon_{BC} = -2000\ R\,\mathrm{J\,mol^{-1}}$.

Composition		Site occupations			
		α	β	γ	δ
A	0.5	1	1	0	0
B	0.25	0	0	1	0
C	0.25	0	0	0	1
A	0.5160	0.9998	0.9998	0.0002	0.0643
B	0.2840	0.0000	0.0000	0.9977	0.1382
C	0.2000	0.0002	0.0002	0.0021	0.7975
A	0.5320	0.9982	0.9982	0.0030	0.1286
B	0.3180	0.0005	0.0005	0.9820	0.2891
C	0.1500	0.0013	0.0013	0.0151	0.5824
A	0.5480	0.9956	0.9956	0.0196	0.1813
B	0.3520	0.0019	0.0019	0.9344	0.4699
C	0.1000	0.0026	0.0026	0.0461	0.3488
A	0.5640	0.9966	0.9966	0.0939	0.1690
B	0.3860	0.0023	0.0023	0.8277	0.7118
C	0.0500	0.0012	0.0012	0.0784	0.1193
A	0.5800	0.9951	0.9951	0.1330	0.1969
B	0.4200	0.0049	0.0049	0.8670	0.8031
C	0.0000	0.0000	0.0000	0.0000	0.0000

B2 and DO$_3$ are the ground states, if $\Delta\varepsilon_{AB}^{(2)}/\Delta\varepsilon_{AB}^{(1)} > 2/3$ B32, DO$_3$ and F$\bar{4}$3m are the ground states.

The first authors to perform CVM calculations using an irregular tetrahedron as basic cluster are Golosov and Tolstik:[19,20] However in the phase diagram calculation for $\Delta\varepsilon_{AB}^{(2)}/\Delta\varepsilon_{AB}^{(1)} = 1$ these authors did not take into account the presence of the F$\bar{4}$3m phase. Later Sluiter et al.[21] studied some prototype bcc phase diagrams but they also did not take into account the presence of the F$\bar{4}$3m phase when this phase is supposed to appear. These authors presented very interesting phase diagrams when the ratio $\Delta\varepsilon_{AB}^{(2)}/\Delta\varepsilon_{AB}^{(1)}$ is around 2/3. Ackermann et al.[22] calculated a series of prototype phase diagrams for different values of the first and second nearest-neighbor pair interactions, however situations where the F$\bar{4}$3m phase is present in the phase diagram have not been studied. One must point out that these authors performed Monte Carlo calculations and found a very nice agreement between phase diagrams calculated by the CVM in the irregular tetrahedron approximation and by Monte Carlo simulations.

The presence of the F$\bar{4}$3m phase was quoted by Inden;[5] phase diagrams including this phase were calculated by this author but using the Bragg and Williams model, unfortunately the point approximation is not at all satisfactory since it does not take into account the correlations among neighboring points.

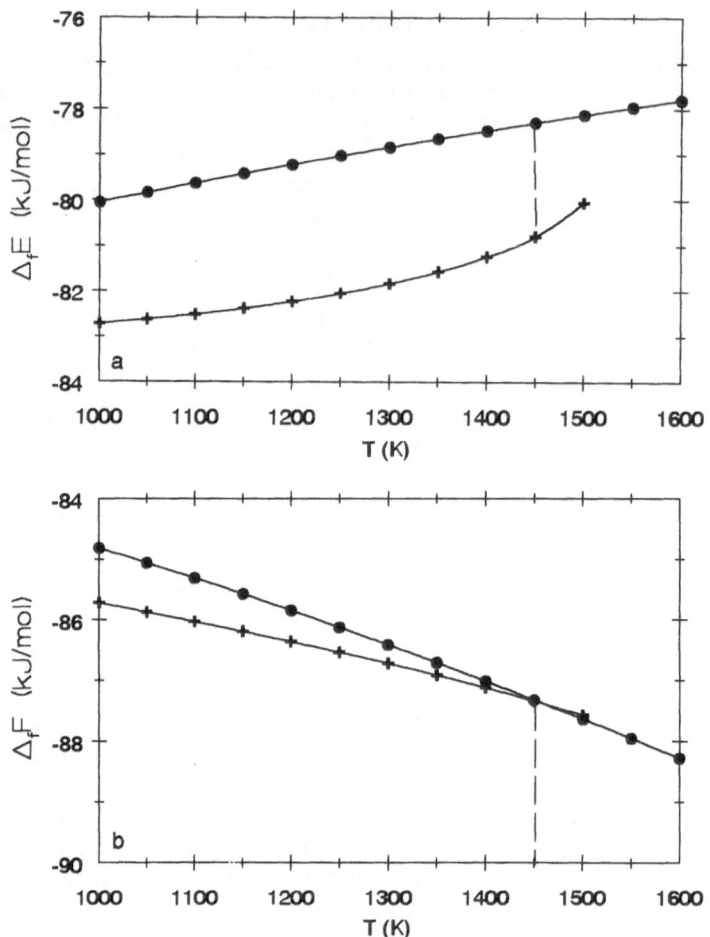

Figure 7. Molar energy (a) and free energy (b) of formation of the L'' (crosses) and A1 (filled-circles) phases as function of temperature in a ternary A-B-C system. The compositions of A, B and C are equal. The interaction parameters in the three limiting binary systems are the same ($\Delta \varepsilon = -2000 \, R \, \text{J} \, \text{mol}^{-1}$).

It is also interesting to quote the work proposed by Bell.[6] This author used a method called Constant Coupling Approximation. A careful analysis of the entropic term shows that it is the same as obtained using a quasi-chemical model using the irregular tetrahedron as the basic cluster, so in the entropic term only tetrahedron and point probabilities appear. This treatment does not take into account the overlapping figures which are the first and second neighbor pairs and the isoceles triangles. Bell[6] calculated a large number of prototype phase diagrams for various ratio of $\Delta \varepsilon_{AB}^{(2)} / \Delta \varepsilon_{AB}^{(1)}$ ($-0.25 < \Delta \varepsilon_{AB}^{(2)} / \Delta \varepsilon_{AB}^{(1)} < 0.85$). The main features of the diagrams presented by Bell[6] are similar with those obtained by Ackermann *et al.*[22] and Sluiter *et al.*[21] except for the value $\Delta \varepsilon_{AB}^{(2)} / \Delta \varepsilon_{AB}^{(1)} = 2/3$; for this value of the ratio of the second neighbor pair interactions and first neighbor pair interactions, the B2 and B32 phases are degenerated. Since frustration effects are not taken into account in the quasi-chemical model, the phase diagram calculated by Bell[6] cannot be correct.

Table V

Site occupations along the line $x_B = x_C$ at $T = 150$ K in the ternary fcc system. The interaction parameters are the same as in Table IV.

Composition		α	β	γ	δ
			Site occupations		
A	0.5	1	1	0	0
B	0.25	0	0	1	0
C	0.25	0	0	0	1
A	0.55	0.9982	0.9982	0.1018	0.1018
B	0.225	0.0009	0.0009	0.8981	0.0001
C	0.225	0.0009	0.0009	0.0001	0.8981
A	0.60	0.9892	0.9892	0.2108	0.2108
B	0.20	0.0054	0.0054	0.7860	0.0032
C	0.20	0.0054	0.0054	0.0032	0.7860
A	0.6500	0.9830	0.9830	0.3170	0.3170
B	0.1750	0.0085	0.0085	0.6530	0.0301
C	0.1750	0.0085	0.0085	0.0301	0.6530
A	0.7000	0.9331	0.9331	0.0008	0.9331
B	0.1500	0.0335	0.0335	0.4996	0.0335
C	0.1500	0.0335	0.0335	0.4996	0.0335

Table VI

Site occupations in the L″ phase as function of temperature for the composition $x_A = x_B = x_C$. The interaction parameters in the three limiting binary systems are the same ($\Delta\varepsilon = -2000\ R\,\mathrm{J\,mol^{-1}}$).

		α	β	γ	δ
			site occupations		
200 K	A	0.3333	0.9760	0.0120	0.0120
	B	0.3333	0.0120	0.9760	0.0120
	C	0.3333	0.0120	0.0120	0.9760
600 K	A	0.3333	0.9717	0.0142	0.0142
	B	0.3333	0.0142	0.9717	0.0142
	C	0.3333	0.0142	0.0142	0.9717
1000 K	A	0.3333	0.9487	0.0257	0.0257
	B	0.3333	0.0257	0.9487	0.0257
	C	0.3333	0.0257	0.0257	0.9487
1400 K	A	0.3333	0.8594	0.0703	0.0703
	B	0.3333	0.0703	0.8594	0.0703
	C	0.3333	0.0703	0.0703	0.8594
1700 K metastable	A	0.3333	0.7653	0.1173	0.1173
	B	0.3333	0.1173	0.7653	0.1173
	C	0.3333	0.1173	0.1173	0.7653

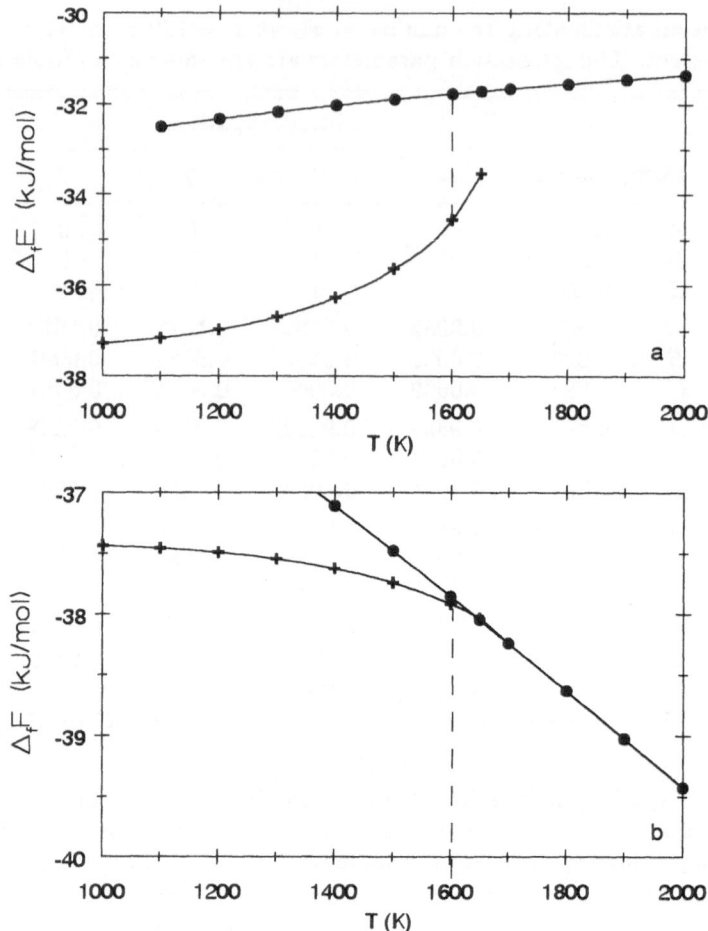

Figure 8. Molar energy (a) and free energy (b) of formation of the stoechiometric DO_3 (crosses) and A2 (filled-circles) phases in a binary *bcc* system. The first and second neighbor interaction parameters are $\Delta\varepsilon^{(1)} = -1500 \ R\,\mathrm{J}\,\mathrm{mol}^{-1}$ and $\Delta\varepsilon^{(2)} = -1000 \ R\,\mathrm{J}\,\mathrm{mol}^{-1}$.

Recently Rubin[7] or Rubin and Finel[8] used the CVM in the irregular tetrahedron approximation to calculate the *bcc* phase diagrams for different values of the ratio $\Delta\varepsilon_{AB}^{(2)}/\Delta\varepsilon_{AB}^{(1)}$. The case where this ratio is equal to 1 was studied carefully. The F$\bar{4}$3m domain was found between the B32 and DO_3 one phase domains. The phase transformations F$\bar{4}$3m $\rightarrow$ B32 and F$\bar{4}$3m $\rightarrow$ DO_3 are both of second order. Rubin and Finel[8] have also shown that the $DO_3 \rightarrow$ A2 transformation is of second order when $\Delta\varepsilon_{AB}^{(2)}/\Delta\varepsilon_{AB}^{(1)} = 1$. In Figs. 8a, 8b, and 9a and 9b the molar energy and free energy of formation of the stoechiometric DO_3 phase have been displayed as function of temperature respectively for the values of the ratio $\Delta\varepsilon_{AB}^{(2)}/\Delta\varepsilon_{AB}^{(1)} = 2/3$ and 1. From these figures it is easy to observe that for the ratio equal to 2/3 the $DO_3 \rightarrow$ A2 transformation is of first order, and that for the ratio equal to 1 the transformation is of second order.

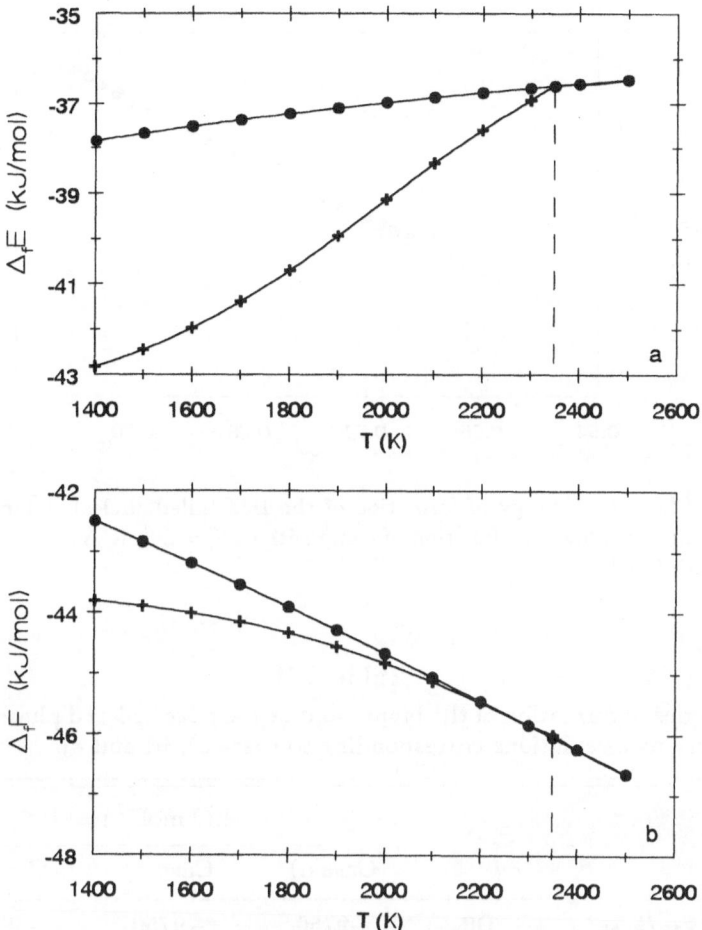

Figure 9. Molar energy (a) and free energy (b) of formation of the DO_3 (crosses) and A2 (filled-circles) phases in a binary *bcc* system. The first and second neighbor interaction parameters are $\Delta\varepsilon^{(1)} = -1500 \ R \, \mathrm{J \, mol^{-1}}$ and $\Delta\varepsilon^{(2)} = -1500 \ R \, \mathrm{J \, mol^{-1}}$.

Considering the prototype phase diagram presented in literature, we decided to look at the stability of the $F\bar{4}3m$ using the CVM in the framework of the irregular tetrahedron approximation. Molar free energy calculations have been performed for various values of the ratio $\Delta\varepsilon^{(2)}_{AB}/\Delta\varepsilon^{(1)}_{AB}$. Figures 10 and 11 present the values of the molar free energy as function of composition for $\Delta\varepsilon^{(2)}_{AB}/\Delta\varepsilon^{(1)}_{AB} = 2.5/3$ and $3.5/3$. The stability of the $F\bar{4}3m$ phase has also been studied for $\Delta\varepsilon^{(2)}_{AB}/\Delta\varepsilon^{(1)}_{AB} = 2/3$, the phase is stable in a very small composition domain inside the B32 and DO_3 phase stability domains.

A last point concerning binary phase diagrams has been studied. Inden[5] and Bell[6] quoted that the DO_3 phase has a particular site occupation for $\Delta\varepsilon^{(2)}_{AB}/\Delta\varepsilon^{(1)}_{AB} = 2/3$,

$$DO_3^* : \qquad x_i^\alpha = x_i^\beta = x_i^\gamma \neq x_i^\delta .$$

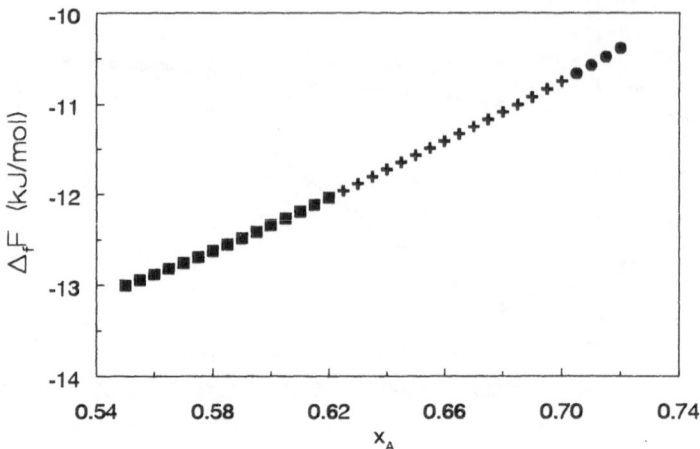

Figure 10. Molar free energy of formation of the B32 (filled-circles), F$\bar{4}$3m (crosses) and DO$_3$ (filled-squares) phases as function of composition. $T = 200$ K $\Delta\varepsilon^{(1)} = -3000$ J mol^{-1}, $\Delta\varepsilon^{(2)} = -2500$ J mol^{-1}.

Table VII

Energies of formation of the binary and ternary *bcc* ordered phases used in various calculations corresponding to cases a), b) and c).

| | | $E(\mathrm{J\,mol^{-1}})$ | | |
		Case a)	Case b)	Case c)
$A_{0.75}B_{0.25}$	DO$_3$	-9750	-9750	-9750
$A_{0.5}B_{0.5}$	B2	-12000	-12000	-12000
$A_{0.5}B_{0.5}$	B32	-13500	-13500	-13500
$A_{0.25}B_{0.75}$	DO$_3$	-9750	-9750	-9750
$B_{0.75}C_{0.25}$	DO$_3$	0	0	-6000
$B_{0.5}C_{0.5}$	B2	0	0	-12000
$B_{0.5}C_{0.5}$	B32	0	0	-6000
$B_{0.25}C_{0.75}$	DO$_3$	0	0	-6000
$A_{0.75}C_{0.25}$	DO$_3$	-11250	-11250	-11250
$A_{0.5}C_{0.5}$	B2	-12000	-12000	-12000
$A_{0.5}C_{0.5}$	B32	-16500	-16500	-16500
$A_{0.25}C_{0.75}$	DO$_3$	-11250	-11250	-11250
$A_{0.5}B_{0.25}C_{0.25}$	L1$_2$	-12000	-12000	-12000
$A_{0.5}B_{0.25}C_{0.25}$	F$\bar{4}$3m	-15000	-18000	-18000
$A_{0.25}B_{0.5}C_{0.25}$	L1$_2$	-11250	-11250	-17250
$A_{0.25}B_{0.5}C_{0.25}$	F$\bar{4}$3m	-9750	-9750	-12750
$A_{0.25}B_{0.5}C_{0.25}$	L1$_2$	-9750	-9750	-15750
$A_{0.25}B_{0.5}C_{0.25}$	F$\bar{4}$3m	-11250	-11250	-14250

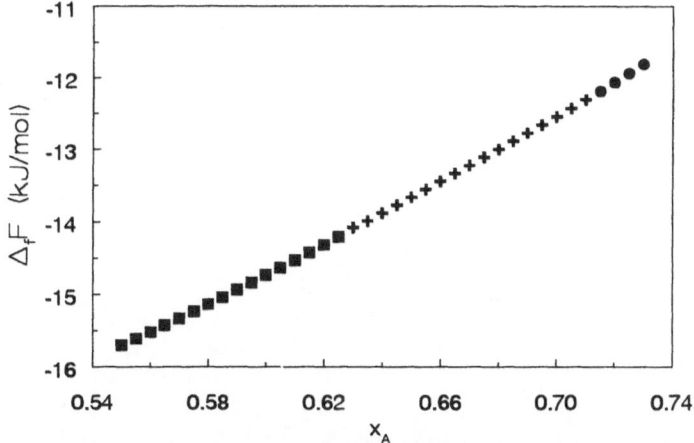

Figure 11. Molar free energy of formation of the B32 (filled-circles), F$\bar{4}$3m (crosses) and DO$_3$ (filled-squares) phases as function of composition $T = 200$ K. $\Delta\varepsilon^{(1)} = -3000$ J mol^{-1}, $\Delta\varepsilon^{(2)} = -3500$ J mol^{-1}.

Table VIII

Site occupations at $T = 200$ K along the line $x_B = x_C$ in the case a).

Composition		Site occupations			
		α	β	γ	δ
A	0.5	1.0000	0	1.0000	0.0000
B	0.25	0.0000	0.5000	0.0000	0.0000
C	0.25	0.0000	0.5000	0.0000	1.0000
A	0.550	0.9999	0.1001	0.9999	0.1001
B	0.225	0.0000	0.4500	0.0000	0.4500
C	0.225	0.0000	0.4500	0.0000	0.4500
A	0.6	0.9993	0.2007	0.9993	0.2007
B	0.2	0.0004	0.3996	0.0004	0.3996
C	0.2	0.0003	0.3997	0.0003	0.3997
A	0.65	0.9998	0.0552	0.9683	0.5768
B	0.175	0.0002	0.4715	0.0165	0.2118
C	0.175	0.0000	0.4733	0.0152	0.2115
A	0.700	0.9997	0.0027	0.9041	0.8934
B	0.150	0.0003	0.4979	0.0483	0.0536
C	0.150	0.0000	0.4995	0.0476	0.0529
A	0.750	1.0000	0.0012	0.9994	0.9994
B	0.125	0.0000	0.4994	0.0003	0.0003
C	0.125	0.0000	0.4994	0.0003	0.0003

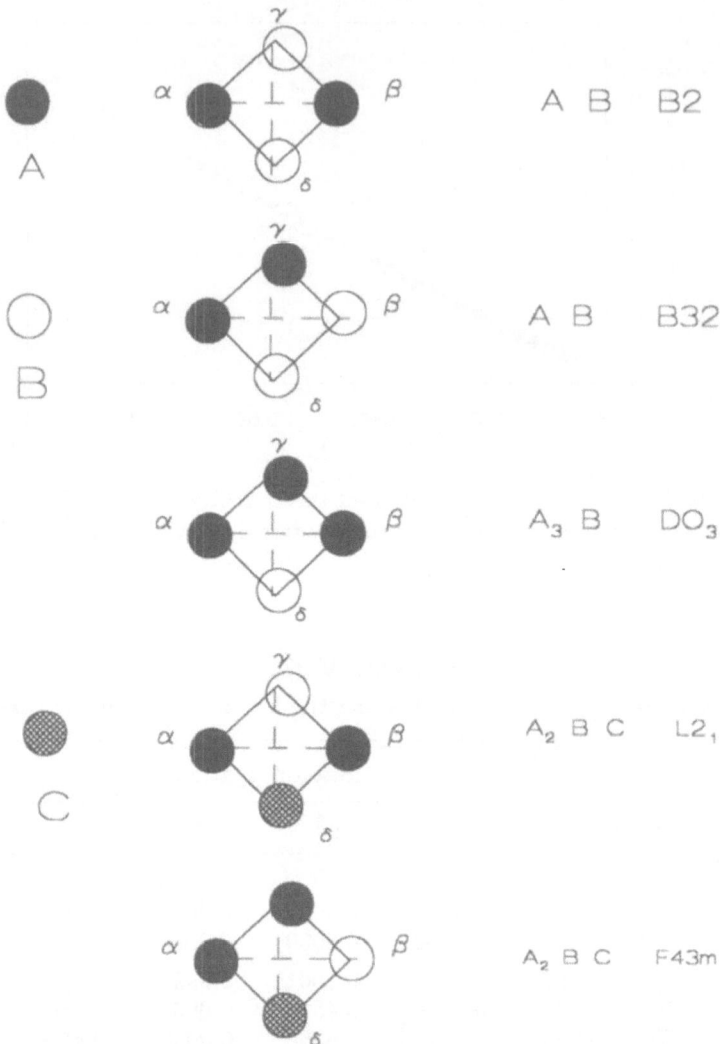

Figure 12. Tetrahedron configurations in binary and ternary *bcc* ordered phases.

We have performed free energy minimization calculations in the D0$_3$ stability domain. For the stoechiometric composition $x_A = 0.75$ $x_B = 0.25$ we observed that the occupations of the α, β and γ sites are the same at low temperature but when the temperature increases it is no more the case.

V.2 Ternary Systems

In ternary *bcc* systems two ternary ordered phases may be observed: the Heusler phase L2$_1$ and the F$\bar{4}$3m phase for the $A_{0.5}B_{0.25}C_{0.25}$ composition. The various tetrahedron configurations obtained in perfectly ordered binary and ternary *bcc* phases are presented in Fig. 12. When only first and second neighbor pair interactions are

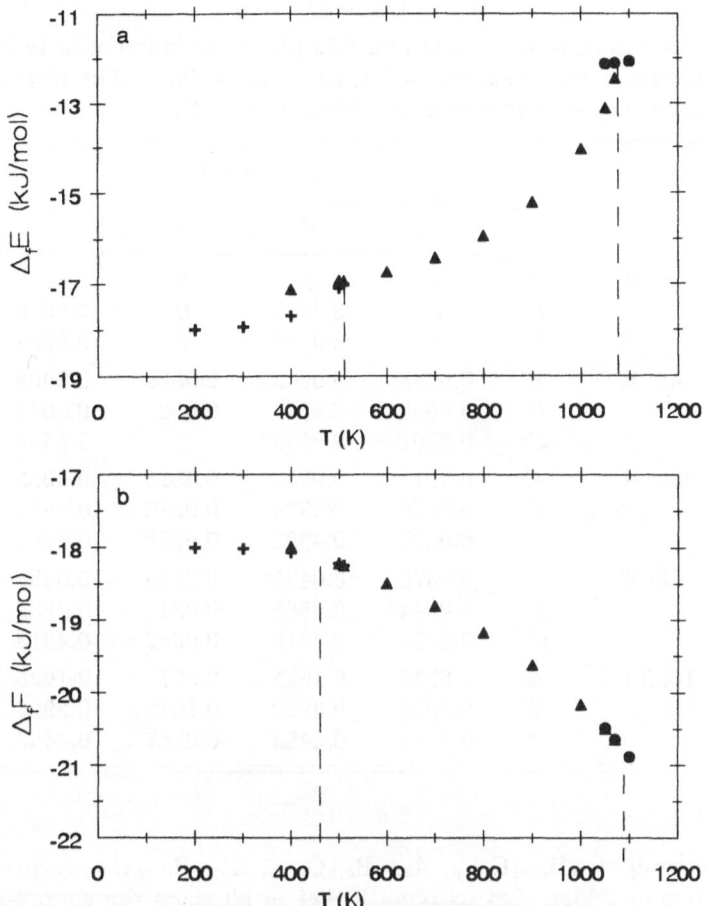

Figure 13. Molar energy (a) and free energy (b) of formation as function of temperature in a ternary *bcc* system at composition $x_A = 0.5$, $x_B = 0.25$, $x_C = 0.25$. F$\bar{4}$3m phase (crosses), B32 phase (up-triangles), A2 phase (filled-circles). The interaction parameters correspond to the case b) of Table VII.

taken into account the energies of formation of the L2$_1$ and F$\bar{4}$3m phases may be calculated as

$$\Delta_f E(L21) = 2\Delta\varepsilon_{AB}^{(1)} + 2\Delta\varepsilon_{AC}^{(1)} + \tfrac{3}{2}\Delta\varepsilon_{BC}^{(2)},$$

$$\Delta_f E(F\bar{4}3m) = \Delta\varepsilon_{AB}^{(1)} + \tfrac{3}{2}\Delta\varepsilon_{AB}^{(2)} + \Delta\varepsilon_{AC}^{(1)} + \tfrac{3}{2}\Delta\varepsilon_{AC}^{(2)} + \Delta\varepsilon_{BC}^{(1)},$$

for the $A_{0.5}B_{0.25}C_{0.25}$ composition. We will considere only systems where the interaction parameters are negative or equal to zero. The F$\bar{4}$3m phase is more stable than the L2$_1$ if

$$\Delta\varepsilon_{AB}^{(2)} + \Delta\varepsilon_{AC}^{(2)} - \Delta\varepsilon_{BC}^{(2)} < \tfrac{2}{3}(\Delta\varepsilon_{AB}^{(1)} + \Delta\varepsilon_{AC}^{(1)} - \Delta\varepsilon_{BC}^{(1)}).$$

Many possibilities may lead to the stability of the F$\bar{4}$3m phase. We have considered three different situations which are called a), b) and c). The values reported in Table VII are the energies of formation of the binary compounds and of the six

Table IX

Site occupations in the F$\bar{4}$3m and B32 phases as function of temperature for the composition $x_A = 0.5$, $x_B = x_C = 0.25$. The interaction
parameters correspond to the case b) of Table VII.

| | | Site occupations | | | |
		α	β	γ	δ
200 K	A	1	1	1	0
	B	0	0.9992	0	0.0008
	C	0	0.0008	0	0.9992
400 K	A	0.9995	0.0003	0.9999	0.0003
	B	0.0005	0.9347	0.0001	0.0647
	C	0.0001	0.0649	0	0.9350
600 K	A	0.9915	0.0085	0.9915	0.0085
	B	0.0076	0.4924	0.0076	0.4924
	C	0.0009	0.4991	0.0009	0.4991
800 K	A	0.9575	0.0425	0.9575	0.0425
	B	0.0344	0.4656	0.0344	0.4656
	C	0.0082	0.4918	0.0082	0.4918
1000 K	A	0.8375	0.1625	0.8375	0.1625
	B	0.1078	0.3922	0.1078	0.3922
	C	0.0547	0.4453	0.0547	0.4453

ternary compounds $A_{0.5}B_{0.25}C_{0.25}$, $A_{0.25}B_{0.5}C_{0.25}$, $A_{0.25}B_{0.25}C_{0.5}$ respectively in the structures L2$_1$ and F$\bar{4}$3m. Let us remark that in all cases the energy of formation has been calculated as a sum of first and second neighbor pair interactions except in the case b) where a ternary effect has been included in such a way that the F$\bar{4}$3m phase of composition $A_{0.5}B_{0.25}C_{0.25}$ has a lower energy (-18000 kJ/mol), therefore in the CVM calculation the energy of the tetrahedron $ABAC$ where A and B as A and C are second nearest neighbor bonds has an energy which is less negative than the energy calculated using the sum of pair interaction.

In the case a) the B32 and F$\bar{4}$3m are degenerated for the composition $A_{0.5}B_{0.25}C_{0.25}$. Calculations performed at low temperature show that the B32 phase forms at low temperature. When increasing the temperature, the B32 → A2 order-disorder transformation occurs at $T = 870$ K. Calculations of the free energy for $x_B = x_C$ and increasing values of the x_A composition show that the F$\bar{4}$3m phase appears; the site occupations are reported in Table VIII. So in the ternary, there is a domain of the F$\bar{4}$3m phase which joins the domains of the F$\bar{4}$3m phase in the AB and in the AC binary systems at low temperature.

In the case b) the F$\bar{4}$3m phase is the ground state for the composition $A_{0.5}B_{0.25}C_{0.25}$ because we have introduced a ternary effect. When increasing the temperature for this composition two transformations occur: F$\bar{4}$3m → B32 at $T = 515$ K and B32 → A2 at $T = 1075$ K. The molar energy and free energy of formation are displayed as function of temperature in Figs. 13a and 13b; in Table IX the sites occu-pations are reported. We have performed calculation along two straight lines which

Table X

Site occupations at $T = 200$ K along the line which joins the point $x_A = 0.5$, $x_B = x_C = 0.25$ in the ternary bcc system to the point $x_A = 0.65$, $x_B = 0.35$ in the binary A-B in the case b) of Table VII.

Composition		Site occupations			
		α	β	γ	δ
A	0.5	0.9998	0.0001	1.0000	0.0001
B	0.25	0.0002	0.9990	0.0000	0.0008
C	0.25	0.0000	0.0009	0.0000	0.9999
A	0.530	1.0000	0.0035	1.0000	0.1166
B	0.270	0.0000	0.9940	0.0000	0.0860
C	0.200	0.0000	0.0025	0.0000	0.7675
A	0.560	0.9999	0.0094	0.9994	0.2313
B	0.290	0.0001	0.9798	0.0001	0.1800
C	0.150	0.0000	0.0109	0.0004	0.5887
A	0.590	0.9998	0.0224	0.9963	0.3416
B	0.310	0.0002	0.9533	0.0014	0.2851
C	0.100	0.0000	0.0224	0.0023	0.3733
A	0.620	0.9993	0.0434	0.9871	0.4503
B	0.330	0.0007	0.9297	0.0079	0.3816
C	0.050	0.0000	0.0269	0.0051	0.1681
A	0.650	0.9980	0.0514	0.9628	0.5878
B	0.350	0.0020	0.9486	0.0372	0.4122
C	0.000	0.0000	0.0000	0.0000	0.0000

respectively joint the point of composition $x_A = 0.5$, $x_B = 0.25$ and $x_C = 0.25$ to the point $x_A = 0.65$, $x_B = 0.35$ in the binary AB system and to the point $x_A = 0.65$, $x_C = 0.35$ in the binary AC system. The site occupations at $T = 200$ K along these lines are presented in Tables X and XI. The site occupations at $T = 200$ K along the line $x_B = x_C$ are presented in Table XII. These calculations show that at low temperature the F$\bar{4}$3m phase has a large domain of stability in the ternary system.

In the case c) the F$\bar{4}$3m is the ground state for the composition $A_{0.5}B_{0.25}C_{0.25}$, this phase is stabilized because in the BC system strong first neighbor pair interactions have been introduced. The conclusions of our calculations are the same as obtained in the case b). For the composition $A_{0.5}B_{0.25}C_{0.25}$ the transformation F$\bar{4}$3m $\rightarrow$ B32 occurs at $T = 545$ K and B32 $\rightarrow$ A2 at $T = 890$ K. Calculations of the molar free energy in the ternary system along the same lines as defined in case b) allow to present the site occupancies. These are displayed on Table XIII for $x_B = x_C$. As in the case b) the domain of the F$\bar{4}$3m phase is large at low temperature in the ternary system.

The cases we have studied are peculiar because the F$\bar{4}$3m appears in two of the limiting binary systems. The F$\bar{4}$3m phase can be a ground state in the ternary system even if the F$\bar{4}$3m is not stable in each of the three limiting binary systems. Some cases have been investigated to considere the evolution of the F$\bar{4}$3m when the tem-

Table XI

Site occupations at $T = 200$ K along the line which joins the point $x_A = 0.5$, $x_B = x_C = 0.25$ in the ternary bcc system to the point $x_A = 0.65$, $x_B = 0.35$ in the binary A-C in the case b) of Table VII.

Composition		Site occupations			
		α	β	γ	δ
A	0.5	0.9998	0.0001	1.0000	0.0001
B	0.25	0.0002	0.9990	0.0000	0.0008
C	0.25	0.0000	0.0009	0.0000	0.9991
A	0.530	1.0000	0.0035	1.0000	0.1166
B	0.200	0.0000	0.0025	0.0000	0.7975
C	0.270	0.0000	0.9940	0.0000	0.0860
A	0.560	1.0000	0.0094	0.9994	0.2312
B	0.150	0.0000	0.0109	0.0004	0.5887
C	0.290	0.0000	0.9798	0.0001	0.1801
A	0.590	1.0000	0.0225	0.9965	0.3410
B	0.100	0.0000	0.0245	0.0023	0.3732
C	0.310	0.0000	0.9530	0.0011	0.2858
A	0.620	1.0000	0.0447	0.9889	0.4465
B	0.050	0.0000	0.0272	0.0049	0.1679
C	0.330	0.0000	0.9281	0.0062	0.3857
A	0.650	1.0000	0.0566	0.9706	0.5728
B	0.000	0.0000	0.0000	0.0000	0.0000
C	0.350	0.0000	0.9434	0.0294	0.4272

perature increases. We have seen that direct $F\bar{4}3m \rightarrow A2$ transition is possible, but $F\bar{4}3m \rightarrow DO_3 \rightarrow B2 \rightarrow A2$ successive phase transitions may also occur, it is the case in the Nb-Ti-Al system.[23] The same holds for the $L2_1$ phase where the direct $L2_1 \rightarrow A2$ transition is possible, however in many cases $L1_2 \rightarrow B2 \rightarrow A2$ successive phase transitions are observed. It is the case in the Fe-Co-Al bcc system for the composition $Co_{0.5}Fe_{0.25}Al_{0.25}$ (Ref. 24).

VI. Conclusions

In binary fcc systems presenting antiferromagnetic first-neighbor interaction Finel and Ducastelle[2] calculated the of L' phase corresponding to three different occupancies of the simple cubic sublattices of the fcc lattice. This L' phase was shown to be a ground state in a ternary system in which all the three binary limiting systems present antiferromagnetic first-neighbor interactions. We have shown that at low temperature there is a large domain of this L' phase extending from the ternary at composition $A_{0.5}B_{0.25}C_{0.25}$ to the limiting AB and AC binaries. The order disorder transformation $L' \rightarrow L1_0$ is of second order in the binaries, in the ternary we observe a first order phase transition $L' \rightarrow A1$ when the interaction parameters are the same in the three

Table XII

Site occupations at $T = 200$ K along the line $x_B = x_C$ in the case b) of Table VII.

Composition		Site occupations			
		α	β	γ	δ
A	0.5	1	0	1.0000	0.0000
B	0.25	0	0.9992	0.0000	0.0008
C	0.25	0	0.0008	0.0000	0.9992
A	0.550	1.0000	0.1000	1.0000	0.1000
B	0.225	0.0000	0.8951	0.0000	0.0049
C	0.225	0.0000	0.0049	0.0000	0.8951
A	0.6	0.9996	0.2004	0.9996	0.2004
B	0.2	0.0004	0.7783	0.0000	0.0213
C	0.2	0.0000	0.0213	0.0004	0.7783
A	0.650	0.9975	0.3027	0.9976	0.3021
B	0.175	0.0020	0.6163	0.0005	0.0812
C	0.175	0.0005	0.0810	0.0019	0.6167
A	0.700	0.8927	0.8927	1.0000	0.0147
B	0.150	0.0539	0.0539	0.0000	0.4922
C	0.150	0.0535	0.0535	0.0000	0.4931

Table XIII

Site occupations at $T = 200$ K along the line $x_B = x_C$ in the case c) of Table VII.

Composition		Site occupations			
		α	β	γ	δ
A	0.5	0.9998	0.0001	1.0000	0.0001
B	0.25	0.0002	0.9990	0.0000	0.0008
C	0.25	0.0000	0.0009	0.0000	0.9991
A	0.550	0.9999	0.1021	1.0000	0.1020
B	0.225	0.0001	0.8929	0.0000	0.0050
C	0.225	0.0000	0.0050	0.0000	0.8929
A	0.6	0.9991	0.2031	0.9995	0.2023
B	0.2	0.0009	0.7751	0.0000	0.0220
C	0.2	0.0000	0.0218	0.0005	0.7757
A	0.650	0.9965	0.3053	0.9974	0.3009
B	0.175	0.0030	0.6145	0.0005	0.0820
C	0.175	0.0005	0.0802	0.0021	0.6172
A	0.700	0.8923	0.8923	1.0000	0.0155
B	0.150	0.0051	0.0051	0.0000	0.4897
C	0.150	0.0526	0.0526	0.0000	0.4948

limiting binary systems. However when it is not the case phase transformations which lead to $L1_2$ or $L1_0$ phases occur before the A1 phase is stable.

In binary *bcc* systems presenting antiferromagnetic first and second neighbor pair interactions the F43m phase corresponding to four different occupancies of the *fcc* sublattices in the *bcc* lattice is observed. This F43m phase may be a ground state in ternary systems. We have shown that the domain of this F43m phase is continuous from the ternary ssytem to the binary systems. Some cases have been studied. However we must point out that the cases we have studied are very peculiar. The F43m phase can be a ground state in ternary systems even this F43m phase can not appear in the binaries. Different phase transitions may occur when increasing temperature for the stoechiometric composition.

References

1. G. Inden and W. Pitsch, in *Materials Science and Technology*, edited by R. W. Cahn, P. Haasen, and E. J. Kramers (VCH Verlagsgesellschaft, 1991) p. 497.
2. A. Finel and F. Ducastelle, *Europhys. Lett.* **1**, 135 (1986).
3. A. Marty, *Thesis*, University Paris 6, Paris, 1990.
4. P. Cenedèse, Y. Calveyrac, and A. Marty, *J. Phys. I. France* **4**, 1063 (1994).
5. G. Inden, *Acta Metallurgica* **22**, 945 (1974).
6. J. M. Bell, *Physica* **142A**, 22 (1987).
7. G. Rubin, *Thesis*, University Paris 6, Paris, 1994.
8. G. Rubin and A. Finel, *J. Phys: Condens. Matter* **5**, 9105 (1993).
9. N. S. Golosov, L. Ya. Pudan, G. S. Golosova, and L. E. Popov, *Soviet Phys. Solid State* **14**, 1280 (1972).
10. C. M. van Baal, *Physica* **64**, 571 (1973).
11. R. Kikuchi, *J. Chem. Phys.* **60**, 1701 (1974).
12. J. M. Sanchez and D. de Fontaine, *Phys. Rev. B* **17**, 2926 (1978).
13. S. K. Aggarwal and T. Tanaka, *Phys. Rev. B* **16**, 3963 (1977).
14. G. Rubin and A. Finel, *J. Phys: Condens. Matter* **7**, 3139 (1995).
15. R. Kikuchi and D. de Fontaine, in *Applications of Phase Diagrams in Metallurgy and Ceramics*, edited by G. C. Carter (NBC Publication N° SP496 1978) p. 967.
16. A. Kikuchi and J. L. Murray, *Calphad* **9**, 311 (1985).
17. R. Kikuchi and H. Sato, *Acta Metall.* **22**, 1099 (1974).
18. R. Kikuchi and C. M. van Baal, *Scripta Metall.* **8**, 425 (1974).
19. N. S. Golosov and A. M. Tolstik, *J. Phys. Chem. Solids* **36**, 899 (1975).
20. N. S. Golosov and A. M. Tolstik, *J. Phys. Chem. Solids* **36**, 903 (1975).
21. M. Sluiter, P. Turchi, Fu Zezhong, and D. de Fontaine, *Physica* **148 A**, 61 (1988).
22. H. Ackermann, G. Inden, and R. Kikuchi, *Acta Metall.* **37**, 1 (1989).
23. V. Jacob, C. Colinet, P. Desre, and F. Moret, *J. Phys. C France*, in press.
24. C. Colinet, G. Inden, and R. Kikuchi, *Acta Metall. Mater.* **41**, 1109 (1993).

CVM and inverse CVM:
Examples in Transition-Metal Alloys and Compounds

M. C. Cadeville,[1] V. Pierron-Bohnes,[1] J. M. Sanchez,[2]
and J. L. Morán-López[3]

[1] Institut de Physique et Chimie des Matériaux de Strasbourg
UMR46 CNRS-Université Louis Pasteur-EHICS 23
67037 Strasbourg
FRANCE

[2] Center for Material Science and Engineering
The University of Texas at Austin
Austin, Texas 78712
U.S.A.

[3] Instituto de Física,
Universidad Autónoma de San Luis Potosí
78000 San Luis Potosí, S.L.P.
MEXICO

Abstract

Some applications of the Cluster Variation Method (CVM) in analysing or interpreting experimental results are presented and discussed. The first example is the experimental determination of the effective pair interactions in alloys from neutron diffuse scattering experiment using the inverse CVM in both direct and reciprocal spaces. Results in $Fe_{80}V_{20}$ and $Fe_{80}Al_{20}$ single crystals are presented. Another example is the modelling of the temperature dependence of the resistivity in ferromagnetic and paramagnetic compounds belonging to Co-Pt, Ni-Pt and Fe-Al systems. This approach uses the chemical (η) and magnetic (σ) long-range order parameter values deduced from the CVM simulations of the corresponding phase diagrams.

Theory and Applications of the Cluster Variation and Path Probability Methods
Edited by J.L. Morán-López and J.M. Sanchez, Plenum Press, New York, 1996

I. Introduction

The primary objective of this paper is to present selected applications of the Cluster Variation Method (CVM) to the analysis and interpretation of experimental results in transition metal based alloys and intermetallic compounds. Through these examples, we show that the increase in the quality of the experimental measurements over the last several years reveals the existence of subtle short-range order effects that, not surprisingly, also require accurate and sophisticated theoretical tools for proper interpretation and understanding. In this regard, the CVM proposed by Ryoichi Kikuchi in 1951,[1] has proven to be invaluable.

In the first example we review a new method[2] to obtain effective atomic interactions directly from the k-space analysis of the short range order (SRO) intensity which is deduced from the diffuse scattering of x-rays or neutrons. As shown in several contributions in this volume, these energies can be used to characterize the thermodynamic properties of alloys, such as relative phase stability, phase diagrams and antiphase boundary energies. Our second example refers to the interpretation of the temperature dependence of the electrical resistivity in terms of simple models that use magnetic and chemical correlation functions calculated from the CVM.[3,4] These two applications of the CVM in various alloy systems are presented, respectively, in sections II and III.

II. Real Space and Reciprocal Space Analysis of SRO Intensities

The diffuse scattering of x-rays and neutrons remains the technique of choice for quantitative studies of short-range order in alloys. In particular, the part of the diffuse intensity related to substitutional SRO has played a prominent role in alloy theory.

Experimental measurements of SRO intensity have evolved significantly from the studies in polycrystalline materials of the early 1950's. At present, accurate measurements are carried out in single crystals at high and low temperature. For example, recent work has focused on the use of high flux neutron and tunable synchrotron X-ray sources.[5-12] On the theoretical side, the study of SRO intensity in alloys and its characterization through diffuse scattering dates back to the pioneering work of Cowley,[13] Krivoglaz,[14] and Clapp and Moss.[15,16] Out of these studies emerged what is presently known as the Krivoglaz-Clapp-Moss (KCM) formula. This formula establishes a simple relationship between the experimentally observed diffuse scattering and the Fourier transform of the effective pair interactions. The KCM formula corresponds to a treatment of concentration fluctuations at the lowest level of approximation for the alloy configurational free energy, namely the single site mean-field approximation, also known as the Bragg-Williams[17] and the molecular field approximation.

More recently, effective pair interactions have been obtained from experimentally determined SRO intensities using higher approximations based on Monte Carlo simulations and the Cluster Variation Method (CVM).[1] The first calculation of SRO intensity carried out directly in reciprocal space using the cluster variation method was done by Sanchez in 1982.[18] This k-space formulation of the CVM provides a significant improvement over the KCM formula, although a price is paid in increased computational effort. Thus, despite the improved treatment of fluctuations by the CVM, the method has been applied only to very few cases.[18,19] An alternative ap-

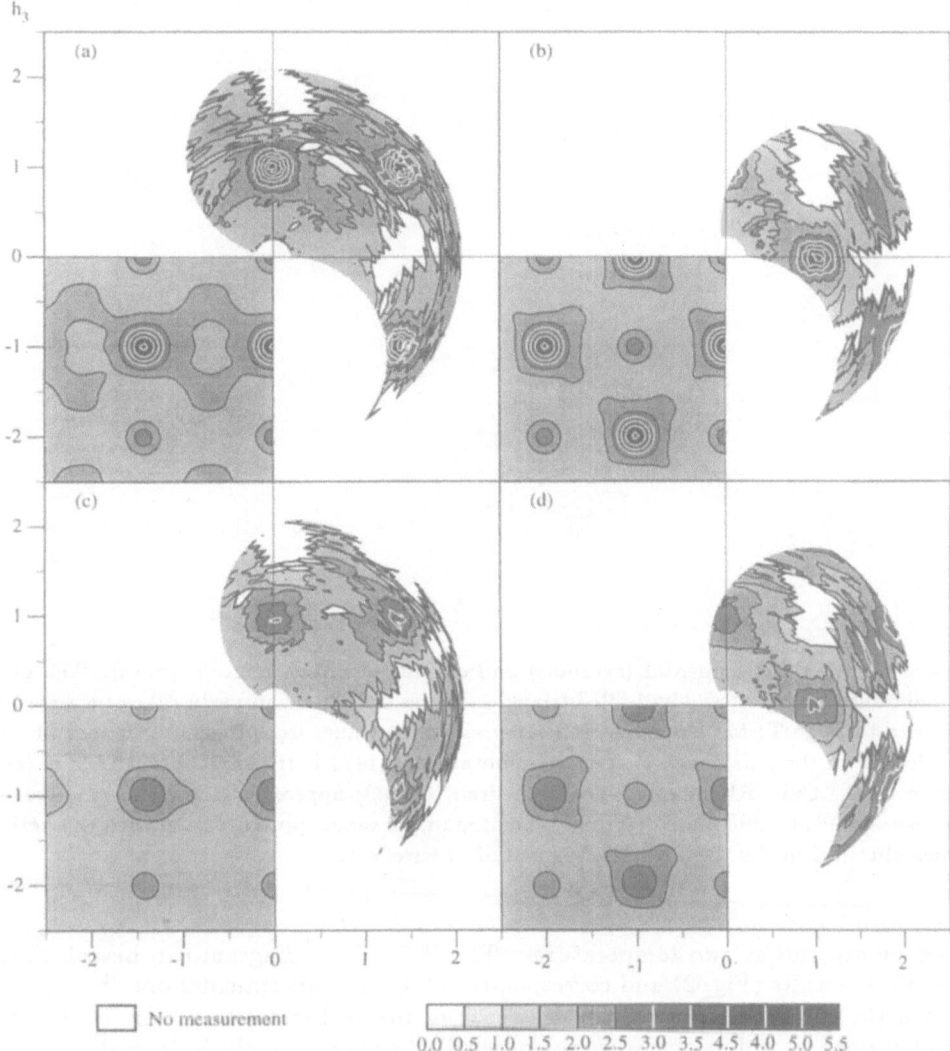

Figure 1. Corrected experimental intensities and calculated ones for Fe-19.6 at.% V in the [110] (a, c) and [100] (b, d) planes, at 1133 K (a, b) and 1473 K (c, d). In each sub-figure, the simulation (down left) has to be compared to the measurement taking into account the symetries of the plane.

proach, which we will refer to as the real space inverse CVM was introduced by Gratias and Cénédèse.[20] In this method, the effective pair interactions are obtained by fitting a few Warren-Cowley SRO parameters which, in turn, are related by a Fourier transform to the experimentally determined SRO intensity. A definite advantage of the real space inverse CVM is that the method involves only a modest computational effort, comparable to that of a regular CVM equilibrium calculation.

This method has been used by different authors[7,8,10,12] to get the effective pair interaction energies. It is illustrated on Fig. 1 for Fe-19.6 at.% V in two reciprocal

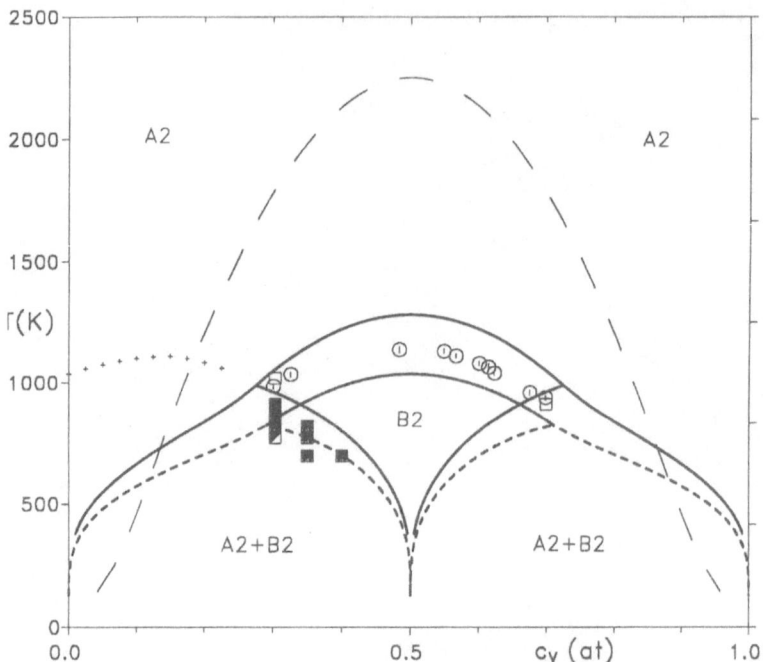

Figure 2. Fe-V experimental (symbols) and calculated phase diagram (lines). The open triangles, full triangles, and half-filled triangles, represent B2, A2 and B2+A2 respectively, as obtained through TEM. The open circles represent the values from Ref. 22. The calculation was done with the pair interaction energies obtained at 1473 K (thick line) and 1133 K (thin line) with the Cube-Rhombohedron-Octahedron (C-R-O) approximation. The long dashed line corresponds to the phase diagram calculated in the same approximation with theoretical values obtained in Ref. 23 with off-diagonal disorder effect.

space planes and at two temperatures. The Fe-V phase diagram can be calculated from these results (Fig. 2) and corresponds well to the experimental one.[21]

On the other hand, the main drawback of the real space technique is that the fitting is done for only a few Warren-Cowley SRO parameters which, typically, are not enough to describe the measured intensities. To a great extent, this problem is solved by applying the inverse method in real space using Monte Carlo simulations, in which longer interaction ranges may be included without any fundamental difficulties.[24,25] A problem of a slightly different nature is the fact that the determination of the Warren-Cowley SRO parameters from the experimental intensities, despite the fact that they are uniquely defined in terms of a Fourier transform, is not without errors. This step in the procedure, which of necessity requires the introduction of a real space cut-off in SRO, is an additional source of uncertainty on the derived effective pair interactions.

Recently[2] has been proposed a new method that carries out the fitting of the SRO intensity directly in k-space using the CVM formulation of the SRO intensity of Sanchez.[18] Since the Warren-Cowley SRO parameters are not required, the k-space fitting circumvents two of the main problems of the real space method, namely the cut-off in SRO and the determination of the SRO parameters which, as mentioned, must be obtained by either fitting or by Fourier transformation of the experimental data. The method has been applied to diffuse neutron scattering data obtained at several

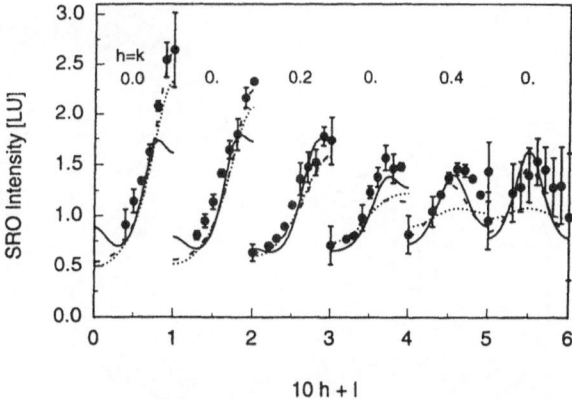

Figure 3. Short range order diffuse intensity obtained using the real space inverse cluster variation method in three approximations: Tetrahedron (dotted line); Cube-Octahedron (dashed line) and Cube-Rhombohedron-Octahedron (full line). The symbols are experimental points.

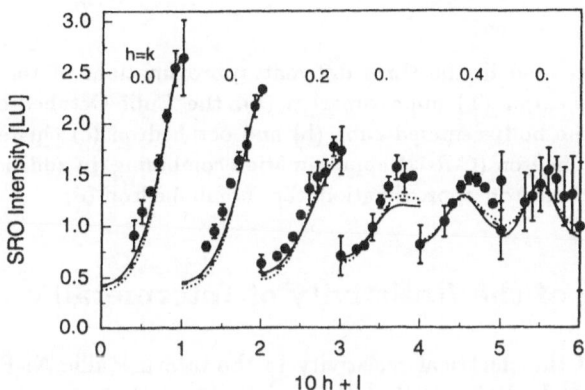

Figure 4. Short range order diffuse intensity obtained using the reciprocal space cluster variation method in three approximations: Tetrahedron (dotted line); Cube-Octahedron (dashed line) and Cube-Rhombohedron-Octahedron (full line). The symbols are experimental points.

temperatures for a Fe-19.5 at.% Al single crystal.[10] A comparison between the real (Fig. 3) and k-space (Fig. 4) methods carried out for three different approximations of the CVM (Fig. 5), shows that the k-space method gives results that are less sensitive to variations in the maximum cluster of the CVM and the temperature. Nevertheless, the corresponding calculation of the Fe-Al phase diagram could not be achieved even with the higher level of approximation due to the presence of the frustrated DO_3 structure. This reciprocal space method has still to be applied to the Fe-V system in which the DO_3 phase does not exist.

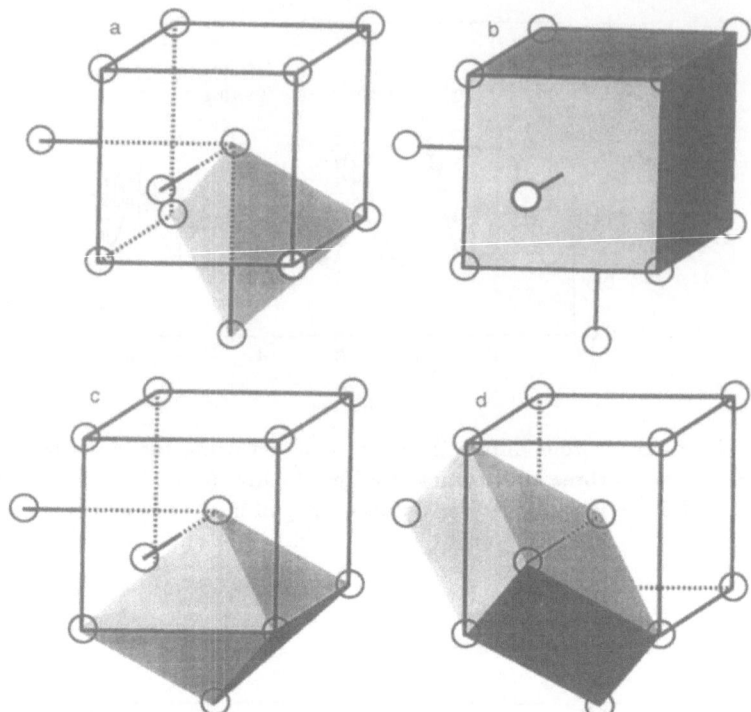

Figure 5. Clusters used in the three different approximations of the Cluster Variation Method: the Tetrahedron (T) approximation (a), the Cube-Octahedron (C-O) approximation combining the body-centered cube (b) and octahedron (c) clusters, and the Cube-Rhombohedron-Octahedron (C-R-O) approximation combining, in addition to the two clusters in the Cube-Octahedron approximation, the rhombohedron (d).

III. Modeling of the Resistivity of Intermetallic Compounds

Previous studies of the electrical resistivity in the intermetallic Ni-Pt, Co-Pt and Fe-Al compounds[3,26,27] had shown that the resistivity of these concentrated alloys can be written, to a first approximation, as the sum of either three or two contributions, in both ordered (O) and disordered (D) states, depending on whether or not the compounds display ferromagnetism at low temperatures

$$\rho_{tot}^{O(D)} = \rho_0^{O(D)} + \Phi^{O(D)}T + \rho_m^{O(D)}, \tag{1}$$

$\rho_0^{O(D)}$ is the residual resistivity or atomic disorder term which depends on temperature in the long range order (LRO) state through the variation of the LRO parameter (η). $\Phi^{O(D)}T$ is the phonon contribution and $\rho_m^{O(D)}$ is the spin disorder scattering term. To express the various contributions in terms of chemical and magnetic LRO parameters, we used the formalism developed by Rossiter[28] for the first two terms Eq. (1), and we adapted that of Kasuya[29] for the last one.

In the Rossiter formalism, the effect of long-range atomic ordering on the scattering of the conduction electrons is treated in an appropriate pseudo-potential model using the Bragg-Williams approximation. The resistivity is described within the simple relaxation time approximation in a nearly free electron model,

$$\rho = \frac{m^*}{n_{eff} e^2 \tau},\tag{2}$$

where τ is order dependent through

$$\tau^{-1} = \tau_0^{-1} \left(1 - \eta^2(T)\right),\tag{3}$$

τ_0 is the relaxation time corresponding to the disordered state, m^* and n_{eff} are, respectively, the effective mass and the effective density of conduction electrons per unit volume. Since the long-range atomic ordering may introduce new gaps into the Fermi surface at the superlattice Brillouin zone boundaries, Rossiter has shown that the value of n_{eff} (or n_{eff}/m^*) will be order dependent according to the following expression

$$n_{eff} = n_0 \left(1 - A\eta^2(T)\right).\tag{4}$$

The coefficient A depends upon the relative positions of the Fermi surface and the superlattice Brillouin zone boundaries. Both its sign and order of magnitude are correlated to the evolution of the electronic band structure near the Fermi level with the formation of the LRO structure. This formalism yields for the residual resistivity term and the phonon contribution the following relations

$$\rho_0^{O(D)} = \frac{\rho_0^D \left(1 - \eta^2(T)\right)}{1 - A\eta^2(T)},\tag{5}$$

where ρ_0^D is the residual resistivity in the disordered state, which is temperature independent, and

$$\Phi^{O(D)}T = \frac{(B/n_0)T}{1 - A\eta^2(T)},\tag{6}$$

where B/n_0 is the temperature coefficient of the phonon contribution in the disordered state. This approach neglects a possible weak dependence of B through the Debye temperature with η.

In the simple model of well-defined local magnetic moments in a single conduction band, as developed by Kasuya,[29] one can write ρ_m as

$$\rho_m = \rho_m^P \left(1 - \sigma^2(T)\right),\tag{7}$$

where σ is the ferromagnetic LRO parameter being equal to 1 at $T = 0$ K and 0 at the Curie temperature (T_{CM}). ρ_m^P is the value of ρ_m in the paramagnetic state

$$\rho_m^P = \text{constant} \times \frac{J^2}{n_{eff}},\tag{8}$$

J is the scalar value of the saturation moment. Using for n_{eff} its formulation through Eq. (4), we obtain for ρ_m a dependence through both chemical and magnetic LRO parameters according to

$$\rho_m = \frac{\rho_m^D \left(1 - \sigma^2(T)\right)}{1 - A\eta^2(T)}.\tag{9}$$

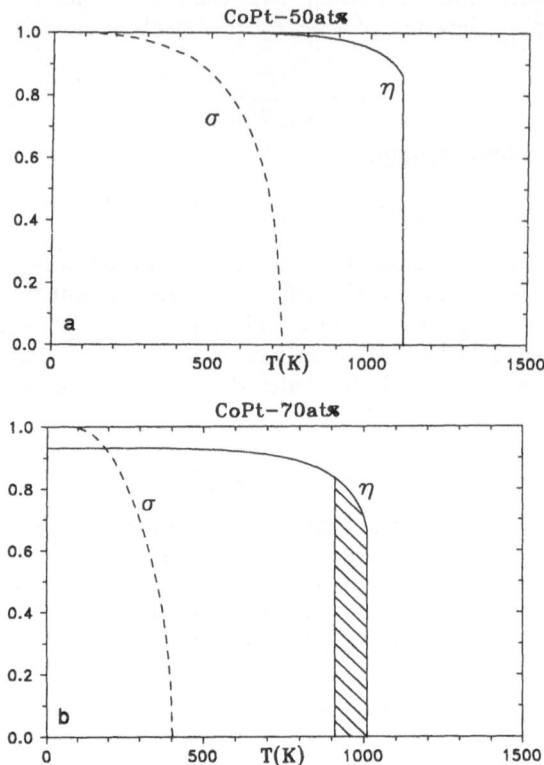

Figure 6. Temperature dependence of chemical (η) and magnetic (σ) LRO parameters in L1$_0$ CoPt (a) and L1$_2$ Co$_{30}$Pt$_{70}$ (b). The hatched domain corresponds to the two-phase region.

Combining Eqs. (1), (5), (6), and (9) we have in the general case, for all concentrations in the L1$_2$, L1$_0$, DO$_3$ and B2 phases

$$\rho(T) = \rho_{\text{APB}} + \frac{\rho_0^D \left(1 - \frac{C_A C_B}{\nu(1-\nu)}\eta^2(T)\right)}{1 - A\eta^2(T)} + \frac{(B/n_0)T}{1 - A\eta^2(T)} + \frac{\rho_m^D \left(1 - \sigma^2(T)\right)}{1 - A\eta^2(T)}. \qquad (10)$$

The constant ρ_{APB} has been added to account for the presence of order defects, like antiphase boundaries, which are always present even in the totally ordered compounds at $T = 0$ K.

Now, to compare this model with experimental data, we use the chemical and magnetic LRO parameter values deduced from previous CVM calculations of the phase diagram, as described in Ref. 30 for the Ni-Pt system and in Ref. 31 for the Co-Pt system, and only fitted to the values of the DO$_3$-B2 and A2-B2 transitions for the non-magnetic Fe$_{70}$Al$_{30}$ alloy. The temperature dependences of the LRO parameters in L1$_0$ CoPt and L1$_2$Co$_{30}$ Pt$_{70}$, and in DO$_3$ and B2 Fe$_{70}$Al$_{30}$ are illustrated in figs. 6 and 7. In Fig. 7, η_{eff} corresponds to

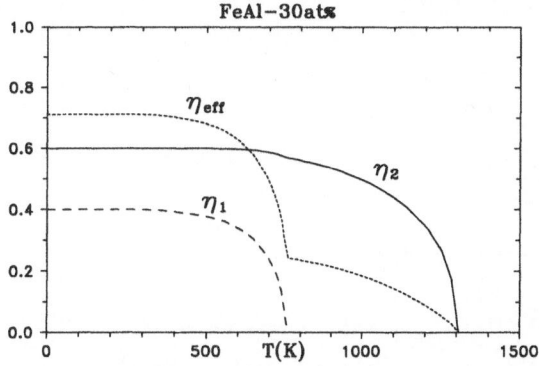

Figure 7. Temperature dependence of LRO parameters in DO_3 and B2 $Fe_{70}Al_{30}$.

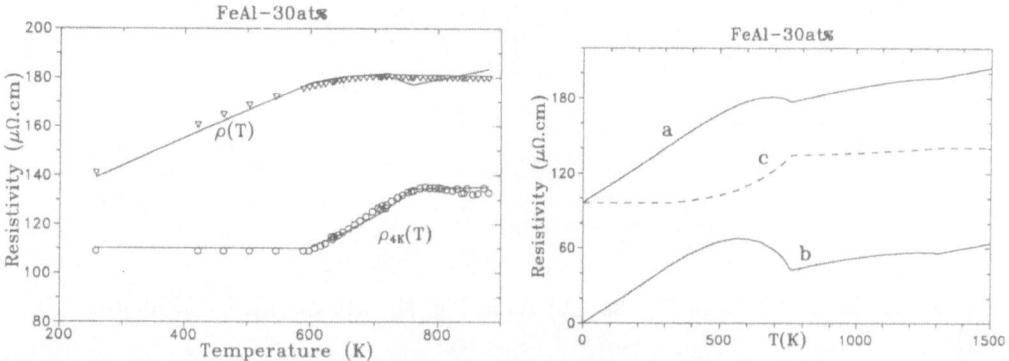

Figure 8. $Fe_{70}Al_{30}$. (a) Comparison between experimental values of ρ_{4K} ($\circ$) and ρ_{tot} ($\triangledown$) and the curves fitted using Eq. 5 over the temperature range of experiments, with η replaced by η_{eff}. (b) Calculated values of ρ_{tot} (a), ρ_{ph} (b) and ρ_{4K} (c) between 0 and 1500 K. The change of resistivity at the transition B2-A2 is also shown.

$$\eta_{eff} = \sqrt{\frac{16}{3}C_AC_B}\left(D\eta_1 + \sqrt{\frac{3}{4}}\eta_2\right), \qquad (11)$$

with η_1 and η_2 the LRO parameters corresponding, respectively, to the B2 and DO_3 phases and with D a parameter which when fitted to the experimental data of Fig. 8, equals 0.81.

The accuracy of the calculated LRO parameters has been shown in the case of $CoPt_3$ compound.[4] One observes a very good agreement between experimental η obtained by NMR at ^{59}Co or by x-ray diffraction and the calculated ones.

In order to test this model, the chemical and magnetic LRO parameters are entered in Eq. (10) and the remaining parameters ρ_{APB}, ρ_0^D, B/n_0, and A are adjusted in order to reproduce the experimental results.

The resistivities of four intermetallic compounds NiPt, CoPt, $Co_{30}Pt_{70}$ and $Fe_{70}Al_{30}$ have been tested through this model, two of them, CoPt and $Co_{30}Pt_{70}$,

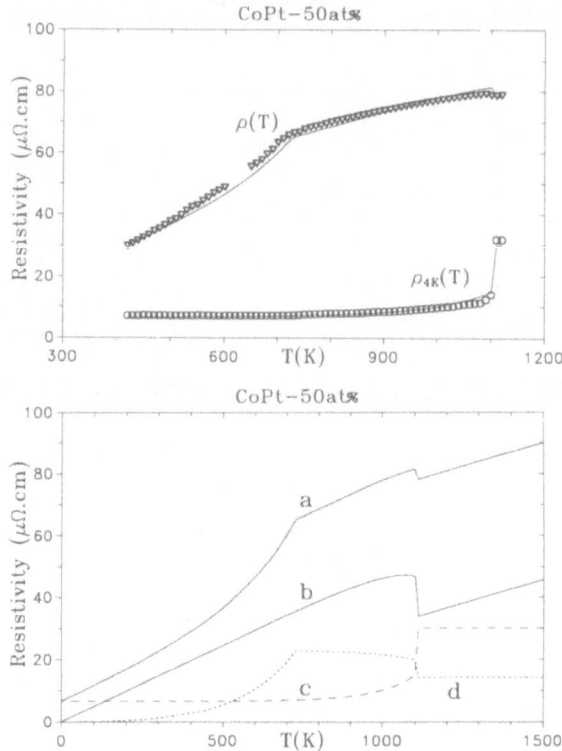

Figure 9. CoPt. (a) As in Fig. 8a. (b) As in Fig. 8b with magnetic contribution (curve (d)).

being ferromagnetic at low temperature. Since in NiPt and $Fe_{70}Al_{30}$ the magnetic contribution ρ_m is zero, it is relatively simple to separate the contribution of the chemical disorder from that of the electron-phonon scattering. In addition to the total resistivity, the residual resistivity due to the chemical disorder, $\rho_0^{O(D)}$, was measured by quenching the sample rapidly to 4 K from an equilibrium state at high temperature, characterized by a LRO parameter η. The phonon contribution is given by the difference

$$\rho_{ph} = \Phi^{O(D)}T = \rho_{tot}(T) - \rho_0^{O(D)}\left(\eta(T)\right), \tag{12}$$

and this experimental determination can be directly compared to its simulation through Eq. (6).

In ferromagnetic compounds, the phonon contribution could not be experimentally separated from the magnetic disorder term. Both quantities were simulated using the values of the constants deduced from the simultaneous fits of ρ_{4K} and ρ_{tot} through Eqs. (5) and (10).

As examples, Figs. 8 and 9 reproduce experimental and fitted values for respectively $Fe_{70}Al_{30}$ and CoPt over the T-range of experiments. At low temperature, the values of η were kept constant and equal to their highest value allowed by diffusion, on account of LRO kinetics data. In addition to the curves fitted to the experimental values, the calculated phonon and magnetic contributions over the 0–1500 K temperature range are shown in Figs. 8b and 9b. For the four investigated compounds, the quality of the fit is satisfactory and the model works well. This approach shows that, besides the expected increase of the resistivity when both chemical and magnetic disorders increase, an important effect is due to the change in the electronic structure, whose consequence is to decrease both the phonon and the magnetic contributions when the chemical disorder increases at the order-disorder transition.

IV. Conclusion

We have illustrated two applications of the CVM in analyzing or interpreting experimental results. The first example is a direct application of the CVM to the analysis of the short range order contribution to the neutron diffuse scattering in $Fe_{80}V_{20}$ and $Fe_{80}Al_{20}$ single crystals. A comparison of the reciprocal space CVM with the real space method in $Fe_{80}Al_{20}$ illustrates the advantages of the k-space versus the real space methods.

The second example is an indirect application of the CVM in the interpretation of physical properties, like the electrical resistivity, which are strongly dependent on the magnetic and chemical LRO parameters. Introducing the LRO parameters obtained through a CVM simulation of the phase diagrams in a simple model derived from the Rossiter and Kasuya formalisms, allows to fit satisfactorily the experimental results in several ordered phases of the Ni-Pt, Co-Pt and Fe-Al systems.

References

1. R. Kikuchi, *Phys. Rev.* **81**, 988 (1951).
2. J. M. Sanchez, V. Pierron-Bohnes, and F. Mejía-Lira, *Phys. Rev. B* **51**, 3429 (1995).
3. M. C. Cadeville, V. Pierron-Bohnes, and J. M. Sanchez, *J. Phys.: Condens. Matter* **4**, 9053 (1992).
4. M. C. Cadeville, V. Pierron-Bohnes, L. Bouzidi, and J. M. Sanchez, *Physica Scripta* **T49**, 364 (1993).
5. M. Bessière, Y. Calvayrac, S. Lefebvre, D. Gratias, and P. Cénédèse, *J. Phys.* **47**, 1961 (1986).
6. B. D. Butler and J.B. Cohen, *J. Appl. Phys.* **65**, 2214 (1986).
7. F. Solal, R. Caudron, F. Ducastelle, A. Finel, and A. Loiseau, *Phys. Rev. Lett.* **58**, 2245 (1989).
8. V. Pierron-Bohnes, S. Lefebvre, M. Bessière, and A. Finel, *Acta Metall. Mater.* **34**, 2701 (1990).
9. L. Reinhard, B. Schönfeld, G. Kostorz, and W. Bürer, *Phys. Rev. B* **44**, 1727 (1990).
10. V. Pierron-Bohnes, M. C. Cadeville, A. Finel, and O. Schaerpf, *J. Phys.* **1**, 247 (1991).
11. L. Reinhard, J. L. Robertson, S. C. Moss, G. E. Ice, P. Zschack, and C. J. Sparks, *Phys. Rev. B* **45**, 2262 (1992).
12. V. Pierron-Bohnes, E. Kentzinger, M. C. Cadeville, J. M. Sanchez, R. Caudron, F. Solal, and R. Kozubski, *Phys. Rev. B* **51**, 5760 (1995).
13. J. M. Cowley, *J. Appl. Phys.* **21**, 24 (1950).

14. M. A. Krivoglaz, *Theory of X-Ray and Thermal Neutron Scattering by Real Crystals*, (Plenum, New York, 1969).
15. P. C. Clapp and S. C. Moss, *Phys. Rev.* **142**, 418 (1966); *ibid* **171**, 754 (1968).
16. S. C. Moss and P. C. Clapp, *Phys. Rev.* **171**, 764 (1968).
17. W. L. Bragg and E. J. Williams, *Proc. Roy. Soc. A* **145**, 69 (1934).
18. J. M. Sanchez, *Physica* **111A**, 200 (1982).
19. T. Mohri, J. M. Sanchez, and D. de Fontaine, *Acta Metall.* **33**, 1463 (1985).
20. D. Gratias and P. Cénédèse, *J. de Physique* **46**, C9-149 (1985).
21. J. M. Sanchez, M. C. Cadeville, V. Pierron-Bohnes, and G. Inden, to be published.
22. J. I. Seki, M. Hagiwara, and T. Suzuki, *J. Mater. Sci.* **14**, 2404 (1979).
23. M. Sluiter and P. E. A. Turchi, in *High-Temperature Ordered Intermetallic Alloys IV.*, edited by L. A. Johnson, D. P. Pope, and J. O. Stiegler *Mater. Res. Soc. Symp. Proc.* **213**, 37 (1991).
24. F. Livet, *Acta Metall.* **35**, 2915 (1987); F. Livet and M. Bessière, *J. Phys.* **48**, 1703 (1987).
25. W. Schweika and H. G. Haubold, *Phys. Rev. B* **37**, 9250 (1988).
26. C. Leroux, M. C. Cadeville, and R. Kozubski, *J. Phys.: Condens. Matter* **1**, 6403 (1989).
27. C. Leroux, M. C. Cadeville, V. Pierron-Bohnes, G. Inden, and F. Hinz, *J. Phys. F: Met. Phys.* **18**, 2033 (1988).
28. P. L. Rossiter, *The Electrical Resistivity of Metals and Alloys*, edited by R. W. Cahn, E. A. Davis, and I. M. Ward (Cambridge, Cambridge University Press, 1987); *J. Phys. F: Met. Phys.* **9**, 891 (1979); *J. Phys; F: Met. Phys.* **11**, 615 (1981).
29. T. Kasuya, *Prog. Theor. Phys.* **16**, 58 (1956).
30. C. E. Dahmani, M. C. Cadeville, J. M. Sanchez, and J. L. Morán-López, *Phys. Rev. Lett.* **55**, 1208 (1985).
31. J. M. Sanchez, J. L. Morán-López, C. Leroux, and M. C. Cadeville, *J. Phys.: Condens. Matter* **1**, 491 (1989).

Application of the Cluster Variation Method to the Image Restoration Problem

Kazuyuki Tanaka[1] and Tohru Morita[2]

[1] Department of Computer Science and Systems Engineering
Muroran Institute of Technology
27-1 Mizumoto-cho, Muroran 050
JAPAN

[2] Department of Computer Science
College of Engineering
Nihon University
Koriyama 963
JAPAN

Abstract

The pair approximation in the cluster variation method is applied to the image restoration problem based on the Q-state Potts model with a local non-uniform external field.

I. Introduction

Recently, some concepts and methods of statistical physics have been applied to information and computer science. Particularly, one of the areas in which equilibrium statistical physics is directly applied to information science is the bayesian approach based on Gibbs (Boltzmann) distribution at a finite temperature. Geman and Geman[1] suggested that the Ising model is applicable to image restoration based on the bayesian formula. This problem corresponds to the one of searching the ground state of a finite-size Ising model under a local non-uniform external field. Geman and Geman[1] recovered an image from a damaged image by using simulated annealing of a spin-S Ising model with a line process. The simulated annealing method needs a large number of Monte-Carlo steps to converge to the thermodynamic equilibrium state in the low-temperature region, resulting in a high-level image. In order to improve this

Theory and Applications of the Cluster Variation and Path Probability Methods
Edited by J.L. Morán-López and J.M. Sanchez, Plenum Press, New York, 1996

Table I

Comparison of the critical temperatures T_c/J
in different approximations.

	T_c/J
Mean field approximation	4.0000
Pair approximation[12]	2.8854
Square approximation[5,13]	2.4257
(2×3) approximation[14]	2.3761
(3×3) approximation[15,16]	2.3462
Exact solution[13,17]	2.2692

difficulty, Gidas proposed a new method based on a combination of the renormalization group technique and the simulated annealing method,[2] and Zhang[3] introduced the mean-field annealing method for studying this problem. Wu and Doerschuk[4] extended the mean-field annealing method into the cluster type mean-field annealing method.

In statistical physics, one of the effective field approximations, into which the mean field approximation is extended, is the Cluster Variation Method (CVM).[5-8] In the CVM an approximation is determined by a set of basic clusters of lattice sites and the entropy of the total lattice system is approximately expressed by a linear combination of the entropies for the basic clusters and their subclusters. The approximate free energy can be expressed in terms of the reduced distribution functions for the basic clusters and their subclusters. The approximate physical quantities are obtained by minimizing the approximate free energy with respect to the reduced distribution functions under some consistency conditions between them. In the Ising model on a square lattice, the CVM can provide a sequence of approximations such that the series of approximate free energies converge to the exact one.[6,9-11] The lowest-level approximation in the CVM is called the pair approximation (PA), which is equivalent to the Bethe approximation.[12] The second-lowest-level approximation in the CVM is called the square approximation, which is equivalent to the Kramers-Wannier-Kikuchi approximation.[5,13] In this approximation, the basic cluster is a (2×2) square plaquette, each edge of which connects a nearest-neighbor pair of lattice sites. In higher-level approximations, the (2×3) clusters or (3×3) clusters are included in the basic cluster set.[14-16] The critical temperatures T_c/J obtained by applying some approximations in the CVM to the spin-$\frac{1}{2}$ Ising model on the square lattice with only nearest-neighbor interaction J are shown in the Table I (in this paper, we put the Boltzmann constant $k_B = 1$). The exact value of the critical temperature was obtained by Kramers and Wannier,[13] and the exact expression for the free energy of this model for zero external field was obtained by Onsager.[17]

In Table I, we see that the critical temperature is improved if a larger basic cluster set is adopted. By combining the CVM and the coherent-anomaly method,[18] we can estimate the critical exponents for many spin systems and the CVM is a powerful method for the study of critical phenomena.[15,16,19-21] The formula of the approximate entropy in the CVM can be expressed in terms of the Möbius function.[28] The results of the variational calculation for the approximate free energy under some consistency conditions between reduced distribution functions of the basic clusters and their subclusters are given in compact form in terms of the Möbius function and in the notation of set theory.[26,27] These formulas are given for the general spin system with local non-uniform interaction J_{ij} and external field y_i. Since the image restoration problem is one of searching the ground state of a finite-size Ising model under local non-uniform external field, it is interesting to provide a general algorithm based on the CVM. The authors applied the pair approximation in the CVM to the binary image restoration.[29]

In studying the multiple-valued image restoration, we can consider the problem in terms either of the spin-S Ising model[22] or of Q-state Potts model,[23] where S is a positive interger or a half odd integer, and Q is a positive integer. Critical properties in these models were investigated by using mean field theory, series expansions, the renormalization group theory, Monte Carlo simulations and so on. The two-state Potts model is equivalent to the spin-$\frac{1}{2}$ Ising model. When we treat a multiple-valued image which consists of faces and have a boundary between faces of different configurations, the Potts model is a powerful approach to recover the damaged image data. In this paper, we adopt the Q-state Potts model as a posterior distribution.

The purpose of this paper is to give a basic algorithm for the application of the PA of the CVM to the restoration problem of multiple-valued images and to study its usefulness. We compare the results obtained by the PA annealing method with the ones obtained by the simulated annealing and mean-field annealing methods. In section II, we construct a statistical lattice model for the multiple-valued image restoration problem based on the bayesian formula. In section III, we give a formulation of the pair approximation in the CVM which is applicable to the image restoration problem. In section IV, we give concluding remarks.

II. Image Restoration and Variational Principle

We consider a square lattice consisting of $L = M \times N$ lattice sites. A set of lattice sites is called a cluster. The whole cluster of all the lattice sites is denoted by the label I. The image data is given by an image consisting of gray faces on the lattice I. The gray configurations of a lattice site is represented by values $0, 1, \ldots,$ or $Q-1$. The black and white configurations are denoted by $Q-1$ and 0, respectively. The original image data is damaged by random noise and then we have a damaged image data different from the original image data. We try to recover the original data from the given damaged data by using the bayesian formula. The conditional probability distribution for an event B under the condition that an event A occurred and the probability distribution for the occurrence of an event A are given by $P(\mathbf{B}|\mathbf{A})$ and $P(\mathbf{A})$, respectively. Then in the bayesian formula the conditional probability distribution for an event A under the condition that an event B occurred, $P(\mathbf{A}|\mathbf{B})$, can be calculated by

$$P(\mathbf{A}|\mathbf{B}) = \frac{P(\mathbf{B}|\mathbf{A})P(\mathbf{A})}{\sum_{\mathbf{A}} P(\mathbf{B}|\mathbf{A})P(\mathbf{A})}. \tag{1}$$

Here, the summation $\sum_{\mathbf{A}}$ is taken over all the possible events for $\mathbf{A}$.

In the original image data, the configuration on the i-th lattice site is denoted by x_i which takes values $0, 1, 2, \ldots,$ and $Q-1$. In the damaged image data, the configuration on the i-th lattice site is denoted by y_i and takes values $0, 1, 2, \ldots,$ and $Q-1$. When an original image data $\mathbf{x_I} = \{x_j; j = 1, 2, \ldots, L\}$ is given, the conditional probability distribution for a damaged image data $\mathbf{y_I} = \{y_j; j = 1, 2, \ldots, L\}$ is denoted by $P(\mathbf{y_I}|\mathbf{x_I})$. The probability distribution for the original image data $\mathbf{x_I}$ is denoted by $P(\mathbf{x_I})$. In this case, the bayesian formula is written as follows

$$P(\mathbf{x_I}|\mathbf{y_I}) = \frac{P(\mathbf{y_I}|\mathbf{x_I})P(\mathbf{x_I})}{\sum_{\mathbf{x_I}} P(\mathbf{y_I}|\mathbf{x_I})P(\mathbf{x_I})} . \tag{2}$$

The notations $\sum_{\mathbf{x_I}}$ and $\sum_{\mathbf{y_I}}$ mean the summations over all possible configurations on the whole lattice sites I as follows

$$\sum_{\mathbf{x_I}} \equiv \sum_{x_1=0}^{Q-1} \sum_{x_2=0}^{Q-1} \cdots \sum_{x_L=0}^{Q-1} , \quad \text{and} \quad \sum_{\mathbf{y_I}} \equiv \sum_{y_1=0}^{Q-1} \sum_{y_2=0}^{Q-1} \cdots \sum_{y_L=0}^{Q-1} . \tag{3}$$

For an original image data $\mathbf{x_I}$, we have a damaged image data $\mathbf{y_I}$, and denote the conditional probability distribution of $\mathbf{y_I}$ for the original image data $\mathbf{x_I}$ by $P(\mathbf{y_I}|\mathbf{x_I})$. We denote the probability that the damaged image data $\mathbf{y_I}$ deviates from the original $\mathbf{x_I}$ at lattice site i by σ_i, so that

$$\sum_{\mathbf{y_I}} (1 - \delta(y_i, x_i)) P(\mathbf{y_I}|\mathbf{x_I}) = \sigma_i , \tag{4}$$

where $\delta(y, x)$ denotes the Kronecker's delta $\delta_{y,x}$. The sum $\sum_{i=1}^{L} \sigma_i$ is the average of the total number of lattice sites i where y_i is different from x_i. When the total deviation $L\bar{\sigma}_1 = \sum_{i=1}^{L} \sigma_i$ is given, we consider a canonical ensemble of systems with given $\mathbf{x_I}$, and the conditional probability distribution $P(\mathbf{y_I}|\mathbf{x_I})$ is determined so as to maximize the entropy

$$S^{(1)} \equiv - \sum_{\mathbf{y_I}} P(\mathbf{y_I}|\mathbf{x_I}) \ln P(\mathbf{y_I}|\mathbf{x_I}) , \tag{5}$$

under the subsidiary condition

$$\sum_{i=1}^{L} \sum_{\mathbf{y_I}} (1 - \delta(y_i, x_i)) P(\mathbf{y_I}|\mathbf{x_I}) = L\bar{\sigma}_1 , \tag{6}$$

and the normalization condition

$$\sum_{\mathbf{y_I}} P(\mathbf{y_I}|\mathbf{x_I}) = 1 . \tag{7}$$

Introducing the Lagrange multipliers $1/T_1$ and $-F^{(1)}/T_1 - 1$, we obtain

$$P(\mathbf{y_I}|\mathbf{x_I}) = \exp\left\{ \frac{1}{T_1} \left[F^{(1)} - \sum_{i=1}^{L} (1 - \delta(y_i, x_i)) \right] \right\} , \tag{8}$$

where $1/T_1$ and $-F^{(1)}/T_1 - 1$ are determined by the conditions (6) and (7).

We next consider the distribution function of the original image data $\mathbf{x_I}$. Here, we assume that the original image consists of faces, where x_i for a lattice site i is usually the same as x_j for j around i as *a priori* bias. We denote the probability that x_i and x_j are different for nearest-neighbors i and j by σ_{ij} in the set of $\mathbf{x_I}$, so that

$$\sum_{\mathbf{x_I}} (1 - \delta(x_i, x_j)) P(\mathbf{x_I}) = \sigma_{ij} , \tag{9}$$

for the nearest-neighbor pair of lattice sites i and j. We use notation $\sum_{(ij)}$ to mean the summation over all the nearest-neighbor pairs of lattice sites i and j. The sum

$\sum_{(ij)} \sigma_{ij}$ is the total number of nearest-neighbor lattice sites i and j, where x_i and x_j are different from each other; that is, the sum of lengths of boundaries between two faces of different darkness in the original image data. As x_I, we shall consider those in which $\sum_{(ij)} \sigma_{ij}$ takes a value given by $L_2 \bar{\sigma}_2$, where L_2 is the total number of pairs of nearest-neighbors and is equal to $2L - M - N$ in the present lattice. With this restriction, we require $P(x_I)$ to satisfy

$$\sum_{(ij)} \sum_{x_I} (1 - \delta(x_i, x_j)) P(x_I) = L_2 \bar{\sigma}_2 , \tag{10}$$

and

$$\sum_{x_I} P(x_I) = 1 . \tag{11}$$

We introduce a canonical ensemble, and make the entropy

$$S^{(2)} \equiv - \sum_{x_I} P(x_I) \ln P(x_I) \tag{12}$$

maximum with the subsidiary conditions (10) and (11), to obtain

$$P(x_I) = \exp \left\{ \frac{1}{T_2} \left[F^{(2)} - \sum_{(ij)} (1 - \delta(x_i, x_j)) \right] \right\} , \tag{13}$$

where $1/T_2$ and $-F^{(2)}/T_2 - 1$ are the Lagrange multipliers to be determined by the conditions (10) and (11). T_2 is the parameter determining the sum of lengths of boundaries of different darkness in the original image data.

By substituting Eqs. (8) and (13) into Eq. (2), we obtain

$$P(x_I|y_I) = \frac{\exp \left[-\frac{J}{T} \sum_{(ij)} (1 - \delta(x_i, x_j)) - \frac{1}{T} \sum_i (1 - \delta(y_i, x_i)) \right]}{\sum_{x_I} \exp \left[-\frac{J}{T} \sum_{(ij)} (1 - \delta(x_i, x_j)) - \frac{1}{T} \sum_i (1 - \delta(y_i, x_i)) \right]}$$

$$= \frac{\exp \left[\frac{J}{T} \sum_{(ij)} \delta(x_i, x_j) + \frac{1}{T} \sum_i \delta(y_i, x_i) \right]}{\sum_{x_I} \exp \left[\frac{J}{T} \sum_{(ij)} \delta(x_i, x_j) + \frac{1}{T} \sum_i \delta(y_i, x_i) \right]} , \tag{14}$$

where J and T are defined by $J \equiv T_1/T_2$ and $T \equiv T_1$. We use the notation $\sum_i$ to mean the summation over all the lattice sites i, such that $\sum_i = \sum_{i=1}^{L}$. T/J, determined by Eq. (4), is a parameter for the sum of lengths of boundaries in the original image data and T is a parameter for the fraction of error due to noise. The problem of determining J and T from the given damaged data is left to a separate work and will not be discussed here. We shall instead try various values of J to see whether or not they give a reasonable result

When we know the values of $\bar{\sigma}_1$ and $\bar{\sigma}_2$, we desire to get x_I consistent with those values. If we adopt the form Eq. (14) for $P(x_I|y_I)$, we require that it satisfies

$$\sum_i \sum_{x_I} (1 - \delta(y_i, x_i)) P(x_I|y_I) = L \bar{\sigma}_1 , \tag{15}$$

$$\sum_{(ij)} \sum_{\mathbf{x_I}} (1 - \delta(x_i, x_j)) P(\mathbf{x_I}|\mathbf{y_I}) = L_2 \bar{\sigma}_2 . \tag{16}$$

The posterior distribution $P(\mathbf{x_I}|\mathbf{y_I})$ of the form (14), satisfying Eqs. (15) and (16), is given by the following variational principle

$$S_I \equiv \max_{P(\mathbf{x_I}|\mathbf{y_I})} S_I\{P(\mathbf{x_I}|\mathbf{y_I})\}, \quad S_I\{P(\mathbf{x_I}|\mathbf{y_I})\} \equiv -\sum_{\mathbf{x_I}} P(\mathbf{x_I}|\mathbf{y_I}) \ln P(\mathbf{x_I}|\mathbf{y_I}) . \tag{17}$$

The variations are taken with respect to the posterior distribution $P(\mathbf{x_I}|\mathbf{y_I})$ under the subsidiary conditions (15), (16) and

$$\sum_{\mathbf{x_I}} P(\mathbf{x_I}|\mathbf{y_I}) = 1 . \tag{18}$$

We introduce the Lagrange multipliers $1/T$, J/T and $-F_I/T - 1$ as follows

$$\mathcal{L}\{P(\mathbf{x_I}|\mathbf{y_I})\} \equiv S_I\{P(\mathbf{x_I}|\mathbf{y_I})\} - \left(-\frac{F_I}{T} - 1\right) \left[\sum_{\mathbf{x_I}} P(\mathbf{x_I}|\mathbf{y_I}) - 1\right]$$

$$- \frac{1}{T} \left[\sum_i \sum_{\mathbf{x_I}} (1 - \delta(x_1, y_i)) P(\mathbf{x_I}|\mathbf{y_I}) - L\bar{\sigma}_1\right]$$

$$- \frac{J}{T} \left[\sum_{(ij)} \sum_{\mathbf{x_I}} (1 - \delta(x_i, x_j)) P(\mathbf{x_I}|\mathbf{y_I}) - L_2\bar{\sigma}_2\right] .$$

From the first variational principle, we have

$$\frac{\partial \mathcal{L}\{P(\mathbf{x_I}|\mathbf{y_I})\}}{\partial P(\mathbf{x_I}|\mathbf{y_I})} = -\ln P(\mathbf{x_I}|\mathbf{y_I}) - 1 + \frac{F_I}{T} + 1 - \frac{1}{T} \sum_i \sum_{\mathbf{x_I}} (1 - \delta(y_i, x_i))$$

$$- \frac{J}{T} \sum_{(ij)} \sum_{\mathbf{x_I}} (1 - \delta(x_i, x_j)) = 0 .$$

As the result, the posterior distribution $P(\mathbf{x_I}|\mathbf{y_I})$ is given by Eq. (14). We can regard that the posterior distribution $P(\mathbf{x_I}|\mathbf{y_I})$ is obtained from the variational principle for the entropy maximum in Eqs. (17) and (18), without using the bayesian formula. It is noted that the conditions determining the parameters J and T are similar, but are not the same in the two approaches.

When $\mathbf{y_I}$ is given, we seek $\mathbf{x_I}$ which makes $P(\mathbf{x_I}|\mathbf{y_I})$ maximum. In statistical physics, this amounts to finding the ground state of the Hamiltonian given by

$$H_I(\mathbf{x_I}) \equiv -J \sum_{(ij)} \delta(x_i, x_j) - \sum_i \delta(y_i, x_i) . \tag{19}$$

If we have a unique minimum configuration, we expect that the ground state determines the low-temperature properties. The average for the quantity $f(\mathbf{x_I})$ in the total system I is defined by

$$\langle f(\mathbf{x_I})\rangle_I \equiv \sum_{\mathbf{x_I}} f(\mathbf{x_I}) P(\mathbf{x_I}|\mathbf{y_I}) . \tag{20}$$

The model whose Hamiltonian is given by Eq. (19) is called the Q-state Potts model in statistical physics.[23] If $\langle x_i \rangle_I \simeq 0, 1, \ldots$, or $Q - 1$, we can use $\{\text{Int}(\langle x_j \rangle_I); j = 1, 2, \ldots, L\}$ in place of the ground state configuration $\mathbf{x}_I$. Here $\text{Int}(x)$ is defined by

$$\text{Int}(x) \equiv n, \qquad n - \tfrac{1}{2} \leq x < n + \tfrac{1}{2}.$$

We shall use the CVM to calculate $\{\langle x_j \rangle_I; j = 1, 2, \ldots, L\}$ at low-temperatures.

In order to recover an original image data from a damaged image data $\mathbf{y}_I$, we search for the configuration $\mathbf{x}_I$ which gives the maximum value of the probability distribution $P(\mathbf{x}_I|\mathbf{y}_I)$ under given $\mathbf{y}_I$. This probability distribution is equivalent to the distribution for a Potts model on a finite square lattice under local non-uniform external field $\mathbf{y}_I$, and we adopt it as the posterior probability distribution giving a posterior image when the degraded image is given. The image restoration problem is reduced to the one of searching for the ground state or the low-temperature averages $\{\langle x_j \rangle_I; j = 1, 2, \ldots, L\}$ of the Potts model. In the Potts model under uniform external field ($y_i = 1$ for any i), on an infinite-size ($L \to +\infty$) regular lattice, the ground state configuration can be exactly obtained when the system have translational and reflectional symmetries.[24] Since the Potts model treated in the image restoration problem has no translational or reflectional symmetries, we must treat all possible Q^L configurations. It is known that low-temperature properties can be achieved by the process of annealing. In this process, we now propose to use an approximation of the CVM.

Hereafter, we write $P(\mathbf{x}_I|\mathbf{y}_I)$ as $\rho_I(\mathbf{x}_I)$. We now note that the probability distribution $\rho_I(\mathbf{x}_I)$ given by Eq. (14) satisfies the following variational principle

$$F_I = \min_{\rho_I} \left(E_I\{\rho_I\} - TS_I\{\rho_I\} \right), \tag{21}$$

where

$$E_I\{\rho_I\} \equiv \text{Tr}_I \ H_I(\mathbf{x}_I)\rho_I(\mathbf{x}_I), \tag{22}$$

and

$$S_I\{\rho_I\} \equiv - \text{Tr}_I \ \rho_I(\mathbf{x}_I) \ln \rho_I(\mathbf{x}_I). \tag{23}$$

Here notation Tr_I stands for $\sum_{\mathbf{x}_I}$. The variations are taken with respect to the distribition function $\rho_I(\mathbf{x}_I)$ under the subsidiary condition

$$\text{Tr}_I \ \rho_I(\mathbf{x}_I) = 1. \tag{24}$$

We introduce the Lagrange multiplier λ as follows

$$\mathcal{L}\{\rho_I\} \equiv E_I\{\rho_I\} - TS_I\{\rho_I\} - \lambda \left(\text{Tr}_I \ \rho_I(x_I) - 1 \right).$$

From the first variational principle, we have

$$\frac{\partial \mathcal{L}\{\rho_I\}}{\partial \rho_I(\mathbf{x}_I)} = H_I(\mathbf{x}_I) + T \ln \rho_I(\mathbf{x}_I) + T - \lambda = 0.$$

If we put $\lambda = F_I + T$, the distribution function $\rho_I(\mathbf{x}_I)$ is given by Eq. (14). We remark that if $\bar{\sigma}_1$ and $\bar{\sigma}_2$ are given, T and J are the unknowns in the variational principle for the entropy maximum given by Eqs. (17) and (18), and that, if T and J are given, $\bar{\sigma}_1$ and $\bar{\sigma}_2$ are the unknowns in the variational principle for the free energy minimum given by Eqs. (21)–(24).

In order to express a function $f(x)$ of an integer variable x, we can use a power series expansion

$$f(x) = a_0 + \sum_{m=1}^{Q-1} a_m x^m, \tag{25}$$

where a_m are constants. We can use an orthonormal set of polynomials $\Phi_m(x)$ of degree m for $m = 0, 1, 2, \ldots, Q-1$, such that

$$\sum_{x=0}^{Q-1} \Phi_m(x)\Phi_n(x) = \delta(m,n), \qquad m, n = 0, 1, 2, \ldots, Q-1. \tag{26}$$

The polynomials are given in Appendix A. We use notation $\langle f(x)\rangle_0$ defined by

$$\langle f(x)\rangle_0 \equiv \sum_{x=0}^{Q-1} f(x). \tag{27}$$

Then $f(x)$ is expanded as

$$f(x) = \sum_{m=0}^{Q-1} b_m \Phi_m(x), \tag{28}$$

where

$$b_m = \langle \Phi_m(x) f_m(x)\rangle_0. \tag{29}$$

If we set $f(x) = \delta(y,x)$, we have $b_m = \Phi_m(y)$ and obtain

$$\delta(y,x) = \sum_{m=0}^{Q-1} \Phi_m(y)\Phi_m(x). \tag{30}$$

$H_I(\mathbf{x_I})$ given by Eq. (19) can be replaced by

$$H_I(\mathbf{x_I}) = -J \sum_{(ij)} \sum_{m=0}^{Q-1} \Phi_m(x_i)\Phi_m(x_j) - \sum_{i} \sum_{m=0}^{Q-1} \Phi_m(y_i)\Phi_m(x_i). \tag{31}$$

III. Pair Approximation

It is difficult to obtain the most probable configuration for the posterior distribution given by Eq. (14) or the ground-state configuration for the Potts model, whose Hamiltonian is given in Eq. (31), exactly. We shall calculate the average of x_i for each lattice site i at the low-temperature region by using the pair approximation in the CVM.

Let cluster i be the one of a single lattice site i. Let cluster ij be the one of a nearest-neighbor pair of lattice sites i and j. Clusters are labeled as α, which represents i or ij in the PA. If the whole cluster of all the lattice sites are as shown in Fig. 1a, the sets of nearest-neighbor pair clusters ij and of single lattice site clusters i are given in Figs. 1b and 1c. The set of nearest-neighbor pair clusters ij is called the basic cluster set and is denoted by **B**. The single lattice site clusters i are all the non-empty clusters consisting of common sites of two or more clusters in **B**.

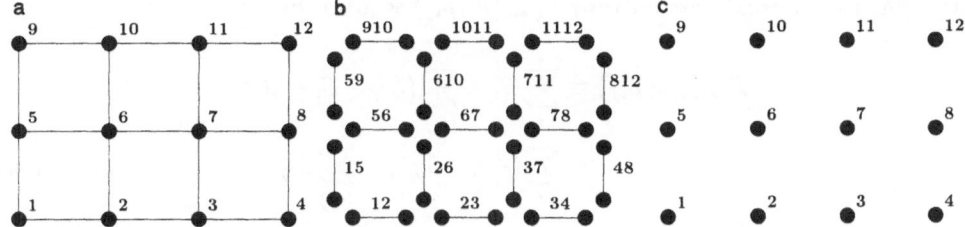

Figure 1. An example of clusters $\mathbf{I}$, ij and i, taken from Refs. 27 and 29. (a) The whole cluster of all the lattice sites $\mathbf{I}$. (b) The nearest-neighbor pair clusters ij. These clusters are subclusters of $\mathbf{I}$. The set of all the clusters is called the basic cluster set and is denoted by **B**. (c) The single lattice site clusters i. These are all the clusters consisting of common sites of two or more clusters in the basic cluster set **B**.

For each cluster α, the set of variables describing the configurations is denoted by $\mathbf{x}_\alpha$. We represent the set $\{x_i\}$ and $\{x_i, x_j\}$ by $\mathbf{x}_i$ and $\mathbf{x}_{ij}$, respectively. We introduce the notations Tr_i and Tr_{ij} defined by $\mathrm{Tr}_i \equiv \sum_{x_i=0}^{Q-1}$ and $\mathrm{Tr}_{ij} \equiv \sum_{x_i=0}^{Q-1} \sum_{x_j=0}^{Q-1}$. The probability distributions for the clusters ij and i, $\rho_{ij}(x_i, x_j)$ and $\rho_i(x_i)$, are defined by the following reducibility equations

$$\rho_{ij}(x_i, x_j) = \rho_{ij}(\mathbf{x}_{ij}) \equiv \mathrm{Tr}_{\mathbf{I}\backslash ij}\, \rho_{\mathbf{I}}(\mathbf{x}_{\mathbf{I}}), \quad \text{and} \quad \rho_i(x_i) = \rho_i(\mathbf{x}_i) \equiv \mathrm{Tr}_{\mathbf{I}\backslash i}\, \rho_{\mathbf{I}}(\mathbf{x}_{\mathbf{I}}), \quad (32)$$

respectively. Here, $\mathrm{Tr}_{\mathbf{I}\backslash ij}$ and $\mathrm{Tr}_{\mathbf{I}\backslash i}$ are the summations over all possible configurations on the whole lattice sites except that the configurations for the nearest-neighbor pair of lattice sites ij and the single lattice site i, respectively, are fixed. We can easily show that

$$\rho_i(x_i) = \mathrm{Tr}_j\, \rho_{ij}(x_i, x_j). \quad (33)$$

This equality is called the reducibility condition from the cluster ij to the cluster i. The averages of a quantity $f(\mathbf{x}_\alpha)$ in the cluster $\alpha = i, j, ij, \mathbf{I}$ are defined by

$$\langle f(\mathbf{x}_\alpha) \rangle_\alpha \equiv \mathrm{Tr}_\alpha\, f(\mathbf{x}_\alpha)\rho_\alpha(\mathbf{x}_\alpha), \quad \alpha = i, j, ij, \mathbf{I}. \quad (34)$$

By using the reducibility conditions, we have

$$\langle \Phi_m(x_i)\Phi_m(x_j)\rangle_{\mathbf{I}} = \langle \Phi_m(x_i)\Phi_m(x_j)\rangle_{ij}, \quad m = 0, 1, 2, \ldots, Q-1,$$

and

$$\langle \Phi_m(x_i)\rangle_{\mathbf{I}} = \langle \Phi_m(x_i)\rangle_i, \quad m = 0, 1, 2, \ldots, Q-1,$$

and the energy $E_{\mathbf{I}}$ can be rewritten as

$$E_{\mathbf{I}}\{\rho_{\mathbf{I}}\} = E_{\mathrm{Pair}}\{\rho_i, \rho_{ij}\}$$

$$\equiv -J \sum_{m=0}^{Q-1} \sum_{(ij)} \langle \Phi_m(x_i)\Phi_m(x_j)\rangle_{ij} - \sum_{m=0}^{Q-1} \sum_i \Phi_m(y_i)\langle \Phi_m(x_i)\rangle_i. \quad (35)$$

In the PA, the approximate entropy $S_{\text{Pair}}\{\rho_i, \rho_{ij}\}$ is given by

$$S_{\text{Pair}}\{\rho_i, \rho_{ij}\} \equiv \sum_i S_i + \sum_{(ij)} (S_{ij} - S_i - S_j)$$

$$= \sum_i (-z_i + 1) S_i + \sum_{(ij)} S_{ij}, \qquad (36)$$

where

$$S_\alpha \equiv - \text{Tr}_\alpha \, \rho_\alpha(\mathbf{x}_\alpha) \ln \rho_\alpha(\mathbf{x}_\alpha). \qquad (37)$$

Here, z_i is the number of nearest-neighbor lattice sites of the site i. For example, we have $z_1 = z_4 = z_9 = z_{12} = 2$, $z_2 = z_3 = z_5 = z_8 = z_{10} = z_{11} = 3$ and $z_6 = z_7 = 4$ in the finite lattice system given in Fig. 1a. If the system has no interaction, the second term in the right-hand side of Eq. (36) is equal to zero. When the entropy S_{ij} is added for the nearest-neighbor pairs of lattice sites ij, the entropies for single lattice sites i and j must be substructed, since the contributions have already been added in the first term. This is the meanings of Eq. (36). A more detailed explanation for the approximate entropy (36) is given in Appendix B.

The variational principle in the PA is given as

$$F_{\text{Pair}} = \min_{\{\rho_i, \rho_{ij}\}} \left(E_{\text{Pair}}\{\rho_i, \rho_{ij}\} - T S_{\text{Pair}}\{\rho_i, \rho_{ij}\} \right). \qquad (38)$$

F_{Pair} is the approximate free energy of the total system in the PA. The variations are taken with respect to the distribution functions $\rho_i(x_i)$ and $\rho_{ij}(x_i, x_j)$ under the subsidiary conditions

$$\text{Tr}_\alpha \, \rho_\alpha(\mathbf{x}_\alpha) = 1, \qquad \alpha = i, j, ij, \qquad (39)$$

and

$$\langle \Phi_m(x_i) \rangle_i = \langle \Phi_m(x_i) \rangle_{ij}, \quad \langle \Phi_m(x_j) \rangle_j = \langle \Phi_m(x_j) \rangle_{ij}, \quad m = 0, 1, 2, \ldots, Q-1, \quad (40)$$

for every pair of nearest-neighbor lattice sites i and j. We remark that the consistency conditions (40) are equivalent to the reducibility conditions $\rho_i(x_i) = \text{Tr}_j \, \rho_{ij}(x_i, x_j)$ and $\rho_j(x_j) = \text{Tr}_i \, \rho_{ij}(x_i, x_j)$.[25] We prove this in Appendix C.

We introduce the Lagrange multipliers λ_i, λ_{ij} and $\lambda_{im, ij}$ as follows

$$\mathcal{L}\{\rho_i, \rho_{ij}\} \equiv E_{\text{Pair}}\{\rho_i, \rho_{ij}\} - T S_{\text{Pair}}\{\rho_i, \rho_{ij}\}$$

$$- \sum_i \lambda_i \left(\text{Tr}_i \, \rho_i(x_i) - 1 \right) - \sum_{(ij)} \lambda_{ij} \left(\text{Tr}_{(ij)} \, \rho_{ij}(x_i, x_j) - 1 \right)$$

$$- \sum_{(ij)} \sum_{m=0}^{Q-1} \lambda_{im, ij} \left(\langle \Phi_m(x_i) \rangle_i - \langle \Phi_m(x_i) \rangle_{ij} \right)$$

$$- \sum_{(ij)} \sum_{m=0}^{Q-1} \lambda_{jm, ij} \left(\langle \Phi_m(x_j) \rangle_j - \langle \Phi_m(x_j) \rangle_{ij} \right).$$

From the first variational principle we have

$$\frac{\partial \mathcal{L}\{\rho_i, \rho_{ij}\}}{\partial \rho_i(x_i)} = -\sum_{m=0}^{Q-1} \Phi_m(y_i)\Phi_m(x_i) + (1 - z_i)T \ln \rho_i(x_i) + (1 - z_i)T - \lambda_i$$

$$- \sum_{\{j|j\leq ij\}} \sum_{m=0}^{Q-1} \lambda_{im,ij}\Phi_m(x_i) = 0,$$

$$\frac{\partial \mathcal{L}\{\rho_i, \rho_{ij}\}}{\partial \rho_{ij}(x_i, x_j)} = -J\sum_{m=0}^{Q-1} \Phi_m(x_i)\Phi_m(x_j) + T \ln \rho_{ij}(x_i, x_j) + T - \lambda_{ij}$$

$$- \sum_{m=0}^{Q-1} \lambda_{jm,ij}\Phi_m(x_i) - \sum_{m=0}^{Q-1} \lambda_{jm,ij}\Phi_m(x_j) = 0.$$

By solving these equations, the one-body distribution functions $\rho_i(x_i)$ and $\rho_j(x_j)$ and the two-body distribution function $\rho_{ij}(x_i, x_j)$ are expressed as

$$\rho_i(x_i) = \exp\left[\frac{1}{T}\left(F_i + \sum_{m=0}^{Q-1} \lambda_{im,i}\Phi_m(x_i)\right)\right],$$

$$\rho_j(x_j) = \exp\left[\frac{1}{T}\left(F_j + \sum_{m=0}^{Q-1} \lambda_{jm,j}\Phi_m(x_j)\right)\right], \tag{41}$$

and

$$\rho_{ij}(x_i, x_j) = \exp\left[\frac{1}{T}\left(F_{ij} + J\sum_{m=0}^{Q-1} \Phi_m(x_i)\Phi_m(x_j)\right.\right.$$

$$\left.\left. + \sum_{m=0}^{Q-1} \lambda_{im,ij}\Phi_m(x_i) + \sum_{m=0}^{Q-1} \lambda_{jm,ij}\Phi_m(x_j)\right)\right], \tag{42}$$

where we put $\lambda_i = (1 - z_i)(F_i + T)$ and $\lambda_{ij} = F_{ij} + T$. The unknown parameters F_i, F_j, F_{ij}, $\lambda_{im,ij}$, $\lambda_{jm,ij}$, $\lambda_{im,i}$ and $\lambda_{jm,j}$ are determined by the conditions (39) and (40) and the following relations

$$(1 - z_i)\lambda_{im,i} + \sum_{\{j|j\leq ij\}} \lambda_{im,ij} = \Phi_m(y_i),$$

$$(1 - z_j)\lambda_{jm,j} + \sum_{\{i|i\leq ij\}} \lambda_{jm,ij} = \Phi_m(y_j). \tag{43}$$

Here, $\sum_{\{j|j\leq ij\}}$ is the summation over all the nearest-neighbor lattice sites j of the fixed i, and we remark that $\lambda_{im,ij} = \lambda_{im,ji}$ for any i and j. For example, in the system given in Fig. 1a, the relations given by Eq. (43) for $i = 1$, 2 and 6 are explicitly expressed by

$$(1 - z_1)\lambda_{1m,1} + \lambda_{1m,12} + \lambda_{1m,15} = \Phi_m(y_1),$$
$$(1 - z_2)\lambda_{2m,2} + \lambda_{2m,21} + \lambda_{2m,23} + \lambda_{1m,26} = \Phi_m(y_2),$$
$$(1 - z_6)\lambda_{6m,6} + \lambda_{6m,62} + \lambda_{6m,65} + \lambda_{6m,67} + \lambda_{1m,610} = \Phi_m(y_6).$$

By using F_i and F_{ij}, the approximate free energy in the total system is expressed as

$$F_{\text{Pair}} = \sum_i (-z_i + 1)F_i + \sum_{(ij)} F_{ij}. \tag{44}$$

We remark that we can regard the CVM as the cluster consistency method (CCP) or as the cluster type effective field theory[8,27,30] by putting $\lambda_{im,i}$ and $\lambda_{im,ij}$ as follows

$$\lambda_{im,i} = \Phi_m(y_i) + \sum_{\{j|j \leq ij\}} \tilde{\lambda}_{ij}, \tag{45}$$

$$\lambda_{im,ij} = \lambda_{im,i} - \tilde{\lambda}_{ij}. \tag{46}$$

We note that the parameters $\tilde{\lambda}_{ij}$ and $\tilde{\lambda}_{ji}$ are different from each other, through $\lambda_{im,ij}$ is equal to $\lambda_{im,ji}$. For example, in the system given in Fig. 1a, the relations given by Eqs. (45) and (46) for $i = 1$, 2 and 6 are explicitly expressed by

$$\lambda_{1m,1} = \Phi_m(y_1) + \tilde{\lambda}_{12} + \tilde{\lambda}_{15},$$
$$\lambda_{2m,2} = \Phi_m(y_2) + \tilde{\lambda}_{21} + \tilde{\lambda}_{23} + \tilde{\lambda}_{26},$$
$$\lambda_{6m,6} = \Phi_m(y_6) + \tilde{\lambda}_{62} + \tilde{\lambda}_{65} + \tilde{\lambda}_{67} + \tilde{\lambda}_{610},$$

and

$$\lambda_{1m,12} = \lambda_{1m,1} - \tilde{\lambda}_{12} = \Phi_m(y_1) + \tilde{\lambda}_{15},$$
$$\lambda_{2m,12} = \lambda_{2m,2} - \tilde{\lambda}_{21} = \Phi_m(y_2) + \tilde{\lambda}_{23} + \tilde{\lambda}_{26},$$
$$\lambda_{6m,56} = \lambda_{6m,6} - \tilde{\lambda}_{65} = \Phi_m(y_6) + \tilde{\lambda}_{62} + \tilde{\lambda}_{67} + \tilde{\lambda}_{610}.$$

In the above equations, $\tilde{\lambda}_{ij}$ can be regarded as the effective field of the lattice site i caused from each of the nearest-neighbor bond clusters which have the lattice site i in common. This interpretation for λ_{ij} was explicitly give in Refs. 31 and 32.

For given parameters T and J, and a given set of $\{y_j; j = 1, 2, \ldots, L\}$, once the unknown parameters F_i, F_{ij}, $\lambda_{im,i}$ and $\lambda_{im,ij}$ are numerically determined by the conditions (39) and (40), the distribution functions for the cluster i and ij, $\rho_i(x_i)$ and $\rho_{ij}(x_i, x_j)$ in Eqs. (41) and (42), can be calculated and the set of $\{\langle x_j \rangle_j; j = 1, 2, \ldots, L\}$ is obtained by Eq. (34). In numerically determining the unknown parameters F_i, F_{ij}, $\lambda_{im,i}$ and $\lambda_{im,ij}$, we apply the iteration method to the simultaneous nonlinear equations given by Eqs. (39) and (40) for the unknown parameters. In order to solve the simultaneous nonlinear equations by the numerical iteration method at an enough small value of T for a fixed parameter J, we start from the large T. We solve the simultaneous nonlinear equations for the next smaller T by adopting the set of the solutions obtained for the last T as the initial value in the iteration for the next smaller T. In this way, we decrease to the final small T. Though the convergence of the numerical iterations in the corresponding simultaneous nonlinear equtions in PA have not been discussed yet, the convergence of the numerical iterations in the mean field annealing was discussed in Ref. 4.

a   b

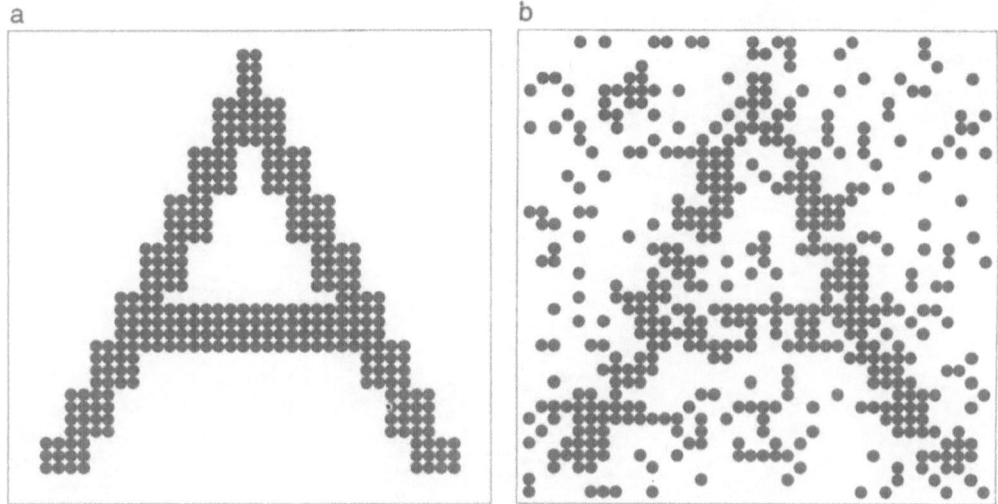

Figure 2. Original binary image data and its damaged image data, taken from Ref. 29. (a) Original image data ($\bar{\sigma}_2 \simeq 0.07539$) (b) Damaged image data ($\bar{\sigma}_1 \simeq 0.21052$).

When the original binary image data given in Fig. 2a is changed to a damaged image data with a random noise as given in Fig. 2b, the process in which the most probable configuration is obtained with the parameter $J = 0.70$ fixed is shown in Fig. 3, as decreasing the parameter T using the PA. In this binary image data, the states 0 and 1 are denoted by blank and a solid circle, respectively. In Fig. 3, the image data obtained by calculating the set $\{\text{Int}(\langle x_j \rangle_j); j = 1, 2, \ldots, L\}$ is shown. In the region of the parameter $T < 0.25$, $\{\text{Int}(\langle x_j \rangle_j); j = 1, 2, \ldots, L\}$ does not change for any value of T and we can regard $\{\text{Int}(\langle x_j \rangle_j); j = 1, 2, \ldots, L\}$ for $T = 0.20$ as the one for $T = 0$, that is the most probable configuration. In the region for the parameter T that the set of $\{\text{Int}(\langle x_j \rangle_j); j = 1, 2, \ldots, L\}$ does not change for any value of T, we regard the obtained set $\{\text{Int}(\langle x_j \rangle_j); j = 1, 2, \ldots, L\}$ as the most probable configuration for a given parameter J. The most probable configurations for various values of parameter J are given in Fig. 4. In this case, we adopt the two-state Potts model, which is a spin-$\frac{1}{2}$ Ising model. When the original three-valued image data given in Fig. 5a is changed to a damaged image data with a random noise as given in Fig. 5b, the most probable configurations for various values of parameter J are given in Fig. 6. In Figs. 5 and 6, the states 0, 1 and 2 are denoted by blank, a cross and a solid circle, respectively. The set of original image data is denoted by $\{x_j^{(o)}; j = 1, 2, \ldots, L\}$. Measures of the restoration rates R, R_1 and R_2 are defined by

$$R \equiv 1 - \sum_i \frac{1}{L} \delta(x_i^{(o)}, \text{Int}(\langle x_i \rangle_i)),$$

$$R_1 \equiv 1 - \sum_i \frac{1}{L} \delta(y_i, \text{Int}(\langle x_i \rangle_i)),$$

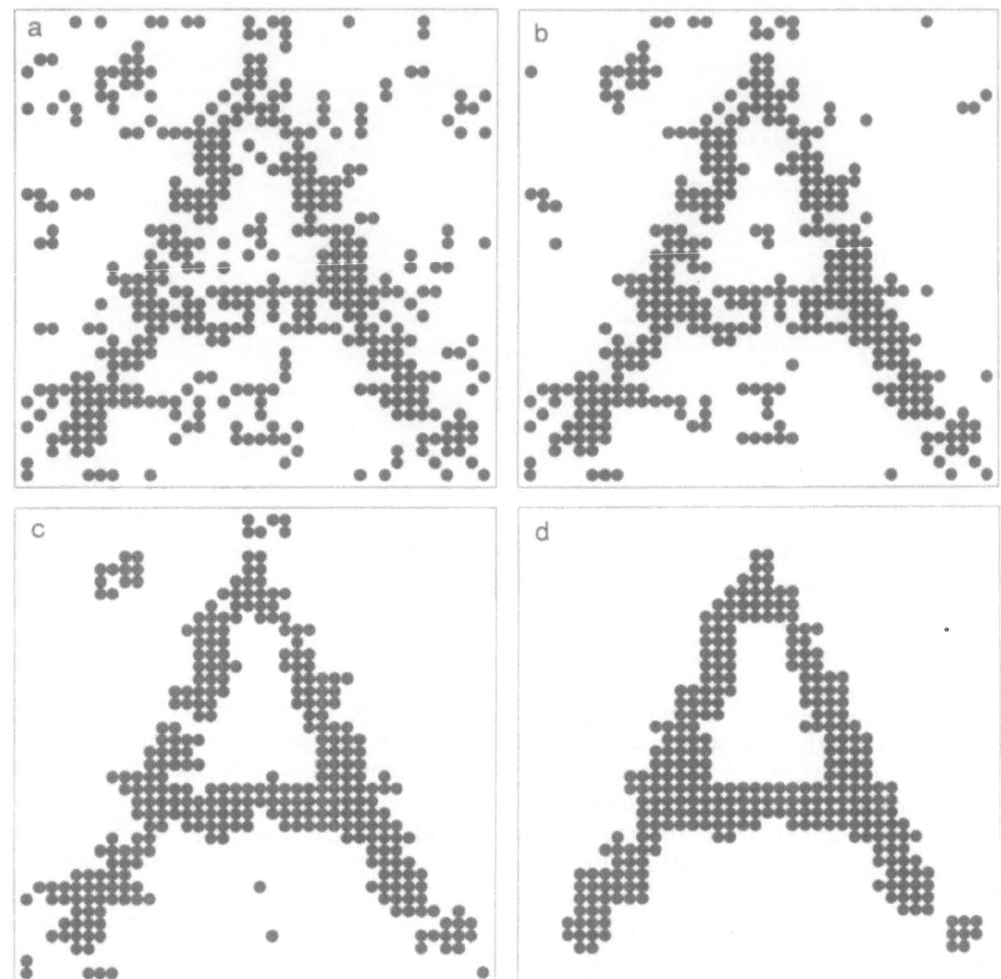

Figure 3. Process that the most probable configuration for a fixed $J = 0.70$ is obtained by PA annealing method in the binary image data, taken from ref. 29. (a) $T = 2.00$ ($R \simeq 0.18213$, $R_1 \simeq 0.02978$, $R_2 \simeq 0.28805$) (b) $T = 1.50$ ($R \simeq 0.12327$, $R_1 \simeq 0.08864$, $R_2 \simeq 0.19275$) (c) $T = 1.00$ ($R \simeq 0.08310$, $R_1 \simeq 0.14404$, $R_2 \simeq 0.12020$) (d) $T = 0.20$ ($R \simeq 0.05956$, $R_1 \simeq 0.18490$, $R_2 \simeq 0.07752$).

$$R_2 \equiv 1 - \sum_{(ij)} \frac{1}{L_2} \delta(\mathrm{Int}(\langle x_i \rangle_i), \mathrm{Int}(\langle x_j \rangle_j)), \tag{47}$$

respectively. When the original image data is perfectly recovered, we have $R = 0$, $R_1 = \overline{\sigma}_1$ and $R_2 = \overline{\sigma}_2$.

We start from the initial value $T = 5.0$ and decrease to the parameter T in the region where the set of $\{\mathrm{Int}(\langle x_j \rangle_j); j = 1, 2, \ldots, L\}$ does not change for any value of T. The upper bound of the region for the parameter T, T_u, increases as the parameter J is increased. This operation is done by an annealing procedure based on the PA, which may be called the PA annealing procedure. We regard the obtained distribution for $\mathrm{Int}(\langle x_i \rangle_i)$ at $T < T_u$ as to the most probable configuration x_I. In the original image

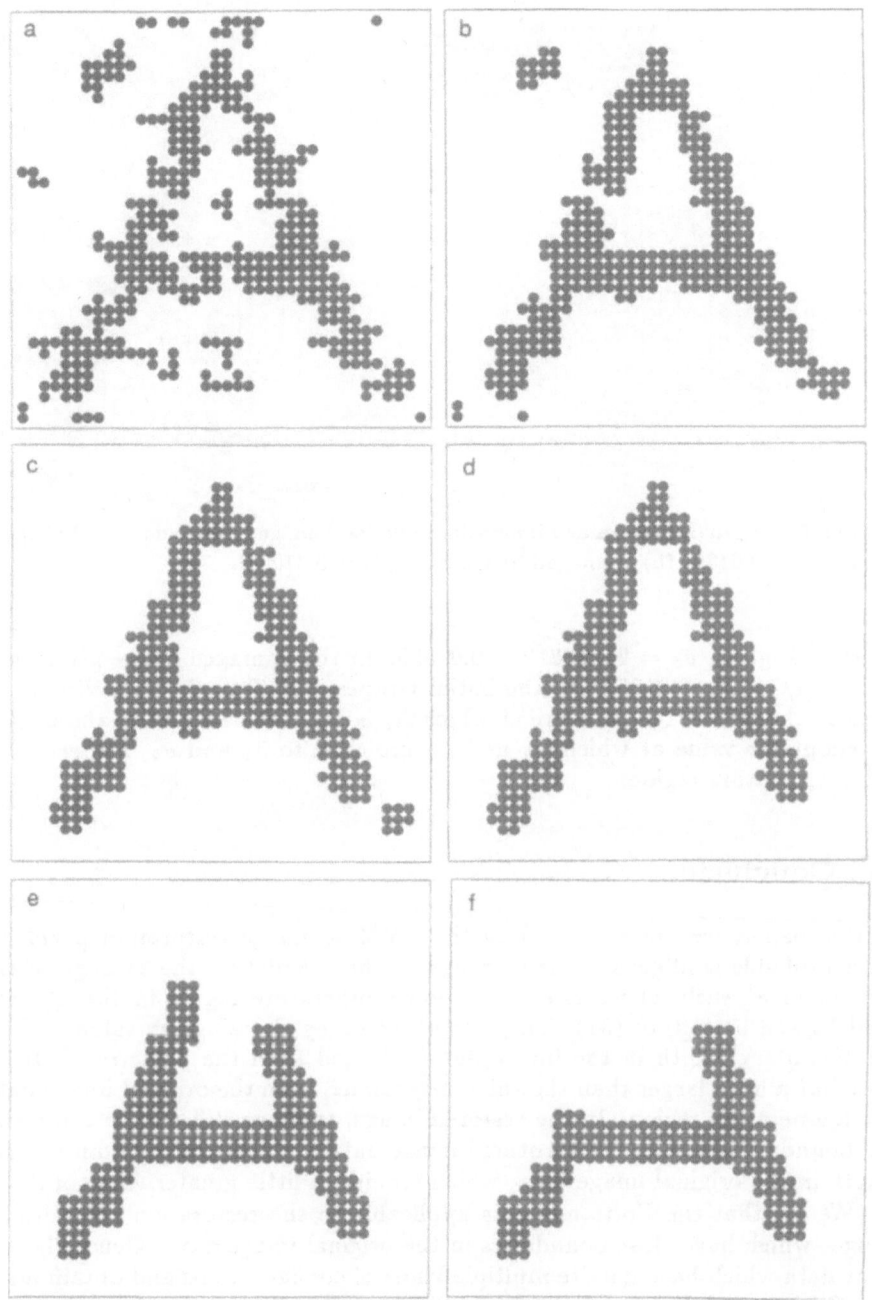

Figure 4. Most probable configurations for various values of the parameter J obtained by the PA annealing method in the binary image data. (c) is taken from Ref. 29. (a) $J = 0.40$ ($R \simeq 0.11219$, $R_1 \simeq 0.10526$, $R_2 \simeq 0.16074$) (b) $J = 0.60$ ($R \simeq 0.06648$, $R_1 \simeq 0.17036$, $R_2 \simeq 0.09033$) (c) $J = 0.70$ ($R \simeq 0.05956$, $R_1 \simeq 0.18490$, $R_2 \simeq 0.07752$) (d) $J = 0.80$ ($R \simeq 0.06371$, $R_1 \simeq 0.19252$, $R_2 \simeq 0.07112$) (e) $J = 1.25$ ($R \simeq 0.08518$, $R_1 \simeq 0.22091$, $R_2 \simeq 0.05334$) (f) $J = 1.40$ ($R \simeq 0.10665$, $R_1 \simeq 0.23961$, $R_2 \simeq 0.04339$).

a b

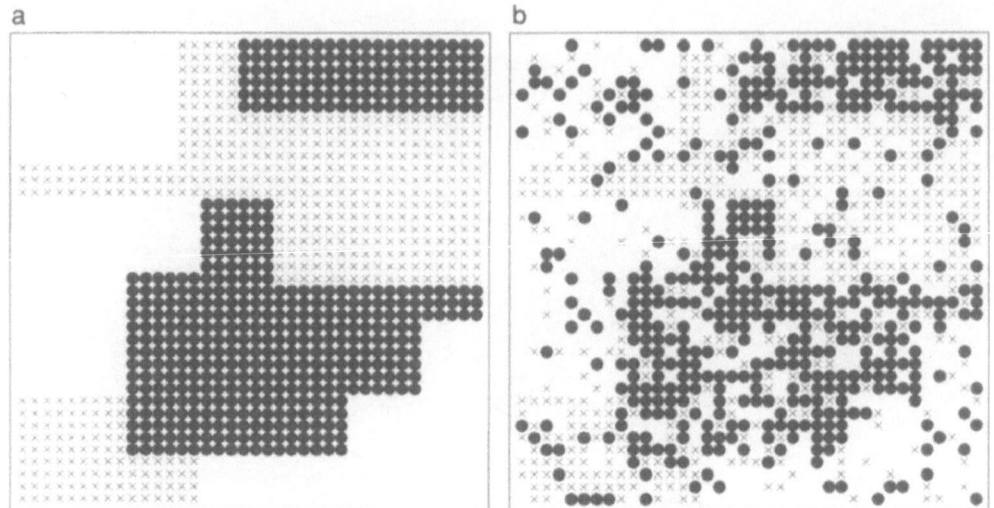

Figure 5. Original three-valued image data and its damaged image data. (a) Original image data ($\bar{\sigma}_2 \simeq 0.06188$) (b) Damaged image data ($\bar{\sigma}_1 \simeq 0.31025$).

given in Fig. 2a, $\bar{\sigma}_2 = 212/2812 \simeq 0.07539$. In the damaged image given in Fig. 2b, $\bar{\sigma}_1 = 304/1444 \simeq 0.21052$. As the initial temperature T in the annealing process, we need to choose the temperature at which R_1 is equal to zero. As to the parameter J, we adopt the value at which R_1 and R_2 are close to $\bar{\sigma}_1$ and $\bar{\sigma}_2$, respectively, in the low-temperature region.

IV. Conclusion

In this paper, we apply the PA of the CVM to image restoration problems. The most probable configuration is searched for by calculating the average of the order parameter at each lattice site in the low-temperature region in the Q-state Potts model given by (19) or (31). They are obtained by choosing the value of J at which the boundary length in the image data obtained from the posterior distribution is close but a little larger than the value determined from the original image data in the low-temperature region. In the restored image, we have still a little error and hence the boundary length in the recovered image data is a little longer than the boundary length in the original image data, which requires a little greater value of J .

We see that the Potts model is applicable to the restoration of multiple-valued images which have clear boundaries in the original image data. Generally, when our treat data which have a finite multiple kinds of configurations and obtain a recovered image data which has a clear boundary between two different configurations, we judge the better recovering data as the recovering image data.

In the CVM, the result is expected to improve successively as we adopt larger basic cluster sets. The generalized algorithm of the CVM for the Ising model with finite-range interactions under a local non-uniform external field is given by Morita.[26,27] In the restoration of more complicated images, a posterior probability distribution must include second-neighbor pair interaction terms and three- or four-body interaction terms. When we treat these models, we need to adopt the CVM with larger basic

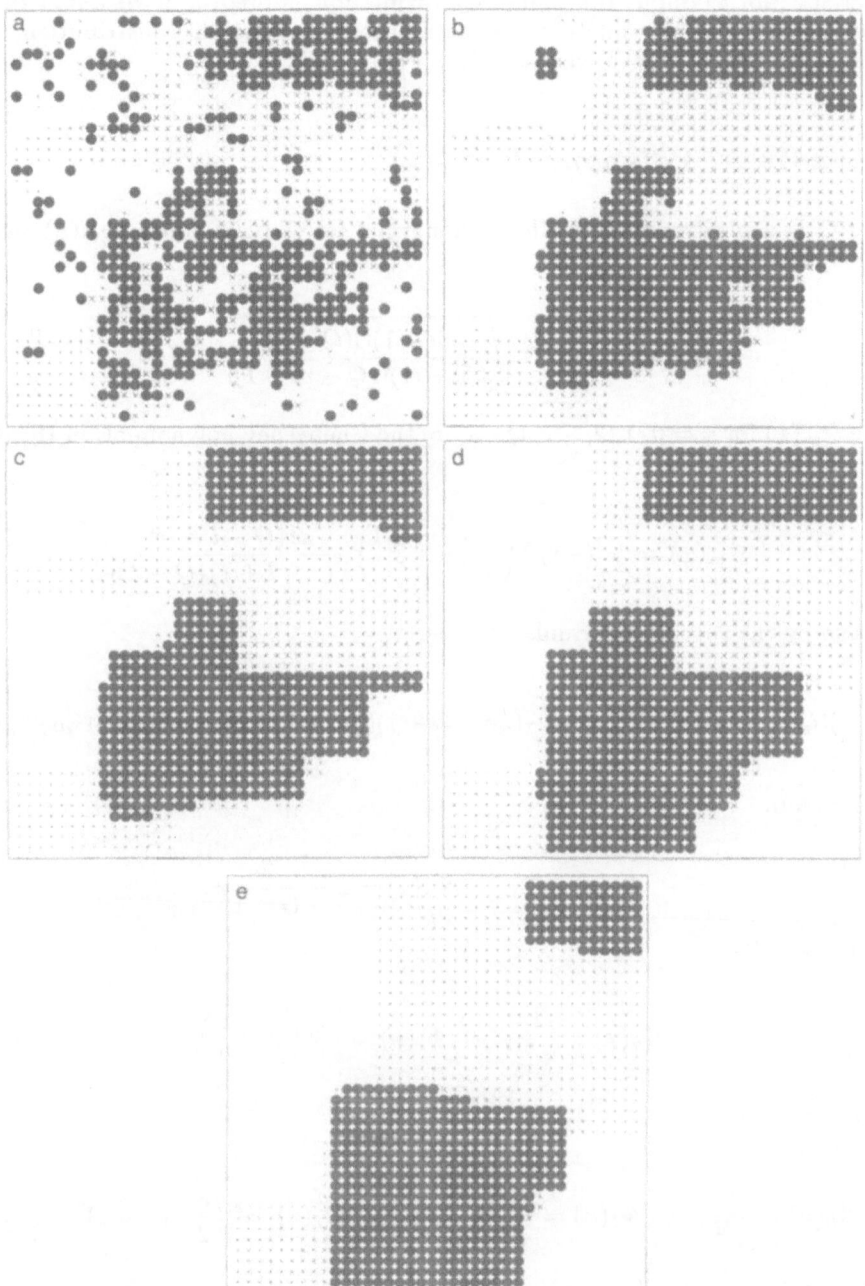

Figure 6. Most probable configurations for various values of the parameter J obtained by the PA annealing method in the three-valued image data. (a) $J = 0.20$ ($R \simeq 0.26524$, $R_1 \simeq 0.26316$, $R_2 \simeq 0.35206$) (b) $J = 0.50$ ($R \simeq 0.09972$, $R_1 \simeq 0.32964$, $R_2 \simeq 0.10597$) (c) $J = 1.00$ ($R \simeq 0.08102$, $R_1 \simeq 0.34349$, $R_2 \simeq 0.07290$) (d) $J = 3.00$ ($R \simeq 0.13227$, $R_1 \simeq 0.38158$, $R_2 \simeq 0.05512$) (e) $J = 5.00$ ($R \simeq 0.19875$, $R_1 \simeq 0.41482$, $R_2 \simeq 0.04623$).

cluster sets, for example, the square approximation. Moreover, if we adopt the Potts model which includes a line process[1] as a posterior probability distribution, we will obtain more clearly recovered image data.

Appendix A: Polynomials $\Phi_m(x)$

When Q is a positive integer, the orthonormal set of polynomials $\Phi_m(x)$ satisfying Eq. (26) are expressed as[33,34]

$$\Phi_m(x) = (-1)^m \sqrt{\frac{(2m+1)!\,[(Q-1)!]^2}{(Q+m)!\,(Q-m-1)!}}\, \Psi_m(x),$$

where $\Psi_m(x)$ for $x = 0, 1, 2, \ldots, Q-1$ are the Chebyshev polynomials in the discrete range, given by

$$\Psi_m(x) = \sum_{l=0}^{m}(-1)^m\binom{m}{l}\binom{m+l}{l}\frac{x!\,(Q-l-1)!}{\Gamma(x-l+1)\,(Q-1)!},$$

and also by the recursion formula

$$(m+1)(Q-1-m)\Psi_{m+1}(x) = -(2x-Q+1)(2m+1)\Psi_m(x) - m(Q+m)\Psi_{m-1}(x),$$

starting from

$$\Psi_0(x) = 1, \qquad \Psi_1(x) = 1 - \frac{2}{Q-1}x.$$

For $Q = 2$,

$$\Phi_0(x) = \frac{1}{\sqrt{2}}, \qquad \Phi_1(x) = \sqrt{2}\left(x - \frac{1}{2}\right).$$

For $Q = 3$,

$$\Phi_0(x) = \frac{1}{\sqrt{3}}, \qquad \Phi_1(x) = \frac{1}{\sqrt{2}}(x-1), \qquad \Phi_2(x) = \sqrt{\frac{3}{2}}\left((x-1)^2 - \frac{2}{3}\right).$$

For $Q = 4$,

$$\Phi_0(x) = \frac{1}{2}, \qquad \Phi_1(x) = \frac{1}{\sqrt{5}}\left(x - \frac{3}{2}\right), \qquad \Phi_2(x) = \frac{1}{2}\left(\left(x - \frac{3}{2}\right)^2 - \frac{5}{4}\right),$$

$$\Phi_3(x) = \frac{\sqrt{5}}{3}\left(x - \frac{3}{2}\right)\left(\left(x - \frac{3}{2}\right)^2 - \frac{41}{20}\right).$$

Appendix B

When the non-interacting Potts model ($J = 0$) has L independent distributions, the entropy S_I is expressed as

$$S_I = \sum_{i=1}^{L} S_i .$$

If only a pair of lattice sites 1 and 2 have some interactions, the entropy is expressed as

$$S_I = \sum_{i=3}^{L} S_i + S_{12}$$

$$= \sum_{i=1}^{L} S_i + (S_{12} - S_1 - S_2) .$$

If only four lattice sites 1, 2, 3 and 4 have some interactions, the entropy is expressed as

$$S_I = \sum_{i=5}^{L} S_i + S_{1234}$$

$$= \sum_{i=1}^{L} S_i + (S_{12} - S_1 - S_2) + (S_{23} - S_2 - S_3)$$

$$+ (S_{34} - S_3 - S_4) + (S_{41} - S_4 - S_1)$$

$$+ (S_{1234} - S_{12} - S_{23} - S_{34} - S_{41} + S_1 + S_2 + S_3 + S_4) .$$

In the system whose Hamiltonian is given by (31), since all lattice sites are not independent of each other, the entropy is expressed as

$$S_I = \sum_i S_i + \sum_{(ij)} (S_{ij} - S_i - S_j)$$

$$+ \sum_{(ijkl)} (S_{ijkl} - S_{ij} - S_{jk} - S_{kl} - S_{li} + S_i + S_j + S_k + S_l) + \cdots$$

$$= \sum_i S_i + \sum_{(ij)} (S_{ij} - S_i - S_j) + \overline{S}_{\text{Pair}}$$

$$= \sum_i (-z_i + 1) S_i + \sum_{(ij)} S_{ij} + \overline{S}_{\text{Pair}} ,$$

where $\overline{S}_{\text{Pair}}$ is a correction term and the notation $\sum_{(ijkl)}$ means the summation over all the square plaquette, each edge of which connects a nearest-neighbor pair of lattice sites. In the PA, we put $\overline{S}_{\text{Pair}} = 0$.

Appendix C

We consider the relation between reducibility and consistency. For two functions $f_\gamma(\{x_\gamma\})$ and $g_\gamma(\{x_\gamma\})$, we introduce the inner product $(f_\gamma(\{x_\gamma\}), g_\gamma(\{x_\gamma\}))_\gamma$ by

$$(f_i(x_i), g_i(x_i))_i \equiv \text{Tr}_i \, f_i(x_i)g_i(x_i),$$

$$(f_{ij}(x_i, x_j), g_{ij}(x_i, x_j))_{ij} \equiv \text{Tr}_{ij} \, f_{ij}(x_i, x_j)g_{ij}(x_i, x_j).$$

The sets $\{\Phi_m(x); \; m = 0, 1, 2, \ldots, Q-1\}$ and $\{\Phi_m(x)\Phi_n(x'); \; m, n = 0, 1, 2, \ldots, Q-1\}$ are an orthogonal system of functions such that

$$(\Phi_m(x_i), \Phi_{m'}(x_i))_i = \delta(m, m'),$$

$$(\Phi_m(x_i)\Phi_n(x_j), \Phi_{m'}(x_i)\Phi_{n'}(x_j))_{ij} = \delta(m, m')\delta(n, n'),$$

and every probability distributions $\rho_i(x_i)$ and $\rho_{ij}(x_i, x_j)$ are expressed in the orthogonal series as

$$\rho_i(x_i) = \sum_{m=0}^{Q-1} \langle \Phi_m(x_i) \rangle_i \Phi_m(x_i),$$

$$\rho_{ij}(x_i, x_j) = \sum_{m=0}^{Q-1} \sum_{n=0}^{Q-1} \langle \Phi_m(x_i)\Phi_n(x_j) \rangle_{ij} \Phi_m(x_i)\Phi_n(x_j).$$

By substituting these into the reducibility $\rho_i(x_i) = \text{Tr}_j \, \rho_{ij}(x_i, x_j)$ and by comparing the coefficients, we can see that the reducibility and the consistency are equivalent to each other[25]

$$\rho_i(x_i) = \text{Tr}_j \, \rho_{ij}(x_i, x_j) \iff \langle \Phi_m(x_i) \rangle_i = \langle \Phi_m(x_i) \rangle_{ij}, \quad m = 0, 1, 2, \ldots, Q-1,$$

$$\rho_j(x_j) = \text{Tr}_i \, \rho_{ij}(x_i, x_j) \iff \langle \Phi_m(x_j) \rangle_j = \langle \Phi_m(x_j) \rangle_{ij}, \quad m = 0, 1, 2, \ldots Q-1.$$

Acknowledgements

The one of authors (K.T.) is grateful to Professors Junji Maeda and Yukinori Suzuki for valuable discussions.

References

1. S. Geman and D. Geman, *IEEE Trans. Pattern Anal. Machine Intel.* **6**, 721 (1984).
2. B. Gidas, *IEEE Trans. Pattern Anal. Machine Intel.* **11**, 164 (1989).
3. J. Zhang, *IEEE Trans. Signal Processing* **40**, 2570 (1992); *IEEE Trans. Image Processing* **2**, 27 (1993); *IEEE Trans. Image Processing* **4**, 19 (1995).
4. C.-h. Wu and P. C. Doerschuk, *IEEE Trans. Pattern Anal. Machine Intel.* **17**, 275 (1995); *ibid.*, p. 391.
5. R. Kikuchi, *Phys. Rev.* **81**, 988 (1951).
6. R. Kikuchi, *Prog. Theor. Phys.* Suppl. No. 115, 1 (1994).
7. T. Morita, *J. Phys. Soc. Jpn.* **12**, 753 (1957).

8. T. Morita, *Prog. Theor. Phys.* Suppl. No. 115, 27 (1994).
9. A. G. Schlijper, *Phys. Rev. B* **27**, 6841 (1983).
10. A. G. Schlijper, *J. Stat. Phys.* **40**, 1 (1985).
11. T. Morita, *Prog. Theor. Phys.* Suppl. No. 80, 103 (1984).
12. H. A. Bethe, *Proc. R. Soc. London* **A150**, 552 (1935).
13. H. A. Kramers and G. H. Wannier, *Phys. Rev.* **60**, 252 (1941); *ibid.*, p. 263.
14. Y. Murai, K. Tanaka, and T. Morita, *Physica* **A217**, 214 (1995).
15. S. Fujiki, M. Katori, and M. Suzuki, *J. Phys. Soc. Jpn.* **59**, 2681 (1990).
16. M. Katori and M. Suzuki, *Prog. Theor. Phys.* Suppl. No. 115, 83 (1994).
17. L. Onsager, *Phys. Rev.* **65**, 117 (1944).
18. M. Suzuki, *Phys. Lett.* **A116**, 3 (1986); *J. Phys. Soc. Jpn.* **55**, 4025 (1986).
19. M. Katori and M. Suzuki, *J. Phys. Soc. Jpn.* **57**, 3753 (1988).
20. K. Wada and N. Watanabe, *J. Phys. Soc. Jpn.* **59**, 2610 (1990).
21. K. Tanaka, T. Horiguchi, and T. Morita, *J. Phys. Soc. Jpn.* **60**, 2576 (1991); *Phys. Lett.* **A165**, 266 (1992); *Physica* **A192**, 647 (1993); *Prog. Theor. Phys.* Suppl. No. 115, 221 (1994).
22. O. Nagai, T. Horiguchi, and S. Miyashita, in *Magnetic Systems with Competing Interactions (Frustrated Spin Systems)*, edited by H. T. Diep (World Scientific, 1994).
23. F. Y. Wu, *Rev. Mod. Phys.* **54**, 235 (1982).
24. T. Morita, *Physica* **A133**, 173 (1985).
25. T. Morita, *J. Stat. Phys.* **34**, 319 (1984).
26. T. Morita, *Phys. Lett.* **A161**, 140 (1991).
27. T. Morita, *Prog. Theor. Phys.* **85**, 243 (1991).
28. T. Morita, *J. Stat. Phys.* **59**, 819 (1990).
29. K. Tanaka and T. Morita, *Phys. Lett.* **A203**, 122 (1995).
30. T. Morita, *J. Math. Phys.* **13**, 115 (1972).
31. S. Katsura and S. Fujiki, *J. Phys.* **C13**, 4711 (1980).
32. K. Tanaka, Y. Takane, H. Ebisawa, and T. Morita, *J. Phys. Soc. Jpn.* **63**, 2909 (1994).
33. M. Abramowitz and I. A. Stegun, *Handbook of Mathematical Functions with Formulas, Graphs, and Mathematical Tables* (Dover Publ. Inc., New York, 1995) p. 788 and p. 791.
34. T. Morita, *J. Phys. Soc. Jpn.* **62**, 4218 (1993), where the polynomials are written but with misprintings: In the equation at the bottom of p. 4222, "$(2k+1)!$" should read "$(2k+1)$", in the equation just above it, "$(x-m)!$" should read "$\Gamma(x-m+1)$", and in the third row in p. 4223, "$=-s$" should read "$=-s/S$".

Note added in proof: After the Workshop on the Cluster Variation Method and the Path Probability Method in Teotihuacan, Mexico, we wrote two related papers. One is entitled as *Determination of Parameters in an Image Recovery by Statistical-mechanical Means*, in which a method of determining the parameters $\bar{\sigma}_1$ and $\bar{\sigma}_2$ is proposed. That paper was published in *Physica* **A223**, 244-262 (1996). In the other, entitled as *Statistical-mechanical Algorithm of Multi-Valued Image Restoration*, the pair approximation in the CVM and the determination method of parameters $\bar{\sigma}_1$ and $\bar{\sigma}_2$ under the periodic boundary condition (on a torus) is applied to the multi-valued image restoration, while all the results in the present paper is calculated under the free boundary condition (on a plane). Figs. 1 and 5 are also used in the paper. That paper is submitted to *Patern Recognition Letters*.

Cluster Approach to Pattern Recognition

Makoto Kaburagi,[1,2] Yasunori Motomura,[2] Qiong Ou,[2]
Kazuhiro Ohtsuki,[1] and Atsuo Ono[1,2]

[1] Faculty of Cross-Cultural-Studies,
 Kobe University,
 Tsurukabuto, Nada, Kobe 657,
 JAPAN

[2] Graduate School of Science and Technology,
 Kobe University,
 Rokkodai, Nada, Kobe 657,
 JAPAN

Abstract

This paper deals with the mathematical morphology as a cluster approach to image processing and the Hough transformation as an object of statistical physics. It is shown that the operation in mathematical morphology is equivalent to a single step of dynamics in a special neural network. We investigate dynamics of an image (the neural network) and examine the relation between the time interval to reach to an associative memory and the threshold. As for the Hough transformation, we derive the *free energy expression* for the transformation and proposed the Gaussian sum method.

I. Introduction

The purpose of this paper is to make a connection between statistical physics and pattern recognition by introducing a cluster approach to pattern recognition.

Usual statistical physics deals with the problem that, for a given model, what is the kind of pattern obtained at thermal equilibrium condition, and to what kind of category, such as the criticality at a phase transition point, *i.e.*, the universality class, the pattern belongs. The typical lattice models used for the studies of the thermodynamic properties of materials are the Ising model, the Heisenberg model and the multicomponent lattice gas model. As is well known, the basic laws of thermodynamics are expressed in terms of a variational (or Mini-Max) principle. One of the most successful methods based on the variational principle of statistical physics is

Theory and Applications of the Cluster Variation and Path Probability Methods
Edited by J.L. Morán-López and J.M. Sanchez, Plenum Press, New York, 1996

the Cluster Variation Method (CVM) which was devised by Kikuchi[1] and generalized by Morita.[2] This method has been successfully applied to a wide variety of fields, magnetism, metallurgy, and surface science.[3]

On the other hand, pattern recognition deals with the problems of how to deduce information from given patterns through construction of a model.[4] One of the most popular model in this field is the neural network[5] and lots of fruitful ideas, such as concept of associative memory,[6-8] have been produced from studies of this model by using techniques of statistical physics. The time of convergence to associative memory is one of the recent issues in the study of the neural network.[6]

An important field related closely to pattern recognition is image processing[9] which includes restoration,[10] reconstruction, matching, detection, representation, and description. This paper deals with two topics; one is the subject concerned with mathematical morphology[11,12] by which recent advance in image processing has been achieved; the other is the subject related to feature detection of a curve by the Hough transformation.[13,14] Mathematical morphology has been applied to extensive image processing and analysis, such as non-linear image filtering, edge detection, noise suppression, shape representation, smoothing and recognition. Therefore, if mathematical morphology can be transformed into another popular method, we may expect that rich results obtained from the popular method can be introduced to the image processing and vice versa. Actually we will show in this paper that the operations in mathematical morphology is equivalent to single a step of dynamics in a special neural network. By using rank order filter in mathematical morphology, as a simple example for an application of this equivalence, we examine the relation between the time interval to reach to an associative memory and the threshold in the neural network system.

The Hough transform is well-known as a method of detecting patterns in image processing.[13,14] In this paper, we propose a new *free energy* expression for feature detection in image processing as an extension of the Hough transformation. To examine the new expression, we apply the square energy case (the Gaussian sum) to a line image with noise and compare the result with that obtained by usual Hough transformation.

The remainder of this paper is organized as follows. After reviewing the mathematical morphology, we demonstrate several examples of image restoration by the method of mathematical morphology in the next section. In section III, we discuss the relation between mathematical morphology and the neural network. Section IV is devoted to the proposal of a new *free energy* expression for feature detection. Finally, supplementary discussions are given in section V.

II. Mathematical Morphology for Image Processing

Mathematical morphology was developed mainly by Matheron and Serra[11] in Europe since the 1960's as an extension of Minkowsky's set-theoretical method for image analysis. Features of this method are based on cluster dependent operators, such as erosion, dilation, opening, closing, and rank order filters. These operations can be defined on binary or grayscale images in any number of dimensions. To be concrete, we consider here only the case of binary signals and review the method.[12]

Let A be the set representation of input signal, and let B a set for a cluster of small size. The set B is called a *structuring element*. Let $A \pm b = \{a \pm b : a \in A\}$ denote the vector translate of A by $\pm b$. The two fundamental operations dilation $\oplus$ and erosion $\ominus$ of A by B are defined as

Figure 1. Images obtained by morphological operations with the cross-type structuring element; (a) original, (b) 10% noise, (c) median filter, (d) erosion, (e) dilation, (f) opening, (g) closing; and those obtained by morphological rank order filters with the 3×3 structuring element; (h) erode, (i) dilate, (j) open, (k) close.

$$A \oplus B = \cup_{b \in B} A + b = \{a + b : a \in A \text{ and } b \in B\}, \tag{1}$$

$$A \ominus B = \cap_{b \in B} A - b = \{c : (B + c) \subseteq A\}, \tag{2}$$

respectively. It is easily seen from these definition that the resultant set of the dilation operator consists of the set of the points such that the translate of the structuring element B has a non-empty intersection with the input set; the resulting set of the erosion consists of the set of translation points such that the translated structuring element is contained in the input set.

Other two important operators, the opening o and the closing • of set A by B, are defined as follows

$$A \circ B = (A \ominus B) \oplus B, \tag{3}$$

$$A \bullet B = (A \oplus B) \ominus B. \tag{4}$$

In order to visualize the behavior of these basic operations, we show in Figs. 1d) to g) the output images of these operations on the input image A shown in Fig. 1b) with the structuring element B of the four neighboring points cross. Figure 1 illustrates that the dilation expands A, whereas the erosion shrinks it; the closing fills in the thin gulfs and small holes, whereas the opening suppresses the sharp capes and cuts the narrow isthmuses.

In image restoration it is important to remove noises in images. For impulsive noises, median filter and rank order filter have been popular techniques. In mathematical morphology, by generalizing the operators defined above, we can construct similar filters, the morphological rank order filters. Following Eqs. (1) and (2), we define two fundamental operations, nth order dilate $\widetilde{\oplus}_n$ and nth order erode $\widetilde{\ominus}_n$ of A by B, as follows

$$A \widetilde{\oplus}_n B = \{c :| A \cap (B + c) | \geq n\}, \tag{5}$$

$$\overline{A \widetilde{\ominus}_n B} = \{c :| \bar{A} \cap (B + c) | \geq n\}. \tag{6}$$

Here $\bar{X}$ represents the inversion of the set X and n is sometimes called threshold. Note that the first order dilate and the last order erode are usual dilation and erosion, respectively. Similarly we define the open $\tilde{o}_{n,m}$ and the close $\tilde{\bullet}_{n,m}$ of set A by B as follows

$$A \tilde{o}_{n,m} B = (A \widetilde{\ominus}_n B) \widetilde{\oplus}_m B, \tag{7}$$

$$A \tilde{\bullet}_{n,m} B = (A \widetilde{\oplus}_n B) \widetilde{\ominus}_m B. \tag{8}$$

In order to demonstrate the effectiveness of these operations for noise reduction, we show in Figs. 1h) to k) the images after operation ($n = m = 4$) on the image with noise A, shown in Fig. 1b), with the structuring element B consisting of eight neighboring points. As shown in Fig. 1, these operators can be used as the filtering technique in image restoration.

III. Neural Network and Morphology

The neural network has been studied extensively in connection with artificial brain and associative memory.[5] As is well known, the general dynamics of the neural network in the Ising variable expression can be represented as

$$S_i(t + 1) = F\big(h_i(\{S_j(t)\})\big), \tag{9}$$

where $S_i(t)$ is the Ising variable representing the state of the ith neuron at time t; $F(h)$ is the filter with sigmoidal shape; $h_i(\{S_j\})$ is the mean field acting on the ith neuron determined by the neural state $\{S_i\}$.

In this section, we will show that the morphological operations are equivalent to a single step of the dynamics, Eq. (9), in a special neural network. We hereafter confine ourselves to the operation in the morphological rank order filter, because it contains the usual morphological operations as a special case. It is easily shown that the operation given by Eqs. (5) and (6) are represented in terms of a set of the Ising variable $\{S_i\}$ as

$$S_{i\,(\text{dilate})} = \big[2\theta(n_i^+ - n + 1) - 1\big]\,\theta(n_i^+ - n + 1) + S_{i\,(\text{input})}\,\theta(n - n_i^+), \tag{10}$$

$$S_{i\,(\text{erode})} = -\big[2\theta(n_i^- - n + 1) - 1\big]\,\theta(n_i^- - n + 1) + S_{i\,(\text{input})}\,\theta(n - n_i^-), \tag{11}$$

where $\theta(x)$ is the heviside step function defined as

$$\theta(x) = \begin{cases} 1 & \text{for } x > 0 \\ 0 & \text{for } x \leq 0, \end{cases} \tag{12}$$

$n_i^{\pm}$ is the number of $\pm$ spins within the ith region specified by the structuring element. $n_i^{\pm}$ is represented as

$$n_i^{\pm} = \sum_{b \in B} \tfrac{1}{2} \left(1 \pm S_{i+b\,(\text{input})}\right), \tag{13}$$

$$= \tfrac{1}{2}\left(1 \pm M_i\right), \tag{14}$$

$$M_i^{\pm} = \sum_{b \in B} S_{i+b\,(\text{input})}. \tag{15}$$

Comparing Eqs. (10) and (11) with Eq. (9), we may say that the morphological operations from the input to the output can be considered as a single step of the neural network.

We have studied the dynamics of this toy neural network and examined the convergence of the image for successive operation of close operator for various threshold n with the same structuring element, as in the previous section. Although there is no learning mechanism in the present case, this convergence might be a kind of associative memory. Figure 2 shows the results of the examination. The time interval t_c to reach convergence increases with decreasing threshold n. In Fig. 3 we plot t_c *versus* n, which implies the relation either $t_c \propto n^{-5}$ or $t_c \propto e^{-\alpha n}$. The mechanism leading to this relation is left for the future studies.

IV. An Extension of the Hough Transformation

The Hough transformation is well-known as a method for detecting patterns in image processing. The original idea was proposed by Hough[13,14] to detect curves in bubble chamber photographs. Later it was introduced to digital image processing community by Rosenfeld.[9]

In feature detection of images, there are many cases in which a given pattern of the image is characterized by small number of parameters. For example, a straight line in 2-dimensional space is characterized by two parameters $(\bar{c}, \bar{m})$ as

$$y = \bar{m}x + \bar{c}. \tag{16}$$

The Hough algorithm is very unique. The idea is based on the fact that the Cartesian coordinates associated with an image can be mapped to the parameter space. In other words, the Hough transformation can be defined as a mapping from the input pattern to the parameter space. To illustrate the Hough transformation, we employ the improved version by Duda and Hart,[15] because the slope-intercept parameter representation (16) used in the original version leads to the practical difficulty of unboundness of parameter space.

Let us consider N points r_i $(i = 1, 2, \ldots, N)$ in Cartesian coordinates on the straight line represented by

$$\bar{\rho} = x \cos \bar{\theta} + y \sin \bar{\theta}. \tag{17}$$

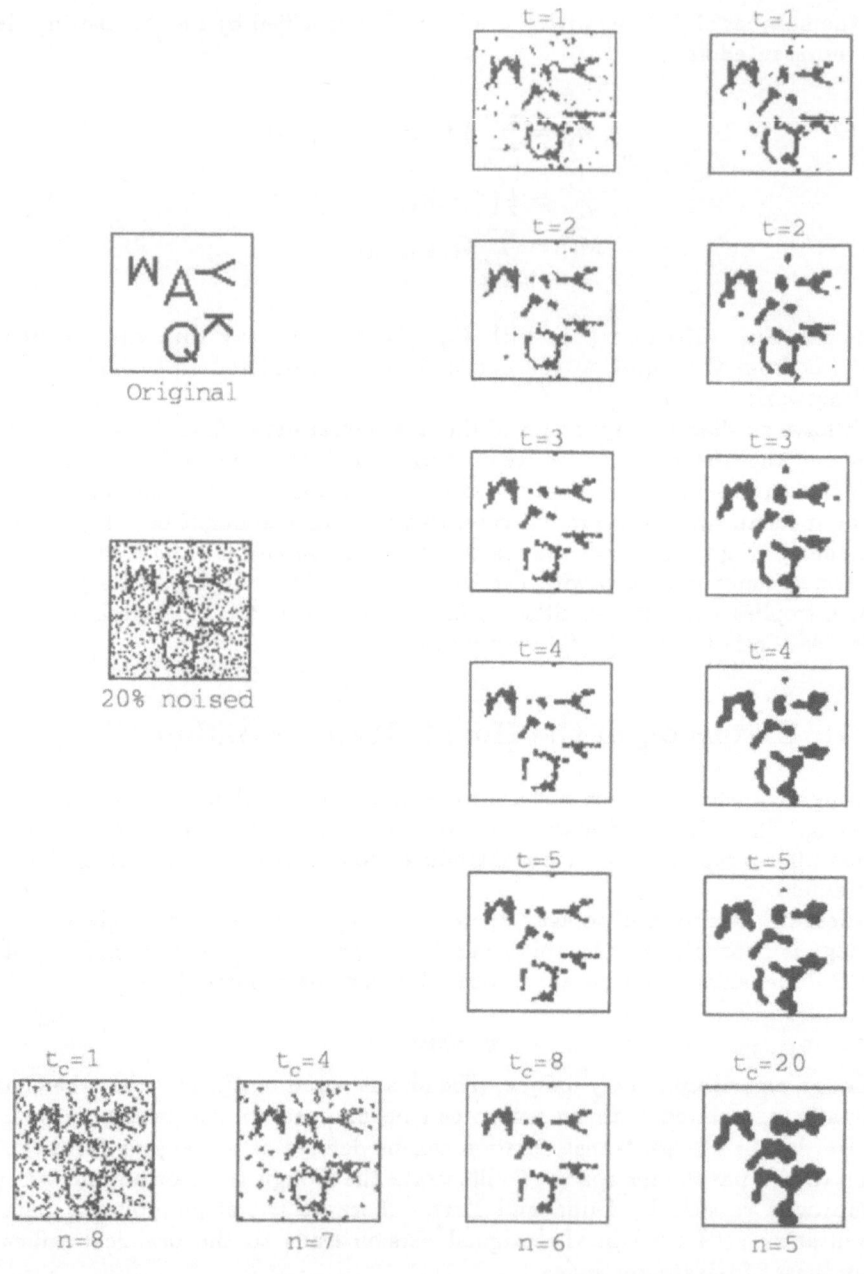

Figure 2. Snapshot and associative memories derived by morphological rank oder filter (close) for the given value of threshold n.

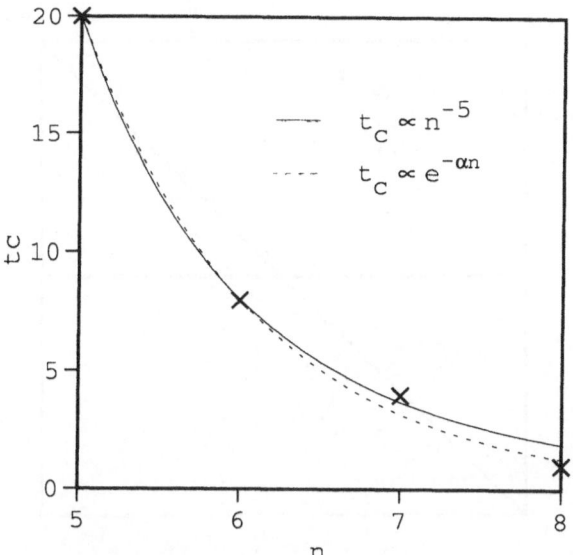

Figure 3. Plots of t_c (time interval to reach convergence) *versus* threshold n (square); full line $t_c \propto n^{-5}$; dashed line $t_c \propto e^{-\alpha n}$.

If we plot the line in the parameter space (ρ, θ) given by

$$\rho = x \cos \theta + y \sin \theta \,, \tag{18}$$

for every point $\mathbf{r}_i$, all of the lines pass through the point $(\bar{\rho}, \bar{\theta})$ in the parameter space. Therefore by searching the intersection point of the lines in parameter space, we can extract feature of the straight line in an image.

The Hough transformation is recognized as a useful tool in detecting straight lines, circles[16] and more complex curves such as ellipses or optional patterns and still more complex 3-dimensional bodies. It is used in a wide variety of fields. For example in high-energy physics experiments, the Hough technique is used for pattern recognition of the layered track chambers.[17]

In order to extend the Hough algorithm, we introduce a *partition function* $Z(\rho, \theta)$ defined as

$$Z(\rho, \theta) = \sum_{i=1}^{N} g\big(v(\rho, \theta, \mathbf{r}_i)\big) \,, \tag{19}$$

$$v(\rho, \theta, \mathbf{r}) = x \cos \theta + y \sin \theta - \rho \,. \tag{20}$$

where $g(v)$ is a filter function with sharp peak at $v = 0$ for the transformation of the image from Cartesian coordinates to parameter space.[18] By searching for peaks of the partition function, feature of the line can be extracted. Note that usual Hough algorithm is realized by setting the filter function as

$$g(v) = \begin{cases} 1 & \text{for } |v| < \delta \\ 0 & \text{otherwise.} \end{cases} \tag{21}$$

O X

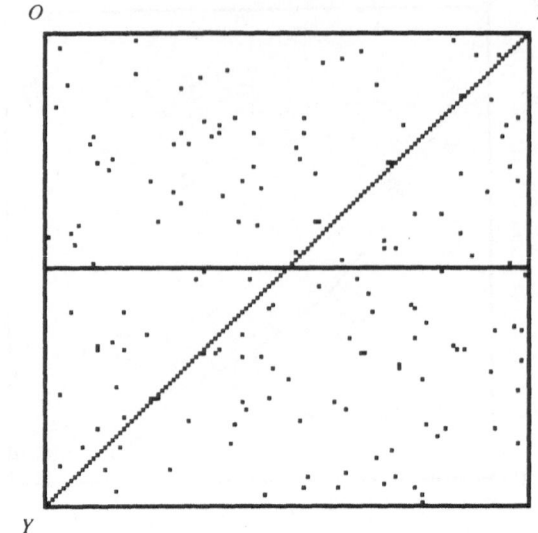

Y

Figure 4. Sample image for the Hough transformation.

We can rewrite Eq. (19) as

$$Z(\rho, \theta) = \int \exp\left[-\beta\varepsilon\big(v(\rho, \theta, \mathbf{r})\big)\right] W(\mathbf{r})\, d\mathbf{r}, \tag{22}$$

where the *energy* function $\beta\varepsilon(v)$ and the density function are given as

$$\exp\big(-\beta\varepsilon(v)\big) = g(v), \tag{23}$$

$$W(\mathbf{r}) = \sum_{i=1}^{N} \delta(\mathbf{r} - \mathbf{r}_i). \tag{24}$$

To examine the dependence of the partition function on the filter function, we calculate it for the line image shown in Fig. 4 for the two cases of the usual voting method and the Gaussian sum method for $\beta = 1$, given by

$$\varepsilon(v) = v^2. \tag{25}$$

In Figs. 5 and 6, we give the results of the calculation for the two cases, which are almost the same to each other. The voting method gives a discrete value for the partition function $Z(\rho, \theta)$, whereas the Gaussian sum method gives continuous values. This will be one advantage of the Gaussian sum method to determine a precise values of the parameters, because we can use an iterative technique, such as Newton-Raphson method, in the determination. The second advantage of the Gaussian sum method is that we can avoid the problem of the slope which gives serious trouble in the voting method. The third advantage is that by changing the value of β we can control the contrast and sharpness of $Z(\rho, \theta)$.

If we define the *entropy function* $S(\mathbf{r})$ by

$$\exp\big(S(\mathbf{r})\big) = W(\mathbf{r}), \tag{26}$$

θ

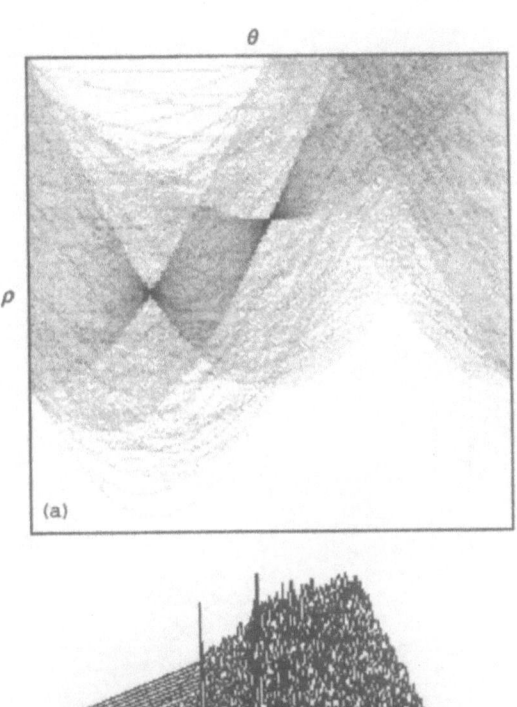

ρ

(a)

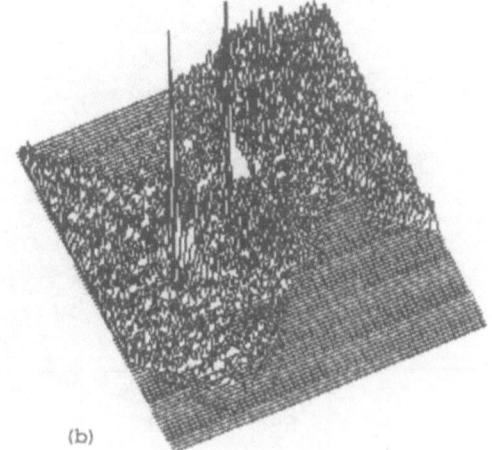

(b)

Figure 5. Results of the Hough transformation; (a) grayscale plot of $Z(\rho, \theta)$, (b) surface plot.

the *partition function* $Z(\rho, \theta)$ is expressed in terms of the *free energy* f in familiar form as

$$Z(\rho, \theta) = \int \exp\left[-\beta f(\rho, \theta, \mathbf{r}, \beta)\right] d\mathbf{r}, \tag{27}$$

$$f(\rho, \theta, \mathbf{r}, \beta) = \varepsilon(v(\rho, \theta, \mathbf{r})) - S(\mathbf{r})/\beta. \tag{28}$$

For estimation of the integral in Eq. (27), we may employ the techniques used in the CVM.

Equation (27) clearly indicates that the Hough transformation corresponds to the point cluster (the mean field) approximation in the CVM. This means that no

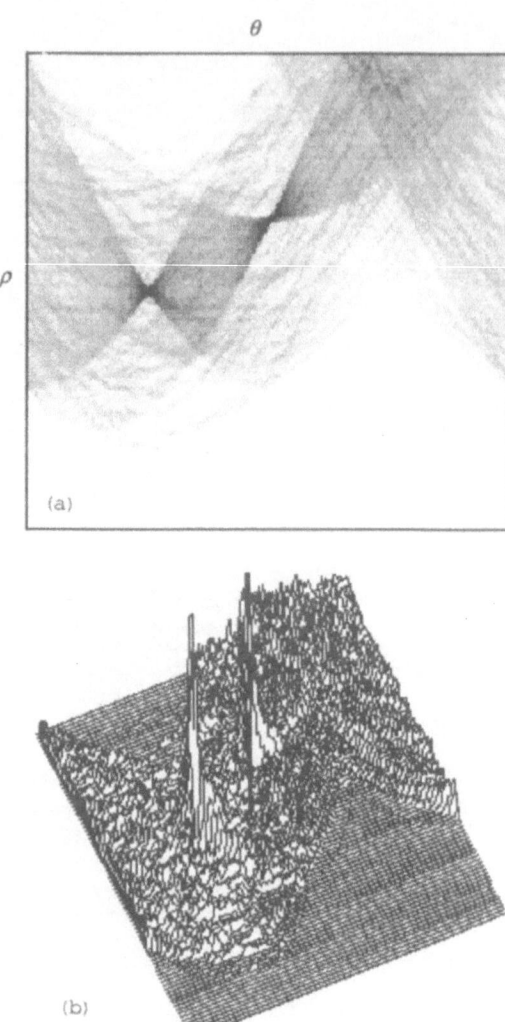

Figure 6. Results of the Gaussian sum; (a) grayscale plot of $Z(\rho, \theta)$, (b) surface plot.

multi-points correlations are taken into account in the Hough transformation and it is desirable to develop the method for dealing with such correlations.

V. Summary

This paper has dealt with two topics, the subject concerned with mathematical morphology and that related to the Hough transformation.

As for the mathematical morphology, we have

1. reviewed the mathematical morphology;
2. demonstrated several examples of image restoration by the method of mathematical morphology;

3. shown that the operations in mathematical morphology are equivalent to a single step of dynamics in a special neural network;
4. studied the dynamics of this toy neural network for successive operation of the close operator for various threshold;
5. examined the relation between the time interval to reach to an associative memory and the threshold in this toy neural network system.

As for the Hough transformation, we have

1. reviewed the Hough transformation;
2. proposed the Gaussian sum method;
3. applied the Gaussian method to an image;
4. derived the *free energy expression* for the transformation.

References

1. R. Kikuchi, *Phys. Rev.* **81**, 988 (1951).
2. T. Morita, *J. Phys. Soc. Jpn.* **12**, 753 (1972).
3. See for example, *Foundations and Applications of Cluster Variation Method and Path Probability Method*, edited by T. Morita, M. Suzuki, K. Wada, and M. Kaburagi, *Progr. Theor. Phys.* Suppl. **113** (1994).
4. K. Fukunaga, *Introduction to Statistical Pttern Recognition*, (Academic Press, New York, 1972).
5. T. Geszti, *Physical Models of Neural Network*, (World Scientfic, Singapore, 1990).
6. S. Amari and K. Maginu, *Neural Networks* **1**, 63 (1988).
7. H. Nishimori and T. Ozeki, *J. Phys. A: Math. Gen.* **26**, 859 (1993).
8. A. C. C. Coolen and D. Sherrington, *Phys. Rev. E* **49**, 1921 (1994).
9. A. Rosenfeld, *Picture processing by computer*, (Academic Press, New York, 1969).
10. K. Tanaka, in *Theory and Application of the Cluster Variation and Path Probability Methods*, edited by J. L. Morán-López and J. M. Sanchez (Plenum, New York, 1996), p. 345.
11. J. Serra, *Image Analysis and Mathematical Morphology*, Vol. 1, (Academic Press, New York, 1989).
12. P. Margos and R. W. Schafer, *Proc. IEEE* **78**, 690 (1990).
13. P. V. C. Hough and B. M. Powell, *Nuovo Cimento* **18**, 1184 (1960).
14. P. V. C. Hough, U. S. Patent 3069654 (1962).
15. R. O. Duda and P. E. Hart, *Commun. ACM* **15**, 11 (1972).
16. Y. Noguchi and A. Ono, *Nucl. Instr.* **A253**, 27 (1986).
17. Q. Ou and A. Ono, preprint.
18. P. L. Palmer, M. Petrou, and J. Kittler, *CVGIP: Image Understanding* **58**, 221 (1993).

On the Superposition of Probabilities

John A. Simmons

Metallurgy Division
National Institute of Standards and Technology
Gaithersburg, MD 20899
U.S.A.

Abstract

Superposition methods—as used in many lattice diffusion calculations to estimate joint probabilities from limited information on lower order correlations—are here reviewed in the general inhomogeneous context. The Kirkwood-Kikuchi-Barker (KKB) superposition principle is shown to be a special case of the application of combinatorial Möbius function techniques over a Boolean lattice of partitions into clusters, or motifs which uses a logarithmic/exponential device. This method is conceptually distinct from the approach utilizing the linear algebra of probabilities over an equal *a priori* probability phase space. It is shown that the probabilities produced by KKB superposition in general fail to satisfy the basic rules of conservation of probability; and a new algorithm is given for embedding the Möbius formalism into probability space. The general problem of superposition of probabilities is formulated in terms of linear convex analysis. The concept of marginal coordinate systems generalizing both the Möbius formalism and the idea of conditional probabilities is introduced and pidgeonhole cluster coordinates developed. The linear convex formalism additionally permits the use of interior point methods for the determination of the permissible polytope in probability space and provides an exact representation of the configurational entropy which incorporating all correlations in finite structures. It is shown that knowledge of the permissible polytope in probability space can restrict the class of permissible structures so severely that entropy maximization is unnecessary; and it is suggested that this is particularly true in the case of ordered structures. The equilibrium configurational free energy computation is discussed in the light of linear convex theory and the ergodic hypothesis.

I. Introduction

Probability superposition has been used extensively in classical statistical mechanics, particularly that devoted to homogeneous systems (For a review, see Ref. 1). It appears to originally been suggested by Kirkwood in 1935 in the pair approximation.

Theory and Applications of the Cluster Variation and Path Probability Methods
Edited by J.L. Morán-López and J.M. Sanchez, Plenum Press, New York, 1996

Table I

Sample pair probability distribution at three sites.

$p_{ij}^{xx'}$	A 2	B 2	A 3	B 3
A 1	0.2	0.1	0.2	0.1
B 1	0.1	0.6	0.1	0.6
A 2	–	–	0.2	0.1
B 2	–	–	0.1	0.6

Although not usually stated in Bayesian terms, it can be described in a Bayesian sense as attempting to incorporate all information known about a probability distribution into one of greater complexity without adding any new information. While the dimensional reduction method of Kikuchi and Brush shows the Kirkwood-Kikuchi-Barker (KKB) superposition approximation to be valid in a lattice context,[2] the KKB approximation does not, in general, conserve probability. In this study we first analyze the Möbius formalism on Boolean partition lattices in detail and develop a consistent alternative to the KKB superposition principle within that context. The Möbius approach, however, does not directly address the general Bayesian statistics goal formulated above. To achieve this one must go beyond the Möbius formalism to look at the convex geometry of probability space. This latter approach will be discussed in the last section, where the connection will also be made with methods used in CVM.

Consider first the following simple example of a probability distribution on three points summarized in the two-point probability Table I. For this distribution the point probabilities for A are 0.3 and that for B are 0.7 at all three points. If we now use the KKB superposition formula to estimate all the three-point probabilities, we find that, among other inconsistencies, probability is not conserved:

$$p_{123}^{AAA} \simeq \frac{p_{12}^{AA}\, p_{13}^{AA}\, p_{23}^{AA}}{p_1^A\, p_2^A\, p_3^A} \simeq 0.2963\,,$$

$$p_{123}^{AAB}\,,\ p_{123}^{ABA}\,,\ p_{123}^{BAA} \simeq 0.0318\,,$$

$$p_{123}^{ABB}\,,\ p_{123}^{BAB}\,,\ p_{123}^{BBA} \simeq 0.0408\,,$$

$$p_{123}^{BBB} \simeq 0.6297\,,$$

$$\sum p_{123}^{xx'x''} \simeq 1.1437\,. \tag{1}$$

Even if we use a graded basis such as the one described in the Appendix to estimate the value of p_{123}^{AAA} (so that the sum of the probabilities is one and the given marginal probabilities are conserved), the value of p_{123}^{AAB} is -0.0963, which is not a robust estimator of the joint probabilities. The presence of negative probability estimates is also discussed in section VI.

On the other hand, if we use the expression

$$p_{123}^{xx'x''} \simeq p_{12}^{xx'}\, p_3^{x''} + p_{13}^{xx''}\, p_2^{x'} + p_{23}^{x'x''}\, p_1^x - 2p_1^x\, p_2^{x'}\, p_3^{x''}\,, \tag{2}$$

Table II

Sample pair probability distribution at four sites.

$p_{ij}^{xx'}$	A 2	B 2	A 3	B 3	A 4	B 4
A 1	0.18	0.02	0.18	0.02	0.18	0.02
B 1	0.12	0.68	0.12	0.68	0.12	0.68
A 2	—	—	0.2	0.1	0.2	0.1
B 2	—	—	0.1	0.6	0.1	0.6
A 3	—	—	—	—	0.2	0.1
B 3	—	—	—	—	0.1	0.6

then all marginal probabilities are conserved, including total probabilities

$$p_{123}^{AAA} = 0.126 \, ,$$

$$p_{123}^{AAB}, \; p_{123}^{ABA}, \; p_{123}^{BAA} = 0.074 \, ,$$

$$p_{123}^{ABB}, \; p_{123}^{BAB}, \; p_{123}^{BBA} = 0.026 \, ,$$

$$p_{123}^{BBB} = 0.574 \, ,$$

$$\sum p_{123}^{xx'x''} = 1 \, . \tag{3}$$

We return to this example in the last section.

Finally, throughout this paper we will illustrate a number of the concepts developed in the context of a four point cluster (with points numbered "1," "2," "3," "4," and species labelled "A" and "B" for which pairwise joint probabilities are given in Table II. These illustrations will be given in italized footnotes.

II. Boolean Partition Lattices

Our discussion will be principally restricted to a finite set of atomic sites,†

$$\Omega = \{\omega_1, \ldots, \omega_w\} \, . \tag{4}$$

Usually we think of the points of Ω as lying on lattice sites, but there is actually no such restriction in this formulation, and Ω merely parametrizes a set of "outcome sites" at each of which one of a fixed finite set, Σ,

$$\Sigma = \{\sigma_1, \ldots, \sigma_s\} \, , \tag{5}$$

---

† A previous manuscript carried out over a countably infinite set of atomic sites was modified when the author underwent a "philosophical" conversion which will not be discussed in this paper. I apologize for any annoying residual traces of that manuscript that occur herein.

of outcomes occurs. In the case of CVM and other applications where periodic boundary conditions are applied, for instance, Ω is topologically equivalent to a three dimensional torus, which, of course, cannot be directly embedded in Euclidean space.

A subset, $S = \{\omega_1, \ldots, \omega_k\}$, of Ω is referred to as a *k-vertex* subset. The order of the vertices is immaterial in describing S, and the empty set, $\emptyset$, is the subset with zero vertices. The set of all subsets of Ω will be denoted by $\mathbf{S}(\Omega)$.

We next want to introduce the concept of *cluster*, by which we mean those subsets of Ω over which joint probability distributions are known, and we call $\mathbf{C}(\Omega)$ the set of all clusters. Because of the ability to calculate marginal probabilities over subsets [as expressed in Eq. (53b)], we postulate for $\mathbf{C}(\Omega)$ the following properties:

$$\bigcup_{C \in \mathbf{C}(\Omega)} C = \Omega, \tag{6a}$$

$$C, C' \in \mathbf{C}(\Omega) \Rightarrow C \cap C' \in \mathbf{C}(\Omega), \tag{6b}$$

$$(C \in \mathbf{C}(\Omega) \,\&\, (\forall i \in \mathcal{I}, C_i \in \mathbf{C}(\Omega) \,\&\, C_i \subseteq C)) \Rightarrow \bigcup_{i \in \mathcal{I}} C_i \in \mathbf{C}(\Omega), \tag{6c}$$

$$(C, C' \in \mathbf{C}(\Omega) \,\&\, C \subseteq C' \Rightarrow (C' - C) \in \mathbf{C}(\Omega), \tag{6d}$$

$$\forall l_1, \ldots, l_k \ni C = [\vec{r}_{l_1}, \ldots, \vec{r}_{l_k}] \in \mathbf{C}(\Omega) \,\&\, \forall \{l_{i_1}, \ldots, l_{i_j}\} \subseteq \{l_1, \ldots, l_k\} \Rightarrow$$

$$\exists C' \subseteq C \ni C' = [\vec{r}_{l_{i_1}}, \ldots, \vec{r}_{l_{i_j}}] \in \mathbf{C}(\Omega). \tag{6e}$$

In finite sets, all of the properties in Eqs. (6) except the first are obvious consequences of the ability to compute marginal probabilities on any subset of a cluster. This is simply accomplished by holding any condition fixed over the subset and summing over all joint probabilities on the cluster for which the fixed condition holds on the subset. Marginal probabilities are different than conditional probabilities which are relative probabilities having the form of a joint probability divided by a marginal probability. The major purpose of Eqs. (6), then, is to define the essential properties of clusters in Ω.

We refer to Eq. (6a) by saying $\mathbf{C}(\Omega)$ *covers* Ω. Eq. (6c) is the condition of *local closure*. A set of clusters is said to be *locally closed* when the union of any set of clusters contained in another cluster is again a cluster. *However, note that the set theoretic union of two clusters need not be a cluster*; it is this fact that forces a lattice structure which is not that of ordinary subsets of Ω. Eq. (6d) is *local complementation*. A set of clusters is said to be *locally complemented* when the set of clusters remains closed under both relative complementation, that is the difference of two clusters is again a cluster. The final condition in Eq. (6), which states that every subset of a finite cluster is a cluster, is referred to as the *richness* of the set of clusters. We note only one simple consequence of Eqs. (6): clusters are closed under arbitrary intersection. This result follows from Eq. (6b) through (6d) by the standard Boolean algebra duality operations of intersection and union interchanging under relative complementation.†

---

† *In our example, the set of clusters is* $\{\{1\}, \{2\}, \{3\}, \{4\}, \{12\}, \{13\}, \{14\}, \{23\}, \{24\},$ $\{34\}\}$. *Note that, for instance, the marginal point probabilities of an atom at point 1, given by* $p_1^A = p_{1234}^{AAAA} + p_{1234}^{AAAB} + p_{1234}^{AABA} + p_{1234}^{AABA} + p_{1234}^{AABB} + p_{1234}^{ABAB} + p_{1234}^{ABBA} + p_{1234}^{ABBB}$, *or at point 2, given by a similar expression, convey less specific information than the marginal pair probability of A atoms at both points 1 and 2, given by* $p_{12}^{AA} = p_{1234}^{AAAA} + p_{1234}^{AAAB} + p_{1234}^{AABA} + p_{1234}^{AABB}$.

A familiarity with two types of mathematical structures, *ordered sets* and *Boolean lattices*, are important for dealing with phase spaces involving joint probabilities. We begin with ordered sets, which are the simpler of the two structures, and which underlie the Möbius formalism. An ordered set is any set with a transitive, reflexive, nonredundant ordering relation as expressed in the first three equations of Eqs. (13). For this short interlude we use the variable, $\mathbf{X}$ for, the set on which the order relation is defined, and $\mathcal{X}$ for an element of $\mathbf{X}$.

The *order tree* segments lying above and below a given element, $\mathcal{X} \in \mathbf{X}$, are defined by

$$\mathbf{T}_\mathbf{X}^\vee(\mathcal{X}) \equiv \{\mathcal{X}' \in \mathbf{X}; \; \mathcal{X} \leq \mathcal{X}'\}, \tag{7a}$$

$$\mathbf{T}_\mathbf{X}^\wedge(\mathcal{X}) \equiv \{\mathcal{X}' \in \mathbf{X}; \; \mathcal{X} \geq \mathcal{X}'\}, \tag{7b}$$

and the order tree between two elements is defined by

$$\mathbf{T}_\mathbf{X}(\mathcal{X}_1, \mathcal{X}_2) \equiv \{\mathcal{X}' \in \mathbf{X}; \; \mathcal{X}_1 \leq \mathcal{X}' \leq \mathcal{X}_2\} = \mathbf{T}_\mathbf{X}^\vee(\mathcal{X}_1) \cap \mathbf{T}_\mathbf{X}^\wedge(\mathcal{X}_2). \tag{8}$$

A *maximal element*, $\mathcal{M}$, is defined by

$$\mathbf{T}_\mathbf{X}^\vee(\mathcal{M}) = \{\mathcal{M}\}. \tag{9}$$

Similarly, a minimal element, or *atomic element*, $\mathcal{A}$, is defined by

$$\mathbf{T}_\mathbf{X}^\wedge(\mathcal{A}) = \{\mathcal{A}\}. \tag{10}$$

For clusters, which are ordered by set inclusion, the set of atoms is Ω.

We are also interested in nonredundant *linear chains* in $\mathbf{X}$, that is, sets $\mathbf{L} \subset \mathbf{X}$ such that if $\mathcal{X}$ and $\mathcal{X}'$ are different elements of $\mathbf{X}$, then $\mathcal{X} \subset \mathcal{X}'$ or $\mathcal{X}' \subset \mathcal{X}$. Transitivity of inclusion implies that finite length linear chains can always be arranged in monotonic order, and we can define the *depth* of a non-empty element, $\phi_\mathbf{X}(\mathcal{X})$, as the maximum length of nonredundant linear chains in $\mathbf{T}_\mathbf{X}^\vee(\mathcal{X})$. Similarly, we define the *height*, $\psi_\mathbf{X}(\mathcal{X})$, of any element as the maximum length of nonredundant linear chains in $\mathbf{T}_\mathbf{X}^\wedge(\mathcal{X})$, and the *relative height* between to elements, $\psi_\mathbf{X}(\mathcal{X}_1, \mathcal{X}_2)$, (which may be zero) as the maximum length of nonredundant linear chains in $\mathbf{T}_\mathbf{X}(\mathcal{X}_1, \mathcal{X}_2)$. The *height* (or *depth*) of $\mathbf{X}$, $\nu_\mathbf{X}$, is the length of the longest linear chains in $\mathbf{X}$.

If the structure $\mathbf{X}$ has depth $\nu_\mathbf{X}$, the *prank* on $\mathbf{X}$ is defined by

$$\mathbf{p}_\mathbf{X}(\mathcal{X}) = \nu_\mathbf{X} - \phi_\mathbf{X}(\mathcal{X}) + 1. \tag{11}$$

Thus, the prank of ⊚ is one. Both the prank and the height are strictly monotonic images of the inclusion relation

$$(\mathcal{X}_1, \mathcal{X}_2 \in \mathbf{X} \; \& \; \mathcal{X}_1 \subset \mathcal{X}_2) \Rightarrow \begin{cases} \mathbf{p}_\mathbf{X}(\mathcal{X}_1) < \mathbf{p}_\mathbf{X}(\mathcal{X}_2), \\ \psi_\mathbf{X}(\mathcal{X}_1) < \psi_\mathbf{X}(\mathcal{X}_2). \end{cases} \tag{12}$$

To see this for prank, for instance, select a maximal chain, $\mathbf{L}$, for $\mathbf{T}_\mathbf{X}^\vee(\mathcal{X}_2)$, of lenght $\phi_\mathbf{X}(\mathcal{X}_2)$ and consider $\{\mathcal{X}_1\} \cup \mathbf{L}$, which is a nonredundant linear chain from $\mathbf{T}_\mathbf{X}^\vee(\mathcal{X}_1)$ and is one longer than $\mathbf{L}$, making $\mathbf{p}_\mathbf{X}(\mathcal{X}_1) > \mathbf{p}_\mathbf{X}(\mathcal{X}_2)$.

For most operations on lattices with finite clusters the height will suffice, but on motif generated lattices—to be introduced shortly—prank is often the preferred way of ordering.

We now construct a number of Boolean lattices. Aside from the crystal lattice which is familiar to most readers, there is the mathematical lattice,[3] which we call a Boolean lattice in that it is one way to abstract the structure of subsets of a set.

A Boolean lattice, as we shall use the term, is a set of objects, $\mathcal{L}$, with an ordering relation, $\leq$, together with a join, $\vee$, and a meet $\wedge$, and a zero element, $\odot$, which satisfy the properties:

$$\mathbb{P}, \mathbb{Q}, \mathbb{R} \in \mathcal{L} \Rightarrow \odot \leq \mathbb{P} \leq \mathbb{P}, \tag{13a}$$

$$\mathbb{P} \leq \mathbb{Q} \ \& \ \mathbb{Q} \leq \mathbb{R} \Rightarrow \mathbb{P} \leq \mathbb{R}, \tag{13b}$$

$$\mathbb{P} \leq \mathbb{Q} \ \& \ \mathbb{Q} \leq \mathbb{P} \Rightarrow \mathbb{P} = \mathbb{Q}, \tag{13c}$$

$$\mathbb{P} \wedge \mathbb{Q} \leq \mathbb{P}, \mathbb{Q} \leq \mathbb{P} \vee \mathbb{Q}, \tag{13d}$$

$$\mathbb{R} \leq \mathbb{P}, \mathbb{Q} \Rightarrow \mathbb{R} \leq \mathbb{P} \wedge \mathbb{Q}, \tag{13e}$$

$$\mathbb{P}, \mathbb{Q} \leq \mathbb{R} \Rightarrow \mathbb{P} \vee \mathbb{Q} \leq \mathbb{R}. \tag{13f}$$

We also note that the term "Boolean lattice" has been used to indicate a special type of mathematical lattice in the mathematical literature.[4] Our use of this term does not convey the concept of complementarity, although the lattices we employ do have universal bounds, a minimum element, and often have relative complementarity.

In the context of multipoint probability distributions, partition lattices provide the natural structure for superposition type approximations. We shall use these "partitions" to break up subsets into clusters over which joint probability distributions are known. $\mathbb{P} \neq \emptyset$ is called a *partition* (or *subset partition*) if $\mathbb{P}$ consists of a non-empty set of subsets of Ω, no one of which is a subset of another. To avoid excessive use of the word "subset," elements of partitions will often be referred to as *motifs*. If all of the motifs are clusters, $\mathbb{P}$ is called a *cluster partition*. Because of the marginal probability formulas, the probability distribution on subclusters is determined from that on the main cluster. There is, thus, no need to include subclusters unless necessary for consistency purposes on cluster overlaps. *One way of thinking about superposition is the extension of a probability distribution over a cluster partition to a probability distribution over a subset partition that includes the cluster partition.*

Two partitions are considered equal if their elements can be put into 1–1 correspondence. If the motifs in a partition are pairwise disjoint, it is called a *disjoint partition*. We denote *partition inequality* by

$$\mathbb{P} \preccurlyeq \mathbb{Q} \iff (S \in \mathbb{P} \Rightarrow \exists S' \in \mathbb{Q} \ni S \subset S'), \tag{14}$$

i.e., every element of $\mathbb{P}$ must be a subset of an element of $\mathbb{Q}$. We typically use the variable, $\mathcal{C}$ to represent clusters; and we will use the variable $\mathbb{C} \equiv \{\mathcal{C}\}$ to represent the partition containing only the cluster $\mathcal{C}$. The minimal partition, $\odot$, is the partition $\{\emptyset\}$. The variables S and $\mathcal{M}$ will typically be used for motifs, which may or may not be clusters.

Partition lattices can be used in two different contexts, that of the Möbius formalism and that of the linear convex formalism (to be introduced briefly at the end of this paper), which strongly rests on the use of marginal probabilities. The former more naturally utilizes motif generated lattices, while the latter uses shadow lattices. Indeed, as was demonstrated in Ref. 5, a motif generated lattice closed under only meet and join is already sufficient (and perhaps the minimum) structure on which the Kikuchi-Barker formalism is valid. Such Boolean lattices need not be locally closed nor locally complemented (closed under partition differences), even though much probability information is, thereby, thrown away. In CVM the clusters are often chosen so that both local closure and complementarity actually obtain, and, indeed, we can modify the definition of motif generated lattice in Ref. 5 to force these conditions; but this distinction, while not compelling, should serve as a warning that the motif lattice formalism, by itself, is not the natural one with which to describe joint probabilities.

In this paper we shall develop a formalism sufficient to describe both of these types of Boolean lattice.†

The *join* of a finite number of partitions, $\bigvee_{\{\mathbb{P}_i; i \in \mathfrak{I}\}} \mathbb{P}_i$ has the usual set theoretic meaning, except that "redundant" subsets—those included in another element of the join—are weeded out. The set, $\bigvee_{i \in \mathfrak{I}} \mathbb{P}_i$, is the smallest set containing all the $\mathbb{P}_i$'s, i.e.,

$$\forall i \in \mathfrak{I}, \; \mathbb{P}_i \preccurlyeq \bigvee_{i \in \mathfrak{I}} \mathbb{P}_i, \tag{15a}$$

$$\forall i \in \mathfrak{I}, \; \mathbb{P}_i \preccurlyeq \mathbb{Q} \Rightarrow \bigvee_{i \in \mathfrak{I}} \mathbb{P}_i \preccurlyeq \mathbb{Q}. \tag{15b}$$

The *meet* of a finite number, m, partitions is defined as the partition made from the set of motifs containing m-wise intersections after weeding out redundant motifs:

$$\mathbb{P}_1 \wedge \cdots \wedge \mathbb{P}_m = \bigvee_{\substack{S_{i_k} \in \mathbb{P}_k \\ i_k \in \mathcal{I}_k, \; k=1,m}} \{S_{i_1} \cap \cdots \cap S_{i_m}\}, \tag{16}$$

where $\mathcal{I}_k$ is the index set for the kth partition. $\mathbb{P}_1 \wedge \cdots \wedge \mathbb{P}_m$, which is a refinement of each of the partitions, is the largest partition which is contained in $\mathbb{P}_1$ to $\mathbb{P}_m$

$$\mathbb{P}_1 \wedge \cdots \wedge \mathbb{P}_m \preccurlyeq \mathbb{P}_1, \ldots, \mathbb{P}_m, \tag{17a}$$

$$\mathbb{Q} \preccurlyeq \mathbb{P}_1, \ldots, \mathbb{P}_m \Rightarrow \mathbb{Q} \preccurlyeq \mathbb{P}_1 \wedge \cdots \wedge \mathbb{P}_m. \tag{17b}$$

Both meet and join are idempotent

$$\mathbb{P} \wedge \mathbb{P} = \mathbb{P} \vee \mathbb{P} = \mathbb{P}. \tag{18}$$

The *complement* of any subset, S, will be designated by $\overline{S} \equiv \Omega - S$, while for partitions, we define

$$\overline{\mathbb{P}} \equiv \{\overline{S}; S \in \mathbb{P}\}, \tag{19a}$$

$$\mathbb{P} - \mathbb{P}' \equiv \mathbb{P} \wedge \overline{\mathbb{P}'}. \tag{19b}$$

$\overline{\mathbb{P}}$ is *not* the Boolean lattice complement of $\mathbb{P}$ nor is it a cluster partition, even when $\mathbb{P}$ is a cluster partition; however, the *difference* of two cluster partitions, $\mathbb{P} - \mathbb{P}'$, *is* a cluster partition.

Under these definitions the *set of all partitions* forms a set, $\mathfrak{F}$, which is closed under finite join and meet. Subsets, $\mathcal{L}$, of $\mathfrak{F}$ which are closed under finite join and meet are called *partition lattices*, written $\mathcal{L} \in \mathsf{L}$. The set-theoretic intersection of any number of lattices in $\mathfrak{F}$ is another lattice of $\mathfrak{F}$:

$$\bigcap_{\substack{\mathcal{L}_i \in \mathsf{L} \\ i \in \mathfrak{I}}} \mathcal{L}_i \in \mathsf{L}, \tag{20}$$

and we can define the meet of two partition lattices, $\mathcal{L}_1 \wedge \mathcal{L}_2$ as the set of partitions made up of finite joins of pairwise meets from $\mathcal{L}_1$ and $\mathcal{L}_2$. Although lattices are not closed under set-theoretic union, we can define the join of two lattices $\mathcal{L}_1, \mathcal{L}_2 \in \mathsf{L}$:

$$\mathcal{L}_1 \vee \mathcal{L}_2 \equiv \bigcap_{\substack{\mathcal{L} \in \mathsf{L} \\ \mathcal{L}_1, \mathcal{L}_2 \subseteq \mathcal{L}}} \mathcal{L}. \tag{21}$$

---

† *A cluster partition* $\{\{1\}, \{2\}\}$, *for instance, conveys the use of marginal point probabilities which are less specific than those obtained from the partition* $\{\{12\}\}$.

The *set of motifs* $S_{\mathcal{L}}$ in a lattice, $\mathcal{L}$, consists of all elements of all partitions in $\mathcal{L}$:

$$S_{\mathcal{L}} = \bigcup_{\mathbb{P}\in\mathcal{L}} \mathbb{P}. \tag{22}$$

By the same arguments, we can define the *set of all cluster partitions*, $\mathfrak{F}_C$, which is also closed under join and meet, and *cluster partition lattices*, written $\mathcal{L} \in \mathbf{L}_C$. Similar statements to the above paragraph hold for cluster partition lattices; $\mathcal{L} \in \mathbf{L}$ is a *cluster lattice* if $S_{\mathcal{L}} \subseteq \mathbf{C}(\mathbf{\Omega})$.

The *extent* of $\mathcal{L}$ or $\mathbb{P}$, denoted, respectively, by $\mathcal{E}[\mathcal{L}]$ or $\mathcal{E}[\mathbb{P}]$ is defined as:

$$\mathcal{E}[\mathcal{L}] = \bigcup_{S\in S_{\mathcal{L}}} S \subseteq \mathbf{\Omega}, \tag{23a}$$

$$\mathcal{E}[\mathbb{P}] = \bigcup_{S\in\mathbb{P}} S \subseteq \mathbf{\Omega}, \tag{23b}$$

and the *set of partitionable subsets of* $\mathcal{E}[\mathcal{L}]$, denoted by $\mathbf{E}(\mathcal{L})$ is defined by:

$$\mathbf{E}(\mathcal{L}) \equiv \{\mathcal{E}[\mathbb{P}]; \mathbb{P}\in\mathcal{L}\}$$

$$= \left\{ S^*; \, \exists\mathbb{P}\in\mathcal{L} \ni S^* = \bigcup_{S\in\mathbb{P}} S \right\} \subseteq \text{Subsets}(\mathbf{\Omega}). \tag{24}$$

Two partitions, $\mathbb{P}$ and $\mathbb{P}'$ are said to be *coextensive* if

$$\mathcal{E}[\mathbb{P}] = \mathcal{E}[\mathbb{P}']. \tag{25}$$

If $S^* \in \mathbf{E}(\mathcal{L})$, then the set of coextensive partitions in $\mathcal{L}$ with extend S^* form a sublattice of $\mathcal{L}$, designated $\mathcal{L}_{S^*}$, which is called tha partition of S *induced by* $\mathcal{L}$.

We say that $\mathbb{P}$ is a *complete* partition of $\mathcal{L}$ if:

$$S \in S_{\mathcal{L}} \ \& \ S \subseteq \mathcal{E}[\mathbb{P}] \Rightarrow S \preccurlyeq \mathbb{P}. \tag{26}$$

This is equivalent to

$$\forall \mathbb{Q}\in\mathcal{L}, \ [\mathcal{E}[\mathbb{Q}] \subseteq \mathcal{E}[\mathbb{P}] \Rightarrow \mathbb{Q} \preccurlyeq \mathbb{P}]. \tag{27}$$

The importance of complete partitions lies in the fact that incomplete cluster partitions throw away information. A simple example of this in the context of the Möbius formalism is given at the end of the next section.†

The *completion* of any partition in $\mathcal{L}$ is defined by:

$$\overset{\bullet}{_{\mathcal{L}}}\mathbb{P} \equiv \{S \in S_{\mathcal{L}}; \ S \subseteq \mathcal{E}[\mathbb{P}]\}. \tag{28}$$

In that case, any partition is complete if $\mathbb{P} = \overset{\bullet}{_{\mathcal{L}}}\mathbb{P}$. Complete partitions are closed under finite meet, but not join. We, also, designate by $^{\bullet}\mathcal{L}$ the *set of complete partitions in* $\mathcal{L}$, which can be made into a Boolean lattice by modifying the definition of join:

$$^{\bullet}\mathbb{P} \overset{\bullet}{\vee} {}^{\bullet}\mathbb{Q} \equiv {}^{\bullet}(^{\bullet}\mathbb{P} \vee {}^{\bullet}\mathbb{Q}). \tag{29}$$

Note that for any partitionable subset of $\mathbf{E}(\mathcal{L})$, there is exactly one complete partition with that extent, and that is the maximal partition with that extent. Thus, $\mathbf{E}(\mathcal{L})$ can be made into a lattice, called the *extent lattice*, where $\leq$ is ordinary set inclusion, $\subseteq$. As such $\mathbf{E}(\mathcal{L})$ is isomorphic to $^{\bullet}\mathcal{L}$ with the modified definition of join.

---

† *The partition* $\{\{1\}, \{2\}\}$ *is not complete, while* $\{\{12\}\}$ *is complete, since higher order probabilities, such as* $\{123\}$ *are not known.*

We distinguish two types of lattices of $\mathfrak{F}$: A *shadow lattice* consists of all partitions in $\mathfrak{F}$ which are "in the shadow" of the partition $\mathbb{P}$:

$$\mathbb{P}' \in \blacktriangle(\mathbb{P}) \iff \mathbb{P}' \in \mathfrak{F} \ \& \ \mathbb{P}' \preccurlyeq \mathbb{P}. \tag{30}$$

Clearly each shadow lattice has a maximal element, $M_{\blacktriangle(\mathbb{P})} = \mathbb{P}$, and we can express the *set of all subsets of* Ω, as $S_{\blacktriangle(\Omega)} \equiv S_{\blacktriangle(\{\Omega\})}$. The meet of any two such lattices is ordinary set intersection, while the join is the shadow lattice of the join of the two maximal elements.

If we restrict the motifs in a shadow lattice to be clusters, the resulting lattice is called the *shadow cluster lattice* of $\mathbb{P}$:

$$\mathbb{P}' \in \triangle(\mathbb{P}) \iff \mathbb{P}' \in \mathfrak{F}_C \ \& \ \mathbb{P}' \preccurlyeq \mathbb{P}. \tag{31}$$

Note here that, since $\mathbb{P}$, itself, need not be a partition cluster, and that while the set of clusters is locally closed, the union of clusters inside a motif need not be a cluster, there may be no maximum element in $\triangle\mathbb{P}$. There is no problem, of course, when $\mathbb{P} = \{S\}$, since, then, $\triangle\mathbb{P}$ is the set of all subsets of atomic clusters.

A *motif-generated lattice* is generated by intersections of the elements of a partition. Let $\mathbb{P} = \{M; i = 1, \ldots, M\}$, and define

$$U_k \equiv \bigcup_{1 \leq i_1 < \cdots < i_k} \{M_{i_1}\} \wedge \cdots \wedge \{M_{i_k}\}, \tag{32a}$$

$$U \equiv \bigcup_k U_k. \tag{32b}$$

Then,

$$S_{\Lambda(\mathbb{P})} \equiv \{\emptyset\} \cup \left\{ \bigcup_{i \in \mathcal{I}} u_i; \ u_i \in U \ \& \ \exists j \ni u_i \subseteq M_j \right\}, \tag{33a}$$

$$\Lambda(\mathbb{P}) = \{\mathbb{P}'; \ \mathbb{P}' \subseteq S_{\Lambda(\mathbb{P})}\}. \tag{33b}$$

Unless otherwise stated the symbol Λ will refer to a motif-generated lattice. As in the case of shadow lattices $M_{\Lambda(\mathbb{P})} = \mathbb{P}$ is the maximal element for its own motif-generated lattice.

By construction, $\Lambda(\mathbb{P})$ consists of all partitions made up from finite meets of the generating motifs. As such $S_{\Lambda(\mathbb{P})} \subseteq E(\Lambda(\mathbb{P}))$ is a set ordered by ordinary inclusion which is closed under finite intersection, but not under local complementation nor local union. By including sets of the type, $\{\overline{M}\}$ in definition (32), one can actually simplify the definition of motif generated lattice since all atomic motifs can be generated by the $2^M - 1$ meets involving all M motifs in either their direct or complemented form with the exception of that involving only complemented motifs. By supplementing this set with all local set theoretic unions of atomic sets (where all atomic sets in the union are contained in some motif of $\mathbb{P}$), and replacing the definition of $S_{\Lambda(\mathbb{P})}$, in (33), the redefined set, $S_{\Lambda(\mathbb{P})}$, would be both locally closed and complemented.†

In a motif lattice a complete partition can have no more elements than there are motifs. Where convenient, we shall use an alternate notation for complete partitions in motif lattices:

$$Q \overset{\circ}{\preccurlyeq} \mathbb{P} \equiv \left(Q = _{\Lambda(\mathbb{P})}\overset{\circ}{Q}\right). \tag{34}$$

---

† *The motif generated lattice for our example is* $\Lambda(\{\{12\}, \{13\}, \{14\}, \{23\}, \{24\}, \{34\}\})$.

It is, also, be useful to define the partitions:

$$\mathbb{U}_k \equiv \bigvee_{1 \leq i_1 < \cdots < i_k \leq M} \{\mathcal{M}_{i_1}\} \wedge \cdots \wedge \{\mathcal{M}_{i_k}\} \,. \tag{35}$$

As seen in Ref. 5,

$$\mathbb{U} = \bigcup_{k=1,\ldots,} \mathbb{U}_k \,. \tag{36}$$

This may not be the case in infinitely generated lattices where an infinite number of motifs may contain the same intersection motif, which would then be left out of $\mathbb{U}$.

III. The Möbius Formalism on Motif-Generated Lattices

Let X be an ordered structure of finite height, ν_X, and so possessing the prank function, p_X. We define the *inclusion function*, η, over $X \otimes X$ by

$$\eta(\mathcal{X}_1, \mathcal{X}_2) \equiv \begin{cases} 1 & \text{if } \mathcal{X}_1 \leq \mathcal{X}_2, \\ 0 & \text{if } \mathcal{X}_1 \nleq \mathcal{X}_2. \end{cases} \tag{37}$$

Clearly, $\eta(\mathcal{X}_1, -)$ is zero outside of $T_X^\vee(\mathcal{X}_1)$, and $\eta(-, \mathcal{X}_2)$ is zero outside of $T_X^\wedge(\mathcal{X}_6)$.†

On the set $T_X(\mathcal{X}_1, \mathcal{X}_2)$, which must be finite, we can order the elements first by prank and then in some lexicographic fashion. Because of Eq. (25), the inclusion function can be represented by a superdiagonal matrix, $\|\eta\|$ with ones on the diagonal, where

$$\|\eta\| = \eta(\mathcal{X}_i, \mathcal{X}_j) \equiv \begin{cases} 1 & \text{if } \mathcal{X}_i \leq \mathcal{X}_j, \\ 0 & \text{if } \mathcal{X}_i \nleq \mathcal{X}_j. \end{cases} \tag{38}$$

Consider the inverse matrix,

$$\|\mu\| \equiv \|\eta\|^{-1} \,. \tag{39}$$

μ is called the *Möbius function* for X as defined on $T_X(\mathcal{X}_1, \mathcal{X}_2)$. We want to show that this definition is independent of the choice of $\mathcal{X}_1$ and $\mathcal{X}_2$.‡

To do this, we make use of the desired properties that $\|\eta\| \, \|\mu\| = I$ and $\mu(\mathcal{X}_1, \mathcal{X}_2) \neq 0 \Rightarrow \mathcal{X}_1 \leq \mathcal{X}_2$ to present an intrinsic construction of μ. Set

$$\mu(\mathcal{X}_1, \mathcal{X}_2) = 0, \quad \text{for } \mathcal{X}_1 \nleq \mathcal{X}_2, \tag{40a}$$

$$\mu(\mathcal{X}, \mathcal{X}) = 1, \tag{40b}$$

---

† *The inclusion matrix for the motif generated partition lattice in our example is that over the set of clusters:*

	{1}	{2}	{3}	{4}	{12}	{13}	{14}	{23}	{24}	{34}
{1}	1	0	0	0	1	1	1	0	0	0
{2}	0	1	0	0	1	0	0	1	1	0
{3}	0	0	1	0	0	1	0	1	0	1
{4}	0	0	0	1	0	0	1	0	1	1
$\eta = $ {12}	0	0	0	0	1	0	0	0	0	0
{13}	0	0	0	0	0	1	0	0	0	0
{14}	0	0	0	0	0	0	1	0	0	0
{23}	0	0	0	0	0	0	0	1	0	0
{24}	0	0	0	0	0	0	0	0	1	0
{34}	0	0	0	0	0	0	0	0	0	1

‡ A similar construction is found in Ref. 6. It is interesting to note that the name M. S. Green is found both in conjunction with Kirkwood's work on the superposition approximation in liquids (which greatly predates the work of Kikuchi and Barker) and with the Möbius formalism in Ref. 7.

and rewrite the expression

$$\sum_{\mathcal{X}_1 \le \mathcal{X}' \le \mathcal{X}_2} \eta(\mathcal{X}_1, \mathcal{X}') \mu(\mathcal{X}', \mathcal{X}_2) = \delta(\mathcal{X}_1, \mathcal{X}_2), \tag{41}$$

with

$$\delta(\mathcal{X}_1, \mathcal{X}_2) = \begin{cases} 1 & \text{when } \mathcal{X}_1 = \mathcal{X}_2, \\ 0 & \text{when } \mathcal{X}_1 \ne \mathcal{X}_2, \end{cases} \tag{42}$$

in the form:

$$\mu(\mathcal{X}, \mathcal{X}_2) = -\sum_{\mathcal{X}'} [\eta(\mathcal{X}, \mathcal{X}') - \delta(\mathcal{X}, \mathcal{X}')] \mu(\mathcal{X}', \mathcal{X}_2). \tag{43}$$

Equation (43) can be used to iteratively define μ over all $\mathbf{T}_{\mathcal{X}}^{\wedge}(\mathcal{X}_2)$ starting after $\mathcal{X}_2$ and using the motif depth as the iteration parameter.

Let S be a k-vertex motif. In the special case that $\mathbf{T}_{\Lambda(\mathbb{P})}^{\wedge}(S)$ consists of all the subsets of S, $\mu(S', S) = (-1)^{k-k'}$, where k' is the number of vertices in S'.†

The Möbius function has a venerable history which is reviewed in Ref. 7 as well as in the graph theory context by Domb[6] and in the cluster context by Guozhong[8] and by Malyshev and Minlos.[9] The results in this section, although independently developed, are also closely related to the combinatorics of sieves.[10]

Let, now, f be any function (with values in an Abelian group) defined over $\mathbf{X}_0 \subseteq \mathbf{X}$. We define the *Möbius transform* of f on $\mathbf{X}_0$ by:

$$\hat{f}(\mathcal{X}) = \sum_{\mathcal{X}' \in \mathbf{X}_0} \mu(\mathcal{X}', \mathcal{X}) f(\mathcal{X}). \tag{44}$$

It, then, follows that for any $\mathcal{X} \in \mathbf{X}_0$,

$$\sum_{\mathcal{X}' \in \mathbf{X}_0} \eta(\mathcal{X}', \mathcal{X}) \hat{f}(\mathcal{X}') = \sum_{\substack{\mathcal{X}' \in \mathbf{X}_0 \\ \mathcal{X}'' \in \mathbf{X}_0}} \eta(\mathcal{X}', \mathcal{X}) \mu(\mathcal{X}'', \mathcal{X}') f(\mathcal{X}'')$$

$$= \sum_{\mathcal{X}'' \in \mathbf{X}_0} \delta(\mathcal{X}'', \mathcal{X}) f(\mathcal{X}'') = f(\mathcal{X}). \tag{45}$$

Equation (45) permits us to formally extend the function f onto $\mathbf{X}$. We can, thus, write Eq. (45) as

$$\sum_{\substack{\mathcal{X} \in \mathbf{X}_0 \\ \mathcal{X}' \le \mathcal{X}}} \hat{f}(\mathcal{X}') = f(\mathcal{X}). \tag{46}$$

If the set $\mathbf{X}_0$ consists of single motif partitions, $\mathbb{S} = \{S\}$, drawn from $\mathbf{S}_{\Lambda(\mathbb{P})}$ where $\Lambda(\mathbb{P})$ is a motif lattice, then we have an additionaly property:

$$\mathcal{X}_0 \in \mathbf{X}_0 \ \& \ \mathcal{X}_0 \le \mathcal{X} \vee \mathcal{X}' \Rightarrow \mathcal{X}_0 \le \mathcal{X} \ \text{or} \ \mathcal{X}_0 \le \mathcal{X}'. \tag{47}$$

---

† *The inverse function to the inclusion function in this example is:*

	{1}	{2}	{3}	{4}	{12}	{13}	{14}	{23}	{24}	{34}
{1}	1	0	0	0	−1	−1	−1	0	0	0
{2}	0	1	0	0	−1	0	0	−1	−1	0
{3}	0	0	1	0	0	−1	0	−1	0	−1
{4}	0	0	0	1	0	0	−1	0	−1	−1
$\mu = $ {12}	0	0	0	0	1	0	0	0	0	0
{13}	0	0	0	0	0	1	0	0	0	0
{14}	0	0	0	0	0	0	1	0	0	0
{23}	0	0	0	0	0	0	0	1	0	0
{24}	0	0	0	0	0	0	0	0	1	0
{34}	0	0	0	0	0	0	0	0	0	1

Equation (47), which is a distinguishing feature of partition lattices then implies that in $\Lambda(\mathbb{P})$:†

$$f(Q_1 \vee Q_2) = f(Q_1) + f(Q_2) - f(Q_1 \wedge Q_2). \tag{48}$$

Equation (48) was called the Kikuchi-Barker approximation in Ref. 5. There, the distinction between the underlying additive Möbius coefficients and the Kirkwood inspired Kirkwood-Kikuchi-Barker approximation were obscured. However, Eq. (48), which is a direct consequence of Eq. (46) and the definition of a motif generated lattices, actually summarizes the Möbius formalism on the motif partition (although not necessarily on the shadow cluster lattice determined by the partition), and has a validity independent of the KKB approximation. This is because Eq. (48) is not only a consequence of the definition in Eq. (44), but determines $\hat{f}$ completely on any partition, $\mathbb{P}$, containing a finite number of motifs when the values of f are known on partitions of the type $\mathbb{S} = \{S\}$, $S \in \mathbf{U}_{\Lambda(\mathbb{P})}$. This proof of this assertion was presented in Ref. 5 for the partition lattices needed in CVM. In that case, due to the use of shift equality to deal with periodic boundary conditions, the meet of two motifs may require redefinition. Nonetheless, the results there, which rested on a key combinatorial lemma whose proof is valid for any finitely generated motif lattice, are still applicable: Let $\mathbb{P}$ contain M motifs. Then, the following result holds in $\Lambda(\mathbb{P})$ for $1 \le j \le M$:

$$f(\mathbb{U}_{M+1-j}) = \sum_{k=1}^{j} (-1)^{k-1} \binom{M-1-j+k}{k-1} \sum_{1 \le i_1 < \cdots < i_{j-k} \le M} f(\{\mathcal{M}_{i_1,\ldots,i_{j-k}}\}), \tag{49}$$

where

$$\mathcal{M}_{i_1,\ldots,i_l} = \begin{cases} \emptyset & \text{for } l < 0, \\ \bigwedge_{j=1,M} \{\mathcal{M}\} & \text{for } l = 0, \\ \bigvee_{\{i_1,\ldots,i_l\} \cup \{j_1,\ldots,j_{M-l}\} = \{1,\ldots,M\}} \{M_{j_1}\} \wedge \cdots \wedge \{M_{j_{M-l}}\} & \text{for } l > 0, \end{cases} \tag{50}$$

from which

$$f(\mathbb{P}) = f(\mathbb{U}_1) = \sum_{k=1}^{M} (-1)^{k-1} \sum_{1 \le i_1 < \cdots < i_{M-k} \le M} f(\{\mathcal{M}_{i_1,\ldots,i_{M-k}}\})$$

$$= \sum_{k=1}^{M} (-1)^{k-1} \sum_{1 \le j_1 < \cdots < j_k \le M} f(\{M_{j_1}\} \wedge \cdots \wedge \{M_{j_k}\}). \tag{51}$$

Equation (51) shows that a function on motifs can be extended to the maximal partition of the lattice generated by the motifs, and contains a simple form of CVM superposition when f is appropriately chosen. As already shown in the introduction, that superposition principle does not, in general, conserve probability if the Kirkwood-Kikuchi-Barker method—which in this additive context rests on the use of logarithms

---

† In our example, for instance,

$$f(\{\{12\}, \{23\}\}) = f(\{12\}) + f(\{23\}) - f(\{2\})$$
$$= [\hat{f}(\{12\}) + \hat{f}(\{1\}) + \hat{f}(\{2\})] + [\hat{f}(\{23\}) + \hat{f}(\{2\}) + \hat{f}(\{3\})] - \hat{f}(\{2\}),$$

with $\hat{f}(\{2\}) = f(\{12\}) - f(\{1\}) - f(\{2\})$, for example, obtained using the Möbius μ matrix. In the Kirkwood-Kikuchi-Barker example shown later, logarithms and exponentiation are used to make the Möbius formalism applicable.

and exponentiation—is used. However, the expansion, itself, is still valid and will be used in modified form in the Bayesian-convex approximation.

Equation (51) implies that Eq. (48) determines f on those partitions of $\Lambda(\mathbb{P})$ which are made up motifs from $U_{\Lambda(\mathbb{P})}$, that is, partitions made up only of finite intersections among the motifs of $\mathbb{P}$. However, without local closure and complementation this approximation has a number of serious failings, and even in the case of cluster partitions must be applied carefully. For example, suppose A_1 and A_2 are two disjoint atomic clusters of $\Lambda(\mathbb{P})$. Casual application of Eq. (48) might lead one to expect that the value associated to $A_1 \cup A_2$ is the sum: $f(\{A_1\}) + f(\{A_2\})$. This is indeed the correct value for $f(\{A_1, A_2\}) = f(\{A_1\} \vee f\{A_2\})$. However, if $A_1 \cup A_2$ is also a cluster of $\Lambda(\mathbb{P})$, then $\{A_1, A_2\}$ is not a complete partition, and one must use $\{A_1 \cup A_2\}$ to get the correct correspondence. Since $f(\{A_1 \cup A_2\})$ contains more information than $f(\{A_1, A_2\})$, the set of clusters of clusters must be extended beyond that of just intersections of clusters. From the above example, we see that, as a minimum, if $\mathbb{P}$ is a cluster partition and $\mathbb{Q} \in \Lambda(\mathbb{P})$, i.e., $\mathbb{Q}$ is a cluster partition of the set $\mathcal{E}[\mathbb{Q}] \in S(\Lambda(\mathbb{P}))$, then $\widehat{f}_{(\Lambda(\mathbb{P})}{}^{\bullet}\mathbb{Q})$ should give the same probability value as that obtained by marginal contraction from $f(\mathbb{P})$. We can be assured that this is the case if we know a joint probability distribution over $\mathcal{E}[\mathbb{Q}]$ which reduces to the joint marginal probabilities assumed for the motifs in $\mathbb{P}$. Unfortunately, such probability distributions are what we are seeking to find by superposition. This issue will be discussed further in section V.

Finally, it is tempting to extend the Möbius approach to lattices which are not locally finite. However, examination of Eq. (44) shows a series with highly oscillatory terms. Equation (51), which uses Eq. (48) to regroup the terms of Eq. (45) in finitely generated motif lattices, clearly shows signs of going out of control as M becomes large, since terms with overlaps are added before corrections due to the overlaps are removed. Thus, for general f, Eq. (48) on locally finite motif lattices appears to offer the best approach.

IV. Distribution Functions

While the partitions in a lattice, $\mathcal{L}$, describe the underlying geometry, actual computations require an assignment, or *mapping*, of physical state quantities to each site of Λ. In the diffusive order case, the state is described by the atomic species, σ_i, belonging to $\Sigma = \{\sigma_1, \ldots, \sigma_s\}$, the set of all *atomic species* types, occupying each site. This total set of mappings is designated by Σ^{Ω}.

However, when dealing with the incomplete marginal information characteristic of superposition problems, we must take into account partition lattice structures. Later, we shall embed these lattice structures into the linear algebra of probability spaces by the use of marginal probabilities. Here, however, we establish the partition lattice structure, $\mathcal{L}$, underlying any assignment on $\mathcal{E}[\mathcal{L}]$:

Let $\mathbb{P} = \{S_1, \ldots, S_k, \ldots\}$, be a partition in the lattice, $\mathcal{L}$, and let $\Sigma = \{\sigma_1, \ldots, \sigma_s\}$ be the set of all *atomic species* types. We refer to $\Sigma^{\mathbb{P}} \equiv \Sigma^{S_1} \times^d \cdots \times^d \Sigma^{S_k} \times^d \cdots$ as the set of assignment mappings, where $\times^d$ represents disordered direct product. Elements of $\Sigma^{\mathbb{P}}$, called *assignments*, are disordered sets, $\Gamma \equiv \{\gamma_1, \ldots, \gamma_k, \ldots\}$, where $\gamma_j \in \Sigma^{S_j}$ is any mapping assigning one element of Σ to each point of S_j. Two assignments are then equal, $\Gamma \equiv \{\gamma_1, \ldots, \gamma_k, \ldots\} = \{\gamma'_1, \ldots, \gamma'_k, \ldots\} \equiv \Gamma'$, if there is a 1-1 correspondence, φ, $\Gamma \overset{\varphi}{=} \Gamma'$ such that $\text{Dom}(\gamma_k) \equiv S_k = S'_{\varphi(k)} \equiv \text{Dom}(\varphi(\gamma_k))$, and the assignments are equal on corresponding vertices. Each assignment, Γ, is directly connected with a partition, $\mathbb{P}_\Gamma$, determined by the domains of the elements of Γ. We shall refer to $\mathbb{P}_\Gamma$ as the *domain partition* of Γ, $\mathrm{Dom}(\Gamma)$.

The physically important assignments are those which represent a partitioned snapshot of an assignment over $\mathcal{E}[\mathfrak{L}]$. Such assignments must agree on the overlaps of the motif domains of each of their elements. Each element, $\gamma_i \in \Gamma$, is itself a set of ordered couples defining a particular atomic assignment over a particular motif, $S_i \in \mathbb{P}$. To describe this we use the notation $\gamma_i \in\in \Sigma^\mathbb{P}$. Clearly, we can use the inclusion notation for such elements: Given $\gamma \in\in \Sigma^\mathbb{P}$ and $\gamma' \in\in \Sigma^{\mathbb{P}'}$, we can write $\gamma \subseteq \gamma'$ when γ as a set of ordered couples is included in γ' as a set of ordered couples. This, of course, implies that $\mathrm{Dom}(\gamma) \equiv S \subseteq S' \equiv \mathrm{Dom}(\gamma')$; and we define *consonant elements*, by which we mean the two elements agree of the overlaps of their domains, *i.e.*, $\gamma_{|S\cap S'} = \gamma'_{|S\cap S'}$. We then say Γ is *consonant* with Γ', or $\Gamma \overset{\cap}{=} \Gamma'$ if every element of Γ is consonant with every element of Γ', and we say that Γ is *self-consonant*, or *coherent*, if Γ is consonant with itself.†

We shall henceforth restrict our discussion to coherent assignments. We designate by $\Sigma^\mathcal{S}$, the *set of mappings* with coextensive domain partitions, $\mathcal{E}[\mathfrak{L}] = \mathcal{S}$, and by $\mathfrak{H}_\mathfrak{L}$, the *set of coherent assignments* with domain partition in $\mathfrak{L}$. This definition requires that $\mathfrak{H}_\mathfrak{L}$ involves all sets of mappings, $\Sigma^\mathcal{S}$, $\mathcal{S} \in \mathbf{E}(\mathfrak{L})$.

It's necessary to pick up in $\mathfrak{H}_\mathfrak{L}$ the partition lattice structure of $\mathfrak{L}$. So, we define an order relation among consonant assignments in $\mathfrak{H}_\mathfrak{L}$: $\Gamma \preccurlyeq \Gamma'$ if and only if $\Gamma \overset{\cap}{=} \Gamma'$, and for each $\gamma \in \Gamma$, there exists a $\gamma \in \Gamma'$ such that $\gamma \subseteq \gamma'$. Similarly, meets and joins of consonant assigments can be defined so that:

$$\Gamma_1 \wedge \Gamma_2 \in \Sigma^{\mathbb{R}_1 \wedge \mathbb{R}_2} \preccurlyeq \Gamma_1, \Gamma_2 , \tag{52a}$$

$$\Gamma_1, \Gamma_2 \preccurlyeq \Gamma_1 \vee \Gamma_2 \in \Sigma^{\mathbb{R}_1 \vee \mathbb{R}_2} . \tag{52b}$$

If $\mathbb{P} \preccurlyeq \mathrm{Dom}(\Gamma)$, we can introduce the notation $\Gamma_{|\mathbb{P}}$, meaning the assignment Γ with domain partition restricted to $\mathbb{P}$, and referred to as Γ *resticted* to $\mathbb{P}$. We also define:

$$\mathbf{G}_{\mathbb{P}\Gamma} \equiv \{\Gamma'; \ \mathrm{Dom}(\Gamma) \preccurlyeq \mathrm{Dom}(\Gamma') = \mathbb{P} \ \& \ \Gamma \preccurlyeq \Gamma'\} , \tag{53}$$

so that $\Gamma_1 \preccurlyeq \Gamma_2 \Rightarrow \mathbf{G}_{\mathbb{P}\Gamma_1} \overset{\supset}{\sim} \mathbf{G}_{\mathbb{P}\Gamma_2}$ and $\mathbf{G}_{\mathbb{P}\emptyset} = \Sigma^\mathbb{P}$.

As stated earlier one way of thinking about superposition is the extension of a probability distribution over a cluster partition to a probability distribution over a subset partition that includes the cluster partition. We can now state exactly what is meant by a probability distribution over a cluster partition:

Definition 1. *For each* $\mathbb{P} \in \mathfrak{F}_C$ *a cluster distribution function,* $\mathbb{p}D$, *written* $\mathbb{p}D \in \mathfrak{D}$, *assigns a probability to any assigment* $\Gamma \in \mathfrak{H}_\Delta\mathbb{P}$ *with the properties that*

$$0 \leq \mathbb{p}D \ \text{ and } \ \mathbb{p}D(\emptyset) = 1 , \tag{54a}$$

$$\mathcal{C}, \mathcal{C}' \in \mathbf{C}_{\Delta(\mathbb{M})}, \ \mathbb{C}' \preccurlyeq \mathbb{C} = \mathrm{Dom}(\Gamma) \Rightarrow \mathbb{p}D(\Gamma) = \sum_{\Gamma' \in \mathbf{G}_{\mathcal{C}'|\Gamma}} \mathbb{p}D(\Gamma') . \tag{54b}$$

Equation (54a), represents the convexity properties for probability distributions. Eq. (54b), which describes *marginal consistency over clusters* expresses the *marginal probability* of Γ on the cluster, $\mathcal{C}$ given all joint probabilities for assignments Γ', on a

---

† *An example of a noncoherent assignment would be:* $\left\{ \left\{ {}^{\mathrm{A\,B}}_{12} \right\} \left\{ {}^{\mathrm{A\,A}}_{23} \right\} \right\}$.

large cluster, $\mathcal{C}'$, where $\Gamma_{|\mathcal{C}} \equiv \Gamma'_{|\{\mathcal{C}\}} = \Gamma$. Equation (54b) can be rewritten in a form more familiar from the Möbius point of view:

$$\mathbb{C} \preccurlyeq \mathbb{C} \ \& \ \Gamma \in \mathbb{C}^\Omega \Rightarrow \mathbb{p}D(\Gamma) = \sum_{\Gamma \preccurlyeq \Gamma' \in (\mathbb{C}')^\Omega} \mathbb{p}D(\Gamma'). \tag{55}$$

Equation (55) is in the dual form of expressions of the type of Eq. (44), where the η matrix here refers to ordering of assignments. By embedding this expression in a linear algebraic framework, we shall show how this—basically tautological—expression provides a simple way of understanding the issues involved in the superposition of probabilities.

Equations (54a) and (54b) imply that

$$\mathbb{C} \preccurlyeq \mathbb{C}' = \mathrm{Dom}(\Gamma') \Rightarrow \mathbb{p}D(\Gamma'_{|\mathbb{C}}) \geq \mathbb{p}D(\Gamma'), \tag{56}$$

so that:

$$\mathbb{p}D'(\Gamma) \leq \mathbb{p}D(\emptyset) = 1. \tag{57}$$

If there is more than one assignment, Γ', with the same restriction, $\Gamma'_{|c} = \Gamma$, then the inequalities in Eq. (55) become strict inequalities in the context of nonzero probability values.†

As was done for assignments we can describe *consonant distribution functions*, ordering among distribution functions, meet and join of consonant distribution functions (all associated to partitions in $\triangle(\mathbb{M})$, $\mathbb{M} \in \mathfrak{F}_\mathbb{C}$):

$$\mathbb{p}D \stackrel{\cap}{=} \mathbb{p}'D' \equiv (\mathbb{p}D)_{|\mathfrak{H}_{\triangle\mathbb{P}} \cap \mathfrak{H}_{\triangle\mathbb{P}'}} = (\mathbb{p}D)_{|\mathfrak{H}_{\triangle(\mathbb{P}\wedge\mathbb{P}')}} = (\mathbb{p}'D')_{|\mathfrak{H}_{\triangle(\mathbb{P}\wedge\mathbb{P}')}}, \tag{58a}$$

$$\mathbb{p}D \preccurlyeq \mathbb{p}'D' \equiv \mathbb{P} \preccurlyeq \mathbb{P}' \ \& \ \mathbb{p}D \subseteq \mathbb{p}'D', \tag{58b}$$

$$\mathbb{p}D \stackrel{\cap}{=} \mathbb{p}'D' \Rightarrow \mathbb{p}D \wedge \mathbb{p}'D' = (\mathbb{p}D)_{|\mathfrak{H}_{\triangle(\mathbb{P}\wedge\mathbb{P}')}} \preccurlyeq \mathbb{p}D, \mathbb{p}'D', \tag{58c}$$

$$\mathbb{p}D \stackrel{\cap}{=} \mathbb{p}'D' \Rightarrow \mathbb{p}D \vee \mathbb{p}'D' = (\mathbb{p}D)_{|\mathfrak{H}_{\triangle(\mathbb{P}\vee\mathbb{P}')}} \succcurlyeq \mathbb{p}D, \mathbb{p}'D'. \tag{58d}$$

In the case of Eq. (58b) we say that $\mathbb{p}'D'$ *subsumes* $\mathbb{p}D$.

V. The Kirkwood-Kikuchi-Barker Formalism

The Kirkwood-Kikuchi-Barker formalism represents an attempt to extend a cluster distribution function associated to a cluster partition, $\mathbb{P}$, with all nonzero probabilities in Eq. (54a) to a motif distribution $\mathbb{P} \succ \mathbb{P}$, where $\mathcal{E}[\mathbb{P}] = \mathcal{E}[\mathbb{P}]$ by the use of a logarithmic/exponential device. In so doing, the set of clusters subsumed by $\mathbb{P}$ must also be extended.

Let $\Gamma \in \mathfrak{H}_{\triangle(\mathbb{P})}$. The approach taken is to ignore the linear convex structure imposed upon multipoint probabilities by Eq. (54a) and (54b), and to force a positive "joint probability" through the use of logarithms and the Möbius construction. Define

$$f(\mathcal{C}) \equiv \ln \mathbb{p}D(\Gamma_{|\mathbb{C}}), \tag{59}$$

---

† *A simple example of marginal consistency over clusters is:*

$$p_1^\wedge = p(\{{}^\wedge_1\}) - p(\{{}^{\wedge\wedge}_{12}\}) + p(\{{}^{AB}_{12}\}) = p_{12}^{AA} + p_{12}^{AB}.$$

over all single cluster partitions and to use the inclusion ordering relation on clusters together with the Möbius formalism given in section III (with $\mathbf{C}(\Omega) \leftrightarrow \mathcal{X}_0$) to define for any partition, $\mathbb{P}' \in \Delta(\mathbb{P})$:

$$\hat{f}(\mathbb{P}') = \sum_{\substack{\{c'\} \preccurlyeq \mathbb{P}' \\ c' \in C(\Omega)}} \ln {}_{\mathbb{P}}D(\Gamma_{|C'}), \tag{60}$$

from which not only:

$$\ln {}_{\mathbb{P}}D(\Gamma_{|C}) = \sum_{\substack{\{c'\} \preccurlyeq \{c\} \preccurlyeq \mathbb{P} \\ c,c' \in C(\Omega)}} \widehat{\ln {}_{\mathbb{P}}D}(\Gamma_{|C'}), \tag{61}$$

follows, but for $\mathrm{Dom}(\Gamma)$, $\mathrm{Dom}(\Gamma') \preccurlyeq \mathbb{P}$,

$$\ln {}_{\mathbb{P}}D^{\mathbf{KKB}}(\Gamma_{|\mathbb{P}'}) \equiv \sum_{\substack{\{c'\} \preccurlyeq \mathbb{P}' \\ c' \in C(\Omega)}} \widehat{\ln {}_{\mathbb{P}}D}(\Gamma_{|C'}) \Longrightarrow$$

$$\Gamma \overset{\Omega}{=} \Gamma' \Rightarrow {}_{\mathbb{P}}D^{\mathbf{KKB}}(\Gamma \vee \Gamma') = \frac{{}_{\mathbb{P}}D^{\mathbf{KKB}}(\Gamma) \cdot {}_{\mathbb{P}}D^{\mathbf{KKB}}(\Gamma')}{{}_{\mathbb{P}}D^{\mathbf{KKB}}(\Gamma \wedge \Gamma')}. \tag{62}$$

Equation (62) defines the *Kirkwood-Kikuchi-Barker pseudoprobability* extension of the cluster distribution function, ${}_{\mathbb{P}}D$. Heuristically, ${}_{\mathbb{P}}D$ expresses an approximation to the "exact" probability, ${}_{\{\Omega\}}D(\Upsilon_{|E[\mathrm{Dom}(\Gamma)]})$, $\Gamma \preccurlyeq \Upsilon \in \Sigma^{\Omega}$.† Eq. (62) embodies the assumption that when no information is available for the probability of the assignment $\Gamma \vee \Gamma'$ on the set $E[\mathrm{Dom}(\Gamma)] \cup E[\mathrm{Dom}(\Gamma')]$ other than that obtainable from ${}_{\mathbb{P}}D(\Gamma)$ and ${}_{\mathbb{P}}D(\Gamma')$, then we can split $E[\mathrm{Dom}(\Gamma)] \cup E[\mathrm{Dom}(\Gamma')]$ into three disjoint parts, $E[\mathrm{Dom}(\Gamma)] - (E[\mathrm{Dom}(\Gamma)] \cap E[\mathrm{Dom}(\Gamma')])$, $E[\mathrm{Dom}(\Gamma)] \cap E[\mathrm{Dom}(\Gamma')]$, and $E[\mathrm{Dom}(\Gamma')] - (E[\mathrm{Dom}(\Gamma)] \cap E[\mathrm{Dom}(\Gamma')])$; and then assume statistical independence over each of these three sets. Thus,

$${}_{\mathbb{P}}D^{\mathbf{KKB}}(\Gamma) = {}_{\mathbb{P}}D^{\mathbf{KKB}}\left(\Gamma_{|(E[\mathrm{Dom}(\Gamma)] \cap E[\mathrm{Dom}(\Gamma')])}\right)$$
$$\cdot {}_{\mathbb{P}}D^{\mathbf{KKB}}\left(\Gamma_{|(E[\mathrm{Dom}(\Gamma)] - E[\mathrm{Dom}(\Gamma)] \cap E[\mathrm{Dom}(\Gamma')])}\right), \tag{63a}$$

$${}_{\mathbb{P}}D^{\mathbf{KKB}}(\Gamma') = {}_{\mathbb{P}}D^{\mathbf{KKB}}\left(\Gamma'_{|(E[\mathrm{Dom}(\Gamma)] \cap E[\mathrm{Dom}(\Gamma')])}\right)$$
$$\cdot {}_{\mathbb{P}}D^{\mathbf{KKB}}\left(\Gamma'_{|(E[\mathrm{Dom}(\Gamma')] - E[\mathrm{Dom}(\Gamma)] \cap E[\mathrm{Dom}(\Gamma')])}\right), \tag{63b}$$

$$\Gamma \overset{\Omega}{=} \Gamma' \Rightarrow {}_{\mathbb{P}}D^{\mathbf{KKB}}\left(\Gamma_{|(E[\mathrm{Dom}(\Gamma)] \cap E[\mathrm{Dom}(\Gamma')])}\right)$$
$$= {}_{\mathbb{P}}D^{\mathbf{KKB}}\left(\Gamma'_{|(E[\mathrm{Dom}(\Gamma)] \cap E[\mathrm{Dom}(\Gamma')])}\right) = {}_{\mathbb{P}}D^{\mathbf{KKB}}(\Gamma \wedge \Gamma'), \tag{63c}$$

$${}_{\mathbb{P}}D^{\mathbf{KKB}}(\Gamma \vee \Gamma') = {}_{\mathbb{P}}D^{\mathbf{KKB}}\left(\Gamma_{|(E[\mathrm{Dom}(\Gamma)] - E[\mathrm{Dom}(\Gamma)] \cap E[\mathrm{Dom}(\Gamma')])}\right)$$
$$\cdot {}_{\mathbb{P}}D^{\mathbf{KKB}}\left(\Gamma_{|(E[\mathrm{Dom}(\Gamma)] \cap E[\mathrm{Dom}(\Gamma')])}\right)$$
$$\cdot {}_{\mathbb{P}}D^{\mathbf{KKB}}\left(\Gamma'_{|(E[\mathrm{Dom}(\Gamma')] - E[\mathrm{Dom}(\Gamma)] \cap E[\mathrm{Dom}(\Gamma')])}\right), \tag{63d}$$

from which Eq. (62) follows. This is certainly true for probability distributions if all the sets involved lie inside the same cluster; *i.e.*, Eq. (62) is a *necessary* condition for any valid superposition extension, but is *not sufficient* to build such an extension.

---

† Actually, as is evident from many considerations, there is, generally, no single "exact" probability.

To gain some understanding of this from the Möbius point of view, let $\mathbb{Q}, \mathbb{Q}' \in \Delta(\mathbb{P})$ with $\mathbb{Q} \preccurlyeq \mathbb{Q}'$, and let $\Gamma \in \mathfrak{H}_{\Delta(\mathbb{P})}$. Define:

$$_\mathbb{p}D^{KKB}(\Gamma_{\mathbb{Q}'|\mathbb{Q}}) \equiv \exp\left(\sum_{\substack{C \preccurlyeq \mathbb{Q}' \ \& \ C \npreceq \mathbb{Q} \\ C \in C_{\Delta(\mathbb{P})}}} \widehat{\ln {}_\mathbb{p}D}(\Gamma_{|C})\right). \tag{64}$$

It is at once evident that (again always using a nonzero $_\mathbb{p}D$):

$$0 < {}_\mathbb{p}D^{KKB}(\Gamma_{\mathbb{Q}'|\mathbb{Q}}), \tag{65a}$$

$$_\mathbb{p}D^{KKB}(\Gamma_{\mathbb{Q}'}) = {}_\mathbb{p}D^{KKB}(\Gamma_{\mathbb{Q}'|\mathbb{Q}}){}_\mathbb{p}D^{KKB}(\Gamma_{\mathbb{Q}}). \tag{65b}$$

In the case that $\mathbb{Q} = C = \{C\}$, $\mathbb{Q}' = C' = \{C'\}$ and $C \subseteq C'$, we also have

$$\sum_{\Gamma' \in G_{C'|\Gamma}} {}_\mathbb{p}D^{KKB}(\Gamma_{C'|C}) = 1. \tag{66}$$

The proof follows from Eq. (61),

$$\sum_{\Gamma' \in G_{C'|\Gamma}} {}_\mathbb{p}D(\Gamma'_{C'|C}) = \sum_{\Gamma' \in G_{C'|\Gamma}} \exp\left(\sum_{C^\star \preccurlyeq C' \ \& \ C^\star \npreceq C} \widehat{\ln {}_\mathbb{p}D}(\Gamma_{|C^\star})\right)$$

$$= \sum_{\Gamma' \in G_{C'|\Gamma}} \exp\left(\sum_{C^\star \preccurlyeq C'} \widehat{\ln {}_\mathbb{p}D}(\Gamma_{|C^\star}) - \sum_{C^\star \preccurlyeq C} \widehat{\ln {}_\mathbb{p}D}(\Gamma_{|C^\star})\right)$$

$$= \sum_{\Gamma' \in G_{C'|\Gamma}} \frac{{}_\mathbb{p}D(\Gamma')}{{}_\mathbb{p}D(\Gamma)} = \frac{1}{{}_\mathbb{p}D(\Gamma)} \sum_{\Gamma' \in G_{\Gamma|\Gamma}} {}_\mathbb{p}D(\Gamma') = \frac{{}_\mathbb{p}D(\Gamma)}{{}_\mathbb{p}D(\Gamma)} = 1. \tag{67}$$

For this condition to hold for partitions, it is necessary that:

$$\mathbb{Q}, \mathbb{Q}' \precdot \mathbb{P}, \ \mathbb{Q}' \succcurlyeq \mathbb{Q} = \mathrm{Dom}(\Gamma) \Rightarrow \sum_{\Gamma' \in G_{\mathbb{Q}'|\Gamma}} {}_\mathbb{p}D^{KKB}(\Gamma'_{\mathbb{Q}'|\mathbb{Q}}) = 1, \tag{68}$$

which is equivalent to

$$\mathbb{Q}, \mathbb{Q}' \precdot \mathbb{P}, \ \mathbb{Q}' \succcurlyeq \mathbb{Q} = \mathrm{Dom}(\Gamma) \Rightarrow {}_\mathbb{p}D^{KKB}(\Gamma) = \sum_{\Gamma' \in G_{\mathbb{Q}'|\Gamma}} {}_\mathbb{p}D^{KKB}(\Gamma'). \tag{69}$$

If Eq. (68) holds, we say that $_\mathbb{p}D^{KKB}$ is *marginally consistent*.[†]

The failure, in general, to satisfy marginal consistency is a significant defect of KKB superposition. It is possible to look for the necessary topological conditions affecting the partition lattice structure for which marginal consistency holds. However, this turns out to be unnecessary, and, in our judgement, leads one in the wrong direction; instead, we shall turn in the next section to the linear algebaic embedding of probabilities. Here, we summarize the important dimensional reduction construction

---

† *An example of the requirement of marginal consistency is:*

$$p\left(\left\{{}^A_1\right\}\right) = p\left(\left\{{}^{AA}_{12}\right\}, \left\{{}^{AA}_{13}\right\}, \left\{{}^{AA}_{23}\right\}\right) + p\left(\left\{{}^{AB}_{12}\right\}, \left\{{}^{AA}_{13}\right\}, \left\{{}^{BA}_{23}\right\}\right)$$

$$+ p\left(\left\{{}^{AA}_{12}\right\}, \left\{{}^{AA}_{13}\right\}, \left\{{}^{AB}_{23}\right\}\right) + p\left(\left\{{}^{AB}_{12}\right\}, \left\{{}^{AB}_{13}\right\}, \left\{{}^{BB}_{23}\right\}\right).$$

This was not true in the case of the example, where the left side is 0.3 and the right side is 0.4703.

of Kikuchi and Brush for which the KKB superposition method *is* valid; and we take advantage of the easily verified fact that whenever all the motifs of a partition, $\mathbb{P}$ are disjoint and the cluster probabilities positive, then the KKB distribution for $\mathbb{P}$ is also marginally consistent.

In the two set case, as originally put forward by Kirkwood, where the KKB approximation takes the form:

$$_{\blacktriangle(\{s\cup s'\})}D^{\text{KKB}}(\Gamma) = \frac{_{\blacktriangle(\{s\})}D(\Gamma_{|\{s\}})\cdot {}_{\blacktriangle(\{s'\})}D(\Gamma_{|\{s'\}})}{_{\blacktriangle(\{s\cap s'\})}D(\Gamma_{|\{s\cap s'\}})}, \tag{70}$$

marginal consistency does hold. In special geometries Eq. (70) can be used as a building block to develop sets for which the overall joint probability distribution is marginally consistent with the probability distributions on each of the motifs. In particular this procedure can be applies to any set, $\mathcal{X}$, which is replicated at all positions on a lattice with a k, l, m (for instance) coordinate system, where $K \leq k \leq K_2$, $L - 1 \leq l \leq L_2$, and $M_1 \leq m \leq M_2$, so that we can refer to $\mathcal{X}(k, l, m)$ with a arbitrarily chosen reference point.

Theorem I. *The dimensional reduction construction of Kikuchi and Brush[2] implies that*

$$\blacktriangle\left(\left\{\bigcup_{K_1, L_1, M_1}^{K_2, L_2, M_2} \mathcal{X}(k, l, m)\right\}\right)^{D^{\text{KKB}}}$$

is marginally consistent.

Theorem I can be usefully generalized to complex unit cells by using shadow cluster lattices.

The second procedure mentioned above gives rise to:

Theorem II. *Let $\mathbb{P} \in {}_{\Delta}(\mathbb{M})$, $\mathbb{M} \in \mathfrak{F}_C$ and consider the set, $\mathbf{X}(\mathbb{P})$, of all partitions, $\mathbb{X} \in {}_{\Delta}(\mathbb{M})$, coextensive with $\mathbb{P}$. Take $\mathbf{X}_0(\mathbb{P})$ be the set of disjoint partitions coextensive with $\mathbb{P}$. The KKB distribution for each of the partitions in $\mathbf{X}_0(\mathbb{P})$ is marginally consistent and defines a probability distribution with domain $\mathfrak{H}_{\Delta(\mathbb{M})}$. Since the set $\mathbf{X}(\mathbb{P})$ is ordered, the construction provided by Eq. (44) can be applied, and will produce a new marginally consistent distribution, which will be a probability distribution provided that all the estimated joint probabilities are positive.*

Proof: The majority of the proof involves the linear algebra of probability space, which is summarized briefly in the Appendix.

Lemma I. *The element of the structure $\mathbf{X}(\mathbb{P})$ with extent, $E(\mathbb{P})$, which is of maximum depth is the partition, $\mathbb{P}_a$, consisting of all atomic clusters, and*

$$\sum_{\mathbb{X}' \in \mathbf{X}_0(\mathbb{P})} \mu(\mathbb{X}', \mathbb{X}) = \delta(\mathbb{X}', \mathbb{X}). \tag{72}$$

Proof: Because the empty partition has empty extent, it is not a member of $\mathbf{X}(\mathbb{P})$, and $\mathbb{P}_a$ is the only element of prank one. The lemma is clearly true for $\mathbb{P}_a$, since $\mathbb{X}'$ can only take the value $\mathbb{P}_a$.

For any other $\mathbb{X}$, we write down the defining Eq. (41) for μ:

$$\sum_{\mathbb{X}_1 \leq \mathbb{X} \leq \mathbb{X}_2} \eta(\mathbb{X}_1, \mathbb{X}') \mu(\mathbb{X}', \mathbb{X}_2) = \delta(\mathbb{X}_1, \mathbb{X}_2), \tag{73a}$$

$$\delta(\mathbb{X}_1, \mathbb{X}_2) = \begin{cases} 1 & \text{when } \mathbb{X}_1 = \mathbb{X}_2, \\ 0 & \text{when } \mathbb{X}_1 \neq \mathbb{X}_2. \end{cases} \tag{73b}$$

If the set $\mathbb{X}_1 = \mathbb{P}_a$ in Eq. (73a), the η term is automatically one, and lemma I follows.

For any assignment, Γ, with domain $\mathbb{X} \in \mathbf{X}_0(\mathbb{P})$, any probability distribution derived from $\vec{P}_{\mathbb{X}} = {}_{\mathbb{X}}D(\Gamma)$ satisfies the relation $\langle \Sigma \mid \vec{P}_{\mathbb{X}} \rangle = 1$ (see Appendix for notation). Accordingly, we set the function f of Eq. (44) to be identically one. We, then, find from lemma I that

$$\hat{f}(\vec{P}_{\mathbb{X}}) = \sum_{\mathbb{X} \in \mathbf{X}_0(\mathbb{P})} \mu(\mathbb{X}_1, \mathbb{X}_2) \, f(\vec{P}_{\mathbb{X}'}) = \delta(\mathbb{P}_a, \mathbb{X}), \tag{74}$$

and from Eq. (46) that:

$$\sum_{\substack{\mathbb{X} \in \mathbf{X}_0(\mathbb{P}) \\ \mathbb{X} \leq \mathbb{X}}} \hat{f}(\vec{P}_{\mathbb{X}}) = f(\vec{P}_{\mathbb{X}}) = 1. \tag{75}$$

If, on $\mathbf{X}(\mathbb{P})$, we now define a non-convex probability distribution by the linear relation:

$$\vec{P}_{\mathbb{X}} = \sum_{\substack{\mathbb{X} \in \mathbf{X}_0(\mathbb{P}) \\ \mathbb{X}' \leq \mathbb{X}}} \overrightarrow{P_{\mathbb{X}'}}, \tag{76a}$$

$$\overrightarrow{P_{\mathbb{X}'}} = \sum_{\mathbb{X} \in \mathbf{X}_0(\mathbb{P})} \mu(\mathbb{X}'', \mathbb{X}') \, \vec{P}_{\mathbb{X}''}, \tag{76b}$$

we see from Eqs. (76) that,

$$\langle \Sigma \mid \vec{P}_{\mathbb{X}} \rangle = 1, \tag{77a}$$

$$\langle \Sigma \mid \overrightarrow{P_{\mathbb{X}}} \rangle = 0, \tag{77b}$$

so that for each disjoint partition beyond the atomic partition, $\vec{P}_{\mathbb{X}}$ lies of the hyperplane and $\overrightarrow{P_{\mathbb{X}}}$ is a differential probability distribution. These latter terms correct the atomic probability distribution for effects due to higher order marginal probabilities.

Theorems I and II are essentially Green's function constructions (although the algorithms for constructing them are nonlinear) when the *complete set of marginals* is given for all clusters in $\Delta(\mathbb{M})$. While certainly useful, they are not sufficient to address the general issue of superposition of probabilities, since neither addresses the solution of the analogous homogeneous linear problem.

VI. The Linear Convex Approach

Theorem II shows how to construct a nonconvexified probability distribution satisfying any complete set of marginality conditions, but, unfortunately, it does not resolve the issue of superposition of probabilities. However, armed with this understanding, Theorem II can lead us to a complete solution of probability superposition.

Let us alter slightly the probability Table I used in the introduction, as shown in Table III. If we now use the expression (2):

$$P_{123}^{\mathbb{X}\mathbb{X}'\mathbb{X}''} \simeq p_{12}^{\mathbb{X}\mathbb{X}'} p_3^{\mathbb{X}''} + p_{13}^{\mathbb{X}\mathbb{X}''} p_2^{\mathbb{X}'} + p_{23}^{\mathbb{X}'\mathbb{X}''} p_1^{\mathbb{X}} - 2p_1^{\mathbb{X}} p_2^{\mathbb{X}'} p_3^{\mathbb{X}''}, \tag{78}$$

Table III

Modified sample pair probability distribution at three sites.

$p_{ij}^{xx'}$	A 2	B 2	A 3	B 3
A 1	0.3	0.0	0.3	0.0
B 1	0.0	0.7	0.0	0.7
A 2	–	–	0.3	0.0
B 2	–	–	0.0	0.7

we find the following set of probabilities,

$$p_{123}^{AAA} = 0.216\,,$$
$$p_{123}^{AAB},\ p_{123}^{ABA},\ p_{123}^{BAA} = 0.084\,,$$
$$p_{123}^{ABB},\ p_{123}^{BAB},\ p_{123}^{BBA} = -0.084\,,$$
$$p_{123}^{BBB} = 0.784\,. \tag{79}$$

The distribution given by Theorem II, while marginally consistent, has a negative probability; $i.e.$, the distribution obtained is nonconvexified. Equation (78) has missed the only correct distribution which has the triple-A probability at 0.3, the triple-B probability at 0.7, and all mixed probabilities zero.

The explanation lies in recognizing that Theorem II only guarantees a nonconvexified distribution on $\mathfrak{P}$ satisfying the given marginality conditions. We have not taken advantage of the structure of $\Delta\mathfrak{P}$. The space, $\mathfrak{RP}$, in this case has dimension $2^3 = 8$ which has been reduced by symmetry. First, if we neglect the symmetry, we can write down a dual basis for $\mathfrak{RP}^\dagger$ which allows us to analyze the situation. It's not difficult to see that the elements of the dual space can be thought of as generated by linear combinations of sets in $\mathbf{E}(\mathbb{P})$, where the value for any one of these sets on a vector in $\mathfrak{RP}$ is obtained by marginal contraction. Accordingly, using Theorem A (see the Appendix), we can write down a basis for $\mathfrak{RP}^\dagger$, say, as:

$$\langle \Sigma|,\ \langle p_1^A|,\ \langle p_2^A|,\ \langle p_3^A|,\ \langle p_{12}^{AA}|,\ \langle p_{13}^{AA}|,\ \langle p_{23}^{AA}|,\ \langle p_{123}^{AAA}|, \tag{80}$$

in terms of the original dual basis:

$$\langle p_{123}^{BBB}|,\ \langle p_{123}^{ABB}|,\ \langle p_{123}^{BAB}|,\ \langle p_{123}^{BBA}|,\ \langle p_{123}^{AAB}|,\ \langle p_{123}^{ABA}|,\ \langle p_{123}^{BAA}|,\ \langle p_{123}^{AAA}|. \tag{81}$$

If we write this expansion in matrix form, we find that it translates the inclusion relation ("$\preccurlyeq$") over assignments into a matrix, η, whose interpretation is exactly that used to define marginal probabilities as expressed, for example, in Eqs. (54). The inverse (Möbius) matrix then expands the joint probabilities in terms of the dual cluster basis, called a *pigeonhole cluster basis*.† The columns of this matrix describe the pigeonhole cluster vectors in terms of the joint probability (pure state) vectors. This basis is called *marginal basis* because $\langle \Sigma \,|$ is one of the dual cluster bases. Such bases have the property that the sum of the probabilities of the basis element dual to $\langle \Sigma \,|$ is one, while the sum for each of the others is zero. Indeed the remaining vectors are a basis for $\Delta \mathfrak{P}$, and are called a *null marginal basis*; they provide a solution to the homogeneous superposition problem.

Theorem III. *We summarize this by writing the general solution to the vector equation for marginal probabilities as:*

$$\langle {}^{0}C \,|^{j} = \eta_{i}^{j} \langle {}^{0}P \,|^{i} \iff {}^{0}\vec{P}_{i} = \eta_{i}^{j} {}^{0}\vec{C}_{j} , \tag{82a}$$

$$\langle {}^{0}P \,|^{i} = \mu_{j}^{i} \langle {}^{0}C \,|^{j} \iff {}^{0}\vec{C}_{j} = \mu_{j}^{i} {}^{0}\vec{P}_{i} , \tag{82b}$$

or, in more familiar terms:

$$\vec{P} = \langle {}^{0}P \,|^{i} \, \vec{P} \rangle \, {}^{0}\vec{P}_{i} \equiv p^{i} \, {}^{0}\vec{P}_{i} = p^{i} \, \eta_{i}^{j} \, {}^{0}\vec{C}_{j} = c^{j} \, {}^{0}\vec{C}_{j} , \tag{83a}$$

$$= \langle {}^{0}C \,|^{j} \, \vec{P} \rangle \, {}^{0}\vec{C}_{j} \equiv c^{j} \, {}^{0}\vec{C}_{j} = c^{j} \, \mu_{j}^{i} \, {}^{0}\vec{P}_{i} = p^{i} \, {}^{0}\vec{P}_{i} . \tag{83b}$$

When $\langle {}^{0}C \,|^{1} = \langle \Sigma \,|$, the basis set, ${}^{0}\vec{C}_{j}$ is called a *marginal basis*. For such a basis, $\sum_{1} \mu_{j}^{i} = \delta_{j}^{1}$, so that ${}^{0}\vec{C}_{j}, j = 2, \ldots , s^{w}$ comprise a null marginal basis for $\Delta \mathfrak{P}$. When the entries of η are all zeroes or ones, the marginal probabilities, $c^{j}, j > 1$, are also conditional probabilities.

The algebraic structure of $\mathfrak{P}$ and pigeonhole cluster probabilities are discussed in more detail in the Appendix.

This elementary result of linear algebra, together with the convex constraints [as contained in Eq. (54a)] that all probabilities must be nonnegative, summarizes the

---

† *In this example, for instance, the η matrix looks as follows:*

η	BBB 123	ABB 123	BAB 123	BBA 123	AAB 123	ABA 123	BAA 123	AAA 123
∅	1	1	1	1	1	1	1	1
A 1	0	1	0	0	1	0	0	1
A 2	0	0	1	0	1	1	1	1
A 3	0	0	0	1	0	1	1	1
AA 12	0	0	0	0	1	0	0	1
AA 13	0	0	0	0	0	1	0	1
AA 23	0	0	0	0	0	0	1	1
AAA 123	0	0	0	0	0	0	0	1

The inverse matrix, μ, is given by:

μ	∅	A 1	A 2	A 3	AA 12	AA 13	AA 23	AAA 123
BBB 123	1	-1	-1	-1	1	1	1	-1
ABB 123	0	1	0	0	-1	-1	0	1
BAB 123	0	0	1	0	-1	0	-1	1
BBA 123	0	0	0	1	0	-1	-1	1
AAB 123	0	0	0	0	1	0	0	-1
ABA 123	0	0	0	0	0	1	0	-1
BAA 123	0	0	0	0	0	0	1	-1
AAA 123	0	0	0	0	0	0	0	1

The rows of η and μ correspond to linear forms (duals vectors) as in (82) or (83), while the columns correspond to vectors. Thus the vector basis is read off the columns of μ with the null marginal basis consisting to the last 7 columns, all of which have zero marginal probability.

linear convex formulation for the superposition of probabilities. Theorem III allows one to write down an *exact* expression for the entropy, which "resides" in the joint *a priori* probability space, in terms of marginal probabilities. Some consequences of this representation will be elaborated further below.

In our example, when we specified the point and pair marginal distributions (as well as the requirement that $\langle \Sigma \mid \vec{P} \rangle = 1$), we left one degree of freedom undetermined in $\vec{P}$, and this can be seen in terms of the first basis set as the value of p_{123}^{AAA}. The set of points of $\mathfrak{RP}$ satisfying the given marginality conditions has a codimension of one, and we find that the value of p_{123}^{AAA} can range between 0.1 and 0.2 to produce a consistent probability distribution. If we adopt the Bayesian point of view that we know nothing about the third order (point probability) distribution, then the use of information theory methods indicates that the information entropy is maximized for the distribution at which p_{123}^{AAA} is 0.16 (which is slightly displaced from the more naively intuitive centroid), *i.e.*,

$$- \left(p_{123}^{AAA} \ln p_{123}^{AAA} + p_{123}^{AAB} \ln p_{123}^{AAB} + p_{123}^{ABA} \ln p_{123}^{ABA} + p_{123}^{BAA} \ln p_{123}^{BAA} \right.$$
$$\left. + p_{123}^{ABB} \ln p_{123}^{ABB} + p_{123}^{BAB} \ln p_{123}^{BAB} + p_{123}^{BBA} \ln p_{123}^{BBA} + p_{123}^{BBB} \ln p_{123}^{BBB} \right), \quad (84)$$

is maximized at the probability distribution:

$$p_{123}^{AAA} = 0.16 ,$$
$$p_{123}^{AAB} , \ p_{123}^{ABA} , \ p_{123}^{BAA} = 0.04 ,$$
$$p_{123}^{ABB} , \ p_{123}^{BAB} , \ p_{123}^{BBA} = 0.06 ,$$
$$p_{123}^{BBB} = 0.54 . \quad (85)$$

The probability distribution given by (85) is the expected (and also the most likely) probability distribution subject to the given marginal constraints. Because the logarithm is only real when all probabilities are positive, in principal one doesn't need to know the exact boundaries of the convex region in order to find this "most likely" probability, if one demands the entropy be real. However, in practice, details concerning the boundaries may be the most efficient way of determining where the entropy is real. The entropy maximum, which identifies the expected probability distribution from among the feasible distributions as delimited by the *probability polytope* in joint probability space, is a function of that polytope. If an additional constraint associated to the expectation of a Hamiltonian is added, the maximum entropy will determine the ensemble state minimizing the appropriate free energy potential.

Techniques of group theory and the convex properties of the entropy function allow—often enormous—simplification in the degrees of freedom associated to the maximum entropy, but it is, generally, impossible to avoid taking into account correlations of all orders without distorting or completely missing the correct probability polytope. Indeed, such entropy approximations are not just slightly off—because of the numerical insensitivity of the logarithm—but completely in error, because in the complete expression, some joint "probabilities" are negative, meaning that the entropy is not real as might be expected in the approximate formulation. The Kikuchi-Brush dimensional reduction scheme, where applicable, avoids the negative probability pitfall, but cannot guarantee that one is on the correct polytope subface.

The importance of the probability polytope must be emphasized. Much of the combinatorics of these types of problems is geometrically summarized by the probability polytope. This polytope incorporates the probability estimates on the feasibility

Table II

Sample pair probability distribution at four sites.

$p_{ij}^{xx'}$	A 2	B 2	A 3	B 3	A 4	B 4
A 1	0.18	0.02	0.18	0.02	0.18	0.02
B 1	0.12	0.68	0.12	0.68	0.12	0.68
A 2	—	—	0.2	0.1	0.2	0.1
B 2	—	—	0.1	0.6	0.1	0.6
A 3	—	—	—	—	0.2	0.1
B 3	—	—	—	—	0.1	0.6

of tailings and on interactions at all levels in a different manner than obtained from graphical expansions.

The statistical linear convex viewpoint given here is completely general, as we shall illustrate with another simple example on the tetrahedron, in which the linear convex analysis can be carried out with desktop software. In that case we can write a dual basis analogous to that given in (80):

$$\langle \Sigma |, \langle p_1^A |, \langle p_2^A |, \langle p_3^A |, \langle p_4^A |, \langle p_{12}^{AA} |, \langle p_{13}^{AA} |, \langle p_{14}^{AA} |, \langle p_{23}^{AA} |,$$

$$\langle p_{24}^{AA} |, \langle p_{34}^{AA} |, \langle p_{123}^{AAA} |, \langle p_{124}^{AAA} |, \langle p_{134}^{AAA} |, \langle p_{234}^{AAA} |, \langle p_{1234}^{AAAA} |. \quad (86)$$

If we assume we know the point and pair distributions, there are 5 degrees of freedom left. We recall Table II of marginal distributions given in Sec. I.

For this example, we reduce that codimension down to three by applying some symmetries; in this case, we can assume that the optimal 123, 124 and 134 three point marginal distributions will be the same. Also, note that symmetries make Table II a function of only 4 variables, such as: p_1^A, p_2^A, p_{12}^{AA}, p_{23}^{AA}.

We begin by presenting the η and μ matrices for the disjoint partitions over the tetrahedron in the case of disjoint partitions as in Theorem II:†

Based on the matrix, μ, of Table IVb , we can use Theorems II and A to write the expressions for nonconvexified distributions satisfying first-order, second-order and third-order marginality constraints:

$$p_{1234}^{xx'x''x'''} \overset{1}{=} p_1^x p_2^{x'} p_3^{x''} p_4^{x'''}, \tag{87a}$$

$$p_{1234}^{xx'x''x'''} \overset{2}{=} p_{12}^{xx'} p_{34}^{x''x'''} + p_{13}^{xx''} p_{24}^{x'x'''} + p_{14}^{xx'''} p_{23}^{x'x''} - 2p_1^x p_2^{x'} p_3^{x''} p_4^{x'''}, \tag{87b}$$

---

† Since we are dealing with disjoint partitions of the numbers from 1 to 4, we use "/" to indicate how the partition is made. This 23/1/4 is {{23}, {1}, {4}}, 1/2/3/4 is {{1}, {2}, {3}, {4}} and 1234 is {1234}. The ordering now is for disjoint partitions (in this case there are 15 such partitions) rather than for clusters.

Table IVa
η matrix for four sites.

η	1/2/3/4	1/2/3/4	1/3/2/4	1/4/2/3	2/3/1/4	2/4/1/3	3/4/1/2	1/2/3/4	1/3/2/4	1/4/2/3	1 2 3/4	1 2 4/3	1 3 4/2	2 3 4/1	1234
1/2/3/4	1	1	1	1	1	1	1	1	1	1	1	1	1	1	1
12/3/4	0	1	0	0	0	0	0	1	0	0	1	1	0	0	1
13/2/4	0	0	1	0	0	0	0	0	1	0	1	0	1	0	1
14/2/3	0	0	0	1	0	0	0	0	0	1	0	1	1	0	1
23/1/4	0	0	0	0	1	0	0	0	0	1	1	0	0	1	1
24/3/1	0	0	0	0	0	1	0	0	1	0	0	1	0	1	1
34/1/2	0	0	0	0	0	0	1	1	0	0	0	0	1	1	1
12/34	0	0	0	0	0	0	0	1	0	0	0	0	0	0	1
13/24	0	0	0	0	0	0	0	0	1	0	0	0	0	0	1
14/23	0	0	0	0	0	0	0	0	0	1	0	0	0	0	1
123/4	0	0	0	0	0	0	0	0	0	0	1	0	0	0	1
124/3	0	0	0	0	0	0	0	0	0	0	0	1	0	0	1
134/2	0	0	0	0	0	0	0	0	0	0	0	0	1	0	1
234/1	0	0	0	0	0	0	0	0	0	0	0	0	0	1	1
1234	0	0	0	0	0	0	0	0	0	0	0	0	0	0	1

Table IVb
μ matrix for four sites.

μ	1/2/3/4	1/2/3/4	1/3/2/4	1/4/2/3	2/3/1/4	2/4/1/3	3/4/1/2	1/2/3/4	1/3/2/4	1/4/2/3	1 2 3/4	1 2 4/3	1 3 4/2	2 3 4/1	1234
1/2/3/4	1	−1	−1	−1	−1	−1	−1	1	1	1	2	2	2	2	−6
12/3/4	0	1	0	0	0	0	0	−1	0	0	−1	−1	0	0	2
13/2/4	0	0	1	0	0	0	0	0	−1	0	−1	0	−1	0	2
14/2/3	0	0	0	1	0	0	0	0	0	−1	0	−1	−1	0	2
23/1/4	0	0	0	0	1	0	0	0	0	−1	−1	0	0	−1	2
24/3/1	0	0	0	0	0	1	0	0	−1	0	0	−1	0	−1	2
34/1/2	0	0	0	0	0	0	1	0	0	0	0	0	−1	−1	2
12/34	0	0	0	0	0	0	0	1	0	0	0	0	0	0	−1
13/24	0	0	0	0	0	0	0	0	1	0	0	0	0	0	−1
14/23	0	0	0	0	0	0	0	0	0	1	0	0	0	0	−1
123/4	0	0	0	0	0	0	0	0	0	0	1	0	0	0	−1
124/3	0	0	0	0	0	0	0	0	0	0	0	1	0	0	−1
134/2	0	0	0	0	0	0	0	0	0	0	0	0	1	0	−1
234/1	0	0	0	0	0	0	0	0	0	0	0	0	0	1	−1
1234	9	0	0	0	0	0	0	0	0	0	0	0	0	0	1

$$p_{1234}^{xx'x''x'''} \stackrel{3}{=} p_{123}^{xx'x''} p_4^{x'''} + p_{124}^{xx'x'''} p_3^{x''} + p_{134}^{xx''x'''} p_2^{x'} + p_{234}^{x'x''x'''} p_1^{x}$$
$$+ p_{12}^{xx'} p_{34}^{x''x'''} + p_{13}^{xx''} p_{24}^{x'x'''} + p_{14}^{xx'''} p_{23}^{x'x''} - 2p_{12}^{xx'} p_3^{x''} p_4^{x'''}$$
$$- 2p_{13}^{xx''} p_2^{x'} p_4^{x'''} - 2p_{14}^{xx'''} p_2^{x'} p_3^{x''} - 2p_{23}^{x'x''} p_1^{x} p_4^{x'''}$$
$$- 2p_{24}^{x'x'''} p_1^{x} p_3^{x''} - 2p_{34}^{x''x'''} p_1^{x} p_2^{x'} + 6p_1^{x} p_2^{x'} p_3^{x''} p_4^{x'''} . \tag{87c}$$

We can now apply Eq. (87b) to the pair probability distribution given by Table II to produce a nonconvexified distribution on $\overline{\mathfrak{P}}$:

$$p_{1234}^{AAAA} = 0.0972,,$$
$$p_{1234}^{AAAB} , p_{1234}^{AABA} , p_{1234}^{AABA} = 0.0148 ,$$
$$p_{1234}^{BAAA} = 0.0288 ,$$
$$p_{1234}^{AABB} , p_{1234}^{ABAB} , p_{1234}^{ABBA} = 0.0532 ,$$
$$p_{1234}^{BAAB} , p_{1234}^{BABA} , p_{1234}^{BBAA} = 0.0592 ,$$
$$p_{1234}^{ABBB} = -0.1012 ,$$
$$p_{1234}^{BABB} , p_{1234}^{BBAB} , p_{1234}^{BBBA} = -0.0272 ,$$
$$p_{1234}^{BBBB} = 0.6752 . \tag{88}$$

To this point on $\overline{\mathfrak{P}}$ we want to add a general vector from $\Delta\mathfrak{P}$, which has been reduced by symmetry to three dimensions.

Complete superposition can be carried out with the use of Theorem III.[†]

We select x be the coefficient of each of p_{123}^{AAA}, p_{124}^{AAA}, and p_{134}^{AAA}, y to be the coefficient of p_{234}^{AAA}, and z to be coefficient of p_{1234}^{AAAA}. The resulting parametric equations describing $\tilde{P}$ are given by:

$$p_{1234}^{AAAA} = z ,$$
$$p_{1234}^{AAAB} , p_{1234}^{AABA} , p_{1234}^{AABA} = x - z ,$$
$$p_{1234}^{BAAA} = y - z ,$$
$$p_{1234}^{AABB} , p_{1234}^{ABAB} , p_{1234}^{ABBA} = 0.18 - 2x + z ,$$
$$p_{1234}^{BAAB} , p_{1234}^{BABA} , p_{1234}^{BBAA} = 0.2 - x - y + z ,$$
$$p_{1234}^{ABBB} = -0.34 + 3x - z ,$$

† Here, η is:

η	BBBB 1234	ABBB 1234	BABB 1234	BBAB 1234	BBBA 1234	AABB 1234	ABAB 1234	ABBA 1234	BAAB 1234	BABA 1234	BBAA 1234	AAAB 1234	AABA 1234	ABAA 1234	BAAA 1234	AAAA 1234
$\emptyset$	1	1	1	1	1	1	1	1	1	1	1	1	1	1	1	1
A_1	0	1	0	0	0	1	1	1	0	0	0	1	1	1	0	1
A_2	0	0	1	0	0	1	0	0	1	1	0	1	1	0	1	1
A_3	0	0	0	1	0	0	1	0	1	0	1	1	0	1	1	1
A_4	0	0	0	0	1	0	0	1	0	1	1	0	1	1	1	1
AA_{12}	0	0	0	0	0	1	0	0	0	0	0	1	1	0	0	1
AA_{13}	0	0	0	0	0	0	1	0	0	0	0	1	0	1	0	1
AA_{14}	0	0	0	0	0	0	0	1	0	0	0	0	1	1	0	1
AA_{23}	0	0	0	0	0	0	0	0	1	0	0	1	0	0	1	1
AA_{24}	0	0	0	0	0	0	0	0	0	1	0	1	0	1	1	1
AA_{34}	0	0	0	0	0	0	0	0	0	0	1	0	0	1	1	1
AAA_{123}	0	0	0	0	0	0	0	0	0	0	0	1	0	0	0	1
AAA_{124}	0	0	0	0	0	0	0	0	0	0	0	0	1	0	0	1
AAA_{134}	0	0	0	0	0	0	0	0	0	0	0	0	0	1	0	1
AAA_{234}	0	0	0	0	0	0	0	0	0	0	0	0	0	0	1	1
$AAAA_{1234}$	0	0	0	0	0	0	0	0	0	0	0	0	0	0	0	1

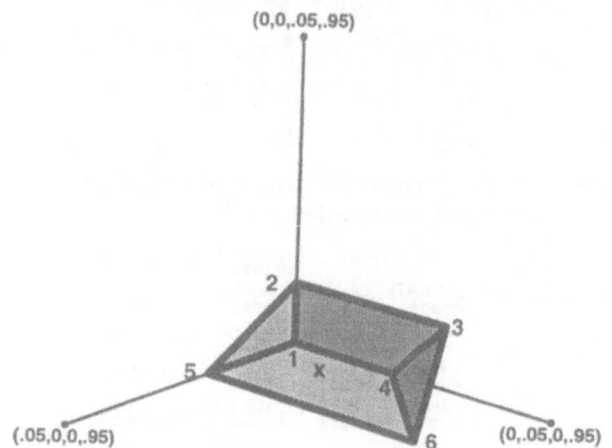

Figure 1. Allowable region in x', y', z', w' probability tetrahedron. The vertices are located in: #1 $= (0,0,0,1)$, #2 $= (0,0,0.01,0.99)$, #3 $= (0,0.03,0.01,0.96)$, #4 $= (0,0.02,0,0.98)$, #5 $= (0.02,0,0,0.98)$, #6 $= (0.02,0.04,0,0.94)$. The × marks the position of the convex Bayesian solution.

$$p_{1234}^{BABB}, p_{1234}^{BBAB}, p_{1234}^{BBBA} = -0.28 + 2x + y - z,$$
$$p_{1234}^{BBBB} = 1.04 - 3x - y + z. \tag{89}$$

Equation (88) can be obtained from Eq. (89) by setting $x = 0.112$, $y = 0.126$, and $z = 0.0972$.

Imposing the convexity restraints that all probabilities be nonnegative delineates a 5 sided prism in x-y-z space, and a corresponding one in $x' = p_{1234}^{AAAB}(= x - z)$, $y' = p_{1234}^{BAAA}(= y - z)$, $z' = p_{1234}^{AABB}(= 0.18 - 2x + z)$ space which is shown in Fig. 1. Note that the prism occupies only a small fraction of the probability tetrahedron composed of $x' + y' + z' + w' = 1$, where w' is the sum of the remaining 13 joint probabilities. Only the probabilities, p_{1234}^{AAAB}, p_{1234}^{BBBA}, p_{1234}^{AABB}, p_{1234}^{ABBB}, p_{1234}^{BBAA}, and their symmetric

and μ together with the set of marginals describing $\vec{P}$ are:

μ	θ	A 1	A 2	A 3	A 4	AA 12	AA 13	AA 14	AA 23	AA 24	AA 34	AAA 123	AAA 124	AAA 134	AAA 234	AAA 1234	$\vec{P}$
BBBB 1234	1	−1	−1	−1	−1	1	1	1	1	1	1	−1	−1	−1	−1	1	1
ABBB 1234	0	1	0	0	0	−1	−1	−1	0	0	0	1	1	1	0	−1	0.2
BABB 1234	0	0	1	0	0	−1	0	0	−1	−1	0	1	1	0	1	−1	0.3
BBAB 1234	0	0	0	1	0	0	−1	0	−1	0	−1	1	0	1	1	−1	0.3
BBBA 1234	0	0	0	0	1	0	0	−1	0	−1	−1	0	1	1	1	−1	0.3
AABB 1234	0	0	0	0	0	1	0	0	0	0	0	−1	−1	0	0	1	0.18
ABAB 1234	0	0	0	0	0	0	1	0	0	0	0	−1	0	−1	0	1	0.18
BAAB 1234	0	0	0	0	0	0	0	0	1	0	0	−1	0	0	−1	1	0.18
BABA 1234	0	0	0	0	0	0	0	0	0	1	0	0	−1	0	−1	1	0.2
BBAA 1234	0	0	0	0	0	0	0	0	0	0	1	0	0	−1	−1	1	0.2
AAAB 1234	0	0	0	0	0	0	0	0	0	0	0	1	0	0	0	−1	0.2
AABA 1234	0	0	0	0	0	0	0	0	0	0	0	0	1	0	0	−1	x
ABAA 1234	0	0	0	0	0	0	0	0	0	0	0	0	0	1	0	−1	x
BAAA 1234	0	0	0	0	0	0	0	0	0	0	0	0	0	0	1	−1	y
AAAA 1234	0	0	0	0	0	0	0	0	0	0	0	0	0	0	0	1	z

Table V

Summary of superposition methods for Table II.

Example summary	Point probability	Kikuchi-Baker	Convex-Bayesian
p_{1234}^{AAAA}	0.0054	1.6	0.151465
p_{1234}^{AAAB}, p_{1234}^{AABA}, p_{1234}^{AABA}	0.0126	0.00816327	0.012872
p_{1234}^{BAAA}	0.0216	0.0296296	0.015735
p_{1234}^{AABB}, p_{1234}^{ABAB}, p_{1234}^{ABBA}	0.0294	0.000499792	0.002791
p_{1234}^{BAAB}, p_{1234}^{BABA}, p_{1234}^{BBAA}	0.0504	0.00770975	0.019928
p_{1234}^{ABBB}	0.0686	0.00367194	0.001546
p_{1234}^{BABB}, p_{1234}^{BBAB}, p_{1234}^{BBBA}	0.1176	0.0240733	0.064409
p_{1234}^{BBBB}	0.2744	0.902012	0.531254
Sum of probabilities	1.0	2.65335	1.0

variants vanish on the boundary of the polytope, so these probabilities prelimit the location of the entropy maximum.

The principle of maximum information entropy can be used to locate the "most likely" *convex Bayesian* or CB probability distribution, as is shown in Table V. As can be seen from this table, the point probability estimates completely fail to predict clustering tendencies, while is it difficult even to give a coherent probabilistic interpretation to the Kirkwood-Kikuchi-Barker approximation.[†] In carrying out estimates involving material systems this simple example already indicates that only in the simplest of physical systems, such as diffuse gases, can one expect to quantitatively predict cooperative phenomena using only point probabilities without invoking corrections for higher order correlations. A discussion of the fluctuation of the CB estimate with sample size (but not with the size of Ω) can be found, for example, in reference 11.

VII. Conclusions

Using any cluster basis, the general problem of superposition of probabilities can be formulated in linear convex form as finding all solutions to:

$$\left\{ \vec{P} = \sum_i p^i \, {}^0\vec{P}_i; \quad \sum_i p^i \eta_i^{jk} = (c^{jk})_0, \quad k = 1, \ldots, k_0 \ \& \ p^i \geq 0 \right\}, \tag{90}$$

[†] I have numerically confirmed that the point probability estimate coincides with the convex Bayesian estimate with only point marginals.

Here η is the marginal probability matrix defining linearly independent cluster variables and the $(c^{jk})_0$ are a set of constants representing the fixed cluster probabilities to be matched.

In the examples given in the previous section we made use of the inverse matrix, μ, in addition to the special Möbius construction of Theorem II and pigeonhole clusters to look at all pair clusters. However, in general, the superposition of probabilities can be carried out over any set of linearly consistent cluster values (conditional probabilities) by using a graded basis set for expressing the conditional probabilities. This generates a polytope formed by the intersection of the subspace of $\Delta\mathfrak{P}$ translated to pass through a point on $\overline{\mathfrak{P}}$ which satisfies the known marginal conditions. In the previous section we gave an explicit, dual, representation of particular points within the polytope. A unique solution is, generally, only determined by the specific conditions and statistics of the probabilities to be estimated. Sometimes a specific solution can be determined either by the dimensional reduction scheme of Kikuchi and Brush or the disjoint partition representation of Theorem II. But, if the polytope is non-empty, a maximum entropy solution always exists, that is:

$$
{}_s\vec{P} = \min_{\substack{\langle \Sigma\,|\,\vec{P}=1,\,p^i\geq 0 \rangle \\ \sum_i p^i \eta_i^{jk}=(c^{jk})_0}} \left[\sum_i p^i \ln p^i \right] . \tag{91}
$$

Linear convex analysis presents a rigorous approach to answering the question of maximum entropy as well as merely finding a consistent answer to (90). Issues of statistics can modify the use of the maximum entropy approximation, but these are a function of the individual application. The critical issue is not one of statistical interpretation, but of the intimidating number of dimensions involved in typical applications. Here, the determination of the probability polytope requires enormous computing resources, although the analytic center and other interior point methods arising from the groundbreaking techniques of Karmarkar may be applicable.[12] The maximum entropy method for selecting the CB joint probability distribution can also be solved using a dual convex algorithm as was done by Eriksson in 1980, but this does not obviate using the full joint probability expression for the entropy, without which an incorrect region in joint probability space may be delineated.[13] However, if one knows the location of the probability polytope, this will often severely restrict the available class of structures even without resorting to entropy maximization. In highly ordered structures, knowledge of the polytope may be more important than the choice of a particular representative probability distribution within that polytope.

Unfortunately, even if we wish to estimate only a limited amount of additional information concerning the probability distribution, the techniques available may be very limited, since Eq. (90) is formulated in terms of the full point probability space whose dimension is usually awesomely large. For instance, suppose we know all nearest-neighbor and second nearest-neighbor pair probabilities and we wish to estimate longer range pair probabilities subject to the nearest-neighbor and second nearest-neighbor constraints. By using a graded basis for all two point probabilities the problem can be formulated in linear convex form over a space of much lower dimension. A maximum entropy solution for this problem can be found, but its physical validity must be carefully examined. Alternatively, in cases of this type, one can throw away some of the two point probability information in order to build a set, $\mathcal{X}$, to which the dimensional reduction construction can be applied; then one can marginally contract to find other pair probabilities. This is tantamount to choosing a number of subclusters of $\mathcal{X}$ for which a special form of superposition can be demonstrated by the above techniques or others to determine the joint probability distributions over $\mathcal{X}$; subsequently applying dimensional reduction. If this solution to (90) is supposed to approximate a solution to (91), one must ask how to justify equal *a priori* probabilities when throwing away certain pair probabilities of the same physical "significance" as those retained.

The linear convex method was principally developed for application to inhomogeneous kinetic problems, but its relationship to configurational statistical equilibrium is evident:

$$
s\vec{P} = \min_{\vec{P}} \left[\sum_i p^i \ln p^i - \frac{1}{kT} \sum_i p^i E_i \right]
$$

$$
= \min_{\substack{\langle \Sigma | \vec{P}=1, p^i \geq 0 \\ \sum_{i=1}^w c^{ir}=\alpha_r,\ r=1,\dots,s-1}} \left[\frac{1}{kT} \sum_i p^i \left(\ln p^i - E_i \right) \right]
$$

$$
= \min_{\substack{\langle \Sigma | \vec{P}=1, p^i \geq 0 \\ \sum_{i=1}^w c^{ir}=\alpha_r,\ r=1,\dots,s-1}} \left[\frac{1}{kT} \sum_i p^i \left(\ln p^i - \eta_i^j \mathcal{E}_j \right) \right]. \tag{92}
$$

Here α_r is the overall concentration of the rth species, and

$$
\mathcal{E}_j = \sum_i \mu_j^i E_i, \tag{93}
$$

represents the formal Connolly-Williams cluster expansion of the energy.[†] Equation (92) is expressed for the Helmholtz' free energy, but can easily be generalized to other forms.

To incorporate symmetry into this formulation, let $^0\vec{P}_i$ be any pure state, and let $\mathfrak{G}$ the intersection of the symmetry group of the Hamiltonian with that of the constraints. Then all states of the set $\mathfrak{G}^0\vec{P}_i$ all have the same energy as $^0\vec{P}_i$. These states form the vertices of a convex polytope all of whose ensemble elements have the same energy. Additionally, because of the convex property of the entropy function, $S = -\sum p_i \ln p_i$, the barycenter will have the greatest entropy of any ensemble in the polytope generated by $\mathfrak{G}^0\vec{P}_i$. Consequently, we can associate one variable to each equivalence class of $\mathfrak{G}$ operating on $\mathfrak{P}$, and be assured that the joint probabilities in the minimum free energy ensemble will be the same for all equivalent pure states. The set of, say n, such variables, one for each equivalence class, are the symmetric variables which can be used to formulate the constraint problem.

Equation (92) has a unique mathematical solution. There are several ways to describe the physical hypotheses underlying this solution, but perhaps the simplest is the ergodic hypothesis, which in a finite system states that there is a finite probability of interchange between any two pure states. As a consequence the Markov process describing the kinetics of the system is irreducible, and the equilibrium point lies in the interior of the polytope, $\mathfrak{P}$. The free energy of that equilibrium ensemble is a

---

† Note that because of the generally highly oscillatory nature of μ, the expression for the cluster energies in terms of the pure state configurational energies is, generally, not physically meaningful. However, if one thinks of the Connolly-Williams "screened potential" coefficients as fundamental, then the complete configurational energies, expressed in terms of η, are well-defined.

lower bound to any other system, even without the ergodic hypothesis. When periodic boundary conditions are applied, translational symmetry of the Hamiltonian forces homogeneity.

The reason for the apparent discrepancy between the idea of a totally interior solution and the idea of a solution (or solutions) lying on subfaces of $\mathfrak{P}$ lies in the fact that as the dimension of the system becomes large (the idea of "large" dimension depends on the temperature), the distance between the "interior" point and the boundary for many material systems becomes meaninglessly small, so that, aside from physical considerations, *no finite computational scheme* can retain the degree of precision needed to retain the interior character of the solution. No matter what numerical method one uses to find the solution to (92), *de facto* linear constraints appear; many of them implying correlation lengths far beyond those of the original Hamiltonian.[16] The numerical solution, then, *appears* to lie in a subspace whose dimension is (usually) far less than that of $\mathfrak{P}$. By forcing the solution to remain homogeneous, the CVM approximation greatly reduces the possibility of ending up in the wrong subface; but as the temperature is reduced, the apparent dimension of the solution to (92) is also greatly reduced and any approximating solution is hard-pressed to even stay on the correct subface. At this point linear convex considerations would seem to be more important than the far simpler problem of minimizing the free energy once one has found the correct subface.

On the other hand, as soon as one admits that numerical considerations force minimizing points to lie on subfaces of $\mathfrak{P}$, the physical meaning of the underlying ergodic hypothesis must be called into question. For it is, then, possible to have multiple minima for (92) which are not metastable, because there is no "fluctuation" connecting them; they are only metastable in a system for which the ergodic hypothesis holds (or, for example, when the size and type of fluctuations are increased by increasing the temperature). These solutions, also, need not be homogeneous if translations are not among the permissible fluctuations. The conflict between these viewpoints concerning the relationship between long term structures and "equilibrium" structures cannot be resolved—other than axiomatically—without the introduction of system kinetics.

Acknowledgements

This work is supported by the ARPA/NIST Program on Mathematical Modeling of Microstructure Evolution in Advanced Alloys. The author thanks R. Kikuchi and John Cahn for inspiration and encouragement; Paul Boggs, Paul Domich, Janet Rogers and Chris Witzgall for insights into the feasibility of convex analysis; Janet Rogers at NIST for determining the prismatic shape of the probability polytope; Pierre Cenedese for invaluable comments on current CVM practice; Ben Burton for organizational criticism and for information on the history of superposition; and Argo Ogawa who instructed me on how to turn an unreadable SGML/TeX mishmash into TeX.

Appendix: Linear Algebra of Probabilities

The s^w assignments in Σ^Ω, can be thought of as generators of a free Abelian group (actually a free vector space), and as such, they comprise a set of basis vectors, called the *pure state* basis ${}^0\vec{P}_i$, $i = 1, \ldots, s^w$ in a Euclidean space, $\Re\mathfrak{P}$ of dimension s^w, called the *embedding* Euclidean space.

We designate the *dual Euclidean space*, or space of *linear forms* as $\Re\mathfrak{P}^\dagger$. $\Re\mathfrak{P}^\dagger$. has an associated *dual pure state* basis:

$$^0\vec{P_i}, \quad \langle^0P|, \quad \langle^0P|^i\,{}^0\vec{P_i} = \delta^i_j, \qquad \text{for } i,j = 1,\ldots,s^w. \tag{A1}$$

A modified Dirac bracket notation has been used in Eq. (A1). In standard matrix terms, dual vectors are usually represented as row vectors, while vectors from $\Re\mathfrak{P}$ are represented as column vectors. *Probability vectors*, $|\vec{P}\rangle$, or just $\vec{P}$, in $\Re\mathfrak{P}$ belong to a set, $\mathfrak{P}$, the *probability space*, and are subject to the condition that the sum of the probabilities is one. This is a linear constraint over $\Re\mathfrak{P}$, and will be expressed as

$$\langle \Sigma \,|\, \vec{P}\rangle = 1, \tag{A2}$$

$$\langle \Sigma| = \sum_{i=1}^{s^w} \langle^0P|^i. \tag{A3}$$

In addition, the probabilities must lie between zero and one. Consequently, $\mathfrak{P}$ is a convex subset of $\Re\mathfrak{P}$ lying on a hyperplane parallel to a hyperspace, $\Delta\mathfrak{P}$, passing through the origin of $\Re\mathfrak{P}$. $\Delta\mathfrak{P}$ is called the *differential probability space* and is defined by

$$\langle \Sigma \,|\, \Delta\vec{P}\rangle = 0. \tag{A4}$$

Because of the linear condition (A3), the inequality demanding that probabilities be no greater than one is not necessary to characterize a probability vector. One need only require the convexity condition that all probability components are nonnegative. No convexity condition is required for $\Delta\mathfrak{P}$, although, clearly, this difference between two probability vectors can never has all components with absolute value bounded by two.

Many physical interpretations can be imposed on both $\Re\mathfrak{P}$ and $\mathfrak{P}$, but we are particularly interested in the *ensemblespace* interpretation. In this interpretation, all pure states must satisfy the *Bayesian hypothesis of equal a priori probabilities*. As is well-known, this idea, which underlies the concept of *phase space* in physics, serves as one of the fundamental concepts in statistical mechanics.[17] The same idea has been translated into mathematics in the fields of probability theory and information theory.[11]† An *ensemble* probability vector (actually only with rational coefficients) can be thought of as an ensemble built from N samples of pure states, where the coefficients of each pure state basis vector is the fraction of the ensemble consisting of that pure state, and the linear dual space over the ensemble space can be used to represent the expectation of any function on pure states (or any physical operator on pure states—no matter how nonlinear) by just adding up the number of pure states in any ensemble vector times the function value for each pure state. The distinction of the Bayesian assumption lies in developing probabilities over the ensemble space, which is itself a probability space. Based on the assumption of equal *a priori* probabilities, one can ask the likelihood of drawing at random a distribution which will give rise to a given ensemble vector. That assumption assigns a multinomial probability to each ensemble probability vector:

$$\frac{N!}{s^{wN}\prod_i(Np_i)!}, \tag{A5}$$

---

† Actually equal *a priori* probabilities are not required, only the assumption of statistical independence is needed to apply the multinomial distribution.

where p_i is the ith component of the ensemble probability vector. The ensemble interpretation of information theory then shows that if one restricts oneself to the set of all ensemble vectors which are subject to a set of linear constraints, the *expected* ensemble probability distribution will be that which maximizes entropy over this set. Because of the properties of the multinomial distribution, this is also the most probable (according to the multinomial distribution) ensemble probability vector in that constrained set.

Returning to $\Delta\mathfrak{P}$, each vertex of $\mathfrak{P}$ is a pure state, and the topology of $\mathfrak{P}$ is that of an $s^w - 1$ hypertetrahedron, or *simplex* with a different facet bounded by every combination of vertices. Any probability distribution which gives zero probability to any pure state, or set of pure states, lies in the boundary of the convex phase space "on the opposite side." In general, though, the points lying on $\overline{\mathfrak{P}}$ satisfy all the marginal constraint conditions, but may have some negative "probabilities." Such points will be called *nonconvexified probability distributions*.

The linear form $\langle \Sigma |$ is the special characteristic of a family of ordered homomorphisms, *marginal projections*, which are of fundamental importance in understanding the structure of $\Re\mathfrak{P}$. Möbius formalism based on the inclusion relation over assignments has been seen in this paper to be embeddable into $\Re\mathfrak{P}$, where the far more powerful tools of linear algebra are available. All of the lattice and related probability concepts translate into operations on the family of marginal projections, which generalize these concepts into the ensemble space.

There are several ways to find a basis for $\Re\mathfrak{P}$, such as those used in the correlation approach by Sanchez *et al.*[14] for a 2 species system. However, I prefer to use a different basis which exploits the underlying ordered structure:

Theorem A. *Let there be s species. Then we can select a basis, called a pigeonhole cluster basis, for all probability distributions on either $\overline{\mathfrak{P}}$ or on $\Delta\mathfrak{P}$ by giving the value of the probabilities over assignments involving only $s - 1$ species.*

To see this, we let the sth species be the one to be eliminated, and we define $\mathbb{p}D(\Gamma) \equiv p_{l_1,\dots,l_k}^{\sigma_{i_1},\dots,\sigma_{i_k}}$, $\Gamma(\vec{\tau}) = \sigma_{i_j}$ to be the probability associated to the indicated species distribution over the cluster $C = [\vec{r}_{l_1},\dots,\vec{r}_{l_k}]$. In the case of a rich lattice, we have, by induction on k:

$$p_{l_1,\dots,l_j,l_{j+1},\dots,l_k}^{\sigma_{i_1},\dots,\sigma_{i_j},s,\dots,s} = \sum_{\substack{q=0\{l_{r_1},\dots,l_{r_q}\}\subseteq\{l_{j+1},\dots,l_k\} \\ 1\leq\sigma_{i'_1},\dots,\sigma_{i'_q}<s}}^{k-j} (-1)^q p_{l_1,\dots,l_j,l_{r_1},\dots,l_{r_q}}^{\sigma_{i_1},\dots,\sigma_{i_j},\sigma_{i'_1},\dots,\sigma_{i'_q}} . \tag{A6}$$

When $k = 1$, Eq. (A6) amounts to Eq. (53b) requiring that the relative probabilities sum to 1 (here the $q = 0$ term, which is one, comes from the empty probability function). Thus, variables of the type

$$p_{l_1,\dots,l_k}^{\sigma_{i_1},\dots,\sigma_{i_k}} , \qquad \text{for } 1 \leq \sigma_{i_1},\dots,\sigma_{i_k} \leq s - 1, \tag{A7}$$

span the space of all probability distributions. Also, since

$$\sum_{q=1}^{k} \binom{k}{q}(s-1)^q = s^k - 1, \tag{A8}$$

where $\binom{k}{q} = \frac{k!}{(k-q)!\,q!}$, there are exactly the right number of variables in (A7) to describe the distribution $p_{l_1,\dots,l_k}^{\sigma_1,\dots,\sigma_k}$, so the variables in (A7) are independent.

The Möbius inclusion relation is easy to understand in terms of pigeonhole cluster bases:

Again, for w the number of sites in Ω, and $0 \le k \le w$, then for any partition of k, $0 \le i_1, \ldots, i_{s-1} \le w$ and $\sum_{j=1}^{s-1} i_j = k$, select $s-1$ disjoint sets,

$$I_1, \ldots, I_{s-1}, \tag{A9}$$

drawn from integers less than or equal to w $\#I_j = i_j$. Think of the set, I_j, as listing the set of site numbers occupied by the species σ_j. Clearly, the joint probability basis,

$$^0\vec{P}_i, \quad \text{for} \quad i = 1, \ldots, \sum_{k=0}^{w} \sum_{\substack{0 \le i_1,\ldots,i_{s-1} \le k \\ \sum_{j=1}^{s-1} i_j = k}} \frac{w!}{i_1! \cdots i_{s-1}!\,(w-k)!}$$

$$= \sum_{k=0}^{w} \binom{w}{k} \sum_{\substack{0 \le i_1,\ldots,i_{s-1} \le k \\ \sum_{j=1}^{s-1} i_j = k}} \frac{k!}{i_1! \cdots i_{s-1}!}, \tag{A10}$$

where the sum for the last index is s^w, can be thought of as indexed by the sets in (A9), since the location of the species σ_s is forced into the remaining $w - k$ sites described by the set of integers, $I - s$, and the joint probabilities can, if desired be written in the forms:

$$p^j = p^{I_1,\ldots,I_{s-1}} \equiv p^{I_1,\ldots,I_{s-1},\bar{I}_s}, \tag{A11}$$

the former form—alternative to that shown in (A6)—being obviously nonredundant. The cluster probabilities:

$$c^j \equiv c^{I_1,\ldots,I_{s-1}} \equiv c^{I_1,\ldots,I_{s-1},\bar{I}_s} = \sum_{l=0}^{w-k} \sum_{\substack{0 \le j_1,\ldots,j_{s-1} \le l \\ \sum_{r=1}^{s-1} j_r = l}} p^{I_1,\ldots,I_{s-1},[J-1,\ldots,J_{s-1},\bar{J}_s]}, \tag{A12}$$

the J_r's comprising a partition of the integers in I_s, are conditional probabilities for occupation of the index set, (A9), by the appropriate species found by allowing any species occupation of the remaining $w - k$ sites in I_s and summing over all relevant point probabilities. Eq. (A12) provides an explicit representation of the matrix, η.

The integer, k, defines the grades. From Eq. (A12) one sees that there are s^k nonzero entries (ones) in the rows of η corresponding to grade k. These rows are linearly independent. Associated to each $k > 0$ there are $\binom{w}{k}$ different bases,

$$c^{I_1,\ldots,I_{s-1},\bar{I}_s}, \tag{A13}$$

obtained by selecting any set, I_s, of $w - k$ distinct sites over which the marginal summation is carried out. Each of these bases has $(s - 1)^k$ independent linear forms, since, by regrouping the terms in (A10), one sees that

$$\sum_{\Omega - \bigcup_{i=1}^{s} K_i = I_s} c^{K_1,\ldots,K_{s-1},\bar{K}_s} = 1, \tag{A14}$$

follows from the marginal condition, $\sum p^i = 1$. When $k = 0$, the single independent form is

$$c^{[\emptyset]\bar{\Omega}} = \langle \Sigma |, \tag{A15}$$

independent of s. For $k = 1$ the format in (A13) can be simply rewritten as

$$c^{ir} \equiv p^{\sigma_r i}, \qquad \text{for } 1 \le i \le w, \ 1 \le r \le s - 1, \qquad (A16)$$

where i_r picks out the one integer in the only nonzero set J_r. When $k = w$, the notation in (A13) is equivalent to that in (A11).

References

1. S. G. Brush, *Kinetic Theory* (Pergamon Press, 1972), Vol. 3, p. 59.
2. R. Kikuchi and S. G. Brush, *J. Chem. Phys.* **47**, 195 (1967).
3. R. Kikuchi, *Phys. Rev.* **88**, 988 (1951).
4. G. Birkhoff and S. MacLane, *A Survey of Modern Algebra* (MacMillan, 1953), p. 351ff; G. Birkhoff and T. Bartree, *Modern Applied Algebra* (McGraw-Hill, 1970), p. 274.
5. J. A. Simmons, *Prog. Theor. Phys.* Suppl. **115**, 351 (1994).
6. C. Domb in *Phase Transformations and Critical Phenomena*, edited by C. Domb and M. S. Green, (Academic Press, New York 1974).
7. R. J. Wilson, in *Combinatorial Mathematics and its Applications*, edited by D. J. A. Welsh, (Academic Pres, New York 1971), p. 315; R. J. Wilson, *Finite Operator Calculus*, (Academic Press, New York 1975), p. 83.
8. Guozhong An, Ph.D. Thesis, The University of Washington, 1989.
9. V. A. Malyshe and R. A. Minlos, *Gibbs Random Fields*, (Kluwer Academic Publishers, 1991).
10. L. Comtet, *Advanced Combinatorics*, (D. Reidel, 1974), p. 176ff.
11. R. D. Levine, in *Maximum Entropy and Bayesian Methods in Applied Statistics*, edited by J. H. Justice, (Cambridge Univ. Press, 1986), p. 85; T. M. Cover and J. A. Thomas, *Elements of Information Theory*, (John Wiley and Sons, 1991).
12. P. Huard, in *Nonlinear Programming*, edited by J. Abadie, (North Holland, 1967), p. 209; N. K. Karmarkar, *Combinatorica*, **4**, 373 (1984).
13. J. Eriksson, *Math. Prog.* **18**, 146 (1980); R. Fletcher, *Practical Methods of Optimization.* Second edition, (John Wiley and Sons, 1987), p. 219.
14. J. M. Sanchez, F. Ducastelle and D. Gratias, *Physica A* **128**, 334 (1984).
15. J. W. D. Connolly and A. R. Williams, *Phys. Rev. B* **27**, 5169 (1983).
16. P. Cenedese, Private Communication.
17. R. Tolman, *The Principles of Statistical Mechanics*, (Oxford, Clarendon Press 1938).

Index